Pictorial communication in virtual and real environments

Pictorial communication in virtual and real environments

Edited by
Stephen R. Ellis
NASA Ames Research Center and University of California,
Berkeley, USA

Section Editor
Mary K. Kaiser
NASA Ames Research Center, USA

Assistant Section Editor
Arthur C. Grunwald
Technion, Israel

Taylor & Francis
London • New York • Philadelphia

UK   Taylor & Francis Ltd, 4 John St, London WC1N 2ET
USA   Taylor & Francis Inc., 1900 Frost Road, Suite 101, Bristol, PA 19007

Copyright © Taylor & Francis Ltd. 1991

All rights reserved. No part of this publication may be reproduced, stored in a retrieval system, or transmitted, in any form or by any means, electronic, electrostatic, magnetic tape, mechanical, photocopying, recording or otherwise, without the prior permission of the copyright owner.

British Library Cataloguing in Publication Data
Pictorial communication in virtual and real environments
1. Graphic displays.
I. Ellis, Stephen R.   II. Kaiser, Mary K.
III. Grunwald, Arthur C.
006.6
ISBN 0-74840-008-7

Library of Congress Cataloging in Publication Data is available

Set in 10/12pt Bembo
by Graphicraft Typesetters Ltd., Hong Kong

Printed in Great Britain by Burgess Science Press, Basingstoke on paper which has a specified pH value on final paper manufacture of not less than 7.5 and is therefore 'acid free'.

Contents

Foreword *J. Koenderink*	viii
Part I Environments	**1**
Prologue *S.R. Ellis*	3
1 Physics at the edge of the Earth *J.P. Allen*	12
2 Pictorial communication: pictures and the synthetic universe *S.R. Ellis*	22
Part II Knowing	**41**
Introduction to Knowing *M.K. Kaiser*	43
3 Perceiving environmental properties from motion information: minimal conditions *D.R. Proffitt and M.K. Kaiser*	47
4 Distortions in memory for visual displays *B. Tversky*	61
5 Cartography and map displays *G.F. McCleary Jr., G.F. Jenks and S.R. Ellis*	76
6 Interactive displays in medical art *D.A. McConathy and M. Doyle*	97
7 Efficiency of graphical perception *G.E. Legge, Y. Gu and A. Luebker*	111
8 Volumetric visualization of 3D data *G. Russell and R. Miles*	131
9 The making of the *The Mechanical Universe* *J. Blinn*	138
Part III Acting	**157**
Introduction to Acting *A.C. Grunwald*	159
Vehicular control	
10 Spatial displays as a means to increase pilot situational awareness *D.M. Fradden, R. Braune and J. Wiedemann*	172

11	Experience and results in teleoperation of land vehicles *D.E. McGovern*	182
12	A computer graphics system for visualizing spacecraft in orbit *D.E. Eyles*	196
13	Design and evaluation of a visual display aid for orbital maneuvering *A.J. Grunwald and S.R. Ellis*	207

Manipulative control

14	Telepresence, time delay and adaptation *R. Held and N. Durlach*	232
15	Multi-axis control in telemanipulation and vehicle guidance *G.M. McKinnon and R.V. Kruk*	247
16	Visual enhancements in pick-and-place tasks: human operators controlling a simulated cylindrical manipulator *W.S. Kim, F. Tendick and L. Stark*	265

Visual/motor mapping and adaptation

17	Target axis effects under transformed visual-motor mappings *H.A. Cunningham and M. Pavel*	283
18	Adapting to variable prismatic displacement *R.B. Welch and M.M. Cohen*	295
19	Visuomotor modularity, ontogeny and training high-performance skills with spatial instruments *W.L. Shebilske*	305
20	Separate visual representations for perception and for visually guided behavior *B. Bridgeman*	316
21	Seeing by exploring *R.L. Gregory*	328

Orientation

22	Spatial vision within egocentric and exocentric frames of reference *I.P. Howard*	338
23	Comments on "Spatial vision within egocentric and exocentric frames of reference" *T. Heckmann and R.B. Post*	359
24	Sensory conflict in motion sickness: an Observer Theory approach *C.M. Oman*	362
25	Interactions of form and orientation *H. Mittelstaedt*	377
26	Optical, gravitational and kinesthetic determinants of judged eye level *A.E. Stoper and M.M. Cohen*	390

27	Voluntary influences on the stabilization of gaze during fast head movements *W.H. Zangemeister*	404

Part IV Seeing 417
Introduction to Seeing 419
S.R. Ellis

Pictorial space

28	The perception of geometrical structure from congruence *J.S. Lappin and T.D. Wason*	425
29	The perception of three-dimensionality across continuous surfaces *K.A. Stevens*	449
30	The effects of viewpoint on the virtual space of pictures *H.A. Sedgwick*	460
31	Perceived orientation, spatial layout and the geometry of pictures *E.B. Goldstein*	480
32	On the efficacy of cinema, or what the visual system did not evolve to do *J.E. Cutting*	486
33	Visual slant underestimation *J.A. Perrone and P. Wenderoth*	496
34	Direction judgement error in computer generated displays and actual scenes *S.R. Ellis, S. Smith, A. Grunwald and M.W. McGreevy*	504
35	How to reinforce perception of depth in single two-dimensional pictures *S. Nagata*	527

Primary depth cues

36	Spatial constraints of stereopsis in video displays *C. Schor*	546
37	Stereoscopic distance perception *J.M. Foley*	558
38	Paradoxical monocular stereopsis and perspective vergence *J.T. Enright*	567
39	The eyes prefer real images *S.N. Roscoe*	577

Index 587

Foreword

Jan Koenderink

Rijks Universiteit Utrecht
The Netherlands

According to the philosopher Kant, space and time are the very form of the human mind. We cannot but perceive and think spatio-temporally. Moreover, we also depend heavily on spatio-temporal expertise for plain survival, something the human holds in common with most animal species. Such vital crafts as orientation, navigation, homing, manipulation, recognition and communication are to a large extent facets of optically guided behavior.

Perhaps paradoxically, modern man relies even more heavily on the senses than his ancestor roving the woods and plains. The reason is the information overload of the senses. The speed and complexity of the required response in such contemporary tasks as driving a vehicle or controlling an industrial plant greatly taxes our sensori-motor coordination. Hence, many professions have developed in which humans are well paid for their well coordinated sensori-motor skills. Many novel cultural artifacts have been invented to help us carry out such tasks. We witness a new technology developing explosively.

It is a curious fact that the science and technology of instrumented aiding of optically-guided spatial behavior is still in its infancy. This field is actually scattered over many scientific and engineering disciplines with only weak mutual couplings. The present volume is indeed a very timely one, and will be useful to both the professional and the newcomer alike. I know of no textbook that covers the area to this breadth. The book presents a veritable bird's-eye view and is a very useful "inventory" for people entering the field, while almost certainly providing some eye-openers even to the pro. It can also be used as a valuable entrance to an extensive, but very scattered literature.

The book has exciting sections on optically guided behavior in exotic and alien environments, such as extraterrestrial space or telepresence, where the human senses cannot rely on evolutionary acquired expertise. In such environments mismatches between the various sensori-motor subsystems are likely, and indeed lead to severe problems. Moreover, some kind of "recalibration" of the senses is called for.

Novel technologies require novel methods of information transfer. Classical methods like maps (Chapter 5) and pictures (Chapter 2) have to be developed much further and have to be augmented with completely novel displays and instruments. It is apparent from the book that this is not just the case in extreme environments (Chapters 12–13), but equally in such settings as medical diagnostics (Chapter 6) or traffic control. The new technology depends heavily on the resources of real time machines that can handle the enormous data volumes pertaining to spatio-temporal structure. The near future will show tremendous progress in machines, and the science of pictorial representation should be ready for it.

In many complex systems the human operator remains a vital link. In practice it is often the weakest in the chain. Now, it is no longer realistic to train a human to fit a machine. Since machines have become so flexible, they can be tailored to exploit the human abilities to the full extent. A fundamental understanding of human abilities and idiosyncrasies is necessary, but only fragments exist. The book has most interesting sections on these issues, hard to find elsewhere. They range from problems of sensory miscalibration, perceptual exploration and misrepresentation to issues of sensorimotor tuning (Chapters 14–18).

A most important issue, also present in the book, is the use of novel technology to improve the data-rate and precision of interhuman communication, where conventional language is supplanted by various pictorial means. The real problems are not of a technological nature, but lie in our limited understanding of human perceptual and conceptual competence. An important example is our limited ability to appreciate the world from other than egocentric frames (Chapter 34).

Aside from being highly informative, reading through the book is also an intellectual pleasure because of the extremely interdisciplinary and richly interconnected material. I wish the reader a good time.

PART I
Environments

Prologue

Stephen R. Ellis

Environments and optimization

Definition of an environment

The theater of human activity is an *environment* which may be considered to have three parts: a *content*, a *geometry*, and a *dynamics*. The content consists of both the actors and the objects with which they interact. The geometry is a description of the properties of the stage of action. The dynamics describes the rules of interaction between the actors and the objects. The following discussion develops an analysis of these parts which is then applied to a concrete example of the unusual environment experienced by astronaut Dr. Joseph Allen while manually flying a Manned Maneuvering Unit in orbit. His experiences provide a reference point for considering the scientific bases of the interaction of human users with the artificial environments presented by computers.

Content

The *objects* in the environment are its content. These objects may be described by *state vectors* which identify their position, orientation, velocity, and acceleration in the environmental space as well as other distinguishing characteristics such as their color, texture, energy etc. The state vector is thus a description of the *properties* of the objects. The subset of all the terms of the state vector which is common to every element of the content may be called the *position vector*. For most physical objects the position vector includes the object's position and orientation as well as the first two time derivatives for each of these vector elements. Though the *actors* in an environment may for some interactions be considered objects, they are distinct from objects in that in addition to characteristics they have *capacities* to initiate interactions with other objects. The basis of these initiated interactions is the storage of energy or information within the actors and their ability to control the release of this stored information or energy after a period of time. The *self* is a distinct actor in the environment which provides a point of view from which the environment may be constructed. All parts of the environment exterior to the self may be considered the field of action. As an example, the balls on a billiards table may be considered the content of the "billiards table" environment and the cue ball combined with the pool player may be considered the "self."

Geometry

The geometry is a description of the environmental field of action. It has a *dimensionality*, *metrics*, and an *extent*. The dimensionality refers to the number of independent descriptive terms needed to specify the position vector for every element of the environment. The metrics are systems of rules that may be applied to the position vector to establish an ordering of the contents and to establish the concept of geodesic or straight lines in the environmental space. The extent of the environment refers to the range of possible values for the elements of the position vector. The environmental space or field of action may be defined as the Cartesian product of all the elements of the position vector over their possible range. The area of a billiard table provides an example of environmental extent. An environmental trajectory is a time history of an object through the environmental space. Since kinematic constraints may preclude an object from traversing the space along some paths, these constraints are also part of the environment's geometric description. The edge bumbers on the billiards table and any other obstacles that may be placed on for variations of the game are examples of kinematic constraints.

Dynamics

The dynamics of an environment are the *rules of interaction* among its contents. An interaction consists of the transfer of energy or information from one element to another. The evidence for this transfer is a change of the state vectors of the interacting elements. All interactions may be reduced for simplicity to asymmetric binary interactions. The directionality of the interaction is determined by two unique elements of the position vector, either the energy state or stored-information state, which allow the definition of an ordering from high to low of all elements. When two elements interact, the element with the higher ranked state may be said to act on that with the lower state. Typical examples of specific dynamical rules may be found in the differential equations of Newtonian dynamics describing the response of billiard balls to impacts initiated by the cue ball. For other environments these rules also may take the form of grammatical rules or even of look-up tables. For example, a syntactically correct command typed at a computer terminal can cause execution of a program with specific parameters. These dynamical rules all provide descriptions of the consequences of the passage of information and energy among the environment's contents.

Life on Earth

Biological optimization through natural selection

Life itself is one notable result of the protracted activity within an environment constrained by the physical laws of nature and the optimizer of natural selection. As a consequence of their ancestors' "practice of physics" in this fixed environment, living creatures often exhibit surprisingly precise and refined mechanical and physiochemical designs. The threshold of absolute human visual sensitivity is, for example, near the limits set by quantum theory (Hecht *et al.*, 1942).

Communications optimization

Some of the more social aspects of human experience have also apparently been optimized for peak performance. The frequency distribution of word lengths in coherent text for almost all alphabetical languages exhibits a characteristic known as Zipf's Law which has been shown by Mandelbrot to be an indication that the encoding system is optimized for transmission over a limited bandwidth communication channel (Mandelbrot, 1953; 1982). The spontaneous patterns of abbreviation that shape Zipf's law are continuously at work in all actively used languages, even computer command languages: for example, as computer users come to be more experienced with a flexible computer operating system such as UNIX, their communications with the system more and more exhibit Zipf's Law (Ellis and Hitchcock, 1986). The driving force behind this change in the frequency distribution of their commands is the desire to maximize the amount of information transmitted over the channel.

As the frequency distribution of lengths of the commands through practice more closely approximates the Zipf distribution, the users come to use their communication channel to the computer more efficiency. One might say that their command language is being matched to their communications environment. This matching of a command language to a communication environment is the kind of optimization that is the goal of good system interface design. It occurs to some extent spontaneously in human behavior. The real challenge of design is to produce interfaces which augment and encourage the natural optimizing tendencies and capacities of the user. This challenge is most vividly evident in the considerations of the problems encountered during interaction with an unusual environment: astronauts orbiting the Earth.

Man in space

The orbital environment is a natural environment but differs from the usual inertial environment with which we are familiar and to which our bodies have been adapted by literally millions of years of evolution. The differences may be separated into those of content, geometry, and dynamics and are most vividly confronted by astronauts during extravehicular activity. A good example is provided by astronaut Dr. Joseph Allen's experiences using a small orbital maneuvering unit called the Manned Maneuvering Unit (MMU), a kind of "space motor scooter" (Chapter 1).

Content

The content of the orbital environment is not radically different from that of the earthbound environment in that it consists mainly of more or less rigid bodies. The fact that many of these objects are pressure vessels designed to fly through the microgravity vacuum of space, of course, introduces a number of special characteristics, i.e., air locks, thruster nozzles, communications antennae, etc., but all of these have direct analogues on earth, i.e., doors, car exhaust pipes, radio antennae, etc. The appearance of the orbiting objects is, however, significantly different due to the absence of the contrast reducing effects of atmospheric diffusion. In this environment all objects appear in sharp, often brilliant contrast interfering with the observer's subjective sense of their distance.

Geometry

The geometric characteristics of the orbital environment are, in contrast to the content, much more unusual. On earth we are primarily surface dwellers with friction and gravity providing considerable constraint on the kinematic freedoms our body has evolved to exploit. With the exception of the privileged few who fly aircraft, our travels are primarily with only 2 degrees of freedom with distinctly limited rotational freedom. A free floating astronaut in space has the full 6 degrees of freedom of translation and rotation and regularly can experience his immediate environment from very unusual attitudes. The relaxed kinematic motion constraints and absence of a gravitation reference axis, when combined with the spacesuit viewing and mobility constraints, can make maintenance of what on earth would be intuitive situation awareness, a significant cognitive activity.

Dynamics

Probably the most unusual aspect of being in orbit is the apparent dynamical environment. Though there is generally no sense of speed, all objects are, of course, in orbits described by Kepler's Laws. These objects are perceived by the actors in the environment in a relative frame of reference often called a local vertical local horizontal (LVLH) frame (Chapter 13). The consequences of imposed thrusts in this reference frame are well known to be counterintuitive and successful maneuvering requires considerable "practice" of the

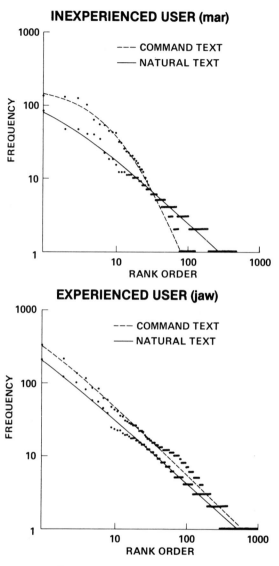

Figure 1. Comparison of frequency of use data from an inexperienced (upper) and an experienced user (lower) of a UNIX operating system. Data gathered separately from natural language texts and collections of system commands in the shell command language are drawn on a Zipf plot of log frequency of use vs. log rank order of use. Since rank order is a transform of frequency of use, the curve must slope downward. Zipf's law asserts that the curve is linear with a slope of -1. The frequency distribution of the experienced user's shell commands more closely resembles a frequency distribution for his natural text than the corresponding comparison for the inexperienced user.

physics, as pointed out by Dr. Allen (Chapter 1). The difficulties in manually conducting orbital maneuvering that are encountered arise not only because the consequences of the state vector changes are counterintuitive from experience in inertially fixed environments, but also because the state vector values are hidden. One cannot see another object's velocity vector or even know from its relative motion in some cases that it has one!

The consequences of the frictionless environment are manifest as dynamical differences. For example, the effects of Newton's Third Law equating action and reaction are constantly evident as unwanted translational or rotational movement which must be explicitly removed by appropriate anchoring or braking thrusts. Consequently, design for effective interaction in this unusual environment requires the development of special purpose tools which through training can become the natural effectors in this unnatural environment.

The considerable work of NASA scientists and engineers and their contractors over the last 30 years has enabled astronauts to traverse the orbital environment. This success has been achieved at the cost of considerable practice of the physics in a variety of simulators for both hardware development and astronaut training. The characteristics of these simulators have been fashioned from a formal and practical understanding of the characteristics of the environment. Similar understanding of the synthetic environments at the human-computer interface will be required for comparable successes.

Pictorial environments for system interfaces

Content

Human-computer interfaces have evolved considerably in the past decade. The teletype interface that characterized operating systems such as RT11 or the original UNIX system has now given way to graphical interfaces such as those developed at Xerox Palo Alto Research Center (PARC) and first used by the Apple Macintosh computer. These interfaces use the computer display screen to present an image to the user in which he may identify separate elements which represent the actors and objects of the interface environment. Being purely symbolic entities the state vectors of these elements may be arbitrarily defined and this arbitrariness is a key source of a design problem. Which elements should be included in the position vector? Which should be excluded?

In natural environments these questions are answered for us by nature. In synthetic environments the answers to them must be determined by the purpose ascribed to the desired interaction. But a successful answer involves the selection of characteristics so that the expectations developed from experience in familiar physical and cultural environments can be tapped through the establishment of a consistent interaction metaphor such as the "Desktop Metaphor" devised at Xerox PARC.

Geometry

Some aspects of the geometry of a graphic computer interface are straightforward. There is clearly a space on the display device that as a first approximation sets the environment's extent. The metric defined on the surface may be arbitrary and determined in a manner contingent upon the mode and nature of interaction. The previous distinction between actors and objects can easily be drawn between the icons representing programs and those representing files. Furthermore, the cursor is a kind of *self* in that it specifies the locus of immediate interaction with the user.

Somewhat less straightforward aspects of the display geometry arise when considering spaces which are 3D virtual spaces inferred from the presentation of depth cues, i.e., perspective, stereopsis, occlusion, motion parallax, etc. Their metrics and other geometric properties will directly depend on the selection and implementation of the depth cues. These displays thus offer significant opportunities for geometric enhancement to emphasize important spatial information.

Probably the most unique aspect of display geometry is what may be called the informational connectivity among the objects and actors. Tree-structured file systems, hierarchically embedded menus, and hypertext documents all represent examples of logically linked complex objects which introduce new dimensions into the position vector. The objects in these environments not only exist in ordinary space and time but also have a dimension of logical proximity determined by a web of connectivity independent of physical space. Defining this web in a way that is useful for a human who wishes to explore the environment is a major design challenge (*ACM Communications*, 1988).

Dynamics

Just as the state vector of an orbiting satellite may be hidden from an observer, so too may the state vector of a synthetic object on a computer display be concealed by unannunciated modes. The synthetic character of such objects, in fact, potentially removes all familiar cues to their current state and can pose would-be actors with serious control problems which may be more constrained by syntactic rules than blocked by physical obstacles. Interaction difficulties are also introduced by the often unnatural mode of interaction required of an operator to interact with the synthetic objects. Typed commands, joystick deflections, mouse rolling, or even hand gestures do not necessarily bear obvious relationships to their designated effects and can require surprising physical effort from the user even for simple activities. In the synthetic universe even the sky is not the limit. The designer of the pictorial interface must learn to appropriately incorporate the constancies, expectations, and constraints from experience in more common environments into the design of the dynamics of the particular artificial environment he is creating. This task requires that he learn to turn our unconscious presumptions of everyday life into conscious assumptions for design. He must learn to make the implicit into the explicit and through this process develop a more complete and coherent description of physical reality which itself is ultimately the source of interface design metaphor.

Pictorial communication

Communication at the object level

Though communication can take place through many senses, human communication with existing computers and computer-based machines is predominantly a visuo-manual activity. The papers in the following sections primarily address the geometric and dynamic issues that arise from analysis of this environment. The analysis adopted by most of the papers may be said to be at the object level in that the units of analysis are clearly identified objects or actors in the environment. Though much of the analysis is visual, there is generally little concern with the properties of the visual image itself, i.e., its brightness, color, or contrast, but interest is focused rather on the spatial interpretation given to the contour patterns formed in the visual image. The focus is on environmental objects not retinal pixels.

The contour patterns of an image are well known to provide only incomplete evidence for the rich spatial environment that we perceive (see Gregory, Chapter 21). Scale and sometimes structure cannot be determined solely from visual information. Accordingly, it is also known that seeing is an active, constructive process utilizing considerable nonvisual information, some of which is in a Kantian sense *a priori* knowledge. As observed by Frank Netter when describing his medical illustrations, we see with more than our eyes.

As a consequence, the following material, which addresses pictorial communication at an object level, is organized into three sections. First, *Knowing*, since our knowledge of our environment provides a filter through which we see it. Second, *Acting*, since our ability to update that knowledge is primarily through observation of the consequences of interaction. And finally *Seeing*, the human capacity that provides the basic data from which our environment is constructed. Those who find this organization backwards need not worry as they may find it possible to sequence the sections in reverse order.

Acknowledgements

The papers in this book have been selected from preliminary drafts that originally appeared as proofs in the Proceedings of the NASA–U.C. Berkeley Conference on *Spatial Displays and Spatial Instruments* (Ellis *et al.*, 1988). This conference was held between August 31 and September 3, 1987, at the Asilomar Conference Center in Pacific Grove, California, and funded by the Division of Information Sciences and Human Factors, NASA Headquarters, through the Aerospace Human Factors Division at the Ames Research Center. Special thanks are due to Dr. Michael W. McGreevy and Dr. David Nagel for supporting the conference. About 40 per cent of the material in the book has been significantly modified since the publication of the *Proceedings* and some chapters have been completely rewritten.

I wish to sincerely thank again all the conference participants and especially my coeditors Mary Kaiser and Art Grunwald, whose assistance and persistent reminders that the paper review must go forward have been helpful. Others who helped with the administrative details of the conference were Fidel Lam, Constance Ramos, Terri Bernaciak, and Michael Moultray. We also should thank the staff at Asilomar. Finally, I wish to thank my family, and especially my wife Aglaia Panos, for the many difficulties they had to endure while I was organizing "the conference" and working on "the book". I do also hope that the personal contacts and interchange of information initiated at the conference is further advanced by the publication of this book.

Stephen R. Ellis
Oakland, California

References

ACM Communications (1988). **31**, (Special issue on hypertext), (7).
Ellis, S.R. and Hitchcock, R.J. (1986). Emergence of Zipf's Law: Spontaneous encoding optimization by users of a command language. *IEEE Transactions on Systems Man and Cybernetics* **SMC-16**, 423–427.
Ellis, S.R., Kaiser, M. and Grunwald, A.J. (1988). *Spatial Displays and Spatial Instruments*, NASA CP 10032, NASA Ames Research Center.
Hecht, S., Schlaer, S. and Pirenne. M.H. (1942). Energy, quanta, and vision. *Journal of General Physiology*, **25**, 819–840.
Mandelbrot, B. (1953). An information theory based on the structure of language based on the theory of informational matching of messages and coding. In *Proceedings of the Symposium on Applications of Communication Theory*, London: Butterworth Scientific Publications.
Mandelbrot, B. (1982). *The Fractal Geometry of Nature*, San Francisco: Freeman.

1
Physics at the edge of the Earth[1]

Joseph P. Allen[2]

*Space Industries, Inc.
Webster, Texas*

I thank you for the opportunity to speak to the International Union of Pure and Applied Physics and the Corporate Associates of the American Institute of Physics. It is particularly appropriate that I voice my gratitude and appreciation for the invitation because I, in many ways, am an outsider to this audience of physicists. For example, I am in no way now a practising physicist. By this, I mean one who studies, teaches, and carries out research on the behavior of matter or one who applies to advanced technical projects the new insights gleaned from such research; and more than 20 years have passed since I was a serious student of physics. Clearly, I am not a practising physicist.

On the other hand, I am not totally without qualification to share with you today some thoughts on the roots of high technology. To begin, I have always been fascinated by the behavior of the world we perceive, a fascination honed by the study of physics through the undergraduate and graduate levels. Secondly, in the last several years I have, in fact, practised physics as a physician would practise medicine. By this I mean "practice" in the literal sense—to go over and over a task in a mechanical way until that task becomes intuitive, second nature, and achievable without great mental effort. I will give an example of practising physics shortly. But, for the moment, let me return to that time I was a university student.

Thirty years ago today, plus or minus 24 hours, I sat in undergraduate Physics 101 at DePauw University and read for the first time in my life of the discoveries and the remarkable concepts of Copernicus, Kepler, and Newton. At that time many of the concepts of these extraordinary thinkers were dramatically demonstrated by the satellite Sputnik as it orbited around our

---

[1] Excerpted with permission from *Physics in a technological world*, (1988) Anthony P. French, Editor, American Institute of Physics, New York. Dr. Allen presented much of this material at the Spatial Displays and Spatial Instruments Conference (see *NASA CP10032*, 1989).

[2] Joseph P. Allen obtained a Ph.D. in physics at Yale University and then went into the United States space program. After retiring from being an astronaut he became Executive Vice-President of Space Industries, Inc., in Webster, Texas

planet Earth. In itself, a satellite was not unique of course, since even in those years we had long been quite accustomed to the Moon. But Sputnik was an artificial satellite, homemade by Russian physicists, engineers, and technicians who were, understandably, quite proud. The rest of us were appropriately impressed at the achievement of its successful launch. A short time later, as I was finishing my graduate studies of physics at Yale, Yuri Gagarin and then John Glenn orbited the Earth in spaceships—artificial satellites large enough to accommodate a person. Suddenly our understanding of the classical laws of nature, and the application of these laws, made it possible for us to travel outside the Earth's atmosphere at speeds far beyond normal human experience. I want to say "at speeds beyond human imagination", but of course a form of imagination led to those very concepts initially envisioned by Copernicus, Kepler and Newton. So clearly, the spaceship velocity of about five miles per second was not beyond human imagination.

Nevertheless, the speed needed to keep homemade satellites in orbit about planet Earth (about 18,000 miles per hour in everyday terms) was, and still is, bold by human standards. Even so, over the last 30 years the combined knowledge, skill, and audacity of thousands of scientists, applied scientists, engineers, and technicians have made possible the construction and orbital flight of many such satellites. What is also remarkable is that scientists as well as pilots, politicians, and others, have been privileged to make space journeys aboard them. Thanks in large part to my interest and education in physics, I have had the good fortune to travel into space on two occasions, have made altogether over 200 orbits of the earth and have reentered the atmosphere twice—landing aboard *Columbia* in California and aboard *Discovery* in Florida. In addition, I have experienced the feeling of floating, untethered, out from the open hatch of an orbiting space shuttle, and have traveled as a satellite myself along one-quarter of a full Earth orbit.

As students of physics we all have carried out *gedanken* experiments, picturing ourselves in our mind's eye as, for example, moving with one reference frame at near-light speed with respect to a second reference frame, then observing meter sticks, clocks, and masses in the two reference frames. The spaceships that orbit Earth do not travel at rates approaching the speed of light, but I assure you that to be a passenger aboard one, as far as classical physics is concerned, is to experience a *gedanken* experiment in real life. To live in the perpetual free fall of orbital flight is to be confronted with example after example of simple classical physics in wonderful and extraordinary detail.

My purpose today is to share with you some of these examples of physics as experienced in orbit—or, as expressed in my title, to reflect on physics at the edge of the Earth. I will use photographs taken from various flights to illustrate the points I wish to make. In that regard, it occurs to me that if Newton or Kepler were somehow to know of our orbiting the Earth, exactly as their equations so elegantly predicted but in artificial satellites, they would not be particularly surprised. Yet they would surely be amazed to see for

Figure 1. Space Shuttle launch.

themselves the views of life aboard these ships that we have captured with the modern invention called the camera. The photography that I use today was taken with quite ordinary cameras and represents to a large degree what your own eyes would see were you to make such a journey.

We are approaching the 500th anniversary of the historic voyage of Christopher Columbus. Although Columbus and his crew were contemporaries of Copernicus, whom I mentioned earlier, the convictions of Copernicus were unknown, or at least unconvincing, to the crew in that they feared, we are told, sailing to the edge of the Earth and then falling beyond. Interestingly enough, each space journey begins precisely that way. A spaceship, modern sailors inside, is propelled by rockets up through the atmosphere and into the vacuum of space with enough velocity to perpetually fall beyond. But there are some dramatic differences as well. The journey from sea level to orbit for a space shuttle, for example, takes 8 minutes, 50 seconds; and contrary to Columbus's direction, space launches are to the east to take advantage of the eastward rotation of the Earth. For the space shuttle the linear outbound acceleration is about 3g's during the launch phase. Of course, at the instant of engine shutdown the acceleration experienced within the spacecraft goes to zero to first order, and the spaceship is then in the silent vacuum of space well beyond the edge of the Earth. As the ship coasts along its orbital path, everything within the ship floats and all human feeling of up or down mysteriously vanishes.

Everything floats—pencils, clothes, cameras, relaxed limbs, and bodies. The appearance of any solid object, of course, does not change, but, because of the floating, ordinary objects can be used for amusing demonstrations. The stability of a book-shaped object spinning around its long axis or its short axis is an easy example. But when gently spun around its intermediate axis, the object rotates, swaps ends, rotates, swaps ends, and so on as though magically demonstrating Euler angles and Euler's equations for an animated videotape on physics instruction.

In contrast to solid objects, the behavior of liquids in microgravity is so different from that of liquids on Earth that the mere sight of water, orange juice, or lemonade out of its proper container borders on comedy. For example, water squeezed from its drink container remains on the end of the container's straw as a perfectly motionless, but glistening globe. When you disturb the straw slightly by shaking or blowing on it, the globe quivers, tiny waves of motion circling its circumference. If disturbed more sharply, the globe shivers, vibrates, and oscillates—veritable Legendre functions of deformation spreading as tidal waves around the sphere.... On touching any relatively flat surface, the liquid changes shape instantly from a sphere to a hemisphere—looking now like weak Jello dessert molded in a round mixing bowl and dumped onto a serving plate. The water will stay there indefinitely unless mopped with an absorbent towel or drunk by a thirsty crewman.

Let me return to the example I mentioned earlier of "practising physics". Consider the satellite (Figure 2) being deployed from the space shuttle Columbia. It is a typical communications satellite—cylindrical, covered on the outside with solar cells with which it generates its needed power as it moves in geosynchronous orbit about 22,000 miles above the equator. Such satellites in space are spin stabilized, that is to say they are their own gyroscopes. As this photograph was taken, the satellite was spinning at 30 rpm and, of course, will continue at this approximate spin rate for its full lifetime of ten years or so. In 1984 two such satellites, properly deployed from the space shuttle *Challenger* and in perfectly good working order, unfortunately later failed to reach their intended destination of geosynchronous orbit because of auxiliary rocket malfunctions.

In the spring of 1984, five astronauts (Hauck, Walker, Fisher, Gardner, and I) were given the assignment of recapturing the lost satellites, if at all possible, and returning them to Earth for refitting and for reuse. The task would clearly be challenging since the satellites were not small (each weighed about 1000 pounds on the Earth's surface), were spinning at an impressive speed and were not equipped with grapple points, handholds, or attachment fittings of any sort for just such a contingency. However, the errant devices were in orbits easily reached by a space shuttle. Moreover, their spins could

Figure 2. Satellite released from the Space Shuttle payload bay.

be slowed by ground commands, and we, as salvage crew members, had for our use the Manned Maneuvering Unit (the MMU—see Figure 3) that could be attached to the back of a suited astronaut and would, in theory, enable the astronaut at least to maneuver up to the satellite. If we could approach the satellite without causing it to wobble or tumble, we could probably seize it through the nozzle of its spent rocket—an opening conveniently aligned through the satellite's center of rotation.

The function of the Manned Maneuvering Unit is to an astronaut what the function of a dory (a small boat) is to a sailor. Each can carry a passenger a short distance out from the mother ship, can move around in the local area and, if handled properly, can return to the ship. The dory, of course, is controlled by the sailor using oars to move the vessel over the two dimensions of the water's surface. The maneuvering unit, on the other hand, is controlled by a spacesuited astronaut using two hand controllers connected to

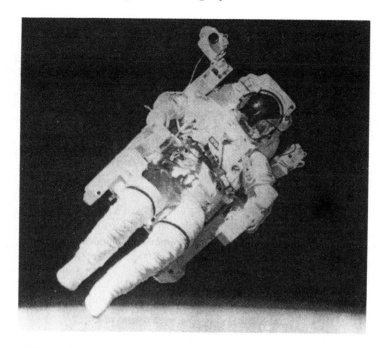

Figure 3. Astronaut in Manned Maneuvering Unit (MMU).

thrusters which move the unit across the three dimensions of space. I do not need to point out that the maneuvering unit is unconstrained by a surface of any kind, nor is it tethered, and thus it may move in all six degrees of freedom, three linear and three rotational. Furthermore, by Newton's first law, any motion imparted along or around any one axis continues unchanged until the astronaut-MMU "cluster" is acted upon by still another force.

Although called at times an "overstuffed rocket chair" because of its appearance, the maneuvering unit is actually a very simple device using compressed gas, typically nitrogen, meted out in bursts through any of 24 finger-sized thrusters mounted on the corners of the back pack. To pilot it, we think of ourselves as wearing a cube-shaped back pack with three perpendicular thrusters at each of the eight corners of the cube. The thrusters are ganged to work in sets of 12. [Any one set of 12 (two thrusters at each of six corners) can provide adequate control to return the astronaut to the mother ship if the other set of 12 should fail.] The astronaut controls the linear motion along the body axes by moving the left-hand controller along those same axes, and the rotational motion around the body axes by twisting, tilting, and rocking the right-hand controller in the direction of the motion to be induced. The hand controllers are coupled electronically to solenoid valves which in turn allot the bursts of compressed gas to the appropriate thrusters. To fly the maneuvering unit in the frictionless, undamped, and silent isolation of space is an extraordinary experience. This is especially the case since there are complications in implementing what I have just described which will, I think, intrigue you. I can testify that these complications indeed do challenge the astronaut piloting the maneuvering unit, particularly when in search of a wayward and potentially skittish satellite. More important to this discussion, I wish to stress that to fly the maneuvering unit, in simulation or in space, is literally to "practise physics", and in this example you are simultaneously the experiment and the experimenter. Let me be more specific.

You will recognize immediately that the center of thrust of the 24 thrusters (the 12 couples) does not correspond exactly with the center of mass of the "cluster" to be maneuvered, i.e., the MMU, the suited astronaut and the attached auxiliary equipment. Of course, this cluster itself is not exactly a rigid body in that the astronaut is not tightly bound by the spacesuit but, rather, floats within this inflated cocoon of air. Some astronauts do qualify as ponderous bodies perhaps, but, in the strict physics sense of the word, certainly not as rigid bodies.

Because of the offset just mentioned, each linear motion commanded by the pilot has an admixture of rotational drift that is added to the resulting linear motion, and for each rotational motion commanded an admixture of linear drift is added to the resulting rotational motion. Put in everyday terms, the MMU "skids" during its translations and "drifts" during its rotations. This skidding and drifting must be imagined as occurring in all three dimensions which is, of course, more complicated than, for example, the simple skidding of a car.

A second perturbation, or complication, in stalking a spinning satellite follows immediately from the undeniable facts of orbital mechanics. The mathematics describing the motions of each satellite—the communication satellite and the astronaut satellite—is straightforward enough, and the position of one object with respect to the other is easy to visualize and predict by, for example, a physics student watching the chase scene from a vantage point above the North Pole. But the relative motion of the two, the change of position of astronaut with respect to satellite, is not necessarily intuitive to the astronaut at the time.

Please consider the following example: the satellite and the astronaut are in identical circular orbits. The astronaut is 20 feet behind the satellite, feet pointed to the Earth's center, attitude inertially stable, with the lance (the capture probe) perfectly aligned with the rocket nozzle of the satellite. (Assume for the moment that the MMU thrusters are not being used.) Then, one-half orbit later (approximately 45 min), the situation will have reversed. The astronaut, head now pointed to the Earth's center, still trails the satellite in the orbital sense, but the lance is pointed away from the satellite which, from the astronaut's perspective, is now behind him. Thus, on a time scale of minutes the astronaut has made a half circle around the satellite, staying always 20 feet from it—a change of relative position due only to the orbital mechanics, not due to motion induced by the MMU.

Now add to this simple example of two objects in identical orbits the

Figure 4. Astronaut practising orbital maneuvers with a MMU simulator (photograph courtesy of Martin-Marietta, Denver).

further complication of the astronaut moving above or below the target satellite and then off to one side (out of the satellite's orbital plane). Finally, recognize that one's field of view from within a space suit is restricted, so that it is difficult to judge simultaneously one's position with respect to the satellite, and the direction to the center of the Earth—information which in theory could enable anticipation of the "Kepler-induced windage" of the situation.

A third complication presents itself, not from physics directly, but rather from the way we humans perceive our surroundings. To attempt to explain this fully would require a long lecture on the physiology and psychology of seeing. An easy description, however, does apply to this situation. In attempting to line oneself up with a large satellite of simple geometric shape, starkly illuminated by direct sunlight, against the velvet black background of deep space, it is very difficult to estimate distances. In fact, it is difficult even to distinguish, for instance, a translation of your position to the right from a rotation of your attitude to the right. In both cases the object against which you are judging moves to the left in your field of vision. Yet, it is important to distinguish the difference since you are commanding, separately, translation and rotation.

Figure 5. An astronaut in a MMU recovering a satellite.

The fourth complication is the most challenging, and in my view the most amusing. The communication satellite spins, and its stability (its resistance to possible wobble induced by plumes of compressed gas from the approaching astronaut and MMU) is a direct function of its spin rate. The suited crewman could, in theory, set up an equal spin rate around the axis of the capture probe jutting out in front of him. Initial contact with the satellite would then be both elegant and the least demanding on the mechanisms to be mated, since sudden torques imposed at impact would be eliminated. It was exactly this spinning capture procedure we planned, and we began to practice this technique in ground-based simulations several months before the mission was to take place (Figure 4).

I suspect that several, if not all, of you now anticipate the results. The seemingly straightforward human task of flying a conceptually simple maneuvering unit in Earth orbit is complicated by the cross-coupling of translation-rotation controls, by the easily predicted but perplexing effect of orbital mechanics on the position of one satellite with respect to another, and by the limitations of human vision in judging $\mathbf{R}$, $\dot{\mathbf{R}}$, and $\ddot{\mathbf{R}}$, where $\mathbf{R}$ is the vector from astronaut to satellite. Each complication is slight, acts slowly and of itself is not insurmountable. But to these three compounded factors we attempted to add a fourth, that is to fly the approach to the target satellite with both astronaut and satellite in a rotating reference frame. We quickly learned from simulation that, even with rotation rates as low as 3 rpm and the approach rate as small as 3 feet per second, the Coriolis force (the $-2m\omega \times \dot{\mathbf{R}}$ term in the force equation) and the centrifugal force (the $-m\omega \times (\omega \times \mathbf{R})$ term) impose a flying task on the human pilots that is confusing in the extreme.

Several test simulations showed that the gas pressure available in the MMU often was depleted before the capture task was accomplished, and the MMU pilot was mentally exhausted by the process. This "practice of physics" clearly indicated to us well before the mission itself that we must fly, inertially fixed, onto the satellite and simply attempt the capture with equipment sturdy enough to withstand the imposed torques of the impact.

Figure 5 shows that our revised flying procedures worked as planned. In November 1984, two satellites (one of them being the astronaut + MMU) were joined on a modern jousting field some 200 miles beyond the edge of the Earth, and both astronaut-satellite collisions were on target and perfectly inelastic, as we had hoped.

I would enjoy digressing from my subject of observing classical physics in action and include several examples of quantum phenomena—plasma physics mostly—which are spectacular from the vantage point of Earth orbit. For example, the northern lights as observed from above or the ion glow that surrounds the orbiter itself on occasion, or perhaps the plasma sheath that spreads out from the shuttle as we start the energy management task of reentry. But these are subjects for another lecture.

I hope you have enjoyed these excerpts from an amateur physicist's logbook of a space voyage. I can assure you that both basic and applied physics are not only essential, but can be great fun when associated with voyages into space. Moreover, a physics education travels very well even over great distances, and an appreciation of physics enhances beautifully these already astounding journeys.

2

Pictorial communication: pictures and the synthetic universe

Stephen R. Ellis

*NASA Ames Research Center
Moffett Field, California
and
U.C. Berkeley School of Optometry
Berkeley, California*

Summary

Principles for the design of dynamic spatial instruments for communicating quantitative information to viewers are considered through a brief review of the history of pictorial communication. Pictorial communication is seen to have two directions: (1) from the picture to the viewer and (2) from the viewer to the picture. Optimization of the design of interactive instruments using pictorial formats requires an understanding of the manipulative, perceptual and cognitive limitations of human viewers.

Pictures

People have been interested in pictures for a long time (Figure 1). This interest has two related aspects. On one hand we have an interest in the picture of reality provided to us in bits and pieces by our visual and gross body orienting systems—and their technological enhancements. Indeed, Western science has provided us with ever clearer pictures of reality through the extension of our senses by specialized instruments.

On the other hand, we also have an interest in pictures for communication, pictures to transmit information among ourselves as well as between us and our increasingly sophisticated information-processing machines. This second aspect will be our prime focus, but some discussion of the first is unavoidable.

It is useful to have a working definition of what a picture is and I will

Pictures and the synthetic universe 23

Figure 1. Prehistoric cave painting of animals from southwestern France. Copyright René Burri/Magnum, New York.

propose the following: a picture is produced through establishment of a relation between one space and another so that some spatial properties of the first are preserved in the second, which is its image. A perspective projection is one of many ways this definition may be satisfied (Figure 2).

The definition may be fleshed out, as cartographers do, by exactly stating what properties are preserved, but the basic idea is that, though the defining relation of the layout of the picture may discard some of the original information, this relation is not arbitrary. The challenge in the design of a picture is the decision what to preserve and what to discard.

Artists, of course, have been making these decisions for thousands of years, and we can learn much from this history. One curious aspect of it, is that early art was not focused on the preservation of spatial properties which have been asserted above to be the essence of a picture.

As art historians have pointed out, early art was often iconographic, depicting symbols, as these Egyptian symbols for fractions illustrate, rather than aspiring to three-dimensional realism (Figure 3) (Gombrich, 1969). This early history underscores a second aspect of pictures which we must consider: their symbolic content. Because of the potentially arbitrary relation between a symbol and what it denotes, a symbol itself is not a picture. Symbols, nevertheless, have from the very beginning found their way into many pictures, and we now must live with both the symbolic and geometric aspects of pictorial communication. Furthermore, focusing on the symbolic content has the useful effect of reminding the viewer of the essentially duplicitous nature of a picture since, though it inherently represents an

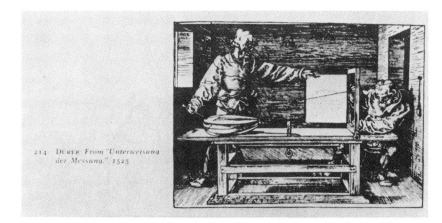

Figure 2. Woodcut by Dürer illustrating how to plot lines of sight with string in order to make a correct perspective projective. Copyright New York City Public Library.

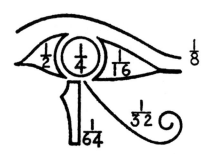

Figure 3. Egyptian hieroglyphic for the Eye of Horus illustrating the symbolic aspect of pictographs. Each part of the eye is also a symbol for a commonly used fraction. These assignments follow from a myth in which the Sun, represented by the eye, was torn to pieces by the God of Darkness later to be reassembled by Thoth, the God of Learning.

alternative space, it itself is an object with a flat surface and fixed distance from the viewer.

The third basic element of pictorial communication is computational. The picture must be created. In the past the computation of a picture has primarily been a manual activity limited by the artist's manual dexterity, observational acumen, and pictorial imagination. The computation has two separable parts: (1) the shaping and placement of the components of the image, and (2) the rendering, that is, the coloring and shading of the parts as is well illustrated by Leonardo's and Escher's studies of hands which use shading to depict depth.

Figure 4. An engraving by Escher illustrating how the ambiguity of depicted height and depicted depth can be used in a picture to create an impossible structure, apparently allowing water to run uphill. Copyright 1988 M.C. Escher heirs/Cordon Art-Baarn-Holland.

While this second part is clearly important and can contribute in a major way to the success of a picture, it is not central to the following discussion. Though the rendering of the image can help establish the virtual or illusory space that the picture depicts and can literally make the subject matter reach out of the picture plane, it is not the primary influence on the definition of this virtual space. Shaping and placement are. These elements reflect the underlying geometry used to create the image and determine how the image is to be rendered. By their manipulation artists can define—or confuse—the virtual space conveyed by their pictures.

While the original problems of shaping, positioning, and rendering still remain (Figure 4), the computation of contemporary pictures is no longer restricted to manual techniques. The introduction of computer technology has enormously expanded the artist's palette, and provided a new 3D canvas on which to create dynamic synthetic universes; yet the perceptual and cognitive limits of the viewers have remained much the same. Thus, there is now a special need for artists, graphic designers, and other creators of pictures for communication to understand these limitations of their viewers. Here is where the scientific interest in the picture of reality and the engineering interest in the picture for communication converge.

Spatial instruments

In order to understand how the spatial information presented in pictures may be communicated, it is helpful to distinguish between images which may be described as spatial displays and those that were designed to be spatial instruments. One may think of a spatial display as any dynamic, synthetic, systematic mapping of one space onto another. A picture or a photograph is a spatial display of an instant of time. A silhouette cast by the sun is not, because it is a natural phenomenon not synthesized by humans.

A spatial instrument, in contrast, is a spatial display that has been enhanced either by geometric, symbolic, or dynamic techniques to ensure that the communicative intent of the instrument is realized. A simple example of a spatial instrument is an analog clock (Figure 5). In a clock the angular positions of the arms are made proportional to time, and the viewer's angle-estimation task is assisted by radial tic marks designating the hours and minutes.

A second aspect of the definition of a spatial instrument, which the clock example also illustrates, is that the communicated variable—time—is made proportional to a spatial property of the display, such as an angle, areas, or length and is not simply encoded as a character string.

The spatial instruments on which we wish to focus attention are generally interactive. That is to say, the communicated information flows both to and fro between the viewer and the instrument. Some of this bidirectional flow exists for practically all spatial instruments, since movement of the viewer can have a major impact on the appearance of the display. However, the displays I wish to consider are those incorporating at least one controlled element, such as a cursor, which is used as a manipulator to extract information from and input information to the instrument.

Spatial instruments have a long history. One of the first ever made, dating from 60 to 80 BC, was an astrolabe-like device uncovered in 1901 near Antikythera, Greece. However, it was not fully described until the late '50's by De Solla Price (1959), who was able to deduce much of its principles of operation by X-raying the highly corroded remains (Figure 6). Here the communicated variables were the positions of heavenly bodies. Nothing approaching the complexity of this device is known until the 16th century. It represents a highly sophisticated technology otherwise unknown in the historical record.

Though many subsequent spatial instruments have been mechanical and,

Pictures and the synthetic universe

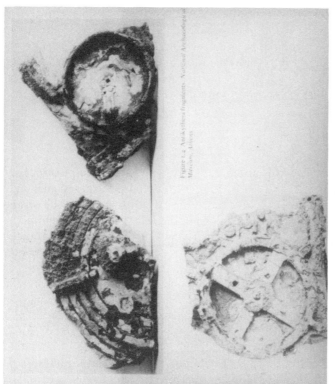

Figure 6. Fragments of an ancient Greek mechanical device used to calculate the display positions of heavenly bodies.

Figure 5. View of the Prague town hall clock, which indicates the positions of heavenly bodies as well as the time.

Figure 7. An old Dutch map of the world from the 17th Century. Copyright New York City Public Library, Spencer Collection, Astor, Lenox, and Tilden Foundations.

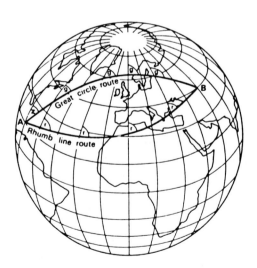

Figure 8. Rhumb-line and great-circle routes between two points on the globe. Note the constant bearing of the rhumb-line route and the constantly changing bearing of the great-circle route. On the globe the great-circle route is analogous to a straight line and direction Z is the azimuth of B from A.

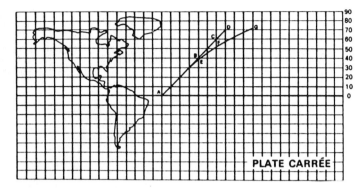

Figure 9. Plate carrée projection illustrating the curved path traced by a rhumb line on this format, i.e., line AEFG.

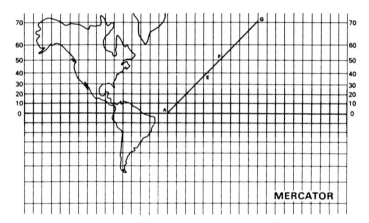

Figure 10. Mercator projection illustrating how a nonlinear distortion of the latitude scale can be used to straighten out the path traced by a rhumb line.

like the Prague town hall clock (Figure 5), have similarly been associated with astronomical calculations (King, 1978), this association is not universal. Maps, when combined with mechanical aids for their use, also certainly meet the definition of a spatial instrument (Figure 7). The map projection may be chosen depending upon the spatial property of importance. For example, straight-line mapping of compass courses (rhumb lines), which are curved on many maps, can be preserved in Mercator projections (Dickinson, 1979; Bunge, 1965). Choice of these projections illustrates a geometric enhancement of the map. The overlaying of latitude and longitude lines illustrates a symbolic enhancement (Figures 8–10). But more modern media may also be adapted to enhance the spatial information that they portray, as illustrated by the reference grids used by Muybridge in his photographs (Muybridge, 1975).

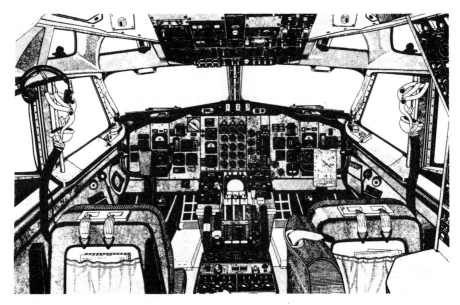

Figure 11. View of the forward panel of a 727 cockpit showing the artificial horizon on the attitude direction indicator. Two of these instruments may be seen directly in front of the control yokes. The lightly colored upper half disk on each shows the aircraft to be in level flight.

Contemporary spatial instruments are found throughout the modern aircraft cockpit (Figure 11), the most notable probably being the attitude direction indicator which displays a variety of signals related to the aircraft's attitude and orientation. More recent versions of these standard cockpit instruments have been realized with CRT displays, which have generally been modeled after their electromechanical predecessors (Boeing, 1983). But future cockpits promise to look more like offices than anything else (Figure 12). In these airborne offices the computer graphics and CRT display media, however, allow the conception of totally novel display formats for totally new, demanding aerospace applications.

For instance, a pictorial spatial instrument to assist informal, complex, orbital navigation in the vicinity of an orbiting spacestation has been described (Figure 13) (Chapter 13). Other graphical visualization aids for docking and orbital maneuvering, as well as other applications, have been demonstrated by Eyles (1986) (Chapter 12). These new instruments can be enhanced in three different ways: geometric, symbolic, or dynamic.

Geometric enhancement

In general, there are various kinds of geometric enhancements that may be introduced into spatial displays, but their common feature is a transformation

Pictures and the synthetic universe 31

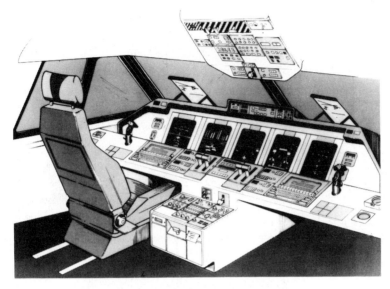

Figure 12. An advanced-concepts commercial aircraft cockpit in the Man-Vehicle Systems Research Facility of NASA Ames Research Center. This artist's conception shows how future cockpits may resemble ordinary offices.

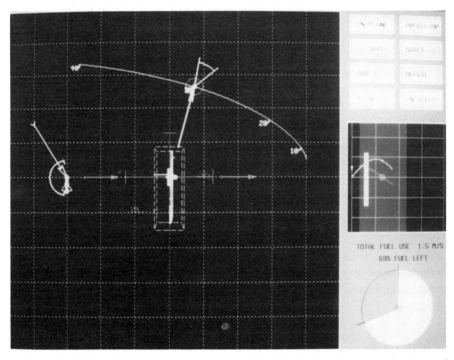

Figure 13. Sample view from an interactive-graphics-based planning tool to be used in assisting informal changes in orbits and proximity operations in the vicinity of a space station.

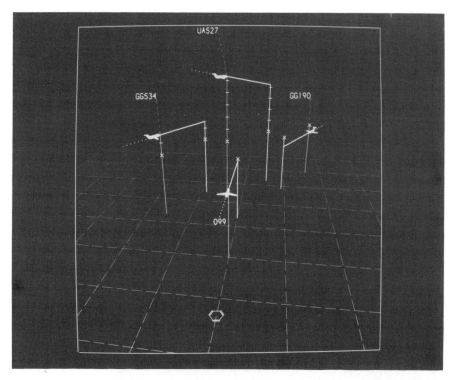

Figure 14. Possible display format for a commercial aircraft cockpit traffic display. The pilot's own craft is shown in the center of the display. All aircraft have predictor vectors attached showing future position and have reference lines to indicate height above a reference grid.

of the metrics of either the displayed space or of the objects it contains. A familiar example is found in relief topographic maps for which it is useful to exaggerate the vertical scale. This technique has also been used for experimental traffic displays for commercial aircraft (Figure 14) (Ellis et al., 1987b).

Another type of geometric enhancement important for displays of objects in 3D space involves the choice of the position and orientation of the eye coordinate system used to calculate the projection (Figure 15). Azimuth, elevation, and roll of the system may be selected to project objects of interest with a useful aspect. This selection is particularly important for displays without stereoscopic cues, but all types of displays can benefit from an appropriate selection of these parameters (Ellis et al., 1985, Chapter 34; Kim et al., 1987; Nagata, 1986; Tachi et al., 1989, Chapter 16).

The introduction of deliberate spatial distortion into a spatial instrument can be a useful way to use geometric enhancement to improve the communication of spatial information to a viewer. The distortion can be used to correct underlying natural biases in spatial judgements. For example, exocen-

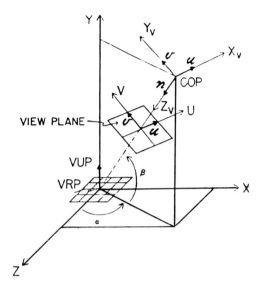

*Figure 15. Illustration of the geometry of perspective projection showing the azimuth and the elevation of the viewing vector **R**, directed from the center of projection COP.*

tric direction judgements (Howard, 1982) made of extended objects in perspective displays, can, for some response measures, exhibit a "telephoto bias". That is to say, the subjects behave as if they were looking at the display through a telephoto lens. This bias can be corrected by introduction of a compensating wide-angle distortion (McGreevy and Ellis, 1986; Grunwald and Ellis, 1987).

Symbolic enhancement

Symbolic enhancements generally consist of objects, scales, or metrics that are introduced into a display to assist pick-up of the communicated information. The usefulness of such symbolic aids can be seen, for example, in displays that present air traffic situation information and that focus attention on the relevant "variables" of a traffic encounter, such as an intruder's relative position, as opposed to less useful "properties" of the aircraft state, such as absolute position (Falzon, 1982).

One way to present an aircraft's position relative to a pilot's own ship on a perspective display is to draw a grid at a fixed altitude below an aircraft symbol and drop reference lines from the symbol onto the grid (Figure 16). If all the displayed aircraft are given predictor vectors that show future position, a similar second reference line can be dropped from the ends of the predictor lines.

The second reference line not only serves to show clearly the aircraft's

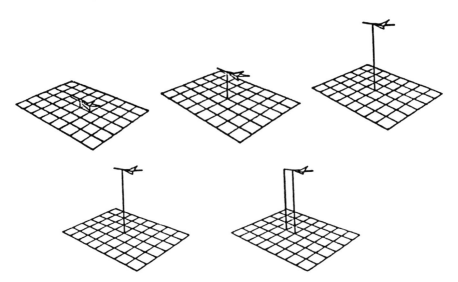

Figure 16. Five views of sample stimuli used to examine the perceptual effect of raising an aircraft symbol above a reference grid. The attitude of the symbol is kept constant. Addition of a second vertical reference line is seen to reduce the illusory rotation caused by the increasing height above the grid.

future position on the grid, but additionally clarifies the symbol's otherwise ambiguous aspect. Interestingly, it can also improve perception of the target's heading difference with a pilot's own ship. This effect has been shown in an experiment examining the effects of reference lines on egocentric perception of azimuth (Ellis *et al.*, 1987a). The experiment may be used as an example of how psychophysical evaluation of images can help improve their information display effectiveness.

In this experiment subjects viewed static perspective projections of aircraft-like symbols elevated at three different levels above a ground reference grid: a low level below the view position, a middle level colinear with it, and a high level above it. The aircraft symbols had straight predictor vectors projecting forward, showing future position. In one condition, reference lines were dropped only from the current aircraft position; in the second condition, lines were dropped from both current and predicted position.

The first result of the experiment was that subjects made substantial errors in their estimation of the azimuth rotation of the aircraft; they generally saw it rotated more towards their frontal plane than it in fact was. The second result was that the error towards the frontal plane for the symbols with one reference line increased as the height of the symbol increased above the grid. Most significantly, however, introduction of the second reference line totally eliminated the effect of height, reducing the azimuth error in some cases by almost 50 per cent (Figure 17).

More detailed discussion of this result is beyond the scope of this introduc-

EGOCENTRIC DIRECTION ERROR

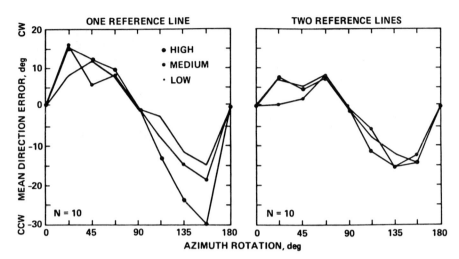

Figure 17. Mean clockwise and counterclockwise egocentric direction judgement for clockwise azimuth rotation of an aircraft symbol. The two panels represent a statistically reliable 3-way interaction ($F = 2.402$, $df = 16144$, $p < 0.003$) from a repeated measures experiment with 10 subjects.

tion; however, these experimental results show in a concrete way how appropriately chosen symbolic enhancements can provide not only qualitative, but quantitative, improvement in pictorial communication. They also show that appropriate psychophysical investigations can help designers improve their spatial instruments.

Combined geometric and symbolic enhancements

Some enhancements combine both symbolic and geometric elements. One interesting example is provided by techniques connecting the photometric properties of objects or regions in the display with other geometric properties of the objects or regions themselves. Russell and Miles (1987) (see also Chapter 8), for example, have controlled the transparency of points in space with the gradient of the density of a distributed component and produced striking visualization of 3D objects otherwise unavailable. These techniques have been applied to data derived from sequences of MRI or CAT scans and allowed a kind of "electronic dissection" of medical images. Though these techniques can provide absolutely remarkable images, one of the challenges of their use is the introduction of metrical aids to allow the viewer to pick up quantitative information from the photometric transformation (Meagher, 1985, 1987).

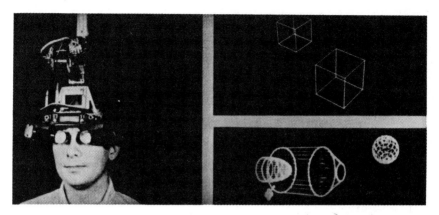

Figure 18. Probably the first computer-driven, head-mounted viewing device. It was developed by Ivan Sutherland to give the viewer the illusion of actually being in the synthetic world defined in the computer. (Used by permission of the University of Utah, Salt Lake City, Utah.)

Dynamic enhancements

While considerable computation may be involved in the rendering and shading of static pictures, the importance of computational enhancement is also particularly evident for shaping and placing objects in interactive spatial instruments. In principle, if unlimited computational resources were available, no computational enhancements would be needed. The enhancements are necessary because resources must be allocated to ensure that the image is computed in a timely and appropriate manner.

An example of a computational enhancement can be found in the selection of a type of geometric distortion to use as a geometric enhancement in a head-mounted, virtual-image computer display of the type pioneered by Ivan Sutherland (1970) (Figure 18) (see also Tachi *et al.*, 1989). Distortions in the imagery used by such displays can be quite useful, since they are one way that the prominence of the components of the image could be controlled.

It is essential, however, that the enhancements operate on the displayed objects before the viewing transformation, because here the picture of reality collides with a picture for communication. The virtual-image presentation makes the picture appear in some ways like a real space. Accordingly, distorting geometric enhancements that are computed after the viewing transformation can disturb visual-vestibular coordination and produce nausea and disorientation. This disturbance shows how different dynamic constraints distinguish head-mounted from panel-mounted formats.

A second example of a dynamic enhancement is shown on the interactive, proximity-operations, orbital planning tool described by Art Grunwald and myself in Chapter 13. When first implemented, the user was given control of

Pictures and the synthetic universe 37

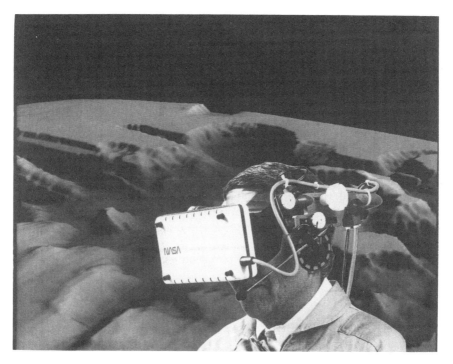

Figure 19. Recent applications of computer-driven, head-mounted viewing devices have benefited from considerable advances in computing power and display technology and may enable new modes of exploring exotic terrain such as planetary surfaces. (McGreevy, 1990).

the direction and magnitude of the thrust vector of the maneuvering system; these seemed reasonable, since they are the basic inputs to making an orbital change. The nonlinearities and counterintuitive nature of the orbital dynamics, however, made manual control of a predictor cursor driven by these variables impossible. The computational trick needed to make the display tool work was allowing the user to command the craft to be at a certain location at a set time and allow the computer to calculate the required burns through an inverse orbital dynamics algorithm. This technique provided a good match between the human user's planning abilities and the computer's accurate computational capacity and speed.

A third example of a dynamic enhancement is shown on the same interactive, proximity-operations, orbital planning tool. Despite the fact that the system has been implemented on a high-performance 68020 workstation with floating-point processor and dedicated graphics geometry engine, unworkably long delays would occur if the orbital dynamics were constantly updated while the user adjusted the cursor to plan a new way-point. Accordingly, the dynamics calculations are partially inhibited whenever the cursor is in motion. This feature allows a faster update when the user is setting a way-point position and eliminates what would otherwise be an annoying delay while adjusting the way-point position.

When Arthur Grunwald finished the first iteration of this display, we decided to name it. Like a dutiful NASA researcher, he searched for an acronym—something like Integrated Orbital and Proximity Planning Systems, or IOPPS for short. This looked as if it might sound like OOPS and I thought we should find a better name. I asked him to find maybe a Hebrew name that would be appropriate. He thought about it for a while and came up with Navie, or "reliable prophet." This is perfect, since that is exactly what the display is intended to provide: reliable prophesy of future position.

But there is another sense in which Navie is a good name. I would like to think that it, and other display concepts developed in our division and elsewhere, also provide a kind of prophesy for the coming displays to be used by NASA during future unmanned, and manned, exploration of air and space.

Like most human activities, this exploration is not an endeavor that can be automated; it will require iteration, trial and error, interactive communication between men and machines and between men and other men. The media for this communication must be designed. Some of them will be spatial instruments.

Bibliography and references

Adams, A. (1975). *Camera and Lens.* New York: Morgan and Morgan.
Arai, H. and Tachi, S. (1989). Development of power assisted head-coupled display system for tele-existence. *Journal of Robotics and Mechanics*, **116**, 36–44.
Bertoz, A. and Melville-Jones, G. (1985). *Adaptive Mechanisms in Gaze Control: Facts and Theories.* New York: Elsevier.
Boeing (1983). 757 Flight deck design development and philosophy, D6T11260-358. Boeing, Seattle.
Bunge, W. (1965). *Theoretical Geography, 2nd ed.* Studies in Geography. The Netherlands, Lund: Gieerup.
De Solla Price, D.J. (1959). An ancient Greek computer. *Scientific American*, **200**, 60–67.
Dickinson, G.C. (1979). *Maps and air photographs.* New York: Wiley.
Ellis, S.R. and Grunwald, A. (1987). A new visual illusion of projected three-dimensional space. *NASA TM 100006.*
Ellis, S.R., Grunwald, A. and Velger, M. (1987a). Head-mounted spatial instruments: synthetic reality or impossible dream. *Proceedings of the 1987 AGARD Symposium on Motion Cues in Flight Simulation and Simulator Induced Sickness*, Brussels, Belgium.
Ellis, S.R., McGreevy, M.W. and Hitchcock, R. (1987b). Perspective traffic display format and airline pilot traffic avoidance. *Human Factors*, **29**, 371–382.
Ellis, S.R., Smith, S. and McGreevy, M.W. (1987c). Distortions of perceived visual directions out of pictures. *Perception and Psychophysics*, **42**, 535–544.
Ellis, S.R., Kim, W.S., Tyler, M., McGreevy, M.W. and Stark, L. (1985). Visual enhancements for perspective displays: perspective parameters. *Proceedings of*

the *International Conference on Systems, Man, and Cybernetics.* IEEE Catalog 85CH2253-3, 815–818.
Eyles, D. (1986). Space stations thrillers unfold at Draper Lab. *Aerospace America,* **24**, 38–41.
Falzon, P. (1982). Display structures: compatibility with the operators mental representations and reasoning processes. *Proceedings of the 2nd European Annual Conference on Human Decision Making and Manual Control,* Wachtberg-Werthoven, Federal Republic of Germany: Forschungsinstitut für Anthropotechnik. 297–305.
Fisher, S.S., McGreevy, M.W., Humphries, J. and Robinett, W. (1986). Virtual environment display system. ACM 1986 Workshop on 3D Interactive Graphics, Chapel Hill, NC.
Foley, J.D. and Van Dam, A. (1982). *Fundamentals of Interactive Computer Graphics.* Boston: Addison-Wesley.
Gombrich, E.H. (1969). *Art and Illusion.* Princeton, N.J.: Princeton University Press.
Gregory, R.L. (1970). *The Intelligent Eye.* New York: McGraw-Hill.
Goldstein, E.B. (1987). Spatial layout, orientation relative to the observer, and perceived projection in pictures viewed at an angle. *Journal of Experimental Psychology,* **13**, 256–266.
Grunwald, A. and Ellis, S.R. (1986). Spatial orientation by familiarity cues. *Proceedings 6th European Annual Conference on Manual Control,* University of Keele, Great Britain.
Grunwald, A. and Ellis, S.R. (1987). Interactive orbital proximity operations planning system. *NASA TP-2839.*
Held, R., Efstathiou, A. and Greene, M. (1966). Adaptation to displaced and delayed visual feedback from the hand. *Journal of Experimental Psychology,* **72**, 887–891.
Helmholtz, H. *Handbook of Physiological Optics (1856–1866).* Southall transcript, Optical Society of America (1924), Rochester, NY.
Herbst, P.J., Wolff, D.E., Ewing, D. and Jones, L.R. (1946). The TELERAN proposal. *Electronics,* **19**, 125–127.
Howard, I. (1982). *Human Visual Orientation.* New York: Wiley.
Ittelson, W.H. (1951). Size as a cue to distance: static localization. *American Journal of Psychology,* **64**, 54–67.
Jenks, G.F. and Brown, D.A. (1966). Three-dimensional map construction. *Science,* **154**, 857–846.
Kim, W.S., Ellis, S.R., Tyler, M. and Hannaford, B. (1987). A quantitative evaluation of perspective and stereoscopic displays in three-axis tracking tasks. *IEEE Transactions on Systems Man and Cybernetics,* **SMC-17**, 61–71.
King, H.C. (1978). *Geared to the Stars.* Toronto: University of Toronto Press.
Marcus, A. (1984). Corporate identity for iconic interface design: The graphic design perspective, *IEEE Computer Graphics and Applications,* IEEE Computer Society, 4, No. 7, New York, 24ff.
Marcus, A., Arent, M. and Brown, B. (1985). Screen design guidelines, *Proceedings National Computer Graphics Association Annual Conference and Exposition,* 14–18 April 1985, 105–137.
McGreevy, M.W. (1990). The exploration metaphor, *NASA TM,* in preparation.
McGreevy, M.W. and Ellis, S.R. (1986). The effects of perspective geometry on judged direction in spatial information instruments. *Human Factors,* **28**, 421–438.
Meagher, D.J. (1985). Surgery by computer. *New Scientist,* **21**.
Meagher, D.J. (1987). Manipulation analysis and display of 3D medical objects using octtree encoding. *Innovative Technology in Biological Medicine,* **8**, 1.
Muybridge, E. (1975). *Animals in Motion,* Lewis S. Brown, (Ed.), New York: Dover.
Nagata, S. (1986). How to reinforce perception of depth in single two-dimensional pictures. Selected papers in basic research, No. 44–51, *Proceedings 1984 SID,* **25**, 239–246.

Parker, D.E., Renschke, M.F., Arrott, A.P., Homick, J. and Lichtenberg, B. (1986). Otolyth tilt-translation reinterpretation following prolonged weightlessness: implications for preflight training. *Aviation, Space, and Environmental Medicine*, **56**, 601–606.

Piaget, J. and Inhelder, B. (1956). *The Child's Conception of Space*. London: Routledge and Kegan Paul.

Roscoe, S.N. (1984). Judgements of size and distance with imaging displays. *Human Factors*, **26**, 617–629.

Roscoe, S.N. (1987). The trouble with HUDs and HMDs. *Human Factors Society Bulletin*, **30**, 1–3.

Russell, G. and Miles, R.B. (1987). Display and perception of 3-D space-filling data. *Applied Optics* **26**, 973–982.

Sutherland, I.E. (1970). Computer displays. *Scientific American*, **222**, 57–81.

Tachi, S., Arai, H. and Maeda, T. (1989). Tele-existence visual display for remote manipulation with a real-time sensation of presence. *Proceedings of 20th International Symposium on Industrial Robotics*, October, 4–6, 1989, Toyko, Japan, 427–434.

Weintraub, D.J., Haines, R.F. and Randle, R.J. (1985). Head up display: HUD utility II: runway to HUD. *Proceedings of 29th Meeting Human Factors Society*, University of Michigan, Ann Arbor.

PART II
Knowing

Knowing

Mary K. Kaiser

Introduction

People are naturally spatial thinkers. We spontaneously encode much of the spatial layout and regularities of our environment. Our perceptual system seems to have incorporated spatial and dynamic properties of the environment, and readily accesses this knowledge to intuitively clarify and interpret visual events. In our perception, action, and cognition, we humans behave as spatially sophisticated beings. As such, we bring a good deal of spatial knowledge (and some biases) to problems couched in visual terms.

The importance that spatial, or visual, thinking has played in gaining conceptual insights has been frequently cited, both in anecdotes and in systematic studies. A well-known example involves Friedrich Kekule's insight concerning the molecular structure of benzene; its ring structure was reportedly inspired by a dream of a snake biting its own tail. More systematic examination of the role visual imagery has played in both the development and intuitive testing of scientific models can be found in a book by Miller (1989). Miller goes so far as to argue that the "visualizability" of proposed models became a significant aesthetic criterion in the late 19th and early 20th century community of German physicists.

If, however, visual-spatial understanding is to function as a truly effective form of knowledge, we must have a way to share this knowledge with others. Thus, an ongoing challenge has been to find effective languages and media for the communication of this knowledge. A whole new discipline devoted to this communication has emerged in the past decade (McCormick et al., 1987; Friedhoff and Benzon, 1989). Called scientific visualization, its goal is to utilize new and emerging computer graphics technology to enable scientists to generate spatially sophisticated but visually comprehensible representations of their data.

The first two chapters in this section consider some of the intuitions and constraints people bring with them to the arena of spatial visualization. Subsequent chapters then describe new technologies and new applications which allow us to see and understand our world in new ways. In some cases, emerging technical developments provide new freedoms and capabilities for well-established visualization disciplines. In other cases, we are able to build visual instantiations of worlds we could previously only imagine. And finally, we are able to develop new mappings between abstract concepts and synthetic visual worlds, to expand our abilities to understand by exploiting our ability to see.

Intuitive understandings and innate biases

What we understand about our environment constrains our visual experiences and memories. In their chapter on perceiving environmental properties, Proffitt and Kaiser (Chapter 3) discuss how observers' expectations and assumptions determine the minimal conditions for specifying environmental properties. Thus, what we know about objects and the way in which they behave influences our interpretation of visual displays. Barbara Tversky (Chapter 4) considers how our memories of spatial layouts are altered and distorted by our tendency to seek alignment and symmetry in visual forms. Understanding such constraints on visual processing assists those who seek to use visual media for information transfer.

Expanding traditional disciplines

Two of the chapters in this section demonstrate how emergent technologies can build upon existing disciplines. Medical illustrators, from the time of Leonardo, have developed many techniques to enhance the spatial information in anatomical illustrations. McConathy and Doyle (Chapter 6) describe these traditional techniques, and then discuss how modern computers can introduce a more convenient level of user interaction, increasing the efficacy of biomedical visualization tools.

The discipline of cartography can also benefit from recent technological developments, as McCleary *et al.* discuss in their chapter (Chapter 5). From the time of his earliest explorations, man has sought ways to record and communicate his discoveries of new lands, trails, and navigational routes. Centuries of insights concerning effective cartographic designs can provide guidelines as modern day pathfinders create new maps of where man and his machines have most recently explored.

Expanding the limits of vision

Exploiting people's visual-spatial talents, some visualization technologies enhance our understanding of physical systems that are in reality impossible to see. Blinn's Mechanical Universe (Chapter 9) invites us to observe electrons and magnetic fields, travel to the outer planets, and ride a railroad flatcar to

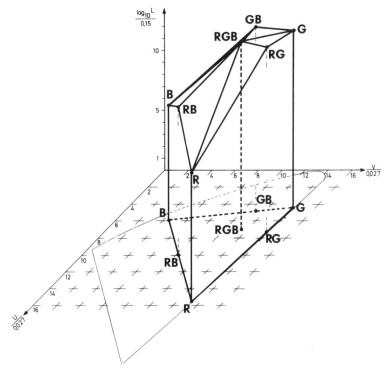

Figure 1. Kraiss and Widdel's (1989) three-dimensional diagram of semantic distances between colors in photo-colorimetric space. The axes are scaled to just noticeable differences, thus equating metric and psychological distances.

understand better time-space relativity. By allowing us visually to explore domains outside our typical perceptual experiences, The Mechanical Universe encourages an understanding of the abstract by grounding physics concepts within common perceptual experience.

Grunwald and Ellis's orbital proximity operations planner (Chapter 13) also takes us to a perceptual scale outside of common experience to give us a synthetic view of an orbital system, and lets us actually view predicted trajectories, force vectors, and plume constraints. Their graphic visualization makes comprehensible the forces and motions which, when experienced in the reference frame of the orbiting spacecraft, are counterintuitive and unexpected. Eyles (Chapter 12) similarly gives us a visualization tool to see the larger scale characteristics of spacecraft orbits.

These three chapters describe unique visualization tools which provide observers with new capabilities by adapting scales of time and space, providing vantage points not actually available to observers, and by making invisible forces visible.

Expanding the realm of visual experience

Visualization can take even broader leaps, lending spatial instantiations to nonspatial information. In their chapter, Russell and Miles (Chapter 8) discuss visualization techniques for three-dimensional data. Here, the issue is how three-dimensional space can best be utilized to clarify the relationships among three metric variables. In another example presented at the conference, Kraiss and his colleagues demonstrated a technique of data representation which utilized color space to map semantic similarities. In order to describe color space, Kraiss used the spatial diagram shown in Figure 1. Color is defined along three orthogonal dimensions: two describe chromaticity (u and v), the third describes luminance. The axes are scaled to just noticeable differences (jnd's), so that equal distances in the depicted space map equal distances in psychological "space"; distance is thus a metric of perceived similarity. Since Kraiss is using a 2D projective drawing of this 3D space, he employs geometric enhancements (e.g., vertical drop lines) to aid viewers' spatial interpretation.

Thus spatial displays and spatial instruments are not only used as a means to communicate knowledge about inherently spatial environments. They are also used to provide views of previously invisible physical systems and allow the perceptual exploration of new worlds. Finally, these displays and instruments can embody spatial metaphors which exploit our propensity for spatial thought to understand structures from nonspatial domains better. We build upon our spatial competence to aid understandings in many disciplines, and in so doing often find that sight lends insight.

References

Friedhoff, R. and Benzon, W. (1989). *Visualization: The Second Computer Revolution.* New York: Harry N. Abrams, Inc.

Kraiss, K.F. and Widdel, H. (1989). The photocolorimetric space as a medium for the representation of spatial data. In *Spatial Displays and Spatial Instruments*, Stephen R. Ellis, Mary K. Kaiser and Arthur J. Grunwald, (Eds). *NASA CP 10032.*

McCormick, B.H., Defanti, T.A. and Brown, M.D. (Eds) (1987). *Visualization in Scientific Computing.* New York: *AMC SIGGRAPH.*

Miller, A.I. (1986). *Imagery in Scientific Thought: Creating 20th-Century Physics.* Cambridge, MA: MIT Press.

3

Perceiving environmental properties from motion information: minimal conditions

Dennis R. Proffitt

University of Virginia
Charlottesville, Virginia

and

Mary K. Kaiser

NASA Ames Research Center
Moffett Field, California

Introduction

Everyday perception occurs in a context of nested motions. Eyes move within heads, heads move on bodies, and bodies move in surroundings that are filled with objects, many of which can themselves move (Gibson, 1966). Motion is omnipresent in perception. Stabilize an image on the retina and it rapidly becomes imperceptible (Pritchard, 1961). Not only is motion a necessary condition for perception, but it is also a sufficient condition for the perception of a variety of environmental properties.

Until recently, spatial instruments had few degrees of freedom with respect to the sorts of motion-carried information that they could provide. With increasing opportunities to employ animation, spatial instruments can be crafted that are tied less to artificial conventions and more to the natural condition of everyday perceptual experience.

The implications of perception research for display design derive from the methods employed by visual scientists in their investigations of how people extract environmental properties from optical information. The approach taken in perception research involves a seeking of *minimal stimulus conditions* for perceiving these properties. Stimuli that typically evoke relevant perceptions are decomposed into minimal information sources, and these sources are evaluated separately. It is almost always found that we humans rely on a large variety of information sources in perceiving any particular aspect of the environment. Knowledge of minimal conditions for perceiving environmental properties can be utilized in the design of effective and technologically efficient spatial instruments.

Since motion information is a minimally sufficient condition for perceiving numerous environmental properties, its use in spatial instruments eliminates the need to employ most of the conventions typically found in static displays. Moreover, in some contexts animated displays can elicit more accurate perceptions than are possible for static displays.

In this chapter, we discuss the status of motion as a minimal information source for perceiving the environmental properties of surface segregation, three-dimensional (3D) form, displacement, and dynamics. The selection of these particular properties was motivated by a desire to present research on perceiving properties that span the range of dimensional complexity.

Surface segregation

Surface segregation refers to the separation of distinct surfaces in depth. In order to represent surface segregation on a two-dimensional (2D) display surface, the surfaces must be distinguished by some apparent optical differences. These distinctions can be achieved with either static images or animated displays; however, only with motion can surface segregation be specified by a single cue without introducing ambiguous depth-order relations. Moreover, the implicit viewer assumptions needed to interpret moving displays are derived from the laws of dynamics, and thus are more fundamental in nature than are those accessed in interpreting static displays.

Perceiving surface segregation in static images

In pictures, surfaces are typically distinguished by color contrasts produced by differences in intensity or wavelength. One surface thereby becomes separated from another at an edge. Figure 1 depicts the familiar faces-vase figure introduced by Rubin (1915). This figure exemplifies the inherent figure-ground ambiguity of all static displays. Here, depending upon which is taken as figure, the vase or the faces, depth-order relations reverse (depth order being a term that refers to what is in front of what).

In order to resolve this depth-order ambiguity, additional cues must be supplied. One effective cue is occlusion. As is shown in Figure 2, having one surface appear to be partially covered by another is an effective convention for specifying depth order. It is important to realize, however, that the disambiguation of Figure 2 is achieved only through the activation of implicit assumptions or biases on the part of the viewer. The viewer must assume that the apparent far surface does not, in fact, have a notch cut out of it. As the Ames demonstrations show, if this assumption is violated, viewers will see erroneous depth-order relations (Ittelson, 1968).

Figure 1. Rubin's (1915) faces–vase figure.

Another static convention that helps to resolve depth-order ambiguity is the use of familiar surfaces. In Figure 3, the "A" is typically seen in front of the background surface. As Figure 1 showed, what is taken as figure—vases or face—is perceived as being in front of the apparent ground (Rubin, 1915). This perceptual bias can be exploited by representing the intended forward surface with a familiar figure. However, as with occlusion, this convention relies heavily on inherent viewer biases. The A is *assumed* to have been placed atop the surrounding surface, as opposed to having been cut out of it. This assumption may be in error.

The inclusion of additional cues, such as shading, perspective, or solid modeling, will further constrain depth-order interpretations. However, so long as the viewer cannot obtain multiple perspectives on the objects depicted, the display remains inherently ambiguous. Again, the Ames demonstrations serve to show that observers can always be made to have erroneous perceptions whenever they are constrained to view an object from a unique perspective.

Intermediate between static and animated displays are those that include flicker. Wong and Weisstein (1987) found that surface segregation is observed in displays consisting of randomly placed dots when a particular region is made to flicker. Moreover, the flickering region usually appears to be behind adjacent nonflickering regions. Spatial instruments have yet to exploit this perceptual influence of flicker.

Perceiving surface segregation in motion displays

The ability of motion information to specify surface segregation without depth-order ambiguity was demonstrated by Gibson *et al.* (1969). They

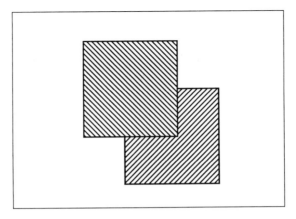

Figure 2. Two surfaces are depicted. The one to the left appears partially to occlude the surface to the right.

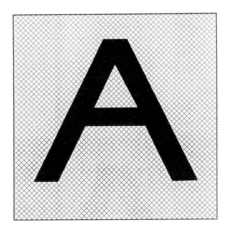

Figure 3. The familiar figure, A, appears to be in front of the background surface.

produced movies of randomly textured surfaces. When the surfaces were superimposed and stationary, segregation could not be achieved. However, when one or both of the surfaces moved, they separated into distinct surfaces and their depth order became unequivocal.

It was thought that the ongoing occlusion of the far surface by the near one served as the essential source of information for the surface segregation demonstration of Gibson *et al*. Recently, however, Yonas *et al*. (1987) showed that surface segregation could be achieved without ongoing occlusion occurring at surface edges. They created a computer-animated display in which surfaces were defined by randomly positioned points of light. As with the original Gibson *et al*. display, when the simulated surfaces were stationary, there was no information suggesting that more than one surface was present; however, when the surfaces moved, their segregation became apparent. In this case, segregation and depth order were specified by the relative motion of point-lights on different surfaces, and by the disappearance of the lights on the far surface when they passed beneath the subjective contour that defined the edge of the close surface.

There are, of course, implicit assumptions that must be made in interpreting moving displays; however, they are of a fundamentally different sort than those that were discussed for static presentations. For static displays, the assumptions are characterized by notions of likelihood and simplicity. It is highly unlikely that anyone would create a display such as Figure 2 with the intent of depicting a square located behind a notched square. Moreover, by any criterion of simplicity, the obvious interpretation of Figure 2 is the simpler of the two (or three) depth-order alternatives (see, for example, Leeuwenberg, 1982). For animated displays, the implicit assumptions reflect fundamental laws of dynamics. Surfaces are not destroyed or brought into being when they pass in front of, or go beyond, more distant surfaces. Unlike those accessed when viewing static displays, the assumptions engaged when perceiving animated displays are based upon dynamical laws.

Three-dimensional form

Any 2D representation of a 3D object is inherently ambiguous. This is true of both static and moving displays. The virtue of animated displays, however, is that time can substitute for the lost spatial dimension.

Implicit viewer assumptions are required to recover 3D relations from either static or moving 2D projections. As was found for perceiving surface segregation, those engaged when viewing animated displays are grounded in the laws of dynamics as opposed to the conventions of artifice.

Perceiving 3D form in static displays

Effective means for representing 3D objects and scenes were discovered by pictorial artists and evolved over time (Gombrich, 1960). Following Berkeley (1709), these pictorial conventions have come to be called secondary or pictorial depth cues. Researchers are still attempting to discover the invented techniques by which artists produced their compelling spatial effects (Kubovy, 1986).

The list of secondary depth cues is a long one; however, all entries share a common origin in the motivation to overcome the ambiguity inherent in 2D representations of a 3D scene. The resolution of ambiguity through the implementation of such conventions as solid modeling, perspective, shading, occlusion, familiarity, and so forth is more apparent than real. Demonstrations, such as those of Ames (Ittleson, 1968), show that perception can always be in error when inferring 3D structure from a single 2D projection. The possibility of such errors reflects, in turn, on the processing assumptions made when interpreting static displays. As with surface segregation, assumptions grounded in likelihood and simplicity are prevalent. To these are added various assumptive geometric conventions (Kubovy, 1986).

Perceiving 3D form in motion displays

The use of geometry can show that the changing spatial pattern, produced when the image of a rotating rigid object is projected onto a 2D surface, uniquely defines the 3D configuration of the object. In addition, three projected images of four non-coplanar points undergoing rotation defines the minimal condition for the recovery of structure from motion (Ullman, 1979).

Wallach and O'Connell (1953) showed that people are able to recover 3D form when viewing 2D projections of rotating objects. They constructed wire forms and projected their shadows onto screens. Viewers of these shadows reported that they saw only 2D configurations of lines when the wire forms were stationary; however, they accurately reported on the 3D configurations when the forms were continuously rotated. Wallach and O'Connell called their demonstration the Kinetic Depth Effect, or KDE.

Interest in KDE has grown over the years. Braunstein (1962), Doner et al. (1984), Todd (1982), and many others have investigated the psychophysics of the phenomenon. Recently, a good deal of research has been directed toward the rigidity assumption.

Recall that transforming a 2D projection of a rotating form is unique to the form's 3D configuration only so long as the form remains rigid. Psychologists are much in doubt as to whether the human perceptual system actually implements a rigidity assumption when extracting structure from motion in KDE (Hochberg, 1986).

When the veracity of interpretive assumptions is evaluated, the issue of whether people utilize a rigidity assumption is less important than that such a dynamical assumption is capable of serving as the sole basis for the recovery of structure from motion. Unlike the assumptions embodied in pictorial depth cues, the rigidity assumption is grounded in the following kinematic law: objects do not distort when rotated. Our perceptual systems were formed in the context of natural constraints. The exploitation of these constraints does not require that they be embodied. The fundamental assumptive nature of the rigidity principle is not based upon whether or not it has been internalized by the perceptual system, but rather upon this fact: vision evolved in a context in which this rigidity assumption is inviolate.

It must be conceded that, in a few known circumstances, the assumptions of picture perception interact with those engaged by motion perception. Ames created a trapezoidal surface that looked like a rectangular window viewed at an angle. When observers viewed it monocularly as it underwent rotation, they typically reported seeing an oscillating rectangular window rather than a rotating trapezoid (Ittelson, 1968). It is important to note that this event's 2D projection is, in fact, inconsistent with the rectangular percept; however, the strong influence of such pictorial assumptions as likelihood and simplicity outweigh, in this case, the motion-carried information defining the actual configuration.

Perceiving 3D structure from motion information has also been shown to occur for joined objects. Johansson (1973) placed point-lights on the joints of people and filmed them as they performed actions in the dark. When shown to observers, these movies were readily perceived as depicting people. It was later found that between 0.1 and 0.2 sec was a sufficient exposure duration for perceiving the human form in these films (Johansson, 1976).

Computational theorists have developed effective algorithms for extracting structure from these jointed events, given certain constraints on the motions of the walkers (Hoffman and Flinchbaugh, 1982; Webb and Aggarwal, 1982). These computational models implement assumptions about the local rigidity of moving limbs. In essence, the models assume that the act of rotating or translating a rod (bones in the case of point-light walkers) does not, itself, change the rod's length. This assumption is based upon a kinematic law of nature. The perceptual system may or may not have internalized this law (Proffitt and Bertenthal, 1988); however, it certainly evolved in a world that is governed by it.

Displacement

The motion of an object relative to an observer is referred to as its displacement. Displacement information can be conveyed in static displays only through the use of very artificial conventions. In moving displays, displacement information is presented directly in the natural medium of time. In addition, the perceptual system effectively segregates those motions specifying form from those that define observer-relative displacement.

Perceiving displacement in static displays

It is not difficult to represent in a static display the fact that an object is moving. What is difficult to represent is the future position that an object will achieve over time. Static representations of motion properties must rely on highly stylized conventions, the most prominent being vector depictions, such as those shown in Figure 4. Interpreting such displays not only requires one effectively to read the intended meaning of the conventions, but he or she must also be able mentally to perform the transformation suggested in the representation. People are not very good at such tasks. In fact, when

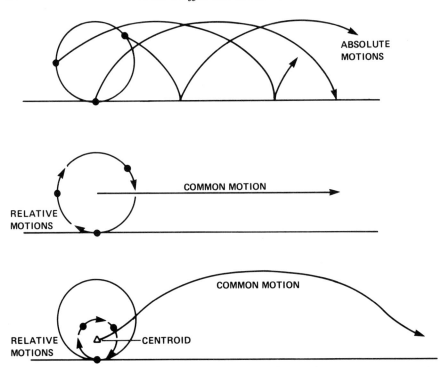

Figure 4. The top panel depicts the absolute motions of three points on a rolling wheel. The middle panel shows the relative and common motions that are perceived in this event. The bottom panel depicts the perceived motions for three points on a rolling wheel in which the configural centroid of the points does not coincide with the wheel's hub.

people attempt to extrapolate the future position of moving objects that become occluded behind barriers, they make sizable errors, particularly for complex motion functions (Jagacinski *et al.*, 1983).

Perceiving displacement in motion displays

It is rare in nature for an object to undergo a pure observer-relative translation such that every object point moves with exactly the same motion. In fact, only when objects move in horizontal circles around the observer do common linear motions project to the observer's point of observation; all nonorthogonal distal translations project a rotational component to the observer's viewpoint. The perceptual system deals effectively with complex motions by analysing them into relative and common motion components (Johansson, 1950). To illustrate this analysis, consider the perception of a rolling wheel.

As is depicted in Figure 4, except for the hub, every point on a rolling wheel follows a complex trajectory belonging to the family of cycloidal curves. These trajectories are referred to as the event's absolute motions. The perceptual system segregates these motions into two components, relative rotations and a common-observer relative displacement (Proffitt et al., 1979). This perceptual analysis selects the configural centroid as the center of relative rotations. Thus, for a rolling wheel, rotations are seen as occurring about the wheel's hub, and the common motion is seen as the hub's translation. However, if point-lights are attached to an unseen rolling wheel and the configural centroid of these lights does not correspond to the wheel's hub, then a different common motion is seen. Again, relative motions are seen as rotations about the configural centroid, but the common motion is, in this case, the prolate cycloidal path followed by this abstract centroid. This perceptual analysis has also been found to occur for configurations moving in depth (Proffitt and Cutting, 1979). It has been proposed that the selection of the configural centroid, as the center for perceived relative motions, reflects a perceptual preference to minimize relative motions; in centroid relative rotations, all instantaneous relative motions sum to zero (Cutting and Proffitt, 1982).

Research findings on the perceptual analysis of absolute motions into relative and common components have two implications for display design. First, object configuration interacts with displacement perception. Whenever an object undergoes a complex motion, its configural properties influence the common motions that are observed. Although the effects are somewhat different, robust configural influences have also been shown to occur in stroboscopically presented apparent motions (Proffitt et al., 1988). Second, relative and common motions have different perceptual significances (Proffitt and Cutting, 1980). As is depicted in Figure 5, relative rotations are used perceptually to define 3D form, whereas common motions are residual to form analysis, and define observer relative displacements.

Dynamics

The laws of dynamics place constraints on the sorts of motions that can occur in nature. Given these constraints, the patterns observed in natural motions reflect back upon underlying dynamical properties. The motions of colliding objects are a good example of this reciprocal specification of dynamic and kinematic properties.

When objects collide, the laws of linear momentum conservation state that post-collision motions must preserve the event's pre-collision momentum. (For the sake of simplicity, we exclude considerations of friction and damping.) Given these laws, it can be shown that the ratio of masses for the objects involved in a collision are specified by ratios in their velocities (Runeson, 1977). It has been found that people are relatively good at judging mass ratios when observing collisions (Todd and Warren, 1982; Kaiser and

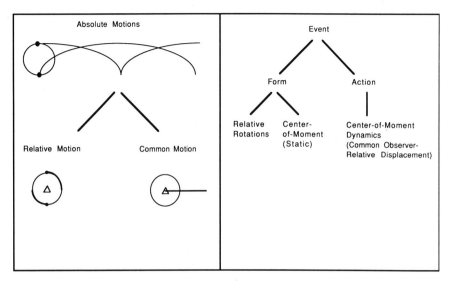

Figure 5. The perceptual system divides absolute motions into relative and common components. The relative rotations are used in form analysis, whereas the form's common motion defines its observer-relative displacement.

Proffitt, 1984). In addition, people are able accurately to discriminate possible collisions from those that violate dynamical principles (Kaiser and Proffitt, 1987a).

These results do not necessarily imply that the human perceptual system has internalized physical conservation laws, and in fact, the results of recent studies strongly suggest that such laws are not inherent to perceptual processing (Gilden and Proffitt, 1989). However, as has been previously discussed for surface segregation and form perception, our sensory systems need not embody natural laws in order to take advantage of the fact that they evolved in an environment in which dynamical laws are always upheld. Motion information is fundamental because dynamical constraints shaped the natural environment in which vision evolved.

The interpretation of static displays requires processing rules shaped in the context of pictorial conventions. The conceptual heritage of static information-processing rules is reflected in their subservience to cognitive beliefs. People hold inaccurate common-sense views about natural dynamics. These erroneous beliefs are reflected in their judgments of static, but not moving, displays.

Perceiving dynamics in static displays

Recently, an intriguing literature has developed on people's naive beliefs about the laws of dynamics. Called "intuitive physics" by McCloskey (1983),

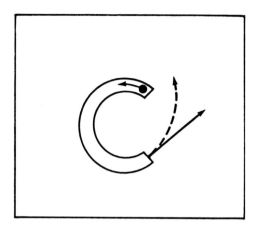

Figure 6. Depicted is a horizontal C-shaped tube through which a ball is rolled. The two drawn trajectories represent the correct path that the ball takes upon exiting the tube, and a frequently drawn erroneous path.

these beliefs influence people's predictions about natural motions; moreover, they are often at odds with the laws of dynamics.

Figure 6 shows one of the problems used by McCloskey *et al.* (1980). Depicted is a C-shaped tube that is lying flat on a horizontal surface. A ball is rolled through the tube and, upon exiting, the ball rolls across the surface. Subjects were asked to predict the path taken when the ball exited the tube. Approximately 45 per cent of the undergraduate subjects who were asked this question incorrectly stated that the ball would continue to follow a curved path. McCloskey and his colleagues have conducted numerous similar experiments, all showing that judgments made about natural object motions often reflect erroneous beliefs.

All of these studies required people to make judgments while looking at pictures. The influence of intuitive physics beliefs is pervasive only in such static contexts. These beliefs have been found to have little or no effect on the perception of animated displays.

Perceiving dynamics in motion displays

We replicated McCloskey *et al.*'s finding with the C-shaped tube problem, using a design in which observers were asked to judge which of a set of drawn trajectories appeared correct. Then, using the same design, we showed observers animated simulations of balls rolling through C-shaped tubes. Upon exiting the tubes, the balls followed a variety of paths. We found that people almost always chose as correct the natural trajectory when viewing these moving displays, and judged their erroneous predictions as being anomalous (Kaiser *et al.*, 1985). We have demonstrated this superiority of motion displays to evoke accurate dynamical judgments in other contexts (Kaiser and Proffitt, 1987b).

Static representations elicit intuitions that reflect cognitive beliefs. Obviously, people would have great difficulty getting about in the world if their perceptions were always tied to their knowledge of physical principles. A baseball outfielder, for example, would probably never succeed in catching a flyball if he was required to plan his pursuit using only his knowledge of physics.

Everyday perceptions necessarily occur in a context of naturally constrained motions. In such circumstances, our perceptual systems can function without recourse to memorial conceptions. Perception is good in motion context because motion is fundamental to the rules of perceptual processing.

Conclusions

Motion is an effective source of information for perceiving a variety of environmental properties. Because it is a minimally sufficient information source, it need not be simply added to the conventions employed in static displays. Rather, motion can replace many of these conventions, and in some contexts, motion can elicit more accurate perceptions than are possible for static displays.

Motion information is fundamental to everyday perception. The interpretive assumptions required to extract structure from motion are based upon the laws of nature—i.e., natural dynamics—whereas those evoked by static displays are based upon the artificial conventions of pictorial representations. The advantage that motion displays have over static ones derives from the heritages of the perceptual processes needed for their interpretation. The perceptual processes required to extract structure from motion information were formed in the context of dynamical constraints. The interpretation of static information relies more on perceptual processes that arise with conceptual development, and thus are grounded in such experientially based notions as simplicity, familiarity, and geometrical conventions.

References

Berkeley, G. (1709). *An Essay Towards a New Theory of Vision*. London: Dent.
Braunstein, M.L. (1962). The perception of depth through motion. *Psychology Bulletin*, **59**, 422–433.
Cutting, J.E. and Proffitt, D.R. (1982). The minimum principle and the perception of absolute, common, and relative motions. *Cognitive Psychology*, **14**, 211–246.
Doner, J., Lappin, J.S. and Perfetto, G. (1984). Detection of three-dimensional structure in moving optical patterns. *Journal of Experimental Psychology: Human Perception and Performance*, **10**, 1–11.

Gibson, J.J. (1966). *The Senses Considered as Perceptual Systems*. Boston: Houghton Mifflin.

Gibson, J.J., Kaplan, G.A., Reynolds, H.N. and Wheeler, K. (1969). The change from visible to invisible: A study of optical transitions. *Perception and Psychophysics*, **5**, 113–116.

Gilden, D.L. and Proffitt, D.R. (1989). Understanding collision dynamics. *Journal of Experimental Psychology: Human Perception Performance*, **15**, 372–383.

Gombrich, E.H. (1960). *Art and Illusion: A Study in the Psychology of Pictorial Representation*. Princeton, NJ: Princeton.

Hochberg, J. (1986). Representation of motion and space in video and cinematic displays. In Boff, K.R., Kaufman, L. and Thomas, J.P. (Eds), *Handbook of Perception and Human Performance*, **1**, 22: 1–63. New York: Wiley.

Hoffman, D.D. and Flinchbaugh, B.E. (1982). The interpretation of biological motion. *Biological Cybernetics*, **42**, 195–204.

Ittelson, W.H. (1968). *The Ames Demonstrations in Perception*. New York: Hafner.

Jagacinski, R.J., Johnson, W.W. and Miller, R.A. (1983). Quantifying the cognitive trajectories of extrapolated movements. *Journal of Experimental Psychology: Human Perception Performance*, **9**, 43–57.

Johansson, G. (1950). *Configuration in Event Perception*. Uppsala, Sweden: Almqvist and Wiksell.

Johansson, G. (1973). Visual perception of biological motion and a model for its analysis. *Perception and Psychophysics*, **14**, 201–211.

Johansson, G. (1976). Spatio-temporal differentiation and integration in visual motion perception. *Psychological Research*, **38**, 379–396.

Kaiser, M.K. and Proffitt, D.R. (1984). The development of sensitivity to causally-relevant dynamic information. *Child Development*, **55**, 1614–1624.

Kaiser, M.K. and Proffitt, D.R. (1987a). Observers' sensitivity to dynamic anomalies in collisions. *Perception and Psychophysics*, **42**, 275–280.

Kaiser, M.K. and Proffitt, D.R. (1987b). *Naive mechanics: Erroneous beliefs, but veridical Perceptions*. Manuscript submitted for publication.

Kaiser, M.K., Proffitt, D.R. and Anderson, K. (1985). Judgments of natural and anomalous trajectories in the presence and absence of motion. *Journal of Experimental Psychology: Learning, Memory, and Cognition*, **11**, 795–803.

Kubovy, M. (1986). *The Psychology of Perspective and Renaissance Art*. Cambridge, England: Cambridge.

Leeuwenberg, E. (1982). Metrical aspects of patterns and structural information theory. In Beck, J. (Ed.), *Organization and Representation in Perception*, 57–71. Hillsdale, NJ: Erlbaum

McCloskey, M. (1983). Intuitive physics. *Scientific American*, **248**(4), 122–130.

McCloskey, M., Caramazza, A. and Green, B. (1980). Curvilinear motion in the absence of external forces: Naive beliefs about the motion of objects. *Science*, **210**, 1139–1141.

Pritchard, R.M. (1961). Stabilized images on the retina. *Scientific American*, **204**, 72–78.

Proffitt, D.R. and Bertenthal, B.I. (1988). Recovering connectivity from moving point-light displays. In Martin, W.N. and Aggarwal, J.K. (Eds), *Motion Understanding*, 298–328. Hingham MA: Kluwer.

Proffitt, D.R. and Cutting, J.E. (1979). Perceiving the centroid of configurations on a rolling wheel. *Perception and Psychophysics*, **25**, 389–398.

Proffitt, D.R. and Cutting, J.E. (1980). An invariant for wheel-generated motions and the logic of its determination. *Perception*, **9**, 435–449.

Proffitt, D.R., Cutting, J.E. and Stier, D.M. (1979). Perception of wheel-generated motions. *Journal of Experimental Psychology, Human Perception and Performance*, **5**, 289–302.

Proffitt, D.R., Gilden, D.L., Kaiser, M.K. and Whelan, S.M. (1988, in press). The effect of configural orientation on perceived trajectory in apparent motion. *Perception & Psychophysics*, **43**, 465–474.

Rubin, E. (1915). *Synsoplevede Figurer*. Copenhagen: Cyldendalske.

Runeson, S. (1977). "On visual perception of dynamic events". Unpublished doctoral dissertation, University of Uppsala, Uppsala, Sweden.

Todd, J.T. (1982). Visual information about rigid and nonrigid motion: A geometric analysis. *Journal of Experimental Psychology, Human Perception Performance*, **8**, 238–252.

Todd, J.T. and Warren, W.H. (1982). Visual perception of relative mass in dynamic events. *Perception*, **11**, 325–335.

Ullman, S. (1979). *The Interpretation of Visual Motion*. Cambridge, MA: MIT Press.

Wallach, H. and O'Connell, D.N. (1953). The kinetic depth effect. *Journal of Experimental Psychology*, **45**, 205–217.

Webb, J.A. and Aggarwal, J.K. (1982). Structure from motion of rigid and jointed objects. *Artificial Intelligence*, **19**, 107–130.

Wong, E. and Weisstein, N. (1987). The effects of flicker on the perception of figure and ground. *Perception and Psychophysics*, **41**, 440–448.

Yonas, A., Craton, L.G. and Thompson, W.B. (1987). Relative motion: Kinetic information for the order of depth at an edge. *Perception and Psychophysics*, **41**, 53–59.

4

Distortions in memory for visual displays

Barbara Tversky

Stanford University
Stanford, California

Abstract

Systematic errors in perception and memory present a challenge to theories of perception and memory and to applied psychologists interested in overcoming them as well. The present paper reviews a number of systematic errors in memory for maps and graphs, and accounts for them by an analysis of the perceptual processing presumed to occur in comprehension of maps and graphs.

Visual stimuli, like verbal stimuli, are organized in comprehension and memory. For visual stimuli, the organization is a consequence of perceptual processing, which is bottom-up or data-driven in its earlier stages, but top-down and affected by conceptual knowledge later on. Segregation of figure from ground is an early process, and figure recognition later; for both, symmetry is a rapidly detected and ecologically valid cue. Once isolated, figures are organized relative to one another and relative to a frame of reference. Both perceptual (e.g., salience) and conceptual factors (e.g., significance) seem likely to affect selection of a reference frame.

Consistent with the analysis, subjects perceived and remembered curves in graphs and rivers in maps as more symmetric than they actually were. Symmetry, useful for detecting and recognizing figures, led to distortions in map and graph figures alike. Top-down processes also seem to operate in that calling attention to the symmetry vs. asymmetry of a slightly asymmetric curve yielded memory errors in the direction of the description. Conceptual frame of reference effects were demonstrated in memory for lines embedded in graphs. In earlier work, the orientation of map figures was distorted in memory toward horizontal or vertical. In recent work, graph lines, but not map lines, were remembered as closer to an imaginary 45 degree line than they had been. Reference frames are determined by both perceptual and conceptual factors, leading to selection of the canonical axes as a reference frame in maps but selection of the imaginary 45 degree line as a reference frame in graphs.

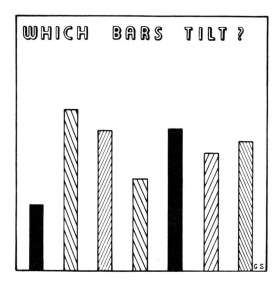

Figure 1. Hypothetical graph taken from Schultz, G.M. (1961). Beware of diagonal lines in bar graphs. The Professional Geographer, **13**, 28–29 (reprinted by Kruskal, (1982)). Reprinted by permission.

Distortions in memory for visual displays

With the best of intentions, scientists, newspaper editors, and textbook authors select graphic displays to present their ideas more clearly and more vividly to their readers. Nevertheless, some of the effects are not only unintended, but unwanted. For example, in Figure 1, presumably the striping on the bars was selected to differentiate the bars, not to instantiate the herringbone illusion, where straight lines are perceived as tilted (this example comes from Schultz, 1961 through Kruskal, 1982). In Figure 2, clipped from the business section of the August 2, 1987 *New York Times*, the graphic artist wanted to contrast two related sets of numbers, the debt and the debt service ratio, year by year. I don't think that the graphic artist intended to create a figure with such a strong tendency to reverse that it makes it difficult to focus on any one section of the graph. The next figure, 3, takes us from the realm of perceptual illusions to experiments in judgment by Cleveland *et al.* (1982). These statisticians asked knowledgeable subjects to estimate correlations from scatter plots and found that higher estimates were given when the point cloud was smaller (or the frame larger). Figure 4, popularized by Tufte (1983) and reprinted by Wainer (1980), is taken from the *Washington Post* of October 25, 1978. Here, the graphic artist probably thought it would be clever to represent the metaphor of the diminishing dollar quite literally. However, only the *length* of the dollar represents the decline of purchasing

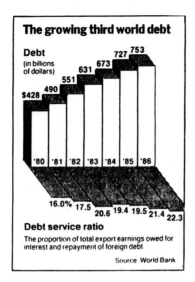

Figure 2. Graph taken from The New York Times, *August 2, 1987. Copyright 1987 by the New York Times Company. Reprinted by permission.*

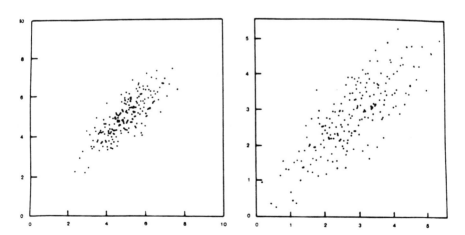

Figure 3. Stimuli used by Cleveland, Diaconis and McGill (1982). Although the correlations in the two scatterplots are the same, the right-hand one in the smaller frame is judged to be higher. Copyright 1982 by the A.A.A.S. Reprinted by permission.

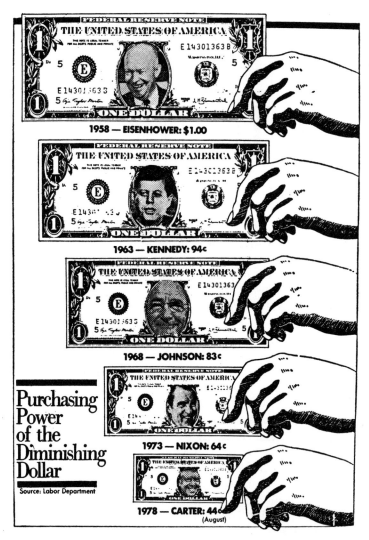

Figure 4. Graph taken from The Washington Post, *October 25, 1978 (reprinted by Tufte (1983) and Wainer (1980)). Copyright 1978 by* The Washington Post. *Reprinted by permission.*

power, not the *area*, yet it is the area that is picked up by the human observer. So, although the Carter dollar purchased a bit less than half of the Eisenhower dollar, the Carter dollar looks less than a quarter of the area of the Eisenhower dollar.

The next example of distorted perception brings me to research in my laboratory. Let me first tell you about a number of different phenomena we have studied, and then I will try to account for them in an analysis of

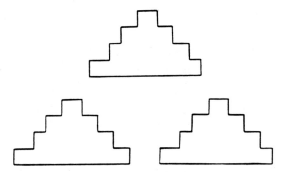

Figure 5. Figures used by Freyd and Tversky (1984). Reprinted by permission.

perceptual organization, where both perceptual and conceptual factors are operative. First, I will discuss examples of perceptual factors. Jennifer Freyd and I (1984) asked subjects to look at figures like that at the top of Figure 5, and then decide whether it was more similar to a slightly more symmetric figure or to an equally different but slightly less symmetric figure. When we selected nearly symmetric figures like that one, subjects nearly always chose the more symmetric alternative as the more similar. What's more, when subjects were asked to select which of the bottom figures was identical to the top figure, subjects were faster to select the identical figure when the alternative figure was *less* symmetric than the original (as in Figure 5) than when it was *more* symmetric than the original. These effects obtained for nearly symmetric figures, but not for less symmetric ones. That was rather complicated, but these experiments, and others like it (see Riley, 1962, and Freyd and Tversky, 1984, for reviews) suggest that there is a symmetry bias in perception. Not only do viewers rapidly detect symmetry, but they also perceive nearly symmetric figures as more symmetric than they are. That is, small deviations from symmetry are overlooked. Human faces, for example, are rarely perfectly symmetric, though we think of them as such. The outer men in Figure 6 (taken from Neville, 1977, p. 335), for example, are actually the same man at the same time. The two outer pictures were constructed by taking the right and left halves of the actual face in the center, and reproducing them in mirror image. It is only by seeing how different the two constructed symmetric faces are that we become aware of the asymmetry of the original face.

Diane Schiano and I (1989) looked for and found distortions toward symmetry in memory for maps and graphs. We presented maps or graphs like those in Figure 7 to different groups of subjects. Sometimes, the subjects were asked to sketch the curves of the graphs or the rivers of the maps, and other times, they were asked questions about the content of the maps or graphs. This was done to induce a natural comprehension attitude toward the figures, and to prevent subjects from simply memorizing line shapes. We then asked judges who knew nothing about the hypotheses to rate whether

Figure 6. Face taken from Neville (1977). *The left and right faces were constructed by taking the left and right halves of the original photograph and reproducing them in mirror image, producing faces that are symmetric, unlike the original. Reprinted with permission from* The Encyclopedia of Ignorances. *Copyright 1977, Pergamon Press.*

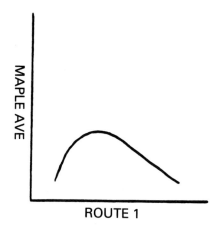

Figure 7. Map curve used by Tversky and Schiano (1987).

the drawn curves and rivers were more or less symmetric than the original ones. The remembered curves, whether in maps or graphs, were judged more symmetric than the originals. These errors in the direction of symmetry, however, apparently occur in perception, not in memory. We asked another group of subjects to copy the curves, and the copied curves were also judged to be more symmetric than the originals, and to the same degree. The first effect to be accounted for, then, is a tendency to perceive nearly symmetric figures as more symmetric than they actually are.

For the next two effects, I turn to maps. In Figure 8 are two maps of the world; which one is correct? If you are like the subjects I have run, most of you will pick the bottom one, that is, the incorrect one. Let me give you

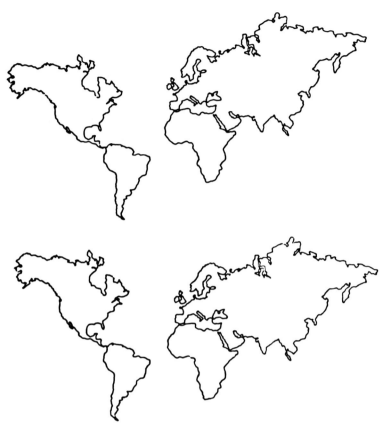

Figure 8. World map stimuli used by Tversky (1981). Subjects incorrectly prefer the lower map, in which the U.S. and Europe, and South America and Africa are more aligned.

another chance. In Figure 9 are two maps of the Americas; my apologies to Central America, which was excised not because of the political situation, but for visual reasons. Again, which map is the correct one? And again, I will predict that most of you will prefer the left, incorrect, one. Why do the incorrect maps look better? Basically, because the incorrect ones are more *aligned*. In the incorrect map of the world, the U.S. and Europe, and South America and Africa are more aligned than they are in the true map. And in the incorrect map of the Americas, North and South America are more aligned. I have found memory errors in the direction of greater alignment for these maps, for directions between major cities on them, for artificial maps, and for visual blobs (Tversky, 1981). Others have found similar results (e.g., Byrne, 1979).

The second widespread error I have found in maps I termed rotation. I

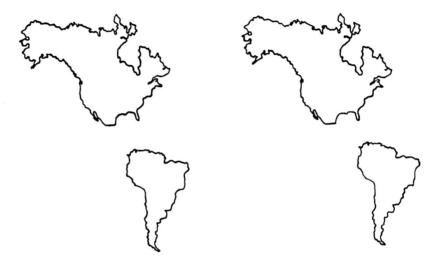

Figure 9. Map of the Americas used by Tversky (1981). Subjects prefer the incorrect left one.

asked a group of subjects to place a cut-out of South America in a frame where the canonical directions, north-south and east-west, corresponded, as usual, to the vertical and horizontal sides of the frame (Figure 10). Although the actual orientation is on the right, most of the subjects uprighted South America to the angle of the left-hand figure, or even more so. Not only South America is perceived as tilted. Those of you who live in the Bay Area, or who arrived from the San Francisco airport may think that you drove southwest to Monterey. Most of my local respondents made mistakes like that, for example, thinking that Berkeley is east of Stanford and Santa Cruz west of Palo Alto. Not so, as this true map of the area shows (Figure 11). Just as for alignment, I have found memory errors of rotation toward the axes for real map figures, for directions between cities on them, for roads, for artificial maps, and for visual blobs (Tversky, 1981). Others have found these effects as well (Byrne, 1979; Moar and Bower, 1983). Unlike the symmetry distortion, the distortions produced by alignment and rotation are stronger in memory than in perception; that is, small tendencies toward alignment and rotation appeared in a copy task, but much greater errors appeared in a memory task.

Until now, we have demonstrated that there is a bias toward symmetry in both maps and graphs that appears in perception and is preserved in memory. I have also demonstrated, primarily in maps, biases toward alignment with other figures and rotation to a vertical/horizontal frame of reference that appear slightly in perception and stronger in memory. Now is the time to start to account for these systematic errors by an analysis of perceptual organization, or more specifically, by the effects of perceptual factors in

Distortions in memory 69

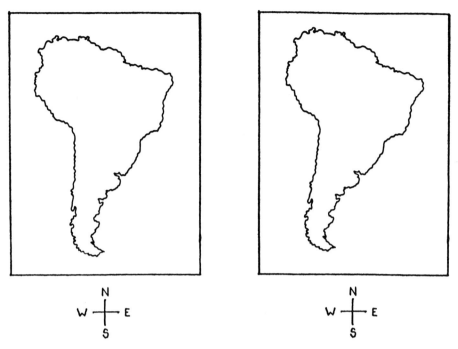

Figure 10. The correct orientation of South America is on the right, but subjects typically upright it, as in the example on the left (from Tversky, 1981). Reprinted with permission of Academic Press.

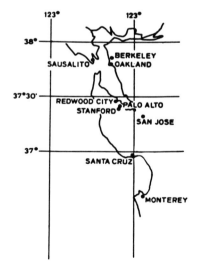

Figure 11. The correct map of the San Francisco Bay area. Subjects erroneously report that Berkeley is east of Stanford and Palo Alto is east of Monterey (from Tversky, 1981). Reprinted with permission of Academic Press.

Process	Perceptual Cue	Perceptually-induced Error	Conceptually-induced Error
Figure segregation/ identification	symmetry	toward symmetry	description enhances or reduces effect
Figure location/ orientation	other figure	toward other figure	
	frame of reference	toward frame of reference	graphic medium (e.g., maps vs. graphs) determines frame of reference (e.g., vertical/ horizontal axes vs. 45° line)

Figure 12. Summary of perceptual analysis and resultant errors.

perceptual organization (Figure 12). One of the earliest forms of spatial organization is distinguishing figures from backgrounds. Because figures are more likely to have symmetry, closure, and other, similar properties than backgrounds, these are valuable cues to figureness (e.g., Hochberg, 1978; Koffka, 1935; Kohler, 1929; Wertheimer, 1958). Symmetry, or near-symmetry, is rapidly and easily detected (e.g., Barlow and Reeves, 1979; Chipman and Mendelson, 1979; Carmody et al., 1977; Corballis, 1976). Thus, because of its usefulness in figure discrimination, symmetry seems to be rapidly detected, and small deviations from symmetry are overlooked so that nearly symmetric figures are coded and remembered as more symmetric than they really are.

Now for anchoring figures in space. In an empty field, figures appear to float, a phenomenon well-known to star-gazers, called the autokinetic effect. In order to perceive and remember the locations of figures, it is useful to anchor them to other figures and/or to a frame of reference. In fact, given that perceivers and the world are rarely static, this seems to be the only way to organize the elements of a scene. Map bodies and graph curves are figures, on backgrounds; they are often nearly symmetric, they appear sometimes with other figures, and typically in a reference frame. Although valuable in locating and orienting figures, anchors pull figures closer to them in memory, yielding systematic errors. Many aspects of a visual scene may serve as anchors: at the same level of analysis, other objects in the scene; at a higher or superordinate level of analysis, a frame of reference. Thus, the analysis of distortion in terms of perceptual organization applies to maps and graphs, and accounts for the errors of symmetry, alignment, and rotation.

This, briefly, is the perceptual analysis. Now, I'd like to present two cases where, we believe, conceptual factors enter into the perceptual analysis of maps and graphs, and yield further distortions. This work was also done with Diane Schiano (Tversky and Schiano, 1989; Schiano and Tversky,

1990). In the first case, we used verbal descriptions to bias the way figures are perceived and encoded. In the second case, we used the graphic medium, maps vs. graphs, to alter the frame of reference selected. The first effect brings us back to symmetry. The graph curves we asked subjects to study were slightly but noticeably less than symmetric. Given that people perceive such curves as more symmetric than they really are, we wondered if we could weaken or strengthen that belief or perception by an accompanying description of the curve, and consequently alter people's memory of the curve. Again, we presented a variety of graphs to subjects to remember, and tested memory either by asking subjects to draw the graphs or to describe some aspect of the relation depicted by the graph. This time the graphs also included descriptions of the functions. For the nearly-symmetric curve of interest, half the subjects received a description emphasizing its symmetry, that is, "Notice that the curve rises smoothly and falls smoothly." The other subjects received a description emphasizing its asymmetry, that is, "Notice that the curve rises sharply and falls slowly." The curves drawn from memory were given to judges who were unaware of the experimental conditions. The results were just as expected: when attention was directed to the symmetry of the curves, remembered curves were drawn more symmetric than when attention was drawn to the asymmetry of the curve. This result is reminiscent of one of the truly classic experiments in psychology, that of Carmichael et al. (1932). In their case, however, subjects were told that a simple picture, such as 0–0, was one of two objects, either eyeglasses or dumb-bells. Subsequent drawings were in the direction of the names. In our case, the verbal component was suggestive yet intended to bias the way subjects perceived and encoded the curves.

The second conceptual factor is more subtle, and addresses the issue of what determines the frame of reference. In the absence of any conceptual or meaningful factors, there are often perceptual factors that provide a frame of reference. The typically horizontal and vertical lines of the actual frame of a picture are one example (e.g., Howard and Templeton, 1971). For an environment, the natural vertical plane, up-down, and the two natural planes, for example, left-right and front-back, form a reference frame; when this is reduced from two to three dimensions, the front-back dimension drops out (e.g., Clark, 1973), usually leaving the horizontal and vertical axes of the picture frame as a reference frame. For maps, there is an additional conceptual factor that is typically perfectly correlated with the perceptually salient axes, namely the cardinal directions, north-south and east-west. Thus far, the evidence for alignment has come either from maps and environments, where both perceptual and conceptual factors suggest the horizontal and vertical as a reference frame, or from visual blobs, where perceptual factors suggest the horizontal and vertical.

Schiano and I wondered if simple straight-line functions at various angles in x–y coordinates would be anchored to those coordinates, and thus distorted toward them. Of course, the x–y coordinates form a natural reference

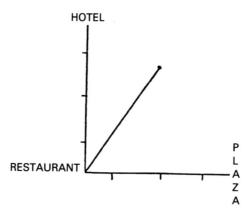

Figure 13. Straight-line maps and graphs used by Tversky and Schiano (1987).

frame for graph functions, but unlike streets, graphed functions are rarely perfectly horizontal or vertical. Moreover, there is another reference frame for graphed lines, the (in this case) implicit forty-five degree line. This is the identity line, where $x = y$, and as such it provides a very important reference point for graphed lines. Above it, rises are steep, and below it, shallow. The experiments we ran were very similar to the previous graph experiments: there were critical stimuli and distractors, and the memory task was designed to elicit comprehension of content, not just remembering the line. The exact same stimuli labelled differently were presented as maps to another group of subjects. Subjects were told that the angled lines were paths or short-cuts; they weren't very convincing maps, as can be seen in Figure 13. In contrast

Table 1. Average deviation toward 45° of drawn (remembered) lines from presented lines in maps and graphs in experiments 5 and 6

Line angle	Angles for which + denotes larger remembered angles				Angles for which + denotes smaller remembered angles			
	20°	25°	30°	35°	55°	60°	65°	70°
Experiment 5								
Graph lines	+3.4	+1.7	+1.0	+1.2	+1.1	+2.9	+2.4	+4.1
Map lines	+0.2	−0.2	+0.4	−2.2	+0.7	+0.5	0.0	+1.5
Experiment 6								
Graph lines	+1.5	+0.1	+1.0	+1.0	+2.8	+2.6	+2.0	+2.3

Copyright 1989 *American Psychological Association*, reprinted by permission of publisher from Tversky and Schiano (1989).

to the prior work on maps showing alignment to the closest axis, horizontal or vertical, the graph lines were remembered as closer to the imaginary forty-five degree line than they actually were. The map lines differed considerably and significantly from the graph lines, yet showed no systematic distortion. We ran this study again, this time using dotted graph lines rather than filled ones. Again, graph lines were remembered as closer to the forty-five degree line. The results of both experiments appear in Table 1. In yet another study (Schiano and Tversky, 1990), we presented lines in axes as a simple memory or copying task, without labels and without calling them graphs. The pattern of errors was systematic, yet very different from the present pattern. However, when we invoked the imaginary forty-five degree line as an anchor by asking subjects to judge whether the line was above or below forty-five degrees before drawing it, remembered lines were drawn closer to the forty-five degree line. This is evidence, we believe, for conceptual factors that influence selection of frame of reference and thereby affect the perceptual analysis, representation, and veridicality of memory of visual displays.

I have presented an analysis of the perceptual processes, particularly figure segregation and organization, that are used in viewing and comprehending visual stimuli. Both these processes can lead to systematic distortions, which were demonstrated in perception and memory of maps and graphs. Thus symmetry, a property more likely to appear in figures than grounds, and therefore useful in figure segregation, is exaggerated in memory for curves in graphs and rivers in maps. Conceptual factors were also shown to affect the perceptual analysis and encoding of visual scenes, and to yield errors of memory, the description of symmetry in one case, and the selection of a frame of reference in the other. The bottom line is "What you see ISN'T what you get".

Acknowledgement

This research was supported by N.S.F. Grant IST 8403273 to Stanford University.

References

Barlow, H.B. and Reeves, B.C. (1979). The versatility and absolute efficiency of detecting mirror symmetry in random dot displays. *Vision Research*, **19**, 783–793.
Byrne, R.W. (1979). Memory for urban geography. *Quarterly Journal of Experimental Psychology*, **31**, 147–154.
Carmichael, L., Hogan, H.P. and Walter, A.A. (1932). An experimental study of the effect of language on the reproduction of visually perceived forms. *Journal of Experimental Psychology*, **15**, 73–86.
Carmody, D.P., Nodine, C.F. and Locher, P.J. (1977). Global detection of symmetry. *Perceptual and Motor Skills*, **45**, 1267–1273.
Chipman, S.F. and Mendelson, M.J. (1979). Influence of six types of visual structure on complexity judgments in children and adults. *Journal of Experimental Psychology: Human Perception and Performance*, **5**, 365–378.
Clark, H.H. (1973). Space, time, semantics, and the child. In T.E. Moore (Ed.), *Cognitive Development and the Acquisition of Language*. New York: Academic Press.
Cleveland, W.S., Diaconis, P. and McGill, R. (1982). Variables on scatterplots look more highly correlated when the scales are increased. *Science*, **216**, 1138–1141.
Cleveland, W.S., Harris, C.S. and McGill, R. (1983). Experiments on quantitative judgments of graphs and maps. *The Bell System Technical Journal*, **62**, 1659–1674.
Corballis, M.C. (1976). *The Psychology of Left and Right*. Hillsdale, N.J.: Erlbaum.
Freyd, J. and Tversky, B. (1984). Force of symmetry in form perception. *American Journal of Psychology*, **97**, 109–126.
Hochberg, J.E. (1978). *Perception*. Second Edition. Englewood Cliffs, N.J.: Prentice-Hall.
Howard, I.P. and Templeton, W.B. (1971). *Human Spatial Orientation*. London: Wiley.
Koffka, K. (1935). *Principles of Gestalt Psychology*. New York: Harcourt, Brace Jovanovich.
Kohler, W. (1929). *Gestalt Psychology*. New York: Liveright.
Kruskal, W. (1982). Criteria for judging statistical graphics. *Utilitas Mathematica*, **21B**, 283–310.
Moar, I. and Bower, G.H. (1983). Inconsistency in spatial knowledge. *Memory and Cognition*, **11**, 107–113.
Neville, A.C. (1977). Symmetry and asymmetry problems in animals. In Duncan, R. and Weston-Smith, M. (Eds), *The Encyclopedia of Ignorance*, 331–338. New York: Pergamon.
Riley, D.A. (1962). Memory for form. In Postman, L. (Ed.), *Psychology in the Making*, 402–465. New York: Knopf.
Schiano, D. and Tversky, B. (1990). Structure and strategy in viewing simple graphs. Submitted for publication.
Schultz, G.M. (1961). Beware of diagonal lines in bar graphs. *The Professional Geographer*, **13**, 28–29.
Tufte, E.R. (1983). *The Visual Display of Quantitative Information*. Cheshire, CO: Graphics Press.

Tversky, B. (1981). Distortions in memory for maps. *Cognitive Psychology*, **13**, 407–433.
Tversky, B. and Schiano, D. (1989). Distortions in memory for graphs and maps. *Journal of Experimental Psychology: General*, **118**, 387–398.
Wainer, H. (1980). Making newspaper graphs fit to print. In Kolers, P.A., Wrolstad, M.E. and Bouma, H. (Eds), *Processing of Visible Language*, 2. New York: Plenum.
Wertheimer, M. (1958). Principles of perceptual organization. In Beardslee, D. and Wertheimer, M. (Eds), *Readings in Perception*. Princeton: Van Nostrand.

5

Cartography and map displays

George F. McCleary, Jr., George F. Jenks
Department of Geography, University of Kansas

and

Stephen R. Ellis
NASA Ames Research Center

Cartographers are creators of maps. While maps have traditionally represented the terrestrial environment, mapping today extends to extraterrestrial objects as well as to artificial (data-space) environments. Maps organize spatial information for use in performing tasks, such as way-finding, and are enhanced symbolically and geometrically to accomplish a wide range of spatial tasks.

Some maps are displays, intended for use without much study. Consider, for example, the paintings by Jasper Johns (Crighton, 1977); these are not intended to communicate accurate spatial information. They are impressionistic and serve the viewer in much the same way as propaganda maps appeal to the emotions. Other maps must be studied, analysed or measured. These are used as instruments, such as the charts used for sea or airborne navigation.

Map characteristics

The characteristics of maps can be divided into two categories: those associated with their geometry, i.e., the scale of the map and the "projection", and those related to the presentation of the content, i.e., the graphic symbols which represent the features of the environment.

The classical focus of cartographic interest in projection geometry has been on the problems associated with the transformation of the spherical earth to the map plane (Chamberlin, 1947). After reduction to some particular scale, the primary considerations have been the geometric properties of the different transformations. Since area, direction and shape cannot all be preserved at

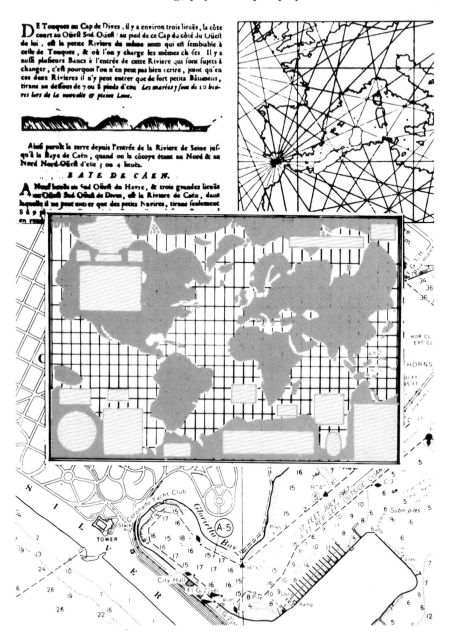

Figure 1. Examples showing the evolution of maps used as navigational instruments. Upper left: portion of a page from a coastal pilot's guide (1691). Upper right: sketch of a segment of Juan de la Cosa's portolan chart (1500). Center: sketch of the Mercator world map of 1569. Lower: portion of a modern nautical chart.

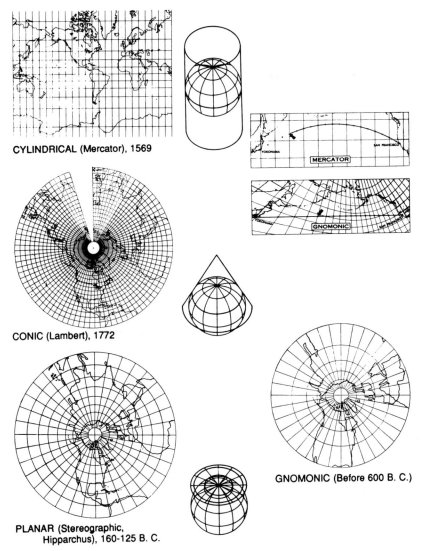

Figure 2. Families of map projections: cylindrical, conic, planar. Most map projections involve a mathematical transformation of the spherical surface onto a plane which may be visualized as a geometric projection onto a cylinder, a cone or a plane in conventional orientation. The earth's graticule then appears in a form illustrated by the three maps on the left. These three are conformal projections, and all angles are shown correctly. On the gnomonic projection, all great circles are represented as straight lines. This property, in concert with the capability of the Mercator projection to show rhumb lines as straight lines, has been of great benefit to the navigator. In the example a great circle path is shown on both the gnomonic and the Mercator.

the same time, tradeoffs must be made. The Mercator projection provides an excellent example of the design of a map. Though there are other projections used for navigation, which like the Mercator are conformal and represent angles correctly, the Mercator projection is unique in that it projects all rhumb lines, lines of constant compass direction, as straight lines. This property was achieved initially by Mercator in 1569 without mathematical analysis by introducing latitudinal scale distortion to compensate for the longitudinal scale distortion which occurs when the spherical surface, with its converging meridians, is transformed with a rectangular projection[1].

There are, however, other aspects of the Mercator projection which make it very important to this discussion. First, it does not show great circles as straight lines. Great circles, the loci of minimum distance between points on a sphere, project as straight lines on the gnomonic projection[2]. Second, in the transformation of the spherical surface to provide conformality, the Mercator projection greatly exaggerates the sizes of the areas in the polar regions. Because the projection came into widespread use for many different purposes, some inappropriate given its properties, it has influenced many map users' subjective senses of geography. This feature has caused great difficulty when it has been used for world maps designed to display statistical data. General Frederick (Lord) Morgan recognized the influence of the great areal size distortion in the middle latitude and polar regions as a key problem in gaining American support for Operation OVERLORD (Morgan, 1950). Saarinen documents the impact which the Mercator and similar projections have had on the development of personal mental maps of the world (Saarinen, 1976; Tobler, 1976; Tversky, Chapter 4).

While for small areas (up to 100 square kilometers) the choice of a map projection is of relatively little consequence, at smaller scales covering large areas, the choice of a projection for any particular mapping task is very significant. If a map is to provide information for visualization about some aspect of the environment, i.e. the geographical range of an animal species, an equivalent (equal-area) projection is the proper choice. The Lambert azimuthal is a good example and contrasts with the Lambert cylindrical

---

[1] In a rectangular projection, the parallels and meridians are laid out at the scale selected, generally using the length of the Equator. Because of the decreasing circumference of the parallels of latitude as they approach the poles, longitudinal scale is increased (it is a function of the cosine of the latitude). The compensating increase in latitudinal scale in the Mercator projection is achieved by setting the vertical map distances from the equator, $y(\phi)$, for each latitude, ϕ, such that

$$y(\phi) = \int_0^\phi \sec(\theta) d\theta$$

[2] The gnomonic is the traditional companion to the Mercator. On it all straight lines are great circles; in traditional navigation procedures, one plots the great circle route between two points, then compiles this path on the Mercator as a set of rhumb lines which are used in the navigation process.

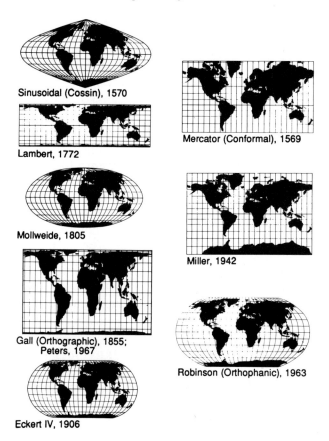

Figure 3. In the left column a series of equivalent area projections are shown. These can be compared with the Mercator, a compromise projection by Miller and the Robinson projection. The Miller is widely used, although it is not an equivalent projection, for it has both significantly less areal exaggeration than the Mercator and a Mercator-like appearance. The projection by Robinson is "orthophanic"—designed to look correct. The projection which has been strongly advocated by Peters was developed earlier by Gall; it does not, as Peters claims, resolve the conflicting demands which are placed on world projections.

(Figure 3). On both the areas on the surface are shown in correct proportions. From the large variety of equivalent projections available, one selects the one with the least angular distortion in the area of most significance. Although there is no substitute for cartographic experience in choosing a map projection, the rule "choose properties first, then minimize distortion" is a reasonable procedure. The Robinson projection, which is neither conformal nor equivalent, was designed with a clear set of properties in mind

(Robinson, 1974). The rational approach taken to the problem of creating a projection which "looks right" contrasts strongly with the claims made by Peters for his equivalent projection (Peters, 1983). Most difficult are the selections of projections which must support a variety of tasks at different scales; this is particularly significant when data are presented for large areas and the map reader has the computer-supported capability to pan across and zoom in to enlarge small areas. No less significant than the choice of projection in these cases are the problems of resolution and generalization[3] in the digital data base.

The graticule is also a significant design element for any map. Its presentation is a form of symbolic enhancement. While the cartographer has the responsibility for presenting information using the appropriate projection for the purpose of the map, the graticule serves to provide a locational context for the map. This context extends beyond simple location, for the graticule itself helps the reader interpret not only the size of the area shown but also the degree of distortion in its representation (Robinson et al., 1984).

Foremost among the problems confronting cartographers has been the mapping of the surface of the land (Harvey, 1980). Long before coastlines had been mapped accurately, cartographers had developed a number of graphic procedures to record the locations and dimensions of hills and mountains, streams and rivers. Over time, as science and industry advanced, the technology of map-making confronted a dichotomy of requirements: maps, on one hand, could be made to portray the surface so that a viewer could visually develop a clear understanding of the characteristics of the land surface. On the other hand, maps could be made to provide scientists or engineers a means for measuring and analyzing the surface (Robinson, 1982).

Among the procedures to be developed were those which produce three-dimensional maps, maps developed from oblique perspectives. Cumbersome and time-consuming to produce manually, there are remarkable examples in the works of Harrison (1944), Imhof (1965/1982), Lobeck (1958), Raisz (1944) and others. The application of the technique to other nonphysical surfaces has been made by Jenks (1967) and Robinson (1961). The potential for the use of this illustrative device is suggested in Figure 5—beyond these examples lie an inexhaustible variety of statistical surfaces as well as a wide variety of features in the natural environment. Any phenomenon which can be conceptualized as a three-dimensional surface can be mapped in this way.

Many maps handle the problems of representing the surface by using different forms of symbolization. Isarithms (contour lines), hachures or shading are used to encode surface features (such as slope, elevation, soil type,

---

[3] Cartographic generalization refers to the rules by which the content of a map is selected based on available space on the map surface. As the scale decreases, the map symbols can become crowded and hard to read. Rules for generalization amount to filters on the map content so as to maintain useful communication to the map user.

Figure 4. Evolution of land surface form mapping. Creation of a map requires a considerable amount of data. For a long time sufficient data were simply not available for detailed maps of land surfaces. As data accumulated, maps with precise graticule coordinate reference systems became possible. Verbal descriptions of the mappable objects or phenomena evolved to graphic map symbols, representations of slope and finally, and with the availability of survey data, to the mapping of elevation by contour lines. More recently shaded relief has been used to assist visualization of land surfaces in situations where contour lines themselves are inadequate.

bearing strength, etc.). Perspective projection, however, provides a more visual portrayal of surface features[4]. With the advent of computer-supported graphics, the process is relatively easy and generally fast enough to allow the map user to redraw the entire map in real time using new display parameters.

While many readers will be familiar with isometric topographic or geological block diagrams, these comprise only a fraction of the three-dimensional maps which have been, or could be, made. This response to the planimetric view, with different and often intellectually abstract forms of symbolization used to convey the topographic or statistical message, has potential utility as an illustrative device (Figure 5). "Why not use maps of this kind to illustrate submarine or fish environments; airport facilities and landing obstacles; the development of hurricanes; intersecting statistical surfaces; the movement of fog, smog, and other air pollutants; and any other areal phenomenon which can be [conceptualized] as a three-dimensional surface" or volume? (Jenks and Brown, 1966).

Basics of three-dimensional map perspective

"Three-dimensional maps present earth surfaces as viewed from a real or assumed position in space. The viewing point can be from any compass position, at any angle above or below the horizon[5]. The map image that is presented for any viewing point is theoretically formed on a transparent perspective plane. This plane is perpendicular to the central line of vision and lies at a point between the eye of the viewer and the surface being mapped. In cartographic presentation the perspective map image, or transformed map, is constructed geometrically, but it can be visualized as being formed by converging projectors (in angular-perspective presentation) or by parallel projectors (in parallel-perspective presentation)" (see Figure 6) (Jenks and Brown, 1966).

The importance of this dichotomy, between the *visual* (*display*) representation, the perspective projection, and the *metric* (*instrument*) representation, the parallel perspective, has not diminished in the more than a quarter-century since this point was made. When "Three-dimensional Map Construction" (Jenks and Brown, 1966) was published, work in computer graphics was in its infancy. Thus, while the techniques explored in their paper advanced the long tradition of manual production, the concepts described expanded the graphic display horizon for people in many different fields. In fact, while some overlook the *visual-metric* distinction in either text or software, others find it fundamental in their discussions. Foley and Van Dam, in their

---

[4] This perspective format is today often described as a 2½D display.
[5] At 90 degrees above the horizon a three-dimensional and a planimetric map would be identical and perspective effects would be minimized since the map would present all information on a plane of constant depth.

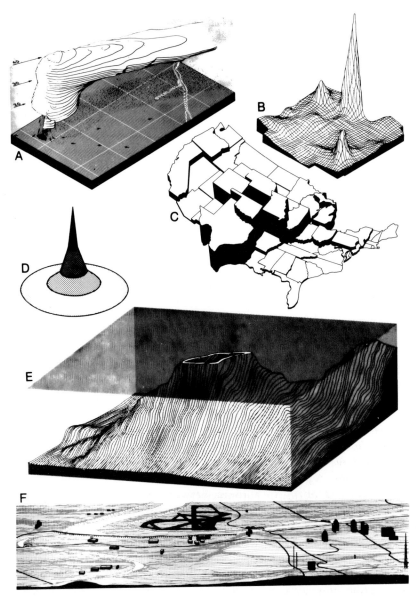

Figure 5. Three-dimensional maps vividly portray physical and conceptual surfaces that are difficult to visualize in the conventional two-dimensional format. Illustrated are (A) a thunderstorm with hail, (B) population density in central Kansas, (C) the value of goods shipped from the conterminous states to Kansas, (D) land values in a hypothetical city, (E) Canton Island, from the sea floor to the sea surface, and (F) the approach to the Kansas City Municipal Airport.

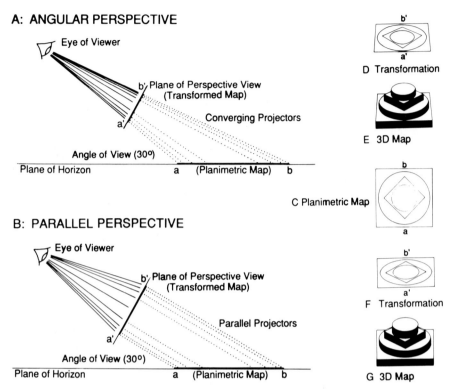

Figure 6. Perspective illustrations can be constructed using either angular or parallel geometric projection. The images are considered to be formed on transparent planes which are perpendicular to the line of sight. In angular perspective (A) an image is formed on this plane by converging projectors, while in parallel perspective (B) the image is formed by parallel projectors. Both projections are useful for three-dimensional maps. A planimetric map (C) can be transformed into either perspective view (D and F). Then, after careful consideration of the viewing axis, the elevation, and the amount of vertical exaggeration, the three-dimensional representations (E and G) can be constructed.

Fundamentals of Interactive Computer Graphics (1982) explore parallel and perspective projection at the beginning of their discussion on "Viewing in three dimensions". Hearn and Baker (1986) begin the same way, and then turn to intensity cueing, hidden-line removal and methods for surface generation, including fractal geometry and octrees. The mechanization of these processes was inevitable, and Tobler led the way with his work on SYMVU in 1965. This software from the Harvard Laboratory for Computer Graphics was in wide use within the decade (Carter, 1984). The computer has made it possible for the series of transformations, which consumed considerable time when handled manually, to be performed almost instantly (Davis and

McCullagh, 1975). The several stages are explicated in Figure 6, with the (planimetric) map first transformed to angular or parallel perspective and the surface features then elevated on the vertical axis.

The properties of both angular or parallel perspective projections may be contrasted. "In terms of visual characteristics, the angular perspective map (Figure 6d) is superior because the convergence cues contribute to the illusion of depth without detracting from the overall configuration" (Vlahos, 1965). The parallel-perspective map (Figure 6f) appears to be a contrived version of the surface because the parallel sides of the map give the illusion of divergence and the background appears to be tilted forward[6]. There is little question that the angular-perspective map more closely simulates the "real" surface than the parallel-perspective map does.

"At this point one might ask, 'Why bother with parallel perspective in three-dimensional mapping?' The question must be answered in two parts: in terms of scaling and in terms of ease of construction. Three-dimensional maps are tools for transferring information, and measurements on the map surface are often a means to this end. On maps constructed in angular perspective, the scale changes continuously over the entire surface of the map in both the vertical and the horizontal planes. Thus, measurements on this changing surface can be made only with a perspective scale constructed to fit a given map[7]. On a parallel-perspective map the scale is constant along any series of parallel lines, and thus measurements can be made along these orientations in both the horizontal and the vertical planes. Because of this, the height or length of a feature can be accurately compared with the height or length of another feature anywhere on the map" (Jenks and Brown, 1966).

There were two problems which had to be confronted before the computer made it possible to develop perspective projections with reasonable ease. "The cartographer selecting the perspective basis for constructing a three-dimensional map must choose between realism and practicality" (Jenks and Brown, 1966). Given the available computer-supported mapping systems, this choice can be made easily, in terms of the problem being resolved by the map which will be produced. While in some cases there will be a clear choice for perspective projection, with its realistic view, parallel (isometric) perspectives will in other situations provide the information more effectively. "In general, parallel-perspective maps with low-to-moderate vertical-scale exaggeration and low angles of view are acceptable[8] while those with great vertical scale exaggeration, or high viewing points, are less desirable" (Jenks and Brown, 1966). With three-dimensional maps now readily available, this problem has been resolved (Yoeli, 1976).

---

[6] See Nagata, Chapter 35, Figure 10.
[7] Angular perspective displays can, however, be designed with appropriate symbolic enhancements to compensate for scale distortion (Ellis et al., 1987).
[8] This comment refers mainly to maps where an angular perspective is expected or required by the map reader.

The second problem confronted is one of continuing interest: what should the viewing angle and elevation be? While this was examined in detail in Jenks's research several decades ago and has been the focus of other research, the questions have not been completely resolved (Jenks and Caspall, 1967; Jenks and Crawford, 1967). Monmonier (1978) explored the problem of "Viewing azimuth and map clarity", concluding that "because trend slope correlates with surface visibility for smooth, regular surfaces, an azimuth in the direction of steepest ascent on the linear trend surface approximates the azimuth with maximum visibility and promotes the cartographic communication of the salient spatial trend" (Monmonier, 1978). While a reader unfamiliar with the area being portrayed might encounter difficulty in handling the map reading task, some additional information, such as a verbal explanation or a version of the map in traditional orientation should alleviate the situation.

Rowles (1978) and Worth (1978) have also provided insight into technical facets of the problem. These studies, by cartographers and relating directly to the three-dimensional map, have been paralleled by the work of others. For example, though digital elevation data can be mapped on three-dimensional maps using contours and hill shading in a perspective view (Stefanovic and Sijmons, 1984; Traylor and Watkins, 1985), these perspective displays may not be perceptually viable. How easy are they to read? How accurately can the reader estimate heights on them and distances across them? For example, Rowles's (1978) study of the perception of height and distance relationships on perspective diagrams makes it clear that it is possible to estimate heights and distances correctly. But some residual judgment errors may remain when interpreting perspective images (Ellis *et al.*, 1987; McGreevy and Ellis, 1986; Chapter 34). Other more general orientation errors not associated with perspective projections have been examined by Baty *et al.* (1974), who studied the effect of map orientation on piloting performance.

As Stevens (Chapter 29) makes clear, in the last quarter century there has been a great increase in our understanding of our perception of three-dimensionality. Tversky (Chapter 4) underscores the need for our understanding of the human information processing system as a whole, for while visual stimuli are organized initially by perceptual processing, the mental image is then affected significantly by a person's conceptual knowledge. Bridgeman (Chapter 20) suggests that the task may serve not only just as a context but also as a control, affecting the human response to visual displays in a number of ways.

The construction procedures outlined in 1966 were essentially adaptations of techniques explicated in detail by Lobeck (1958). Lobeck had developed his work on the foundation begun by geographers and geologists a century earlier. The method of drawing vertical profiles without constructing numerous verticals is unique, quick, and reliable. Even though each vertical profile is a simple uniform line, the combination of profiles gives the final map an appearance of a continuous surface. Furthermore, the technique makes it possible to render minor nuances in surface configuration which are frequently lost in other forms of symbolization. None of this conceptual framework has disappeared. Nevertheless, while our knowledge base has expanded, there are a great many problems and prospects to be addressed.

Speculations on three-dimensional maps

"Three-dimensional maps are useful tools which have been neglected for some time" (Jenks and Brown, 1966). This statement was true in 1966, and it was true as well in the 1980's. While the earlier article continued, "They should be more commonly used, and familiarity with the techniques discussed in this article should dispel any qualms anyone might have about needing artistic talent to construct them", it is clear now that they have, with the available computer support, become quite commonplace[9].

We are a decade beyond Moellering's consideration of "The Real-time Animation of Three-dimensional Maps" (1980). This older work has been surpassed in many ways, but many of the same conceptual problems still remain to be solved. Digital Elevation Models (DEMs), also called Digital Terrain Models (DTMs), are integral aspects of Geographical Information Systems (GIS). Burrough (1986) devotes a chapter to their development and use. They have become interchangeable with the planimetric perspective, exploring and visualizing the data surfaces which are the substance of a GIS (see Tomlin, 1990). An entire volume, *Three Dimensional Applications in Geographical Information Systems* (Raper, 1989), explores a number of the issues, both conceptual and technical, which have remained unresolved for a long time. Also see *Analytical and Computer Cartography* (Clarke, 1990) for similar material. These volumes discuss the "...infinite number of ways to produce nonparallel transformations". The illustrations in Figure 7, produced using a slit-line anamorphoser 25 years ago, could be produced now using computer graphics systems. All the illustrations in Figure 7 are transformations of the planimetric weather map shown in Figure 7a. "These are only a few of the many transformations which can be made with an anamorphoser or computer, but they do point to some interesting possibilities which today can be achieved more generally by computer graphics techniques. For example, it appears that maps based on one projection might be altered to satisfy the coordinates of a completely different projection. Note, for example, the change of parallels from concave to convex curves and the change from converging meridians to diverging meridians. Similarly, the grids of maps approximate projections which are quite different from the original" (Jenks and Brown, 1966).

---

[9] Animated procedures of this sort have in fact been used to visualize planetary and terrestrial surface databases in the JPL productions of *LA: The Movie* and *Miranda: The Movie*.

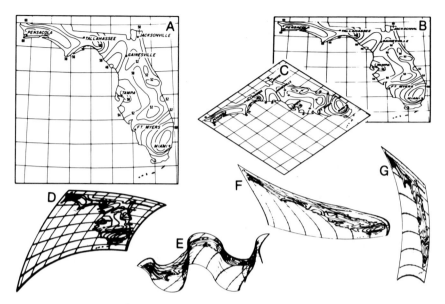

Figure 7. A series of experimental transformations of a map (A). The optical transformations, (B) and (C), both have a viewing angle of 40 degrees, (B) in a south-north orientation and (C) is rotated to a southwest-northeast orientation, are relatively common. The other examples are nonparallel transformations suggesting some unusual possibilities. Such maps could fit into non-traditional formats or provide distinctive emphasis in the visualization of a geographic situation.

Cartograms

Another interesting "innovation" in map making has been the development of cartograms, topological transformations of geographic space on the basis of geographically indexed statistical data. For example, in a cartogram the areas of countries can be made proportional to their populations, economic levels, or some other statistical measure such as the ratio of their immigration to emigration (Tobler, 1963, 1976). Cartograms of this type are not a recent invention. While Raisz (1934) is an early American resource, Levasseur appears to have employed the technique in French textbooks more than a century ago. Navigational counterparts were, in fact, first used by the Romans. Some of the other early cartograms, the T-in-O maps (*orbis terrarum*), date from the Crusades and were based on a structure of a conceptual space. These medieval maps, generally considered to be based on myth and dogma, simply reflect a view of the world organized more on the basis of theology than geography (Wilford, 1981). Modern cartograms in frequent use include automobile strip maps, the distorted maps used by many rapid transit systems, and the schematic maps used by airlines. These modern examples are often much easier to understand or use than their geographically correct

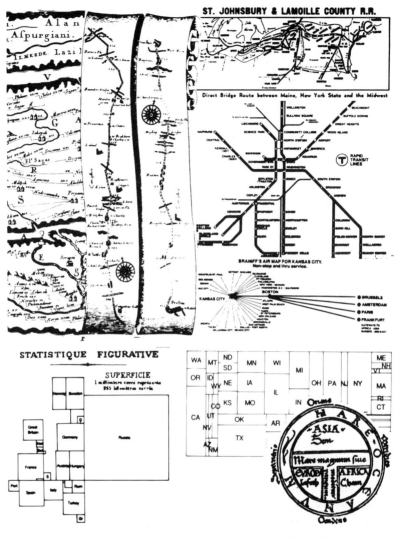

Figure 8. Spatial transformations, cartograms, can be used both for navigation and for managing environmental information. Top left: section of the Peutinger Table, a road map of the Roman Empire, ca. 450 AD. Top center left: portion of a seventeenth century English road map. Top right: portions of maps of a railroad line, a rapid transit system and an airline service network. Bottom left: a graphic statistical display, redrawn from an example by the French educator-economist-geographer Levasseur (1870). Bottom right: a rectangular statistical cartogram in the style of those made by Raisz to associate statistical variables with states as areas. These examples may be compared with the first printed map, a biblical view of the world prepared by a seventh-century scholar which also is rather schematic.

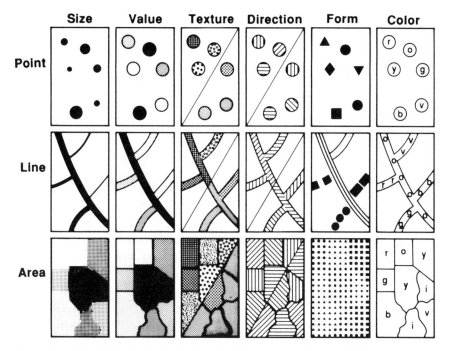

Figure 9. The Visual Variables described by Jacques Bertin (1967; 1983). Size and value are efficient for portraying quantitative data, while form and color are more suited for qualitative data. Texture and direction can be used for both types of data. Direction is not particularly visually effective and the total range of possibilities for texture have yet to be fully explored.

counterparts. The key question asks if these displays are understood more accurately than alternative descriptions?

Jacques Bertin (1967, 1983) has attempted to answer such questions and has examined the six visual variables that contribute to graphic communication. His approach, though systematic, is not experimental. It is, nonetheless, supported by a considerable body of research (Stevens, 1975). This approach appears to offer a viable avenue leading to a development of an empirically based science of pictorial communication (McCleary, 1983).

Any map is a display for a reader, for a user. From it one must garner information and ideas. Often one must use a map as the basis for making a decision. This theme is found throughout the history of human activity, and the maps which exist at any particular time and in any particular culture mirror the environmental concerns, activities and behaviors of those who created and used them. Only within the last half century has the map become more than a repository for information. Only for a very short extent of the human race's fascination with maps and mapping have those who assumed the responsibility for making maps seriously created displays and instruments

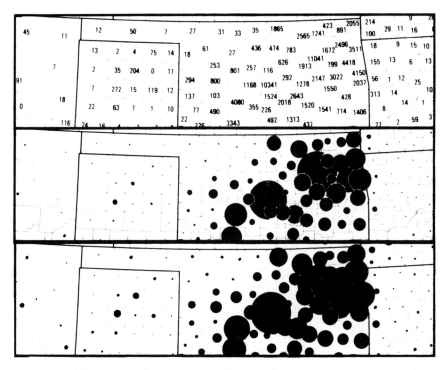

Figure 10. Three maps showing the distribution of Native Americans (Indians) in a portion of the Great Plains (Arkansas, Kansas, Missouri, New Mexico, Oklahoma and Texas). The actual values are shown in the top map. In the middle map, the sizes of the circles are directly proportional to the values; a circle representing 1000 is half the size of one representing 2000, and a tenth the size of one representing 10,000. A large number of studies have shown that the human visual system does not perceive circle sizes linearly, but rather underestimates size differences (Stevens, 1975). The map at the bottom compensates for this visual underestimation; the smallest circle is the same in size as the other map, but the sizes of the other circles have been rescaled to compensate for the perceptual underestimation (Robinson et al., 1984; McCleary, 1983).

which would work well for those who had to manage or navigate through the environment. The clear explication of the vocabulary, the visual variables of Bertin, the growing series of studies of the perceptual and cognitive characteristics of maps, the more complete understanding of map reading emerging from studies in psychophysics, all provide the cartographer with a considerably better set of rubrics on which to base a map. But the task is far from complete, and more remains to be done to provide an adequate understanding of the effectiveness of the different types of displays used to represent surface form.

Access to cartography

The statistical textbook by Schmid and Schmid (1979) provides a perspective on the traditional principles for the design of graphic displays. Dickinson (1973) focuses directly on the merger of statistics and maps. Monkhouse and Wilkinson (1971), on the other hand, provide an in-depth exploration of mapping techniques. Their encyclopedic approach contrasts with the technical approach used in nearly all of the other cartographic textbooks available; see, for example, *Elements of Cartography* (Robinson *et al.*, 1984) and Dent (1990). Cotter's (1966) *The Astronomical and Mathematical Foundations of Geography* provides a very accessible but rigorous introduction to map projections and their use in navigation. More detail about map projections can be found in Snyder (1982, 1987); also see Maling (1973) and Snyder and Voxland (1989) for a detailed analysis of over a hundred projections. An introduction to map projections is available in Hypertext on Macintosh Hypercard (Alpha *et al.*, 1988). Lockwood (1969) considers a wide variety of maps and graphs, while Fisher (1982) focuses on fundamental facets of the mapping problem. Herdeg (1981) has collected a wide variety of material from an even wider array of resources. Southworth and Southworth (1982) focus on *Maps* using a scrapbook approach. See also *Map Appreciation* by Monmonier and Schnell (1988), a volume which focuses on the types of maps and their uses. *Map Use* by Muehrcke (1986) provides a significantly different perspective, one which is mirrored in Campbell (1990).

References

Alpha, T.R., Virgil, J.F. and Bucholz, L. (1988). A guide to the commonly used map projections for use with Hypercard, USGS Open-file report 88364, Menlo Park, CA.

Barfield. W. and Robless, R. (1989). The effects of two or three dimensional graphics on problem solving performance of experienced and novice decision makers. *Behavior and Information Technology*, **8**, 369–385.

Baty, D.L., Wempe, T.E. and Huff, E.M. (1974). A study of aircraft map display and orientation. *IEEE Transactions in Systems, Man, and Cybernetics*, **SMC-4**, 560–568.

Bertin, J. (1967/1983). *Semiology of Graphics: Diagrams, Networks, Maps.* William J. Berg, Trans. Madison: University of Wisconsin Press.

Bridgeman, B. (1989). Separate visual representations for perception and for visually guided behavior. In Ellis, S.R., Kaiser, M.K. and Grunwald, A. 1989, *Spatial Displays and Spatial Instruments*. NASA Conference Publication 10032; Moffett Field, CA: NASA Ames Research Center, 14-1-2-15.

Burrough, P.A. (1986). *Principles of Geographical Information Systems for Land Resources Assessment.* Oxford: Clarendon Press.

Campbell, J. (1990). *Map Use and Analysis.* Dubuque, IA: Wm. C. Brown.

Carter, J.R. (1984). *Computer Mapping: Progress in the '80's.* Resource Publication in Geography (1983/4) Washington: Association of American Geographers.

Chamberlin, W. (1947). *The Round Earth on Flat Paper: Map Projections Used by Cartographers.* Washington: National Geographic Society.

Clarke, K.C. (1990). *Analytical and Computer Cartography.* Englewood Cliffs, NJ: Prentice-Hall.

Cotter, C.H. (1966). *The Astronomical and Mathematical Foundations of Geography.* New York: American Elsevier.

Crighton, M. (1977). *Jasper Johns.* New York: Harry N. Abrams.

Davis, J.C. and McCullagh, M.J. (1975). *Display and Analysis of Spatial Data.* London: John Wiley and Sons.

Dent, B.D. (1990). *Cartography: Thematic Map Design.* Second edition; Dubuque, IA: Wm. C. Brown.

Dickinson, G.C. (1973). *Statistical Mapping and the Presentation of Statistics.* Second edition; London: Edward Arnold.

Ellis, S.R. (1989). Pictorial communication: pictures and the synthetic universe. In Ellis, S.R., Kaiser, M.K. and Grunwald, A. (Eds) 1989, 1-1–1-23.

Ellis, S.R., Kaiser, M.K. and Grunwald, A. (1989). *Spatial Displays and Spatial Instruments.* NASA Conference Publication 10032; Moffett Field, CA: NASA Ames Research Center.

Ellis, S.R., McGreevy, M.W. and Hitchcock, R. (1987). Perspective traffic display format and airline pilot traffic avoidance. *Human Factors,* **29,** 371–382.

Fisher, H.T. (1982). *Mapping Information: The Graphic Display of Quantitative Information.* Cambridge, MA: Abt Books.

Foley, J.D. and Van Dam, A. (1982). *Fundamentals of Interactive Computer Graphics.* Reading, MA: Addison-Wesley.

Harrison, R.E. (1944). *Look at the World: The Fortune Atlas for World Strategy,* New York: Knopf.

Harvey, P.D.A. (1980). *The History of Topographical Maps: Symbols, Pictures and Surveys.* London: Thames and Hudson.

Hearn, D. and Baker, M.P. (1986). *Computer Graphics.* Englewood Cliffs, NJ: Prentice-Hall.

Herdeg, W. (Ed.) (1981). *Graphic Diagrams: The Graphics Visualization of Abstract Data.* Zurich: Graphis Press.

Hodgkiss, A. (1981). *Understanding Maps: A Systematic History of their Use and Development.* Folkstone: Dawson.

Imhof, E. (1965/1982). *Cartographic Relief Presentation.* Steward, H.J. (Ed.), Berlin: Walter de Gruyter.

Jenks, G.F. (1967). The data model concept in statistical mapping. *International Yearbook of Cartography,* **7,** 186–189.

Jenks, G.F. and Brown, D.A. (1966). Three dimensional map construction. *Science,* **154,** 857–864.

Jenks, G.F. and Caspall, F.C. (1967). Vertical exaggeration in three-dimensional mapping. Technical Report 2 (Nr 389-1246, Nonr 583-15); Washington: Office of Naval Research.

Jenks, G.F. and Crawford, P.V. (1967). Viewing points for three dimensional maps. Technical report 3 (Nr 389-146, Nonr 583-15); Washington: Office of Naval Research.

Lobeck, A.K. (1958). *Block Diagrams and Other Graphic Methods Used in Geology and Geography.* Amherst, MA: Emerson-Trussell.

Lockwood, A. (1969). *Diagrams: A Visual Survey of Graphs, Maps, Charts and Diagrams for the Graphic Designer.* London.

Maling, D.H. (1973). *Coordinate Systems and Map Projections.* London: George Philip.

McCleary, G.F., Jr. (1983). An effective graphic vocabulary. *IEEE Computer Graphics and Applications,* **3** (2), 46–53.

McGreevy, M.W. and Ellis, S.R. (1986). The effect of perspective geometry on judged direction in spatial information instruments. *Human Factors*, **28**, 439–456.
Moellering, H. (1980). Strategies of real-time cartography. *American Cartographer*, **7**, 12–15.
Moellering, H. (1980). The real-time animation of three dimensional maps. *American Cartographer*, **7**, 67–75.
Monkhouse, F.J. and Wilkinson, H.R. (1971). *Maps and Diagrams*. Third edition; London: Methuen.
Monmonier, M.S. (1978). Viewing azimuth and map clarity. *Annals*, Association of American Geographers, **68**, 180–195.
Monmonier, M.S. (1982). *Computer-Assisted Cartography: Principles and Prospects*. Englewood Cliffs, NJ: Prentice-Hall.
Monmonier, M. and Schnell, G.A. (1988). *Map Appreciation*. Englewood Cliffs NJ: Prentice-Hall.
Morgan, Sir F. (1950). *Overture to Overlord*. Garden City, NY: Doubleday.
Muehrcke, P.C. (1986). *Map Use: Reading, Analysis, Interpretation*. Second edition; Madison, WI: JP Publications.
Peters, A. (1983). *The New Cartography*. New York: Friendship Press.
Raisz, E. (1934). The rectangular statistical cartogram. *Geographical Review*, **24**, 292–296.
Raisz, E. (1944). *Atlas of Global Geography*. New York: Global Press.
Raper, J., ed. (1989). *Three Dimensional Applications in Geographical Information Systems*. London: Taylor & Francis.
Robinson, A.H. (1961). The cartographic representation of the statistical surface. *International Yearbook of Cartography*, **1**, 53–63.
Robinson, A.H. (1974). A new map projection: its development and characteristics. *International Yearbook of Cartography*, **14**, 145–155.
Robinson, A.H. (1982). *Early Thematic Mapping in the History of Cartography*. Chicago: University of Chicago Press.
Robinson, A., Sale, R., Morrison, J. and Muehrcke, P. (1984). *Elements of Cartography*. Fifth edition; New York: John Wiley and Sons.
Rowles, R.A. (1978). Perception of perspective block diagrams. *American Cartographer*, **5**, 31–44.
Saarinen, T.F. (1976). *Environmental Planning: Perception and Behavior*. Prospect Heights, IL: Waveland Press.
Schmid, S.F. and Schmid, S.E. (1979). *Handbook of Graphic Presentation*. Second edition: New York: John Wiley and Sons.
Snyder, J.P. (1982). *Map Projections Used by the U.S. Geological Survey*. U.S. Geological Survey Bulletin 1532; Washington: U.S. Government Printing Office.
Snyder, J.P. (1987). *Map Projections—A Working Manual*. U.S. Geological Survey Professional Paper 1395; Washington: U.S. Government Printing Office.
Snyder, J.P. and Voxland, P.M. (1989). *An Album of Map Projections*. U.S. Geological Survey Professional Paper 1453; Washington: U.S. Government Printing Office.
Southworth, M. and Southworth, S. (1982). *Maps: A Visual Survey and Design Guide*. New York Graphic Society Book; Boston: Little, Brown.
Stefanovic, P. and Sijmons, K. (1984). Computer-assisted relief representation. *ITC Journal*, 40–47.
Stevens, K.A. (1989). The perception of three-dimensionaity across continuous surfaces. In Ellis, S.R., Kaiser, M.K. and Grunwald, A. (Eds), 6-1–6-8.
Stevens, S.S. (1975). *Psychophysics: Introduction to its Perceptual, Neural and Social Prospects*. New York: Wiley-Interscience.
Tobler, W.R. (1963). Geographic area and map projections. *Geographical Review*, **53**, 59–78.

Tobler, W.R. (1976). The geometry of mental maps. In Golledge, R.G. and Rushton, G. (Eds), *Spatial Choice and Spatial Behavior*. Columbus: Ohio State University Press.

Tomlin, C.D. (1990). *Geographic Information Systems and Cartographic Modeling*. Englewood Cliffs, NJ: Prentice-Hall.

Traylor, C.T. and Watkins, J.F. (1985). Map symbols for use in the three dimensional graphic display of large scale digital terrain models using microcomputer technology. *Proceedings, AUTO-CARTO 7: Digital Representations of Spatial Knowledge*. Washington, 526–531.

Tversky, B. (1989). Distortions in memory for visual displays. In Ellis, S.R., Kaiser, M.K. and Grunwald, A., 12-1–12-17.

Vlahos, P. (1965). Three dimensional display: its cues and techniques. *Information Displays*, **2** (6).

Wilford, J.N. (1981). *The Map Makers*. New York: Alfred A. Knopf.

Worth, C. (1978). Determining a vertical scale for graphical representations of three-dimensional surfaces. *Cartographic Journal*, **15**, 31–44.

Yoeli, P. (1976). Computer-aided relief presentation by traces of inclined planes. *American Cartographer*, **3**, 75–85.

6

Interactive displays in medical art

Deirdre Alla McConathy and Michael Doyle

University of Illinois, Department of Biomedical Visualization
Chicago, Illinois

Medical illustration is a field of visual communication with a long history. Leonardo DaVinci, inventor, scientist, and illustrator, is perhaps the best known pioneer of medical art, but many other individuals, such as the famous anatomist Vesalius, also contributed to the development of the profession. Understandably, many factors have impacted the field throughout its growth, but the primary goal of a medical artist—to explain visually information about the health sciences—has remained unchanged. Other goals such as marketing and advertising of products are subsidiary to this central objective of presenting educational imagery to health science professionals and patients alike.

Traditional medical illustrations such as the one shown in Figure 1 are static, two-dimensional, printed images—highly realistic depictions of the gross morphology of anatomical structures (Netter, 1948; Pernkopf, 1963, 1977). Coincidental with technological advances in both medicine and image production, however, is the expansion of the role of medical art. Today medicine requires the visualization of structures and processes that have never before been seen. Complex three-dimensional spatial relationships require interpretation from two-dimensional diagnostic imagery. Pictures that move in real time have become clinical and research tools for physicians.

Medical artists are uniquely qualified to plan and produce visual displays for use in health communication. Basic science courses taken within a medical school curriculum prepare them to be content experts. Prerequisite life drawing, painting, color theory, graphic design and other fine art courses, and subsequent graduate coursework including anatomical drawing and surgical illustration imbue artistic skills. Using instructional design theory, artists plan goals and objectives, perform critical analyses of task and learning performance, and evaluate products and procedures. Medical artists are media technologists as well. They must choose from a plethora of media the appropriate mode of presentation for the specific content being represented. The objective in medical art is to incorporate new technologies as both

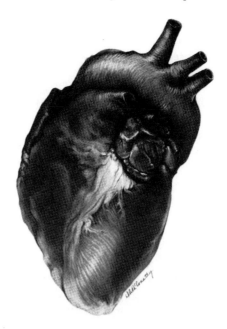

Figure 1. Traditional medical illustration by Deirdre McConathy, depicting gross morphology of a cadaver heart.

production tools and modes of final presentation. The artists are therefore knowledgeable of a wide variety of media, including printed images in line, continuous tone or color; projection media such as slides, video, film, and animation; computer graphics; and three-dimensional models and simulators.

In addition to formal instruction, medical artists possess those abilities often attributed to the mystical realm of art. Perhaps because of their comprehensive knowledge base relevant to problems of visual representation, for artists an iterative problem-solving process often becomes automatic to the point of appearing to be intuitive. Previsualization of visual solutions by the artist allows exploration to occur in an effective, if not well-understood, manner. For example, Ansel Adams, renowned for his development of the zone system in black and white photography, was consciously aware of the limitations of film for representing the range of values we are able to see with the human eye. He could, however, mentally image how a landscape would be recorded by film, and thereby "see" a predictable translation to guide him. In a similar manner, medical illustrators use a combination of factual, theoretical and artistic knowledge to previsualize.

Clients and content experts need to be involved in the process of preparing visuals, but many important production decisions pertaining to the final appearance of the image are solely the domain of the artist. Artists are able to identify and manipulate many variables with predictable results and recognize the contributions of unpredictable "happy accidents".

The most fundamental decisions upon initiating a drawing involve characteristics of the light source portrayed. The importance of direction of a light source is well documented. Perceptual psychologists have demonstrated that an upper-left light source is generally the default assumption for a viewer, but direction is only one variable to be considered. Two other important considerations are color temperature and intensity, as each of these conveys information about spatial relationships and can be used to invoke affective reactions. The artist sometimes needs to invent the light source, creating an unreality that is more effective than reality. For example, operating room lights provide very diffuse, even lighting of the surgical field to avoid fatigue to the surgeon's eyes; therefore photographs appear to be flat spatially. Surgical illustrators enhance the impression of space by creating an imaginary, directional light source, with strong highlights and cast shadows. Many other artistic decisions, such as viewer station point, composition, and color harmony, all impact the final results, and should be entrusted to professional communicators and qualified artists.

The medical artist embodies a link between the technical and aesthetic realms of visual communication. The skills exemplified by medical artists for the health sciences community can demonstrate an appropriate model for other fields that need to make judgments about visuals from a holistic viewpoint.

The importance of a qualified consultant and producer of visuals cannot be overemphasized. In their report to the National Science Foundation ("Visualization in Scientific Computing") McCormick et al. (1987) comment that "Because of inadequate visualization tools, users from industry, universities, medicine, and government are largely unable to comprehend the flood of data produced by contemporary sources such as supercomputers, satellites, spacecraft, and medical scanners. Today's data sources are such fire hoses of information that all we can do is warehouse the numbers they generate, and there is every indication that the number of sources will multiply". The authors suggest that interactive graphics are the best available solution to managing this information deluge. They go on to recommend that interdisciplinary teams of computer scientists, engineers, cognitive scientists, systems support personnel, and artists be enlisted to attack the visualization challenge.

One inevitable question for all types of pictorial displays is how realistic should the image be? Much debate exists as to the appropriate amount of realism it is necessary to include in different types of visuals. Research into the realism continuum and its effect on learning has not, however, established usable guidelines to be implemented. The current trend in educational resources is toward editing of information within pictures to a more diagrammatic style, whereas efforts to improve simulators are toward maximizing realism. Interactive displays may prove to be a reasonable solution to

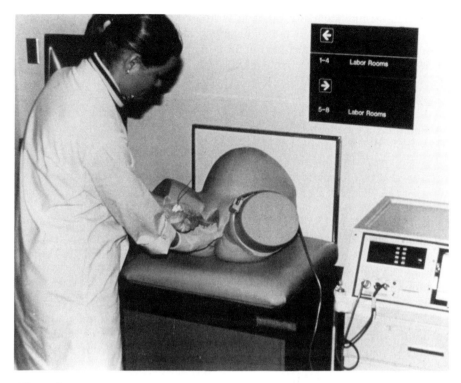

Figure 2. Interactive patient simulator developed by Evenhouse used to teach instrumentation for fetal monitoring procedure.

the editing question by providing users the flexibility of controlling the variable of realism and detail themselves. In reality, however, the issue of optimal levels of detail to include in a particular illustration is most often settled by budgetary constraints or subjective client preferences.

Medical illustrators are involved with the development of interactive visual displays for three different, but not discrete, functions: as educational materials, as clinical and research tools, and as databases of standard imagery used to produce visuals.

Health education visuals are required for a diverse audience including patients, medical students in training, and experienced surgeons. The information depicted may be factual, theoretical, abstract, or motor-skill training. Patient simulators are, for example, important methods for training manual skills because they offer the greatest breadth of learning experience with no risk of damage or discomfort to the patient. A successful simulator should provide a high degree of procedural realism. A three-dimensional model (Figure 2) used to train personnel in the procedure for fetal monitoring exemplifies a traditional type of interactive teaching display.

Monitoring a fetal heart rate during labor requires the insertion of an intra-uterine pressure catheter and the attachment of a scalp electrode to the baby. Placement of the instruments is critical since misapplication can result in devastating damage to the newborn. Correct positioning of the instruments requires the technician to palpate anatomical landmarks and visualize spatial relationships.

To satisfy these requirements in the simulator, medical sculptor Ray Evenhouse mimics soft and bony tissues with layers of synthetic materials. The structures are made from casts of bones and sculptures of soft tissues based on morphometric data. The completed simulator consists of a fetal head that is positioned within the maternal torso by an instructor in a variety of presentations. Visual and tactile realism is essential so that underlying structures such as the anterior and posterior fontanelles and facial features can be palpated to orient the trainee. In addition, the motivational factor induced by a highly aesthetic simulator contributes to the overall success of the model (Evenhouse and McConathy, 1989).

A quite different simulation is represented by an electronic textbook recently developed by Doyle *et al.* (1987) as a tool for teaching histology, the study of cell and tissue biology, to medical students (Figure 3). This prototype system operates on an IBM PC microcomputer fitted with both a high-resolution graphics display, capable of 256 on-screen colors, and a separate monochrome text display. The textbook uses the interactive digital video (IDV) interface, a device-independent process for user interaction with digital video images. This process uses a novel technique for the manipulation of color look-up tables to identify features of an image on a video display. For a technical description of this process see U.S. Patent #4,847,604. Using this system, a medical student can call up from a menu a microscopic image from one of the body's organ or tissue systems. This image is then displayed on the video monitor with no labels or identifying structure names shown. The student can then use a mouse to indicate a particular image that he or she would like more information about. Clicking one of the mouse buttons causes the computer to display a screen of explanatory text concerning the particular histological structure indicated (e.g., an individual cell in an image of a group of cells). Pressing the other mouse button would cause the display of a higher-magnification image of that image element (histological structure) selected. The student is then free to interact with this higher-magnification image to obtain further textual explanation or to see even higher magnification views. It should be noted here that this "zooming" capability does not merely involve the higher-magnification display of the same digital image (with the resultant loss of resolution), but rather causes the display of an entirely different image with no decay in resolution or image quality. For example, if the on-screen image was of a 1000× light microscopic view of some tissue, selecting the "zoom" feature would cause the display of a low-magnification (3500×) electron microscopic image of that particular type of structure. These correlations can be caused to run in reverse, so that the student could zoom from high

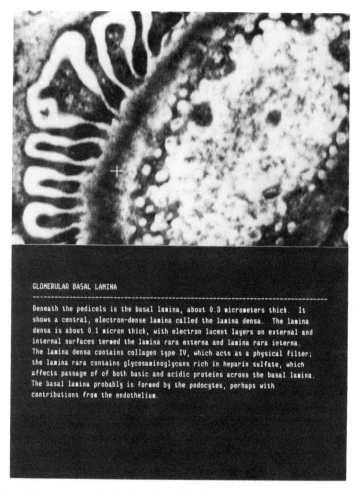

Figure 3. Electronic textbook developed by Doyle (1987) used to teach medical students about cells and tissues.

magnifications to lower magnification views or he or she could enter the name of a structure from the keyboard with the resultant display of an image containing the highlighted structure on the video display.

Image databases adapted for the IDV interface are extremely memory-efficient. The data storage load for a single image and correlation mechanism is less than 1 per cent larger than the original compressed image file before adaptation to the system. It is therefore practical to include all of the 1200 or so images needed for a complete histology atlas on a single CD-ROM disk. Another advantage to the process is that it runs very quickly, and this speed is not affected by the resolution of the image. The histology atlas runs very fast on an unadorned IBM PC (4.77 Mhz) with the appropriate graphics controller and disk storage device. Although the images in the atlas are only

512 × 484 pixels in resolution, the program would achieve the feature identification just as quickly if the image resolution were 4000 × 4000 pixels.

A specific objective in the development of the *Interactive Atlas of Histology* was to eliminate the distraction of having all of the important discrete elements within an image labeled on the screen and yet maintain the capability for immediate access to the exact descriptive textual information which the student desires.

In some situations, computer graphic images can contain so much information that it is not practical or not necessary to see all of the text-based information relevant to a particular image. Such a case exists in the graphic display of supercomputer-level image output. The IDV interface could be of great practical value in allowing the scientist to interact directly with the graphic display of, for example, a complex biological process simulation. A custom-designed interface could allow the researcher direct and immediate control over program flow for a simulation while it is executing, or immediate textual elaboration on an interesting feature of the simulation output display.

Head-up displays are currently of great interest in the aerospace industry. These displays have the effect of placing the user within the virtual environment of the computer image. A great deal of research is being done toward making the user interface for such a display as intuitive as possible. Techniques such as retinal scanning are being investigated as possible means to achieve a very natural-feeling way to specify a location within the display. The IDV interface would be an effective way to correlate this intuitive locator mechanism with desired relevant computer responses.

Other possible applications for the process are numerous: computer-aided education for information-intensive fields such as medicine or the military, for the earliest educational levels or for remedial or special education; image-based reference works such as atlases, catalogs, maps or navigation systems; cognitive rehabilitation systems, for head injury or Alzheimer's patients, to build associative relationships and still allow a controllable degree of freedom of interaction; interactive art displays; foreign language education systems; and entertainment programs or games.

These examples highlight the range of possibilities for teaching with interactive visuals. The opportunities for students to learn in real time, encounter variations, self-edit information, and adopt learning strategies best suited to their own needs represent a major advancement in education.

Another burgeoning area of interactive displays involves visuals as clinical and research tools. The advent of computer technology in combination with new technologies of diagnostic imaging has provided physicians and researchers with new methods of visualization.

The imaging modalities of computed transmission and emission tomography, magnetic resonance imaging and ultrasound are revolutionizing medicine. "Improved 3D visualization techniques are essential for the comprehension of complex spatial and, in some cases, temporal relationships between anatomical features within and across these imaging modalities"

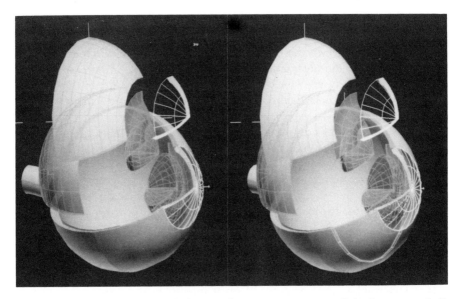

Figure 4. Stereoscopic exploded view showing components of the human eyeball. This geometric construction, developed by R.F. Parshall and L.L. Sadler (1990), from statistical norms of wall thickness, radii of curvature, etc., was produced on a CADD workstation using ICEM software by Control Data Corporation. This "theoretical eye" model is to be docked within a serial section reconstructed orbital (of the eye) environment based on cadaver specimens. For unaided stereoviewing, relax your eyes and focus out beyond the the page. The left and right eye images will converge yielding a triple image, the center one appearing three-dimensional. Images like this are to be part of the DaVinci Project's reconstructed three-dimensional computer database of standard anatomy.

(Pernkopf, 1987). For example, using the computer a plastic surgeon can modify a patient's features to simulate postoperative results. Such manipulation, based on each patient's diagnostic imagery, can be a powerful tool to help plan a surgery and also allay a patient's anxiety about the outcome. Another emerging application of computer visualization is the custom design of orthopedic reconstructions such as knee replacements through noninvasive 3D imaging.

Such developments in diagnostic imagery dictate a radical departure from conventional methods of teaching and communicating anatomical information. Medicine has traditionally relied on frontal, anterior-posterior views, but this flattened perspective is not sufficient. The explosion of diagnostic imagery has shattered conventions of orientation and requires visualization of oblique, cross-sectional and other unique viewpoints. Using computer-aided design software, students can rotate structures to improve their spatial understanding.

These major changes in spatial representation require heightened attention

Interactive displays in medical art 105

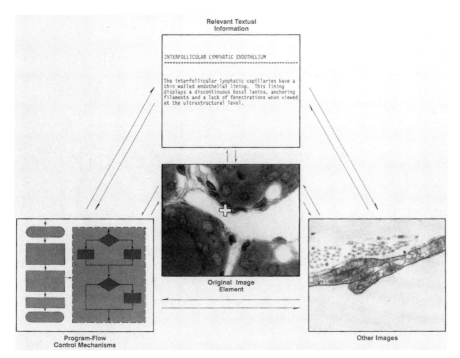

Figure 5. The IDV interface allows very sophisticated interrelationships to be set up between images, text and program-control mechanisms.

to fundamental aspects of preparing visuals, such as orienting the viewer. The impression of space can be enhanced by unusual oblique views, but is useful only when the user is properly oriented. Failure to establish the viewer's orientation seriously compromises the communication of the visual, yet we continue to see slides flashed with little or no orienting landmarks or graphic elements. This leaves the viewer with orientation as a first cognitive task rather than proceeding to the intended task of information processing.

Research concerning orientation and mental rotation of figures has provided a body of theory which can potentially be used to solve questions of orientation; however, application of these theories is still sorely lacking. In surgical illustration it is unclear whether it is better to depict a procedure from the surgeon's point of view during the surgery, or whether a view of the patient in anatomical position (upright, anterior-posterior orientation) is best.

Another problem that plagues visual communicators is a lack of standardization of both verbal and visual symbols. Specialty areas often develop representations that are learned by users over time, but comprehensive "dictionaries" of graphic elements would be helpful to assist the new learner and to assure consensus of interpretation.

Standardization of graphic elements would also maximize the amount of

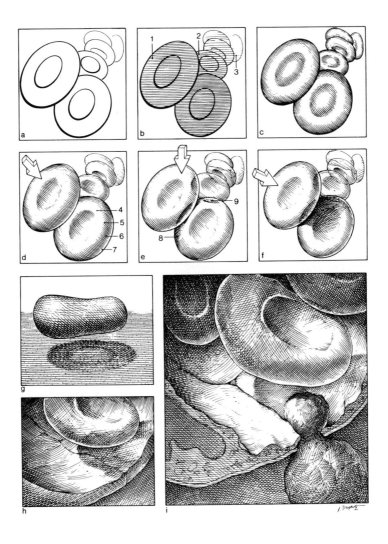

Figure 6. The representation of three-dimensional form using lines and values is paramount to creating a sense of space. This series of red blood cells demonstrates fundamental decisions and cues available and their cumulative effect in the finished rendering of a hypothetical environment.

(a) In addition to position depth cues such as overlap and linear perspective (objects become smaller as they recede towards a vanishing point), subtle reinforcement of aerial perspective can be obtained by using variable line weights both within the same object and between objects. Heavier lines suggest a foreground and lighter lines simulate atmospheric interference for distant objects.

(b) A more obvious demonstration of aerial perspective is obtained by assigning categorical values for objects. Artists subjectively manipulate intervals between values to "stretch or compress" space. For example, the difference between values measured on an arbitrary grey scale for red blood cells 1 and 2 is two steps, whereas for cells 2 and 3 the difference is eight.

(c) Axial lighting (as though the light source is from the viewer's eye) produces a symmetrical distribution of continuous tones within objects. Even with the addition of highlights and shadows, however, a relatively flat field is produced.

(d) A directional light source (arrow) enhances the impression of dimensionality substantially by creating areas of highlight (1), mid-tone (2), form shadow (3) and reflected light (4) that are defined by the geometry of the object.

(e) Cast shadows (6) and phantoming of an object behind an overlapping transparent object (7) further integrate spatial relations between cells.

(f) Although it is well established that the default expectation for direction of light source in pictures is upper left, alterations may produce more meaningful information because cast shadows conform to the surface upon which they fall. By redirecting the light source (arrow) slightly the viewer is given information from the cast shadow about the convex and concave surface of the cell.

(g) In some instances a cast shadow may be more descriptive than the object itself. Shape information about the cell, obscured in the lateral viewpoint, is revealed by the cast shadow. In addition, variation of value within the cast shadow gives form information. The dark edges define both refractive properties and the density of the object casting the shadow.

(h) The cast shadow simultaneously gives information about the surfaces of the cell and the capillary wall—a visual interaction.

(i) The cumulative effect of line and value cues used to render the migration of a blood cell through the wall of a capillary. The simultaneous display of a full complement of depth cues and their subtle variations, presented here, represents a significant obstacle for dynamic interactive computer technologies. Furthermore, imaging techniques such as electron microscopy lack decision rules and editing capabilities to produce these images that are more "spatial than life."

information which could be encoded into graphic symbols. For example, illustrators often employ arrows as devices for portraying the idea of direction, movement, or force. What do different types of arrows mean? In medical art there is a tendency to use simple, two-dimensional arrows to imply direction of movement. A three-dimensional arrow can also encode information about force, and can be made more or less monumental to correlate with the amount of force produced. Arrows drawn in perspectives that seem to pierce space can give information about complicated movements such as spirals or rotations. Unfortunately, no standardized vocabulary for graphic elements exists for medical art or for most specialties.

Standardization of data used to construct images would also be a boon to improving accuracy and production efficiency. At present, as each artist begins an illustration he or she must subjectively synthesize information from many resources. A database of morphometric information would assist the artist by providing measurements for an idealized form that can be manipulated, rotated, and embellished using the computer. Following the approach of human factors specialists in the design of tools and environments, the artist would have data sets of measurements to describe the range and standard for forms. Image banks would alleviate the necessity of "reinventing the wheel" (or kidney, brain, or heart in the case of medical art!) every time a new illustration is requisitioned. This way of thinking is somewhat antithetical to the traditional illustrator's mode of thinking, in which the product of artistic labor is considered to be a personal, unique interpretation of the subject matter—a problem that may impede acceptance of stock supplies of imagery.

A project that addresses the issues raised thus far is under way at the Department of Biomedical Visualization at the University of Illinois at Chicago. Aptly named The DaVinci Project, the interdisciplinary research group, consisting of experts from engineering, institutional computing, educational development, supercomputing, urban planning, architecture, medical imaging, and medical illustration, aims to create a *Resource center for anatomical imaging*. Using methods traditionally employed at a microscopic level, the DaVinci Project will establish a comprehensive, accurate description of standard human gross anatomy and its development through time, based on quantitative and qualitative data gathered from diagnostic images and actual specimens (Figure 4). Morphometric analysis and stereology will be used to develop a computer-based stereoanthropomorphic database which can be manipulated, analysed, and enhanced for various visualization purposes. The database will benefit diverse fields including medical education, bioengineering, anatomical simulator design, forensic science, biological process simulation, surgical instrument design, pharmaceutical research and development, military technology, sports equipment design, and missing persons research.

The DaVinci Project will contribute to teaching efforts, provide a research tool to clinicians and basic scientists, serve as a production tool for artists,

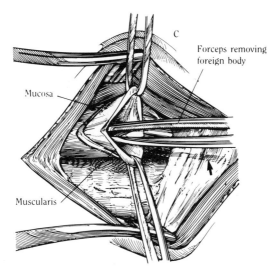

Figure 7 This surgical illustration depicting removal of a foreign body from the esophagus of a horse demonstrates application of depth cues to produce a cohesive spatial scene with far greater clarity than the physical scene itself. Structures are delineated with strong contour lines, inherent local values and shading of forms to convey dimensionality. While the shadow cast by the forceps (arrow) is inconsistent with the even, diffuse lighting that actually occurs in the operating room, this element can spatially separate elements and give information about the angle of insertion. In addition, this illustration points out that while continuous tones may be used to increase realism they are not necessary. The "loose-jointed" graphic style using bold lines in this rendering are effective for creating the "sense" of realism sufficient for the student surgeon. (Illustration by Tom McCracken in Equine Surgery Advanced Techniques, by permission of Lea & Febiger.)

integrate diagnostic imagery, and utilize computer technology to standardize and visualize information. Such an endeavor summarily represents the trend toward approaching visual information interdisciplinarily, interactively, and electronically.

References

Doyle, M.D. (1989). *Method and Apparatus for Identifying Features of an Image on a Video Display*, United States Patent #4,847,604.

Doyle, M.D., O'Morchoe, P.J. and O'Morchoe, C.C. (1987). A microcomputer based histology atlas. *Proceedings of the American Association of Anatomists.*

Evenhouse, R. and McConathy, D.A. (1989). Development and construction of a fetal monitoring simulator. *The Journal of Biocommunication*, **16** (1), Winter.

Netter, F. (1948). *Ciba Collection of Medical Illustrations*. Ciba Pharmaceutical Products, Inc.; Colorpress.

Parshall, R.F. and Sadler, L.L. (1990). CADD visualization of anatomical structure of the human eye and orbit. *SPIE Conference on Biostereometrics Technology and Applications*, accepted for publication.

Pernkopf, E. (1963). *Atlas of Topographical and Applied Human Anatomy, I, II.* Philadelphia: W.B. Saunders Co.

Pernkopf, E. (1987). *The Urban and Schwarzenberg Collection of Medical Illustrations Since 1896.* Vienna: Urban and Schwarzenberg.

Pernkopf, E. (1987). Visualization in scientific computing, *Computer Graphics*, **21** (6).

7

Efficiency of graphical perception[1]

Gordon E. Legge, Yuanchao Gu and Andrew Luebker

Department of Psychology
University of Minnesota
Minneapolis, Minnesota

Summary

Graphical perception refers to the part played by visual perception in analysing graphs. Computer graphics has stimulated interest in the perceptual pros and cons of different formats for displaying data. One way of evaluating the effectiveness of a display is to measure the efficiency (as defined by signal-detection theory) with which an observer extracts information from the graph. We measured observers' efficiencies for detecting differences in the means or variances of pairs of data sets sampled from Gaussian distributions. Sample size ranged from 1 to 20 for viewing times of 0.3 or 1 sec. The samples were displayed in three formats: as numerical tables, as scatterplots, and as luminance-coded displays. Efficiency was highest for the scatterplots (~60 per cent for both means and variances) and was only weakly dependent on sample size and exposure time. The pattern of results suggests parallel perceptual computation in which a constant proportion of the available information is used. Efficiency was lowest for the numerical tables and depended more strongly on sample size and viewing time. The results suggest serial processing in which a fixed amount of the available information is processed in a given time.

Introduction

Computer graphics has stimulated lively interest in the design of new formats for displaying data. A set of numerical data can be displayed in

---

[1] This chapter has appeared previously in *Perception and Psychophysics* (1989), **46**, 365–374 and is reprinted in slightly modified form by permission.

countless ways, but only some of these formats are well suited to the information-processing capacities of human vision. The phrase "graphical perception" has been coined to refer to the role of visual perception in analysing graphs (Cleveland 1985, Chapter 4; Cleveland and McGill, 1985). These authors have studied several elementary visual tasks relevant to graphical perception, such as discrimination of slopes or lengths of lines. Cleveland and McGill's Table 1 ranks several such tasks in order of the accuracy of perceptual judgment.

Cleveland and McGill point out that the great advantage of graphical displays over numerical tables is due to the capacity of human vision to process global pattern features at a glance. Julesz (1981) has used the term "preattentive" to refer to perceptual operations that can be carried on in parallel across the visual field.

While it is clear that we can make relative statements concerning the perceptual superiority of one type of data display over another, can we also make absolute statements about the perceptual effectiveness of a given type of display? Efficiency, as defined by signal-detection theory, provides a measure on an absolute scale of how effectively information is used. Efficiency ranges from zero to one and represents the performance of a measuring instrument or real observer relative to the performance of an ideal observer. The ideal observer makes optimal use of all available information. H.B. Barlow and colleagues have measured human efficiency for several visual information-processing tasks, including detection of mirror symmetry (Barlow and Reeves, 1979), discrimination of dot density (Barlow, 1978), detection of modulation of dot density (van Meeteren and Barlow, 1981), and judgments of the number of dots in displays (Burgess and Barlow, 1983).

Our purpose was to use the concept of efficiency to study graphical perception. Two common graphical tasks are the estimation of means and variances in sets of noisy data. We have measured observers' efficiencies for estimating these two statistical parameters from samples of data drawn from Gaussian distributions.

We compared efficiencies for data displayed in three different formats: as numerical displays, as scatterplots, and as luminance-coded displays. Numerical tables are the traditional means for displaying data. How well can observers estimate the means and variances of columns of numbers? Scatterplots are representative of pictorial methods for displaying data in which the substantial capacities for spatial vision can be used. Luminance-coded displays are ones in which salient differences are conveyed by luminance contrast. A large body of recent research points to the importance of contrast coding in vision.

Burgess and Barlow (1983) described two generic ways in which human performance can be suboptimal. Observers might simply fail to use some of the information available to them, but optimally process the remaining information. On the other hand, observers might use all the information, but

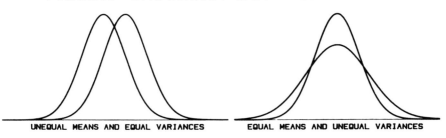

Figure 1. Stimuli were based on values sampled from two Gaussian distributions. In one series of experiments, these Gaussians had equal variances, but means that differed by one standard deviation (right). In a second series, the Gaussians had equal means, but their variances differed by a factor of two (left).

contribute imprecision (intrinsic noise) as a result of errors of internal representation. Burgess and Barlow (1983) showed how these two factors—incomplete sampling and internal noise—can be teased apart by measuring discrimination thresholds as a function of the level of externally added visual noise (see also Barlow, 1977; Pelli, 1981; Burgess et al., 1981; and Legge et al., 1987). We used this technique to evaluate the roles of the two factors in limiting perceptual estimation of means and variances.

Preliminary reports of this work were given at the 1987 annual meeting of the Association for Research in Vision and Ophthalmology, and the Spatial Displays and Spatial Instruments Conference, sponsored by NASA Ames Research Center and UC Berkeley, at Asilomar, CA, Aug. 31–Sept. 3, 1987. The research was supported by U.S. Public Health Service Grants EY02857 and EY02934 and AFOSR Grant 86–0280.

Method

The stimuli were derived from numbers drawn at random from two parent Gaussian distributions (Figure 1). In one series of experiments, the two Gaussians had identical variances, but their means differed by one standard deviation. In a second series, the two Gaussians had identical means, but their variances differed by a factor of two.

In an experimental trial, an observer was shown N random samples from each distribution for a time T. The observer was instructed to choose which set of samples was drawn from the distribution with the higher mean (or larger variance). An observer's performance was measured as percent correct in blocks of 300 trials. At least two blocks, collected in separate sessions, were averaged to estimate performance levels.

In order to compute efficiency, percent correct was transformed to accuracy D using formulae from signal-detection theory. For discrimination of means, an observer's accuracy D_{obs} is given by

$$D_{obs} = -\sqrt{2Z} \tag{1}$$

where Z is the standardized normal deviate corresponding to proportion correct in the two-alternative, forced-choice procedure (Green and Swets, 1974). D_{ideal} is the accuracy obtained if an optimal strategy is used. In the case of means, the optimal strategy is straightforward; compute the means of the two sets of samples and choose the one with the higher value. When there are N samples

$$D_{ideal} = \sqrt{(N)}(M_1 - M_2)/sd \tag{2}$$

where M_1 and M_2 are the means of the two parent distributions and sd is the common standard deviation of the parent distributions. In our experiments, the difference of the means was equal to the standard deviation so

$$D_{ideal} = \sqrt{N} \tag{3}$$

In the case of equal means, but unequal variances, the optimal strategy is to compute the sum of squared deviations of sample values from the parent mean for each set of samples, and choose the sum with the higher value. When there are N samples, D_{ideal} is given by

$$D_{ideal} = \sqrt{(N)} \, (sd_1/sd_2 - sd_2/sd_1) \tag{4}$$

where sd_1 and sd_2 are the standard deviations of the parent distributions (Egan, 1975, p. 136). The accuracy of the real observer is obtained from proportion correct as follows. Using formulae given by Egan (1975, pp. 239–240) receiver-operating characteristics (ROC) curves can be computed for different values of sd_1 and sd_2. Equation (4) associates a value of D with each ROC curve. The area under the ROC curve is the proportion correct in forced choice (Green and Swets, 1974). Therefore, the ROC forges the link between the observed proportion correct and a value of accuracy attributable to the real observer.

An observer's efficiency E is defined to be (Tanner and Birdsall, 1958)

$$E = (D_{obs}/D_{ideal})^2 \tag{5}$$

Efficiency ranges from zero to one and provides an absolute scale for judging the effectiveness of an observer's performance.

As illustrated in Figure 2, the stimuli were displayed in three formats: numerical displays, scatterplots, and luminance-coded displays. Figure 2 shows examples in which the sample size N was 10. In each case, the

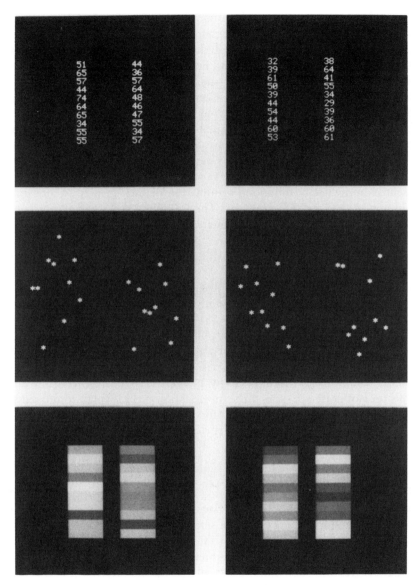

Figure 2. The three display formats. The column of three panels on the left shows the same trial display as numbers, scatterplots, and luminance-coded bars. The subject's task is to decide which set of samples was drawn from the Gaussian distribution with the higher mean. The three panels on the right depict a trial in which the subject's task was to decide which set of samples came from the distribution with a larger variance.

observer was asked to choose which member of the stimulus pair was drawn from the parent distribution with higher mean or larger variance. In the case of scatterplots, the observer judged the mean or variance in the vertical position of the dots on the screen. For luminance coding, the judgment referred to luminance levels of the bars.

All stimuli were displayed on a Conrac SNA 17/Y monochrome video monitor with P4 phosphor at a viewing distance of 64 cm. For numbers and scatterplots, the stimuli had a luminance of 300 cd/m^2 on a black background. Stimuli were generated using an LSI-11/23 computer and Grinnell GMR274 frame buffer with a display resolution of 512 × 480 pixels. Stimuli were presented for either 300 or 1000 ms. A masking pattern of X's was presented immediately following termination of each display so afterimages could not be used by the observers. The following three paragraphs contain details of three types of display formats.

Numerical displays

The parent Gaussian distributions had means of 45 and 55 and standard deviations of 10. Sampled values were rounded off to the nearest integer before being displayed. Accordingly, there was a quantization error corresponding to about 5 per cent of the standard deviation. (Quantization error is a form of "noise" that will reduce the accuracy of the ideal observer. Accordingly, our estimates of efficiency are slightly low. The issue of quantization noise is dealt with in detail by Burgess (1985). In general, the effects of this type of noise are small if the quantization error is small compared with the standard deviation of the noise process.) The center-to-center spacing of the digits was 0.51° and the empty space separating two columns subtended 4.5°.

Scatterplots

Numerical values drawn from the Gaussian distributions were transformed linearly to a vertical position on the screen. A "+" symbol centered on the screen specified the vertical position midway between the two means. One standard deviation of the Gaussian subtended 2.6°. Each sample was displayed as a "*" symbol subtending 0.51°, placed on the screen with an accuracy of 2 per cent of one standard deviation. The horizontal spacing between samples was 0.51° and the horizontal separation between the two sets of samples was 4.5°.

Luminance-coded displays

The numbers drawn from the Gaussian distributions were mapped to luminance and displayed as bars on the screen. The digital-to-analog converter of the frame buffer quantized values to 256 levels. A look-up table was used to correct for the video monitor's nonlinear transformation from voltage to luminance. The two distributions had mean luminances of 150 cd/m² and 185 cd/m² and the standard deviation was 35 cd/m². In terms of luminance, the quantization error was 2 per cent of one standard deviation near the mean luminance levels of the two distributions. Each bar subtended 3.5° × 0.86°. The columns of bars were separated by 1.8°.

In a separate experiment, we measured thresholds for discriminating differences in mean values. A threshold was found by reducing the difference between the means of the parent Gaussian distributions until a criterion accuracy was achieved. The criterion was 75 per cent correct, corresponding to D_{obs} = 0.95. The QUEST procedure was used to find the threshold difference in means (Watson and Pelli, 1983). As discussed in the results, the threshold data were used to distinguish between internal noise and inappropriate sampling as reasons for an observer's deviation from ideal performance.

Two highly practiced observers participated in the experiments.

Results and discussion

Effect of sample size

Figure 3a illustrates subject YG's performance with scatterplot means for sample sizes ranging from 1 to 20. Accuracy D_{obs} is plotted on the left vertical scale and corresponding values of percent correct on the right. The dashed line shows the performance of an ideal observer. Notice that when the sample size is 1, both YG and the ideal observer have accuracies near 1.0 (per cent correct near 76 per cent). When only one sample is drawn from each of the overlapping parent Gaussian distributions, there will be occasions when the sample from the distributions with lower mean is greater than the sample from the distribution with higher mean. The ideal observer will select this value and be counted wrong. In this way, the accuracy of the ideal observer is limited by the variability (noise) inherent in the parent distributions. As the sample size increases, the standard deviations of the sampling distributions decrease in proportion to $\sqrt{N}$, and the performance of the ideal observer steadily improves. Figure 3a illustrates that the performance of the real observer also improves, but not as fast as the ideal. YG's efficiency can be computed from D_{obs} and D_{ideal} using equation (5). Figure 3b shows the data of Figure 3a transformed to efficiency on the vertical scale. Even though YG's accuracy increased with sample size (Figure 3a), his efficiency declined slowly for the same range of sample size (Figure 3b). This decline reflects the growing vertical separation between the curves for YG and the ideal observer in Figure 3a as sample size increases. In subsequent figures, we plot efficiency as the dependent variable.

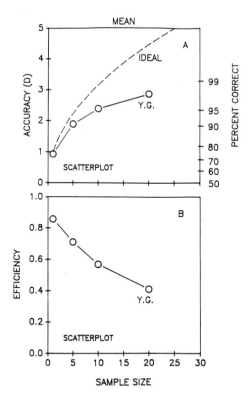

Figure 3. Use of efficiency as a performance measure. In panel A, YG's performance for discriminating scatterplot means is plotted as a function of sample size. Accuracy, D, is plotted on the left vertical scale and corresponding values of percent correct on the right. The performance of the ideal observer, equation 3, is also shown. Panel B shows corresponding values of YG's efficiency, computed using equation 5.

In Figure 4, efficiency is shown for the discrimination of means. The viewing duration was fixed at 0.3 sec. Data are shown for the three display formats. Best-fitting straight lines have been fitted through the sets of data. Because both scales are logarithmic, a slope of −1.0 would represent inverse proportionality between efficiency and sample size. Values of the slopes are shown in the figure and are summarized in Table 1.

Efficiencies are highest for scatterplots, >50 per cent. These values are a little higher than the 25 per cent efficiencies obtained by Barlow and Reeves (1979) for the detection of mirror symmetry, and equivalent to the 50 per cent efficiency observed by Barlow (1978) for the discrimination of dot density. The scatterplot data show a weak dependence on sample size (slopes of −0.23 and −0.36). Had the slopes been zero (constant efficiency), subjects would have been processing new samples with equal effectiveness. Instead, both observers made substantial, but incomplete, use of additional information available from additional samples.

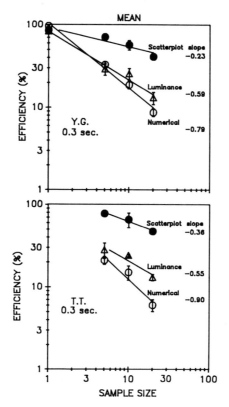

Figure 4. Efficiency for discriminating means is plotted as a function of sample size on log-log coordinates. Data are shown for the three display formats. Slopes are given for the best-fitting lines through the sets of data. These slopes are listed in Table 1. The two panels are for observers YG and TT.

Table 1. Slopes

Efficiency as a function of sample size, t = 0.3 sec:

	SCATTERPLOTS		LUMINANCE		NUMERICAL	
	Mean	Var.	Mean	Var.	Mean	Var.
YG	−0.23	0.26	−0.59	−0.37	−0.79	−1.00
TT	−0.36	0.03	−0.55	−0.22	−0.90	−1.38

Efficiency as a function of viewing time, sample size = 10:

TT	0.08	0.08	0.67	0.55	0.70	1.13

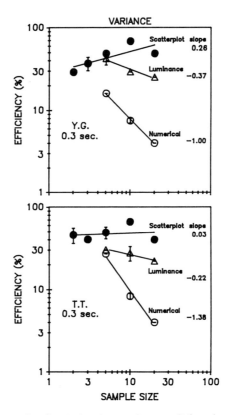

Figure 5. Efficiency for discriminating variances. Other details as in Figure 4.

Efficiencies are lowest for numerical displays and show a more pronounced decline with increasing sample size. The decline means that subjects are less able to use the additional information provided by the extra samples.

The results for luminance lie intermediate between those for scatterplots and numerical displays.

Notice that YG's efficiencies for sample sizes of one were close to 100 per cent. The $N = 1$ condition is really a control to verify that performance is not limited by visual resolution or other forms of sensory discrimination. We believe that departures from 100 per cent efficiency are not sensory in origin, but result from more central processes. Brief experiments with two low-vision observers—one with central-field loss caused by optic atrophy and the other with severe corneal vascularization (Snellen acuity = 20/1000)—confirm this point. Both had about the same efficiencies for scatterplots as normal observers.

Figure 5 shows comparable data for discrimination of variance. On the whole, the pattern of results is very similar to those for means. Observers are as efficient at estimating variances as means. This may be surprising because,

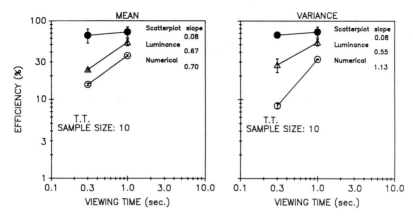

Figure 6. Efficiency is plotted as a function of viewing time for observer TT. The two panels show results for discrimination of means and variances.

as every neophyte statistics student knows, numerical calculation of variance is much harder than means. Unlike the data for means, the scatterplot curves for variance in Figure 5 are nonmonotonic with a peak at a sample size of 10. We return to this point below when we discuss algorithms that subjects may use in estimating variances.

Effect of viewing time

Figure 6 shows the effect of viewing time on efficiency with a fixed sample size of 10. Data are shown for subject TT. Once again, the results are similar for means and variances.

There is only a slight increase in efficiency for scatterplots as exposure time increases from 0.3 to 1.0 sec. Little increase is to be expected because efficiency is already high for the shortest exposure. In comparison, efficiencies for numerical displays rise more rapidly, with log-log slopes much nearer one. For viewing times even longer than 1 sec, we might expect efficiencies for numerical displays to keep rising. Eventually, they might catch up with the scatterplots. In the extreme, a subject might consciously compute statistics on the columns of numbers and possibly achieve very high efficiencies (limited only by skill at mental arithmetic).

In visual search tasks, perceptual processing is said to be parallel if performance is independent of exposure time and the number of elements. For our discrimination task, parallel processing would mean constant efficiency. It is not quite the case that scatterplot perception is a purely parallel process. Efficiency rises slowly with viewing time and drops slowly with sample size. Relatively speaking, however, scatterplots are processed in a more parallel manner than numerical displays. This difference between the two types of displays quantifies and supports Cleveland and McGill's (1985) view that the preattentive (and hence parallel) processing of graphs is the most fundamental reason for their superiority to tabular displays.

In visual search, processing is said to be serial when performance increases linearly with time and declines in inverse proportionality to the number of elements. We can see that these relations apply to the efficiency of a serial processor as follows. Suppose that an observer processes samples serially at a rate of N_o elements in time T_o. Suppose that this observer makes optimal use of the available information and is limited only by the rate at which samples can be processed. In time T such an observer can process $(N_o/T_o)T$ samples. Substituting this for N in equation (3), we obtain the accuracy of the serial processor for discriminating means:

$$D_{serial} = \sqrt{[(N_o/T_o)T]}$$

Substituting this expression in equation (5) for efficiency, we find that

$$E = (D_{serial}/D_{ideal})^2$$
$$= (N_o/T_o)(T/N) = k(T/N)$$

where k is a rate constant. (E can never exceed 1.0.) Such an observer's efficiency is linearly related to viewing time T and is inversely proportional to sample size N. This is roughly the pattern of results we found for numerical displays.

The results for luminance-coded data lie between those for scatterplots and numerical displays, both in values of efficiency and in slopes. In the experiments with means, the task amounted to discrimination between the average luminances of two patches of static noise. We are not aware of any experiments in which this capacity has been studied. The variance experiment amounts to discrimination of the r.m.s. contrasts (contrast power) of two displays of one-dimensional static visual noise. The efficiencies we found—typically 30 per cent—are in good agreement with values measured by Kersten (1987). He measured efficiencies for the detection of static visual noise on uniform fields or on fields of dynamic visual noise.

Sorted and unsorted samples

Our data indicate that perceptual analysis is serial for numerical displays, but more parallel in character for scatterplots. A serial processor can handle only a few samples in a brief exposure. Its performance might be enhanced, however, if those few samples were well chosen. For example, if only one of N samples can be processed in a brief exposure, the median, maximum, or minimum value would be more useful than a value selected at random. We can make it easier for a real observer to use such values by presenting displays in which the samples are sorted (in ascending or descending) order. Then, in a brief exposure, an observer knows where to look in the set of samples for the extrema or median values. Because a parallel processor has

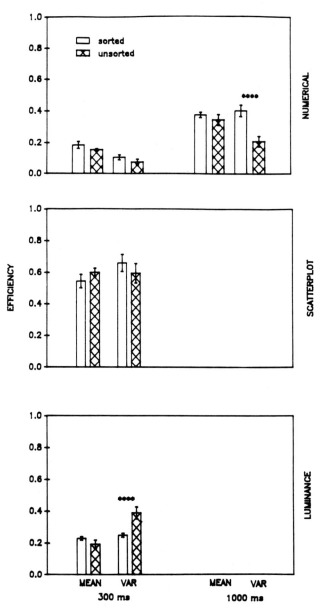

Figure 7. Comparison of efficiencies for sorted and unsorted samples.

simultaneous access to all values, sorting should be of less advantage. Sorting has no effect on the performance of the ideal observer who uses all information in the set of samples regardless of presentation order.

These considerations led us to predict that sorting should benefit perceptual analysis of numerical displays more than scatterplots. We measured efficiencies for sorted and unsorted sets of 10 samples for observer TT. The open and cross-hatched bars in Figure 7 compare efficiencies for sorted and unsorted displays. In all four comparisons with numerical displays, efficiency was higher for sorted samples, but the advantage was significant only for discrimination of variances with 1000-ms exposures ($p < 0.01$). Sorting had no systematic effect on efficiency for scatterplots. As predicted, sorting was more beneficial for numerical displays than for scatterplots, but the effect was relatively weak. For luminance-coded displays, sorting did not produce significant changes in efficiency for discriminating means, but actually hampered performance on variance. Perhaps the r.m.s. contrast (corresponding to variance in the luminance-coded displays) was harder to estimate in the sorted displays because local-contrast steps were much smaller.

Perceptual strategies

How efficient would we expect subjects to be if they based their decisions on median values or extrema? It is possible to suggest a variety of strategies using these values and to compute the efficiencies associated with each. If we find that an observer's efficiency exceeds that of the strategy, we can be sure that the observer is not using the strategy. On the other hand, an observer whose efficiency is less than that of the strategy may be attempting to use the strategy, but failing to execute it perfectly.

The curves in Figure 8 show efficiency as a function of sample size for two strategies for discriminating means and three strategies for discriminating variances. The strategies are described in Table 2. Data have been replotted from Figures 4 and 5. Efficiencies for the five strategies were derived from Monte Carlo simulations programmed in Pascal and run on a Sun 3/160 with a floating-point accelerator board. Gaussian random numbers were obtained from a uniform random distribution (the RANDOM function under Sun UNIX) then transformed to a Gaussian random variable based on the cumulative normal distribution (Abramowitz and Stegun, 1972, p. 933). The curves in Figure 8 are based on simulated experiments run with sample sizes ranging from 1 to 50 in integer steps. For each sample size there were 100,000 trails.

The median strategy for discriminating means is efficient. In the limit of large sample size, its efficiency drops to $2/\pi$ which is close to 64 per cent (Freund, 1962, p. 219). Less efficient is the maximum strategy, especially as sample size grows large. Neither strategy provides a good fit to any of the data sets, including the sorted-list data.

We considered three strategies for discriminating variances. The absolute-value strategy is efficient and outperforms real observers even for scatterplots.

Efficiency of graphical perception

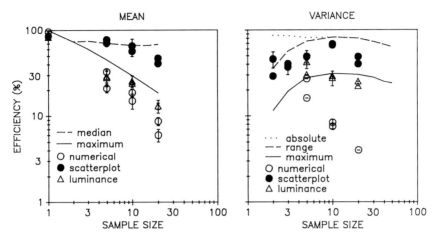

Figure 8. The five curves, two for means and three for variance, show how efficiency depends on sample size for five strategies (Table 2). The curves were derived from results of computer simulations. Data for observer YG have been replotted from Figures 4 and 5 for comparison.

Table 2. Some possible perceptual strategies

	Strategy	Description of Statistics*
Discrimination of Means		
	Median	Median value
	Maximum	Maximum value
Discrimination of Variances		
	Absolute value	Sum of the absolute values of the maximum positive and negative deviations from the mean
	Range	Difference between maximum and minimum values
	Maximum	Maximum

* In a trial, the subject is presented with two sets of N samples, each set drawn from a parent Gaussian distribution. The subject attempts to identify the set of samples that came from the parent distribution with higher mean (or larger variance). The subject who adopts one of these strategies picks the set of samples having the higher value of the indicated statistic.

The range strategy (Table 2), however, shows the same sort of nonmonotonic dependence on sample size as the scatterplot data of real observers. This strategy has a peak efficiency of 85 per cent at a sample size of 11, a little better than the real observers. The rough correspondence, however, suggests that real observers may rely heavily on extreme values in their estimates of dispersion of data depicted by scatterplots. As might be anticipated, the curve for the maximum strategy roughly parallels the curve for the range strategy, but with overall lower efficiency.

As statisticians have long known, order statistics like medians or maxima or minima, are particularly informative. The curves in Figure 8 indicate that means and variances can be efficiently discriminated with simple strategies based on these values. Only in the case of the range strategy for variance discrimination, however, is there evidence that subjects actually adopt one of the strategies we propose.

Thresholds for discriminating means

The two panels of Figure 9 show threshold data rather than efficiency. For a given value of variance (the same for the two-parent Gaussian distributions), the difference between the means was adjusted by a forced-choice staircase procedure to find the threshold difference. The squared difference between the means at threshold (see next paragraph) is plotted as a function of variance. Thresholds are shown for a sample size of five and a viewing time of 0.3 sec. Each point represents the mean of two to six separate threshold estimates. Best-fitting straight lines have been fitted to the data.

These threshold plots are analogous to plots of signal energy vs. noise spectral density (see Burgess et al., 1981; Legge et al., 1987). Slopes and intercepts of lines through the data can be used to distinguish between two generic forms of inefficiency: nonoptimal sampling and internal noise. The slope of a straight line through such data is related to the efficiency with which the observer processes stimulus samples. The greater the slope, the lower the observer's efficiency. Burgess et al. (1981) have defined sampling efficiency as the ideal observer's slope divided by the real observer's slope. Values less than 1.0 reveal nonoptimal sampling. The magnitude of the internal noise is proportional to the x-axis intercept's distance to the left of the origin. The farther to the left of the origin the intercept, the greater the internal noise. The ideal observer's straight line passes through the origin (no internal noise) and has a slope of 0.18^2. These concepts are described in more detail by Legge et al. (1987).

---

[2] The slope value of 0.18 can be derived from equation (2). Squaring both sides and denoting the threshold difference of means by Δm, we have:

$$(\Delta m)^2 = (D_{ideal}^2/N)sd^2$$

A threshold criterion of 75 per cent correct corresponds to a value of D_{ideal} of 0.95. For $N = 5$, the slope of the relation between squared difference of means and variance is $(0.95)^2/5 = 0.18$.

Efficiency of graphical perception

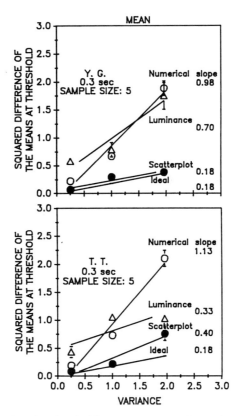

Figure 9. The results of a threshold experiment are shown in which the difference between the means of the parent Gaussian distributions was reduced until the observers were 75 per cent correct in the forced-choice procedure. The squared difference of the means at threshold is plotted as a function of the common variance of the Gaussian distributions. The intercepts and slopes of straight lines fitted to the data are used to partition losses of efficiency into two sources. Results are shown for the three display formats and for the ideal observer. The two panels give results for observers YG and TT.

For both subjects, the scatterplot lines have slopes and intercepts close to the ideal. This is not surprising because efficiencies for scatterplots are very high.

Both subjects had slopes that were much higher than the ideal, but intercepts near zero, for numerical displays. This reveals that the source of inefficiency is almost entirely due to incomplete sampling, confirming the serial-processing interpretation. In the brief exposure interval, the subjects could sample only a small fraction of the available information, but this they processed flawlessly.

For both subjects, the luminance displays had intercepts substantially to the left of the origin, suggesting the existence of internal noise. This noise might

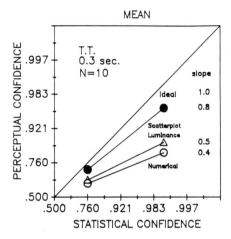

Figure 10. How perceptual confidence can be related to statistical confidence.

have a sensory origin. It might be related to the precision with which contrast is coded in the nervous system (Legge et al., 1987).

Perceptual confidence

When people "eyeball" two sets of data, how is their confidence in discerning differences in the means related to the confidence of a statistical test? Because people are not 100 per cent efficient in using available information, perceptual confidence should be less than statistical confidence.

When discriminating between the means of two normal distributions of equal variance, statistical confidence can be indexed by d', the difference between the means divided by the common standard deviation. The d' value, representing statistical confidence, is what we have referred to above as the accuracy of the ideal observer D_{ideal}. The accuracy, D_{obs}, of a real observer can be estimated easily from forced-choice data using equation (1). Equation (5) defines efficiency in terms of the accuracies of the real and ideal observers. If we equate perceptual confidence with perceptual accuracy, we can use equation (5) to relate perceptual confidence to statistical confidence:

$$\text{perceptual confidence} = \sqrt{E} \times \text{statistical confidence}$$

From this equation we see that perceptual confidence is linearly related to statistical confidence with the constant of proportionality being the square-root of efficiency.

Figure 10 plots perceptual confidence against statistical confidence for subject TT for one set of conditions. The axes are linear in d', but the numerical labels are the corresponding values of proportion correct. These are the numbers typically used in expressing confidence. For example, when the data support 0.95 statistical confidence, the corresponding perceptual confidence associated with numeric displays is 0.74.

Conclusion

Efficiency provides an absolute measure of perceptual performance. We have used efficiency to study graphical perception. Our findings extend those of Cleveland (1985) and Cleveland and McGill (1985) by quantifying the superiority of graphs over numerical tables. Specifically, our results tell us how well real observers can estimate statistical parameters—means and variances—compared with an ideal observer (or statistical test) who uses all information optimally.

Perceptual efficiencies are very high for scatterplots, often 60 per cent or more. Efficiencies are much lower for numerical tables (<10 per cent for a moderate number of samples and short exposures). Efficiency for the luminance-coded displays lies intermediate between those for scatterplots and numerical tables.

Performance with scatterplots has the earmark of a parallel process: weak dependence on sample size and viewing time. This confirms the view that the major contributor to the superiority of graphical displays is the preattentive, parallel-processing capacity of spatial vision. Our simulations of possible perceptual strategies, however, indicate that relatively high efficiencies can be achieved without analysing all the samples. Instead, observers may use a parallel spatial process quickly to identify particularly informative samples—e.g., minimum and maximum values. Their decisions may depend only on these values.

Real observers appear to process tables of numbers in a much more serial fashion. Their efficiencies drop roughly linearly with increasing sample size and increase in rough proportion to time. A plausible interpretation is that entries in tables are processed sequentially at a fixed rate. Given enough time, efficiencies might become quite high even for numerical displays.

References

Abramowitz, M. and Stegun, I.A. (1972). *Handbook of Mathematical Functions*. Washington: US Government Printing Office.

Barlow, H.B. (1977). Retinal and central factors in human vision limited by noise. In Barlow, H.B. and Fatt, P. (Eds) *Vertebrate Photoreception*. New York: Academic Press.

Barlow, H.B. (1978). The efficiency of detecting changes of density in random dot patterns. *Vision Research*, **18**, 637–650.

Barlow, H.B. and Reeves, B.C. (1979). The versatility and absolute efficiency of detecting mirror symmetry in random-dot displays. *Vision Research*, **19**, 783–793.

Burgess, A.E. (1985). Effect of quantization noise on visual signal detection in noisy images. *Journal of the Optical Society of America*, **A2**, 1424–1428.
Burgess, A.E. and Barlow, H.B. (1983). The precision of numerosity discrimination in arrays of random dots. *Vision Research*, **23**, 811–820.
Burgess, A.E., Wagner, R.F., Jennings, R.J. and Barlow, H.B. (1981). Efficiency of human visual signal discrimination. *Science*, **214**, 93–94.
Cleveland, W.S. (1985). *The Elements of Graphing Data*. Moneterey, CA: Wadsworth.
Cleveland, W.S. and McGill, R. (1985). Graphical perception and graphical methods for analyzing scientific data. *Science*, **229**, 828–833.
Egan, J.P. (1975). *Signal Detection Theory and ROC Analysis*. New York: Academic Press.
Freund, J.E. (1962). *Mathematical Statistics*. Englewood Cliffs, NJ: Prentice-Hall.
Green, D.M. and Swets, J.A. (1974). *Signal Detection Theory and Psychophysics*. Huntington, NY: Robert E. Krieger Publishing Company.
Julesz, B. (1981). Textons, the elements of texture perception, and their interactions. *Nature*, **290**, 91–97.
Kersten, D. (1987). Statistical efficiency for the detection of visual noise. *Vision Research*, **27**, 1029–1040.
Legge, G.E., Kersten, D. and Burgess, A.E. (1987). Contrast discrimination in noise. *Journal of the Optical Society of America*, **A4**, 391–404.
van Meeteren, A. and Barlow, H.B. (1981). The statistical efficiency for detecting sinusoidal modulation of average dot density in random figures. *Vision Research*, **21**, 765–777.
Pelli, D.G. (1981). 'The effects of visual noise.' Ph.D. Thesis, Physiology Department, Cambridge University.
Tanner, W.P. Jr. and Birdsall, T.G. (1958). Definitions of d' and η as psychophysical measures. *Journal of the Acoustical Society of America*, **30**, 922–928.
Watson, A.B. and Pelli, D.G. (1983). Quest: A Bayesian adaptive psychometric method. *Perception & Psychophysics*, **33**, 113–120.

8

Volumetric visualization of 3D data[1]

Gregory Russell[2] and Richard Miles

Princeton University
Princeton, New Jersey

Introduction

In recent years, there has been a rapid growth in the ability to obtain detailed data on large complex structures in three dimensions. This development occurred first in the medical field, with CAT scans and now magnetic resonance imaging, and in seismological exploration. With the advances in supercomputing and computational fluid dynamics, and in experimental techniques in fluid dynamics, there is now the ability to produce similar large data fields representing 3D structures and phenomena in these disciplines.

These developments have produced a situation in which currently we have access to data which is too complex to be understood using the tools available for data reduction and presentation. Researchers in these areas are becoming limited by their ability to visualize and comprehend the 3D systems they are measuring and simulating.

History

In response to this, there is growing activity in the area of visualization of 3D data. Some early work in this area was done by Harris *et al.* (1979) at the Mayo Clinic and Herman *et al.* (1984) at the University of Pennsylvania in the area of medical imaging. In 1983, Jaffey *et al.* (1984) approached the subject from a different direction. They developed the "source-attenuation" model, and used holograms to visualize 3D subjects. More recently, there is stronger emphasis on interactive visualization, and concentration on techni-

---

[1] This work was supported by Princeton University School of Engineering. Additional support for G. Russell was provided by the Office of Naval Research through their Graduate Fellowship program.
[2] G. Russell is currently at IBM T.J. Watson Research Lab, Yorktown Heights, NY 10598.

ques and systems for general use and commercial products (Goldwasser et al., 1985; Hunter, 1984).

Much of the recent activity is directed toward improving and extending the use of graphics techniques for interactive visualization of data based on surface representations. The groundwork for this was done by Herman et al. (1984). Work in this area is continuing both in academic groups (Herman at the University of Pennsylvania (Herman et al., 1984) and Fuchs at North Carolina (Fuchs et al., 1985), and in several commercial ventures (notably CEMAX)). Also, graphics projects at NASA, JPL, and aerospace corporations have been providing increasing support for visualization tasks based on conventional graphics concepts.

The more interesting projects involve departures from conventional graphics. By careful use of transparency, it is possible to produce images of 3D systems which provide true volumetric visualization, rather than surface projections. We have been working on this type of system for the past three years (Russell and Miles, 1987), concentrating on techniques which are efficient enough to be used interactively on existing computer systems. Pixar Corporation has recently been developing a package to support volumetric visualization, including an approach called Volume Rendering Technique, which they developed with Phillips Medical Systems and Dr. E. Fishman (1987) of Johns Hopkins University. This package is perhaps the most comprehensive image-based system commercially available at this time.

An approximation to volumetric imaging is also provided in PLOT3D, a graphics software system developed at JPL. This package includes a facility for producing nested transparent contour surfaces from a volumetric data base, which provides surprisingly good visualization of the data. Its primary limitations are data size (about 100,000 data points) and the number of contours it can support. Also, since this is a rather symbolic representation, it must be interpreted with care.

Volumetric vs. 2½D visualization

Normal pictorial illustration (stills), and most widely used 3D graphics techniques are limited to providing 2½D surface images. That is to say, along any line of sight there is only one object or surface visible. This usually produces pictures from which a rough idea of the three-dimensional structure of the original scene can be deduced. In contrast, X-ray images generally do not have a unique interpretation as projections of some three-dimensional subject, and even X-ray stereo pairs are insufficient to provide an unambiguous interpretation without *a priori* knowledge about the subject.

This is a computational constraint which applies not only to visual observation of pictures, but to interpretation of volumetric projections in general. Vision, however, is capable of limited volumetric perception and comprehension, if given adequate stimulus.

In order to achieve effective volumetric perception, it is necessary to

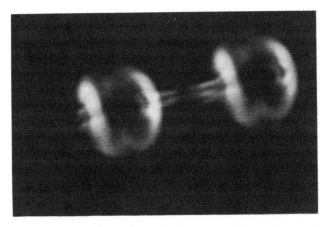

Figure 1. Vortex rings resulting from the Crow instability. Navier-Stokes simulation data provided by Dr. Michael Shelley, Princeton University.

present volumetric data in a form that vision is accustomed to dealing with. While cross sections are often useful for detailed study of internal features, it is difficult or impossible to comprehend fully the 3D structure of an object in this manner. Instead, data must be presented as we would see a real object. Natural visual processing transforms this information back into a mental structural model. Volumetric characteristics of the data are conveyed by making the projection TRANSPARENT, as implied in the earlier discussion.

The requirements for volumetric perception are basically the same as for computed axial tomography. A set of projection images from many different viewpoints is computationally sufficient to reconstruct the internal details of a subject. Visual reconstruction has several added constraints: the images must be presented as an ordered sequence of closely spaced views, and they must be shown at a rate of at least 8–10 frames/sec. These constraints are dictated by the temporal character of visual perception.

For perception of volumetric structure (rather than surface structure), complex optical phenomena such as lighting and shading, specular (surface) reflections, and diffraction and diffusion are not useful. In fact, these effects generally make the basic structure of volumetric scenes more difficult to understand, overwhelming the viewer with fine details and optical distortions. Simple luminance and opacity are adequate for volumetric visualization.

System implementation

We have developed a system at Princeton which implements this approach to volumetric visualization on a PC/AT (Russell and Miles, 1987). The algorithms upon which it is based are efficient enough to provide a usable off line visualization system on the AT (precomputed images take approximately 1 min/view for 2 million data points) and they are suitable for development into a real-time interactive visualization system using current state-of-the-art commercial hardware (AT&T Pixel Machine, for example).

The model for the system has the following characteristics:

(1) Data consists of samples on any regular 3D lattice (e.g., simple cubic, face-centered cubic, hexagonal close packed).
(2) The data elements are treated as nebulous, fuzzy regions localized around the sample coordinates (i.e., no subvoxel definition—consistent with proper sampling technique).
(3) Optical model includes luminance and opacity control at each data point, with the possibility of handling a light source (no refraction or specular reflection).
(4) Views are computed directly from the data, without any intermediate representation. This reduces the risk of artifacts and avoids simplification of the data that may lead to the loss of features.
(5) Perspective is not supported (this is subordinate to motion).

This combination of characteristics yields a model which is well-behaved and computationally efficient, with enough flexibility to provide a broad range of visual effects.

The implementation on the PC/AT operates in a two-step process. For a given data base, a sequence of views is computed, based on a selected set of optical characteristics onto which the data are mapped, and a viewpoint and axis of rotation for the data. Each image takes about 60 to 75 sec, for a typical data base of 2 million samples (e.g., $32 \times 256 \times 256$ or $128 \times 128 \times 128$), and we usually generate anywhere from 15 images (for a restricted range of views) to 120 images (for a full rotation of the data). The images are stored on a disk as they are generated. When a sequence is complete, the images are loaded by a second program for viewing. Up to 180 clipped images (176×176) may be loaded into 6 Mbytes of RAM on the PC/AT. They may then be viewed as a movie on a full-color, 8-bit greyscale display at frame rates up to 15 frames/sec. The viewpoint is controlled interactively using a mouse, within the precomputed range.

Evaluation

This method of visualization provides good comprehension for a range of subjects and optical characteristics. Its most significant advantage is that it is very robust. There is little or no preprocessing of the data, so there are generally no computational artifacts. Even data containing no distinct surfaces can be accurately visualized, since this method does not rely on surfaces as the fundamental elements of a scene. The use of motion as the means of communicating structure allows all the data to be made visible through the use of transparency. This provides a high degree of confidence in the resulting visualization. It is also robust in the sense that an informative set of images can be produced using simple optical characteristics (luminance = data value, high transparency) with little or no *a priori* knowledge about the data itself.

The motion/transparency approach is most effective with scenes of moderate complexity (such as that shown in Figure 1), that is, scenes whose structure can be largely comprehended as a whole. With very complex scenes, containing perhaps hundreds of detailed components (e.g., a video cassette recorder guts), this type of visualization suffers from showing too much information, which cannot be fully comprehended as a single entity.

Complexity

The issue of complexity arises in visualization for two distinct reasons. The first is the visual limitation just mentioned. The mind is incapable of performing a complete internal reconstruction of a volumetric scene, as is done in a CAT scan, for example. We have observed that beyond a certain level of complexity in depth (apparently three to four layers of structure), the mind's ability to maintain a conceptual model of a scene begins to fail.

In addition to the visual/conceptual limitation, there is an optical constraint which limits the degree of complexity which is practically acceptable. There is a tradeoff between the amount of transparency used (which affects the visibility of embedded structures) and the amount of contrast available in small features. This is directly related to signal-to-noise (S/N) ratio. Vision does not have particularly large S/N ratio, so fine details quickly lose definition as transparency is increased. This is also a limiting factor in CAT scans, but the devices used have much higher S/N ratios, so much lower contrast can be tolerated in CAT-scan source images than is detectable visually.

These considerations provide strong motivation to develop means of reducing and controlling the level of complexity in volumetric visualization.

The role of binocular vision

From a very early point in our investigation of visualization, it was clear that stereo pairs were inadequate as illustration of volumetric scenes. Once we had a working visualization system based on motion, it was easy to see how much more comprehensive this approach is than static stereo viewing. For some time, we assumed that adding stereopsis to the motion-based system would not be worthwhile, since static experiments suggested that stereopsis would not work well on precisely those scenes where some improvement was needed. Specifically, scenes with extensive volumetric content and high complexity, such as medical data, generally have low contrast and few clearly defined, unique features on which stereopsis can operate. For scenes which are visualized with low transparency, which provides more distinct features, stereopsis is not really needed since these scenes are generally quite easily understood with only the motion-based visualization.

When we actually were able to try out stereo and motion together, the results were somewhat surprising. With scenes of medical data with moderate to high transparency, static stereo viewing is relatively ineffective, as expected. However, when motion and stereo viewing are used together, the stereopsis provides noticeable enhancement to the visual perception of the structure over motion alone. There is apparently some interaction between the visual mechanisms which use stereo and motion to deduce structure. The combined effectiveness suggests that stereopsis is facilitated by information made available by motion, which perhaps allows better feature matching between images, resulting in more and better disparity measurements.

This strong interaction between stereopsis and motion perception means that stereopsis must be considered as an important part of any visualization system. Though motion is very powerful alone, considerable enhancement is possible through the use of binocular vision.

Conclusions

This approach to visualization, using transparency and motion in an image-based system, has significant advantages over systems based on solid rendering or graphical modeling. Most significant are the broader range of volumetric structure which can be visually represented and the robustness and freedom from artifact which volumetric visualization provides. A comprehensive visualization facility should certainly include the ability to perform both image-based and graphical rendering, and in the future these techniques should be increasingly integrated to allow both graphical and image-based components in a single visualization.

Computers are now becoming available which will be capable of performing visualization tasks interactively. This will dramatically change the way in which visualization is used, particularly for very complex subjects. As interactive visualization becomes more practical, the current emphasis on development of techniques for data reduction and rendering should be supplanted by the need for means of controlling and interacting with the visualization process. As the potential degrees of freedom for controlling a visualization increase with the complexity and size of scenes, the design of effective control mechanisms will be a difficult endeavor.

Some simple control mechanisms, such as clipping, spatial editing tools, and 3D cursors, are relatively easy to implement. However, for complex data, control mechanisms should parallel the way in which structures are decomposed and manipulated conceptually. This means providing the capability to specify the structural components of a scene and control their visual characteristics by referring to them as objects. Automated or computer-aided object segmentation is required to make this practical, but for the purpose of interactive control of visualizations, the accuracy and reliability of segmentations need not be as high as it must be for conventional, noninteractive visualization.

Additionally, it may be useful to be able to produce geometric distortions of data in order to push obstructing objects out of the way without separating them altogether from the region of interest. The net effect would be to produce the equivalent of an exploded view for structures of nondiscrete components. This would be particularly useful in medical applications. If information about connectivity and stiffness can be incorporated into the process, this could make the visualization system even more useful in surgical training or preoperative planning environments, where the mechanical properties of tissue structures are very important.

Advanced modes of interaction will become more and more important as volumetric display is applied to more ambitious problems of data interpretation.

References

Fishman, E., Drebin, B., Magid, D., Scott, W.W., Jr., Ney, D.R., Brooker, A.F., Jr., Riley, L.H., Jr., St. Ville, J.A., Zerhouni, E.A. and Siegelmann, S.S. (1987). "Volumetric rendering techniques: applications for three dimensional imaging of the hip". *Radiology*, **163** (3), 737.

Fuchs, H., Goldfeather, J., Hultquist, J.P., Spach, S., Austin, J.D., Brooks, F.P., Jr., Eyles, J.G. and Poulton, J. (1985). "Fast spheres, shadows, textures, transparencies, and image enhancements in pixel planes". *ACM Computer Graphics*, **19** (3), 111.

Goldwasser, S., Reynolds, R.A., Bapty, T., Baraff, D., Summers, J., Tatton, D.A. and Walsh, E. (1985). "Physician's workstation with real-time performance". *IEEE Computer Graphics Applications*, **5** (12), CGA.00, 44.

Harris, L.D., Robb, R.A., Yuen, T.S. and Ritman, E.L. (1979). "Display and visualization of three-dimensional reconstructed anatomic morphology: experience with the thorax, heart, and coronary vasculature of dogs". *Journal of Computer Assisted Tomography*, **3**, 439.

Herman, G.T., Reynolds, R.A. and Udupa, J.K. (1984). "Computer techniques for the representation of three-dimensional data on a two dimensional display". *Proceedings from the Society Photo-Optical Instrument Engineering*, **507**, 3.

Hunter, G. (1984). "3D frame buffers for interactive analysis of 3D data". *Proceedings from the Society Photo-Optical Instrument Engineering*, **507**, 178.

Jaffey, S., Dutta, K. and Hesselink, L. (1984). "Digital reconstruction methods for three-dimensional image visualization". *Proceedings from the Society Photo-Optical Instrument Engineering*, **507**, 155.

Russell, G. and Miles, R. (1987). "Display and perception of 3D space-filling data". *Applied Optics*, **26** (6), 973.

9
The making of The Mechanical Universe

James Blinn

JPL Graphics Laboratory
Pasadena, California

Editorial foreword

The Mechanical Universe is a two-semester, introductory level, television-based physics course. In the fall of 1985 the first semester of *The Mechanical Universe* was released to the academic community and public broadcasters. The two semesters of the course, *The Mechanical Universe* and *Beyond the Mechanical Universe*, consist of 26 half-hour television lessons and two versions of a text, one for science and engineering majors and the other for nonmajors. The course is scientifically sophisticated and mathematically rigorous, teaching and using calculus. The lecture programs contain computer animation used as a primary tool for instruction in physics. Each program begins and ends with Professor David Goodstein providing philosophical, historical, and often humorous comments from his lectures at Caltech.

The television series is not only the basis for a college course, but it is also suitable for a general audience interested in stimulating and challenging science programming. *The Mechanical Universe* television series and college course were funded by The Annenberg/CPB Project and The National Science Foundation (California Institute of Technology, 1986).

The following sections excerpt a number of design considerations regarding the dynamic computer graphics used to communicate physical phenomena and mathematical principles included in *The Mechanical Universe* (Blinn, 1987). The specific recommendations were not intended to be freely extended to other graphics interface applications, but do represent the considered judgment of a pioneer of computer graphics and certainly identify design issues that are faced in all attempts to use computer graphics as a medium for communication of spatial information. *The Mechanical Universe* represents a pioneering example of the use of computer graphics techniques to assist the visualization and comprehension of both physical phenomena and unusual environments.[1]

Overview

The Mechanical Universe project required the production of over 550 different animated scenes, totaling about 7½ hours of screen time. The project required the use of a wide range of techniques and motivated the development of several different software packages. This paper is a documentation of many aspects of the project, encompassing artistic/design issues, scientific simulations, software engineering, and video engineering.

My interest in *Mechanical Universe* is twofold. One, to produce the material and two, to see what tools need to be developed. It is hard to develop tools if you don't know what they are supposed to do. Having a large animation project provides a lot of experience of what the problems really are, instead of what somebody thinks they might be. This is a somewhat empirical approach to systems design. That is, several special-case systems are built, motivated just by the needs of some particular project. They are then analysed to see what things they seem to have in common. In doing this sort of examination, it is important to realize that you cannot prove that your assertions are correct in the same sense that you can prove a mathematical theorem. The best that can be said is that the mechanisms described here seem to work well for the problems to which they have been applied.

In this section I will discuss a few ideas on graphical design in general. The emphasis will be on concepts that are not specifically for scientific animation, but those that may be applied to other uses of visual communication.

I haven't learned this by formal training. It has come by practice, intuition, and perhaps genetics (I come from a family of artists). I learned to solve design problems by being presented with them and by being forced to think about the implications of color and shape choices. The results are what made sense to me at the time.

Graphical design (static)

Static design refers to the appearance of a single frame. The concept of motion design is discussed later.

What is a design problem?

Let us begin with the question, "what is a design problem?". It can be likened to pantomime. You must present some information that, perhaps, could be described in words, but you are required to use only pictures.

---

[1] See also Hussey (1989) for visualization of planetary surfaces, and Kaiser *et al.* (1990) for instruction in some phenomena discussed in this book.

Some examples:

(1) The Voyager spacecraft approaches a planet. A moon is off to the side. You must pan across to see it, but still give the viewers some idea of context of where they are now looking, compared with where they were looking before.

(2) How about a more detailed example? We will take an example from program 5, Vectors. The idea is to list the various types of vector expressions and to give an idea of whether the result is a vector or scalar. New items are added to the list as the program proceeds. The whole list may not fit entirely on the screen. In addition, as a new item is added, some geometric demonstration is needed to show what it is.

Let's look at a solution to this last example. We represent an abstract "space" where the vectors live as a kind of vector land. There is a river running down the middle separating it from scalar land. This allows us to display the lists in perspective receding into the distance. As each new object is introduced, it is added to the front of the list and the list recedes farther into the distance. Old list items may no longer be legible, but the memory of them is enough to remind the viewer of what they are. The key elements are to (1) differentiate between vectors and scalars and (2) give an impression of three-dimensional (3D) space, but not to make it look too realistic.

An oblique view of the ground plane must appear to recede into the distance. This can be shown by texture. An obvious texture is a grid which shows perspective very well. However, at this point in the academic development, the notion of a coordinate system has not yet been presented. Some other textural effect must be used. Texture mapping a random, say pebbly, texture would be slow. The resolution is to place a randomly scattered group of lines looking like grass across the plane. Just a few such lines can give a very cheap impression of receding ground plane. Also, the color of the plane is made to get bluer and paler as it moves into the distance.

Drop shadows help to bring out the 3D quality and make the vectors seem to hover above the plane, giving an interesting surreal effect.

Later in the program, when unit vectors and coordinates are introduced, the grid is placed on the plane (but only a small piece of it). Grids are a bit overused in computer graphics, but for much of what we do at *Mechanical Universe*, they are necessary because we are actually plotting graphs.

When we introduce unit vectors $\hat{c}$ and $\hat{\jmath}$, they tip their hats. When we show the construction of a vector product, the term for *vector add* and *vector multiply* are slid down close to the grid.

Direction of attention

It is necessary to direct the attention of the viewer to the important parts of the picture. Scenes are shown on television in fairly brief bursts, so the important parts must stand out. One good trick for doing this is to look away from the screen and look back quickly; determine what you see first when looking back. Is that the important part of the picture? If not, change the picture to make it so.

This means avoiding gaudy backgrounds; the background should not look more interesting than the foreground. In one example I had an equation over a dark blue background that graded into orange, giving a sort of sunset effect. It was very pretty, but the problem was that when you first looked at the screen, all you saw was the orange. I changed the background to a more neutral color and now the first thing you see is the equation.

Avoiding information overload

I consciously avoid trying to "dazzle" the viewers. Dazzling implies an overload or numbing of the senses. The idea is to communicate and draw the viewers in instead of making them tip backwards off their chairs.

For the same reason, I don't use lots of spinning or tumbling of 3D objects. It's distracting. There is a trade-off here between not giving your audience enough views of an object to be able to understand its 3D shape versus making it confusing by spinning it around too quickly.

One important trick to encourage simplicity is to arrange for the designs to be done while viewing a monitor from across the room. If the image can be made legible at a distance of 10 ft, it's about right. This discourages putting in too much small detail.

Color selection

Given the color television medium, we have both the opportunity to make scenes in color and the responsibility to make the colors look good. There are a few tricks to use in color selection.

I have favorite colors; I lean toward blues and greens. However, I don't like purple. I once used it purposely to break out of a rut, as a background in the scene on conic sections. I originally wanted to put a red cone in front of it, but I couldn't get a red that didn't disappear into the purple in dark areas (as seen in black and white). Finally, I went to a brighter yellow cone.

Make it work in black and white

When designing, look at the picture with the color turned off and see if it "reads" (to use a designer term). Reads in this context means "can you tell what is going on; do the appropriate things stand out?".

While color is important in the *Mechanical Universe* animations, it is not the only thing that differentiates items on the screen. It's not crucial. I have made consistent color decisions, but the viewer is not expected to remember color schemes to understand a scene.

Context

Color selection programs are minimally useful because colors always look different in context. The only real way to see how they look is to make an actual picture of the scene.

Distance cues

Distance can be represented by making things disappear into a fog. This was done literally in a scene of the molecular arrangement of a salt crystal.

Other color cues: the color of things gets bluer and paler with distance.

Field lines are a complex set of 3D curves. They can look like a pile of spaghetti if you're not careful. The distance effect is aided by three things: (1) normal depth cueing (things get darker—i.e., less luminance contrast—with distance); (2) drawing them in depth order so a closer (brighter) line will overlay a farther line; (3) making the intensity of the line darker at the edges than in the middle. This gives a slight "cylindrical" solid quality to the lines.

Not too many

Don't use too many colors.

There is a problem with running out of colors. There are more physical quantities to represent than there are easily distinguishable colors. You can't use saturation or value to distinguish things because sometimes these need to be adjusted depending on context, e.g., energy.

Consistency

Consistently use color schemes to recall previous results as well as to differentiate things. We will discuss the color scheme later:

(1) But the color scheme wasn't always consistent
(2) Paler colors for mass multiplied by something
(3) Colored backgrounds for two integrations of gravity law
(4) Colored backgrounds for bringing external equations to prove Kepler's third law
(5) Blue texture for energy equation

2D/3D Considerations

Two-dimensional diagrams are easier to understand than 3D, especially when they are in motion. This is partly because labels keep getting in the way of 3D diagrams in some views. Most of the physics of the first term of *Mechanical Universe* is essentially 2D problems (like Keplerian orbits). These remain 2D. The inherently 3D concepts are torque and angular momentum. The punch line is, use 3D only when absolutely needed.

Figure 1. Computer animation dissects the forces and motions that make a gyroscope do its tricks.

In fact, some 3D situations were simplified to 2D. For example, I used 2D for the Lennard-Jones atomic motion simulation and the ideal gas simulation. The actual physics is 3D, of course, but 2D shows the phenomena adequately and 3D would be really confusing.

In the second term there were more inherently 3D problems. You must use 3D for electromagnetic fields. Many textbooks use 2D for fields, but much is lost.

Three dimensions are also used as a trick to put more text on the screen. As the screen tilts back, more text fits. The top row might not remain legible, but we can remember what it was.

Making things stand out from the background

Drop shadows help make things stand out from the background. While they are good for labels on graphs, don't put a drop shadow on the plotted graph line because it detaches it from the grid.

Put 3D shadows for 3D vectors even if there are abstract shapes with no light source. One can more easily see a 3D shape by simultaneously having two views of the object, a 3D view and a projection of that view on the x,y plane. This is what the cubists were trying to do—show many views of an object at once. The shadow technique is more the way we are used to seeing and interpreting things.

Make the background a different value; use pale colors.

Realism versus abstraction

Images representing some real, physical object are often overlaid with labels, vectors, etc. For such scenes, the real object is rendered with a simulated light source and shading (usually with a simple polygon rendering program).

The mathematical abstractions are overlaid with a line drawing program (lines don't change thickness as they get closer or farther from viewer).

Graphical design (dynamic)

From reading Thomas and Johnson's book (1981), you are left with the impression that animation is the highest form of human art. It encompasses all aspects of static art and adds timing and motion, too. Motion design may well be the next great research topic in computer graphics. Results shown here are very preliminary.

Interpolation

It is the popular wisdom in animation that spline interpolation is better than linear interpolation. It is smoother. Most of the animations were done with splined motion. However, later in the series I began experimenting with linear interpolation and found it quite pleasing. Let's face it, the algebraic motions represent mechanical operations, so why not make them mechanical looking? In this case non-natural (jerky) motion sometimes looks more interesting than smooth motion because it's different and contains more high frequencies at the key frames.

Incorporation of "classic" techniques

There are various classic techniques that are found in "conventional" animation that apply here.

Squash/stretch

Squash and stretch refer to a distortion applied to the shape of an object when it undergoes acceleration. This is easily done by animating the x and y scale factor of an object. Before it begins to move, it gathers itself up by shrinking in x, then it stretches out in x as it is moving, and when it stops it shrinks briefly and returns to its normal size. This wasn't done in the *Mechanical Universe* as much as it should have been.

Overlapped motion

The concept of overlapped motion states that motion 2 should start before motion 1 is completed. This works well with character animation, but I found it of limited use in algebraic animation. In algebra there is just too much to follow as it is, without having the individual steps of a derivation merge into each other. Making the steps disjointed in time gives the viewer a chance to absorb one step before another begins. I did make the x and y

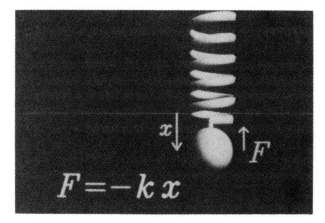

Figure 2. The spring force, or Hooke's law, is described in this animated scene from the Harmonic Motion episode.

motion of an object overlap, but this just rounds off the corners of the motion.

Perceptions of speed

I found it interesting to discover how limited our perception of velocity is. Given two successive scenes, where an object moves, say, one and a half times as fast in the second scene, it is very hard to tell which is which. This was proven because we were showing velocity changes in a lot of the physics. Most of the solutions involved representing velocity spatially as well as temporally by adding streaks or velocity vectors to moving objects.

Another interesting speed-perception discovery concerns double framing. One would think that all animation is ideally single framed. Double framing is just an economy measure if you don't have the computer time to do all the frames. Double framing looks jerkier. But there's another perceptual effect of double framing—double-framed motion looks faster than single-framed motion.

That is, if an object moves across the screen in 1 sec, it will look like it is moving faster if it is animated as 15 frames double-framed rather than 30 frames single-framed. This was alluded to in Thomas and Johnson's book (1981) on Disney animation. They said that motion was sometimes purposely double-framed to give it a "jaunty" look.

Audience

When doing something of this nature, it is important to keep the audience in mind. I had a very specific audience in mind when I designed these animations before I understood the concepts. For the most part, these are the explanations that I would have liked to have had, and that would have made the most sense to me when I was learning physics.

Roots

We are all products of our environment. I would like to mention some previous experiences that have affected my design motions here.

Lillian Lieber and Hugh Lieber are a mathematician/artist team that produced a series of charming books in the 1940s. Hugh, the artist, had a very surreal sense of making mathematical symbology visually interesting.

Various Disney animations were produced for science and mathematics. Among these were "Man in Space" and "Donald Duck in Mathemagic Land."

George Gamow (1967) wrote several books popularizing physics. His best creation is the Mr. Tompkins series. In these books, Mr. Tompkins attends a physics lecture and falls asleep. In his dreams the physical point of the lecture is illustrated, usually by exaggerating the effects so they were more noticeable in daily life. Particularly memorable was a scene in the "Old Woodcarver's Shop" where a sculptor makes atoms out of little green marbles (electrons) and little red marbles (protons).

The "Chem-studies" series of films were made for high school use. These had several conventionally done animations of molecular dynamics during chemical reactions. The motion of the atoms in these animations beautifully gives a sense of the energetics of atomic bonding. These were produced by David Ridgeway, who is on the national advisory committee to the *Mechanical Universe*.

The Bell Labs produced science films such as *The Unchained Goddess* and *Our Mr. Sun*. These were directed by Frank Capra, a Caltech graduate, and also a member of the advisory committee to the *Mechanical Universe*.

Finally, a telecourse from the past: "Continental Classroom"; this was a for-credit course offered on television in about 1960. It had classes in mathematics, physics, and chemistry. When I was young I was interested in this stuff, but I didn't know where to go for information. When I found this course I got up religiously each morning at 6 a.m. to watch it. I understood only about half of it, but it kept my interest in the subject alive. I hope that, with the *Mechanical Universe*, I might be making a series that generates similar interest in a new generation of students.

Visual metaphors (design)

In this section I will discuss visual metaphors for physics, grouped by design concepts.

Color

A normal textbook diagram has shapes, lines, and text. In video we have, in addition, color and motion. The challenge is using them. Motion usage is, for the most part, more obvious than color usage. Where there is some previous convention for color assignment, I tried to use it. Where there was none, I had to invent one.

When referring to explicit color values, I will use the notation developed by Alvy Smith. Color is represented by three numbers signifying:

(1) Value or brightness (0–1).
(2) Hue going around the color wheel. Numerical quantities go from 0–5 for one cycle: 0 = red, 1 = yellow, 2 = green, 3 = cyan, 4 = blue, 5 = magenta.
(3) Saturation. 0 = neutral, 1 = fully saturated.

Written, as an expression, (i, j, k) (i.e., (1, 0, 1)) would be a red of maximum brightness and saturation.

Many different ideas were keyed to colors. Much of this was subtle, and the animations never relied solely on the color to be understandable. I was left with the impression, however, that there simply aren't enough colors to have a unique one for *everything*.

For dimensional analysis

When physical abstractions such as acceleration or torque are represented in vector diagrams or algebraic labels, there must be some color. Rather than just making all vectors and labels white, I chose to institute a color scheme that is keyed to the units in which the quantity is measured. These color schemes are maintained throughout the series. This provides for a sense of continuity and also gives the viewer a sense for dimensional analysis.

Also, I tried to avoid the temptation to get overly cute with the colors. Colors are used primarily for labels. Terms in equations are usually white; otherwise, the equation tends to look like confetti. A term is shown in color only if the dimensions are important for a particular derivation.

Position, velocity, and acceleration are the most commonly used quantities. Position was a green = (1, 1.8, 1); velocity was a yellow = (0.7, 1.2, 1); and acceleration was a red = (1, 0.2, 1).

There are several motivations for this general color scheme. As successive derivatives are taken, the color shows a smooth progression along the color wheel from green to red so there is a visual progression between the colors. (Actually, the reddening applies not so much to derivatives as to the division by time.)

Acceleration is the most "active" of the three concepts. But red means "stop," not a very dynamic idea (although it takes deceleration to stop). This might be a counter-argument for the use of this color. But red is also the most exciting, attention-getting color. It shows that something is going on, and thus looks dynamic.

Green (as in grass) shows a static "place-like" effect.

This color scheme worked well when applied to a scene showing an abstract bicycle rider. The intent was to show elevation and slope. The normal color for informational traffic signs (green) was used to label the elevation. The normal color for warning traffic signs (yellow) then labeled the slope.

Note that the colors chosen are not pure; the hue values are not integers. The exact hues were selected visually to look nice together. Exact primary colors tend to look boring.

Mass times acceleration gives force. Mass times velocity gives momentum. Force and momentum were given the same colors as acceleration and velocity except that the saturation was reduced. I think of mass as a sort of dark grey color, looking solid, like lead or iron. So adding grey to the above colors desaturates them.

Energy is a dark blue color. This was chosen to look sort of like a lightning bolt. Energy's color is (0.2, 4, 1).

Angular momentum is a sort of rotational concept. I toyed with the idea of giving angular momentum vectors a sort of barber-pole effect, but it seemed too busy. Angular momentum is also mass times velocity times distance. Maybe a sort of pale yellowish-green? But that would not make it distinguishable enough from the other two. Finally, I decided to take off in a new direction and make it a pale blue. Torque, the derivative of angular momentum, is lavender (blue with red added to it).

Area and volume were made variants on the green color. Area is a slightly bluer shade. Volume is a still bluer shade. Maybe I was getting too subtle here, but you have to pick *some* color, and it might as well be for *some* reason.

Actually this choice was not entirely conscious, and as a result, the color for area is not exactly consistent through the entire series. For example, the color of Gaussian surfaces in the electricity programs was the position color, not the area color. This led to some problems when showing surface integrals. You do your best, but sometimes mistakes creep in.

Solid, liquid, gas

In the thermodynamics discussion there is a section on the states of matter. In particular, a PVT diagram is separated into regions where a substance is a solid, a liquid, and a gas. These regions were colored as follows:

> Solid—medium brown; an earth color, designates the solidity of ground.
> Liquid—bluish; like the color of water.
> Gas—white; a transparent color.

In the PVT diagram there is a region above the critical point where the distinction between liquid and gas disappears. Van der Waals' equation was used to find the degree of liquidity and to calculate a saturation value smoothly grading from blue to white for this region.

Electric charge

Positive and negative charges are shown in many scenes. There has been a sort of convention for some time in engineering to make the positive leads red. In addition, the books by George Gamow represented electrons as green marbles. So a similar color scheme was chosen for the *Mechanical Universe*.

But there are two problems here. First, not everyone has a color television set. So the colors were chosen so that, in black and white, they would still have enough difference in brightness to be distinguishable. Second, although red and green are complementary colors visually, in video it is red and cyan (a sort of pale blue). In some instances a neutral charge (e.g., for neutrons) is shown as, obviously, white. It would seem best to make the plus and minus colors add up to white. So a more bluish hue was chosen for negative charge. The exact value was actually changed during the second half of the series to be exactly cyan. This seemed necessary to make plus and minus add up to neutral, but I'm not sure it was a good idea in retrospect.

Electric and magnetic fields

I've always thought of magnetic fields as blue, and many published diagrams have shown it as blue. In fact, in an earlier project showing the magnetic field of Jupiter, I made the field lines blue. The question is, what color are electric fields? Since they are lines between positive (red) and negative (greenish blue), I decided to make it the color halfway between them, yellow. Note again that this is a different yellow than is used for velocity.

Relativity coordinate systems

There were many scenes in the relativity section that illustrated events as seen from two different reference frames. The two frames were usually those of a cartoon Albert Einstein and a cartoon Henry Lorentz. When they first appear, Albert is wearing a tan suit and Henry is wearing a blue suit. Thereafter, any algebraic or pictorial reference to Albert's frame is drawn in tan and any reference to Henry's frame is blue. These colors were initially selected as typical colors that suits come in, but they were fine-tuned to show up distinctly in black and white and when placed on a common background. Actually, when I first decided to do this, I had made Henry's suit dark grey. But dark grey didn't look good as a comparison color to tan—tan and blue are more balanced complementary colors. I had to remake one of the first animations just to change the color of Henry's suit. The production people probably thought I was nuts.

Wave/particle duality

The last three programs of the series begin to touch on quantum mechanics. Several of the scenes depicted wave-particle duality. Complementary background colors were selected to represent particles and waves. All particle equations appeared over dark pale green; all wave equations and plots of wave functions appeared with a dark pale magenta background.

Literal versus schematic

My tendency is to be too literal. The sizes and timings of some phenomena sometimes have too big a range to make this easy. But, because this is computer animation, the viewer expects precision and accuracy. When sizes or timings must be distorted into schematic diagrams, it is important to give some visual cues that this is being done. One way to do this is to have the schematic scenes drawn with sketchy or irregular lines. This removes the precision effect of perfect lines.

Literal

Some things were done geometrically correctly, even though it was difficult. For example, the radii of the orbits of the Bohr atom are proportional to the perfect squares (1, 4, 9, 16, ...). To see as many as four orbits, the scale must be too small to make the first orbit clear. This was usually solved by having the camera pull back when discussing the larger and larger orbits. This is a useful general principle, as it was described earlier concerning a list receding in perspective. If some things are too small, start close up and pull back.

Schematic

When force laws are introduced, we needed to show the operation of gravitational and electric forces. At this point, the magnitudes weren't important, only the signs. Crude schematic faces were used as mass particles (grey faces) and as positive and negative charges (red and cyan faces). The motion was sketchy, showing only attraction versus repulsion, and the faces were sketchy, with irregular and comical lines. This visual signaling was not done enough in the series.

Other scenes with schematized motion included:

(1) A depiction of resistance in metals. The normal velocity of electrons in a metal is far greater than the drift velocity, which is the electric current. Therefore, an accurate depiction of current wouldn't look much dif-

ferent than random thermal motion. The relative velocities were made more equal for illustration purposes. Also, resistance is caused by collisions of electrons with imperfections and thermal motions of the atoms in the metal lattice. These are usually too few and far between to be easily noticeable. They were made more obvious by flagging some metal atoms a different color and having the electrons bounce off them elastically, while not being affected by the positions of all of the nonflagged atoms.

(2) An electrical spark is generated by a chain reaction. Electrons are accelerated by an electric field and build up enough kinetic energy to knock other electrons off atoms. Again the typical spacing and frequency of the real situation would not fit on the screen. Some exaggeration was done.

Visual metaphors (physics)

Here are some more visual metaphors, this time grouped by subject matter, rather than by design issues.

Algebraic ballet

To make the science respectable we had a lot of algebra to present. Algebra, however, can be a bit draggy. We decided to liven it up by animating the algebraic transformations that the equations go through. These animations usually go by quickly. In fact, it is unlikely that the viewer will be able to follow all the steps upon first viewing. The speed was a concern, but we felt that making it slower would slow down the programs too much. The idea is to get a feel for what is going on and be able to look at a videotape shower to get the detail later if desired.

Transforming algebraic operation into motion proved to be an interesting exercise. Many of the motions seemed pretty obvious to me, but they are listed here for completeness.

Term labeling

It's easy to lose track of what different symbols in an equation represent. This was addressed by having the symbols identify themselves with English words popping out and shrinking back into them.

Balancing act

Simple algebraic operations to move terms around were animated literally.

(1) Terms moving to the opposite side of the = sign. Adding on one side means subtracting on the other so a + or − sign flips its identity as the term hops over the =.

(2) Factors moving to the opposite side of the = sign. Multiplying on one side means dividing on the other. When a factor jumps over the =, it lands below or above a division bar according to whether it came from above or below.
(3) Distribution: $a(b + c)$ becomes $ab + ac$ by having the *a* jump up, split in two, and each copy land next to the appropriate term.
(4) Squaring: Either two 2s come down from above and land on each side of the =, or a 2 on one side of an = sails over and changes to a $\sqrt{}$ sign on the other side.

Canceling

This applies to the removal of identities like $a - a$ or a/a. Some ways used to depict this were:

(1) A lightning bolt zaps the two terms and they disappear.
(2) An eraser appears and erases the terms.
(3) The two terms turn red and fall off the bottom of the screen together.
(4) A video-game-style spaceship flies in and fires a missile to explode the term.
(5) A Monty Python-style foot stomps out the terms.
(6) The Hand of God touches the term and it becomes a puff of smoke. This was used in the program that derived Kepler's first law (orbits are ellipses) from Newton's laws. The program made comparisons between the accomplishments of mathematics and physics and the accomplishments of art, drama, and music. Art was represented by the Sistine Chapel of Michaelangelo with the Hand of God giving life to Adam. The essential cancellation in the math that makes the derivation work is r^2/r^2; this is done by the Hand of God, too.
(7) Multiplication sign snipping out a term: The expression $\mathbf{v} \times \mathbf{v}$ is equal to 1. When this appears, the cross product sign magnifies around the surrounding $\mathbf{v}$'s and then squashes rapidly in y, snipping out the terms.
(8) Simply fading the terms out: This, of course, was the simplest and was done the most often.

Recalling old results

When a result from a previous program, or from a previous course is introduced, some effort was made to indicate to the viewer where it came from. Some examples are:

(1) A trigonometry book flies in, opens, and trig. identities fly out.
(2) A head with a hinged lid opens to receive some intermediate results; later it returns and the intermediate results fly out.
(3) A hand pulls down a window shade with old energy equations.
(4) An entire scene is reprised from a previous program.

(5) Some results were derived against a background image of some distinctive color. Later, when the results are needed, a slide comes in containing the equation with the same background as the old scene.

Substitution

Substitution involves taking an equation, defining some variable and replacing occurrences of that variable into another equation. Some examples:

(1) Vertical shrinking. A term is replaced with a number by shrinking the term vertically to zero and having the number expand up from zero in its place.
(2) Vacuum cleaner. The identity equation appears above the main equation. The replaced term from the lower equation moves up to the identity to merge with its copy there. The other side of the identity equation moves down to the empty spot left in the original equation.
(3) Several calculus identities (such as turning dr/dt into v) were shown by rotating the dr/dt about the y axis and having it become v when the other side appeared.

Jokes

The program on wave motion shows some approximate relations between wave speed and various physical parameters. The $\approx$ sign ripples like a propagating sine wave while these equations appear. This was done by modeling the lines of the $\approx$ sign with a one-cycle helix. Rotating it about x and then scaling by 0 in z made it ripple.

Calculus

A few algebraic operations on calculus notation:

(1) d/dt flies in from left and impacts f to form df/dt.
(2) The f slides up and down to form $(d/dt)\,f$ from df/dt.
(3) The two symbols $\int$ and dt move in on either side of f and clamp it together to form $\int f(x)dt$.
(4) A simple differential equation like $(dx/dt) = y$ is solved by moving dt to the other side to make $dx = y\,dt$. Then the left-hand d hops over the equal sign and changes into a $\int$ sign, to make $x = \int y\,dt$.
(5) Integration is done by the $\int$ sign ratcheting across an expression, sort of like a credit card imprinter.
(6) $\oint$ is formed by drawing the circle on the $\int$ as the path of integration is traced out in a geometric diagram in the background.
(7) $\oiint$ is formed by revealing the circle on $\iint$ as a Gaussian surface is spread out around a volume in a parallel diagram.

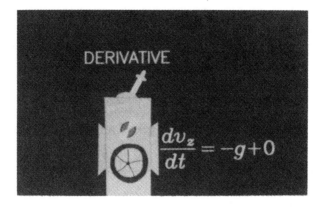

Figure 3. The Mechanical Universe *derivative machine has become a legend in its own time.*

Calculus

Limits

Use explosion to express the limiting process when Δ turns into d. The explosion was generated by a simple 2D pattern scaled up and faded out simultaneously.

Symbolic derivative machine

Because we evaluate derivatives and integrals symbolically many times in the series, we developed a quick way to do it—the derivative machine (Figure 3).

Design
The derivative machine is an expression transformer. It has two functions—differentiation and integration. An expression goes in one end and comes out the other end, so it needed to be thin in the x direction so there would be plenty of room on each side to show the inputs/outputs. When the derivative machine is first introduced, it comes in a crate marked "ACME Derivative Machine" (a hat tip to the old Chuck Jones Roadrunner movies). A crowbar shaped like an integral sign opens the crate.

Some random wheels and lights made it look Rube Goldbergish. The sides are not exactly straight and the wheels are not exactly round.

Internals
When the derivative machine is introduced in program 3, the internals are shown two ways:

(1) As various elementary operations are introduced, they shrink down into a sort of circuit board that is plugged into the machine, the door slams, and a new light blinks on on the front panel.

The making of the The Mechanical Universe 155

(2) An alternative view of the internals was given briefly, showing the details of how the elementary operations are applied to take the derivative of the simple expression x^2. This was intended to be something of a metaphor on how symbolic derivative computer programs work. The input function comes in on a conveyor belt. An eyeball on a stalk comes down and looks at it. (This is indicated by a dotted line running from the eyeball to the function.) This is the pattern recognizer. The derivative operation is basically one of matching the desired function against a list of known patterns which are pulled down into the scene like window shades. Then the proper pattern is found and checked. There will be some dummy parameters in the pattern which need to be filled in with the specific terms from the equation. The eyeball observes these and some handles come down and simultaneously turn all occurrences of the dummy parameter into the specific term needed. Identities such as $x + 0$ or $x \times 1$ are removed by an eraser. The expression $x + x$ is turned into $2x$ by a vise-like adder. The final expression is carried out on a conveyor belt.

Operation
The lever on the top controls the operation of the derivative machine. When you throw the lever to the right, it takes an expression in the left hopper and spits the derivative out of the right hopper. When you throw the lever to the left it takes an expression in the right hopper and spits out the antiderivative (integral) on the left. Sometimes the expression stays put and the derivative machine passes over it. Note: it doesn't evaluate integral expressions, it just takes the antiderivative (i.e., you don't feed $\int x^2$ in to get $\frac{1}{3} x^3$, you just feed in x^2). As it operates, the horizontal and vertical scales cycle up and down a bit to give it a squash and stretch look.

References

Blinn, J.F. (1987). The mechanical universe: an integrated view of a large animation project. (Course Notes: Course #6) *Proceedings 14th Annual Conference on Computer Graphics and Interactive Techniques*, ACM SIGGRAPH and IEEE Technical Committee on Computer Graphics. Anaheim, Calif.
California Institute of Technology (1986). *The Mechanical Universe ... and beyond*. Corporation for Community College Television: Pasadena, Calif.
Gamow, G. (1967). *Mr. Tompkins*. Cambridge: Cambridge University Press.
Hussey, K.J. (Producer/director) (1990). *Mars, the movie*, (video). Visualization and earth science applications group, observational system division, JPL Audiovisual Services, Pasadena, Calif.
Kaiser, M.K., MacFee, E. and Proffitt, D.R. (1990). "Seeing beyond the obvious: understanding perception in everyday and novel environment". NASA Ames Research Center, Moffett Field, Calif.
Thomas, F. and Johnston, O. (1981). *Disney Animation: The Illusion of Life*. New York: Abberville Press.

PART III
Acting

Acting

Arthur J. Grunwald

Introduction

The papers in the following section generally deal with the problems of human operators who interact with complex machinery or work in unfamiliar, counter-intuitive, strange and sometimes hostile environments. It deals with those characteristics of the human visual and motor systems which allow us to perceive and compose spatial maps of our environment on one hand, and guide visuomotor behavior on the other hand. It investigates the visual system in terms of environments, unnatural to the human, such as virtual environments for telepresence, or head-mounted displays, and discusses the human ability to adapt to mismatches in the visuomotor system. Such mismatches might occur in remote manipulation under conditions in which, due to time delays, system noise or distortions, a stable or normal correlation between the motion of the hand and of the visual image is lacking. Finally, the chapter discusses how the various sensory inputs interact in the human spatial orientation process, how sensory conflict can result in motion sickness, how orientation interacts with our perception of form, and how the visual ocular reflex can be voluntarily counteracted.

Professor Koenderink, in his foreword, has remarked that in complex systems the human operator often remains the weakest link in the chain, and that thus the machinery, the control task and operating environment should be tailored to the inherent human abilities and limitations. Here, spatial instruments play a vital role. The key to a successful tailoring however, is a thorough understanding of the highly complex activities taking place in the human controller. The papers in this section constitute an available basic body of knowledge, composed of a broad range of different approaches, which might guide the reader in the design of exotic systems such as virtual environments for teleoperation, helmet-mounted displays in Nap-of-the-Earth[1] flight, orbital maneuvering displays, or just novel head-up displays for automobiles. Figures 1 and 2 show examples of spatial instruments for telemanipulation from Hannaford et al. (1987). These displays are spatial

---

[1] Very low altitude flight following the contours of the ground

instruments in the sense that the perspective view at the "wrist" of the manipulator is visually augmented by the display of non-visual physical quantities, here the three forces (bars) and three torques (dials) acting on the manipulated object. Thus, the displays do not only relocate the operator in the remote environment, they also communicate to him task-relevant non-visual quantities. Although, for this particular application, the authors have chosen to communicate these quantities by visual means, other ways might be equally useful and are left to the imagination of the reader.

The detailed empirical analysis of experimental psychology used by most of the authors in this section to unravel the mysteries of the human sensory–motor system greatly compliments the engineering approach usually taken by aeronautical control system designers. The following discussion provides an alternative analysis of human sensory–motor performance in the simplified framework of control system theory.

The human in the framework of control system theory

Three stages of a control action

A human operator, interacting in a complex environment, can be considered as the controller of a dynamical system. In general, whatever the complexity of the system is, any control action can be subdivided in three stages: (1) Sensing: measurement of the system state by means of sensors, (2) Computation: processing of the measurement, estimation of the actual system state, comparison of actual and desired state, and definition of a control strategy to bring the system to the desired state, and (3) Actuation: execution of the strategy through an actual control action.

These three stages can be identified not only for humans, controlling complex machinery, but also for humans, engaged in everyday activities like walking, picking up and placing objects, playing tennis or riding a bicycle. The bicycle rider for instance, processes a complex blend of visual, vestibular, kinesthetic, proprioceptive and tactile measurements in order to create a control command to the handlebar (Van Lunteren and Stassen, 1969).

Overview of man-machine models

In vehicular control, for instance, control models have been formulated for describing, analysing and understanding the human response in controlling road vehicles or aircraft. Early classical control theory man-machine models, McRuer's "crossover" model (McRuer and Weir, 1969), and later optimal control models for human response (Kleinman et al., 1970) have compared the human response in an instrument tracking task with the control actions of a well designed servo-system. In these simplified models the human is considered to perceive an explicitly displayed, well defined control error, i.e., the deviation of a cross from a square in a flight director or the tilting of a horizon bar in instrument flight, and to utilize this error to create a unique control action, like the displacement of a control stick. In both the classical and the optimal control models, the three stages of control: sensing, computation and actuation are represented by simple linear, scalar dynamical functions. The breakthrough with the optimal control framework, however, is the assumption that the human controller contains an "internal dynamic model" of the controlled element, of the forcing functions and disturbance noises. This model plays the key role in the process of estimating the controlled element state from incomplete noisy observations and in defining an optimal control strategy for closing the loop. Oman, in his paper presented in this section about sensory conflicts in motion system, has used this optimal control framework as a model for motion sickness.

Later models have analysed the more complex situation of controlling land vehicles or aircraft by visual field cues. In contrast to a well defined control error shown on an instrument display, the control-oriented information has to be extracted from the complex visual motion pattern (Grunwald and Merhav, 1976; Zacharias and Caglyan, 1985). In this case a complex set of measurements has to be processed to reconstruct state information which can be used in a control strategy. The sensing and processing stages of this task are thus far more complex.

A third group of models deals with supervisory control and decision making (Van de Graaf and Wewerinke, 1983; Yajima et al., 1983). Here the processing stage involves complex non-linear and multidimensional optimizations. In contrast to the limited tracking models which involve motor skills and regard the human controller as a servo system, these models attempt to reproduce the intelligent behavior of human operators.

It is clear from the above examples that the complexity of the human control action can be rooted in each one of the three control stages. The complex process of spatial orientation is a prerequisite for any egomotion or vehicular manipulation or control task. It includes the collection of a set of noisy, sometimes incomplete or conflicting measurements (sensing stage) and a computational process aimed at reconstructing the human's position and orientation with respect to an outside world (computation stage).

Human spatial orientation and state estimators

There is an important analogy between the human spatial estimation process and "state estimators" as known in control engineering for the last four decades. The "Kalman estimator", for instance, has found its use in a broad range of control applications. In inertial navigation it is used to compute an optimal estimate of the vehicle position from noisy accelerometer and rate gyro measurements. The basic assumption with the Kalman estimator is that the dynamics of the system and the statistical properties of the measurement and system noises, are known and modelled by a set of dynamic equations.

This model is used to compute expected output values for each one of the sensors. These values are compared with the actual noisy measured values and the difference is used cleverly and in an optimal manner, to retune the parameters of the estimator. A weakness of the approach, however, is evident when the system dynamics or noise processes are not exactly known or turn out to be different from what was expected. In this case the filter might produce far from optimal and sometimes totally erroneous results.

The key to a successful engineering design of state estimators is twofold: (1) insensitivity of the estimator performance to inaccuracies or errors in the assumed system model and noise processes, known as "robustness" and (2) the ability to adapt to a wide range of varying conditions. The techniques for obtaining these desired properties vary from using redundant or complementary measurements, i.e., using more measurements than are strictly needed for obtaining an "observable" system, to incorporating on-line learning schemes, i.e., algorithms which readjust the estimator parameters automatically by considering the statistics of the signals involved. Some advanced filters include "banks" of models for different conditions, like a set of wrenches in a tool shop. Clever identification schemes are able to identify the condition and choose the correct model from the bank, very much like the mechanic, choosing the right tool for the job. However, accuracy and adaptability seldom come together. An estimator doesn't behave differently in this sense from a human. In the course of action, estimators manage to "learn" the statistics of more or less stationary processes. In the beginning of its "life span" the estimator is still young and enthusiastic and easily adapts its parameters to a varying process. However, being quite inexperienced, it might make occasional large errors. The longer the estimator is active, the better it is able to tune itself to the particular situation, but the less it is willing to change its parameters if a sudden change in conditions occurs.

It is fascinating to note that these dynamic characteristics are inherently also present in the human observer. Adaptability is apparent when considering that the human spatial orientation works fine both when strolling easily through the park or when driving at high speed on a curved road. The "redundancy" in measurements in vehicular control (visual, vestibular, kinesthetic, proprioceptive, tactile, audio, etc.) becomes vital when visual cues are largely impaired, as is the case in night driving on a wet road. The role of nonvisual cues is clearly demonstrated in McGovern's chapter (Chapter 11) about teleoperated land vehicles in this section. He reports that operators, remotely controlling small vehicles by means of a vehicle-mounted TV camera ("inside-out" view), seriously underestimate terrain roughness, resulting in a large number of roll overs.

Estimators with erroneous models

As is the case with artificial state estimators, the human spatial estimation process will also break down when basic assumptions about the visual environment (the "model") are in error with reality. This breakdown might lead to to surprising visual illusions or erroneous control actions. Consider, for example, the well known "Ames Room" demonstration (Ittelson, 1952). In this demonstration a monoscopically viewed person walking up and down a room with a floor consisting of non-parallel, converging lines, apparently expands and shrinks in size as a result of the subject's erroneous assumption that the floor is made up out of square tiles instead of converging ones. This produces a bias in the floor slant angle estimate, which, in turn, produces an error in estimated distances.

A second case of "breakdown" of the human estimation process occurs when the various measurements do not match with an expected motion pattern or model, or are in conflict with each other[2]. Oman, in this section, bases his theory of motion sickness on these conflicts. Virtual environments in which the visual image is decoupled from the inputs felt by the vestibular system, are sensitive in particular to these conflicts. Well known examples are head-mounted displays, in which the head rotation is slaved to a remote camera or a computer generated image system. Scaling errors and time lags between the motions of the head and the corresponding motions on the display will cause visual inputs and vestibular inputs to be in conflict with each other and cause nausea or motion sickness.

Other typical examples are flight simulators, which attempt to create a realistic illusion of the motions as felt in the aircraft. The cabin is closed so that a visual scene can be generated independent of the inertially stable outside room. Ideally, three rotational and three translational motions have to be generated. Since the simulator stroke is limited, and thus is able to sustain the acceleration for a short time only, longer lasting accelerations are simulated by using a component of gravity by tilting the cabin without the pilot noticing this. Likewise the component is removed by slowly rotating the cabin back to the horizontal position. This is known as "washout" motion. For moderately moving simulators, like an airline trainer, the washouts can be well below the human threshold of detection of rotational motion and the results can be satisfactory. However, for a violently maneuvering fighter aircraft, the limited simulator stroke might be inadequate to simulate the required angular and linear accelerations and washouts can not be designed below human thresholds. This will lead to erroneous vestibular inputs.

Brief overview of the section

Vehicular control

The vehicular control part of this section presents ideas for spatial displays for atmospheric flight, for orbital maneuvering in space and for teleoperation

---

[2] Other analyses of the breakdown of human perceptual estimation processes are considered by Ellis et al. (Chapter 34) and Foley (Chapter 37).

of land vehicles. They are spatial displays in the sense that they are enhanced by geometric, symbolic and dynamic techniques to ensure that the task-relevant information is communicated properly.

Fadden et al. (Chapter 10) use separate displays for the horizontal and vertical aircraft control task, i.e. a map and a vertical situation display. This choice was motivated partly by the huge difference in tracking accuracy requirements between the lateral and the vertical task. They also discuss the scaling differences between longitudinal and vertical axes. Although in the vertical situation display the vertical path angle had to be exaggerated by a factor of 20:1 to ensure proper communication of the vertical aircraft motion, this enhancement did not have a negative effect. It was also important to keep the longitudinal scaling of the map and of the vertical situation display congruent.

McGovern (Chapter 11) discusses his experiences in the development of displays for teleoperated terrain vehicles. He reported operators being uncomfortable with limited field-of-view systems, problems in distance estimation and, most of all, a tendency of over-controlling as a result of the lack of vestibular cues.

Eyles (Chapter 12), and Grunwald and Ellis (Chapter 13) propose orbital maneuvering displays for carrying out unanticipated missions, when already in space. The visualization system of Eyles allows the astronaut virtually to move around in space freely by flexibly changing the display coordinate system, viewpoint and field-of-view. This capacity can be seen as a way of achieving telepresence, with the purpose of identifying with crucial parts of the mission. The orbital maneuvering display of Grunwald and Ellis addresses the problem of the counter-intuitive and unfamiliar relative motion between coorbiting spacecraft. A dynamic enhancement is introduced in which the dynamics are solved in an "inverse manner", i.e. instead of having the astronaut control maneuvering burns he will control the final desired position which is directly mapped to a hand controller. The establishment of a direct correspondence between movements of the hand and the controlled element greatly simplifies the design procedure.

Manipulative control

In this section typical problems of remote manipulation, like choice of display, choice of hand controllers and correlation between motion of the hand and of the display, are discussed.

Held and Durlach (Chapter 14) discuss the notion of telepresence in teleoperated systems. Anthropomorphically designed teleoperators offer the best means of transmitting man's adaptive and creative problem solving into the remote environment. The crucial factor in creating telepresence is to establish a high and familiar correlation between the movements of the operator, sensed directly via proprioceptive/kinesthetic cues, and the actions seen on the display. Time delays destroy this correlation. The degree of telepresence that can be achieved depends on the adaptability of the operator. They composed an interesting experiment of eye-hand coordination in which, in addition to prismatic displacements, a time delay was introduced between the actual motion of the hand and the image. The adaptation to the prismatic displacement was found to break down with larger time delays, because the motions of the hand could not be identified with the motion of the display. They feel therefore that adaptation requires a certain degree of identification.

McKinnon and Kruk (Chapter 15) developed a multi-axis, six degrees-of-freedom (DOF) hand controller for teleoperated systems in space. The design objective was standardization of systematic modes of control in order to reduce training time and improve efficiency. They review the Space Shuttle robot arm control, which has two three DOF devices: Translational Hand Control for the left hand and Rotational Hand Control for the right hand. The Shuttle arm controls are shown to be operable but not optimal. Their shortcoming is a lack of direct correspondence between the axes of control and of visual displays. Their objective was to demonstrate the feasibility of integrating the three rotational and three translational degrees of freedom in one hand controller. The only way to avoid device-induced cross-coupling in such a controller and still have direct correlation between control inputs and resulting actions, was to center all axes at the geometric center of the cupped hand. This utilizes the human ability to make complex multi-axis movements with one hand using conscious position sensing. They conclude that consistent orientation between controller axes and visual feedback is essential.

Kim et al. (Chapter 16) investigated the contribution of stereo cues in a simulated pick-and-place task with a perspective display. In the presence of the proper visual enhancements such as a ground grid and projecting lines, the stereo display had no clear advantage over the mono display. However, without display enhancements, the stereo display was far superior. This shows that natural visual cues, when lacking in the spatial instrument, can be restored by proper augmentation. Interestingly, the relative performance with the different display conditions reported for this pick-and-place task is almost identical to subjects' relative 3D tracking performance with an identical set of display conditions (Kim et al., 1987).

Visual/motor mapping and adaptation

The papers in this section discuss the complex processes underlying eye-hand coordination. The understanding of these processes is relevant in particular for teleoperated devices and virtual environments, in which the motions of the hand and the motions of the manipulator are not congruent. The lack of congruency might be due to reversible systematic distortions like a rotated frame of reference, a fixed bias in the displacement, or a fixed scale factor. It might also result from irreversible distortions, like pure time delays or random noise. Through a learning process, the human is able to adapt to systematic distortions. The success of a teleoperated device depends on

Figure 1. Displays of force/torque information for telerobotics. Several monochrome and color formats have been developed and experimentally evaluated at JPL for the display of forces and torques encountered by a remote manipulator (left panels) to the controlling operator. The right panels show bargraph displays of force (upper three bars in each video screen) and torque (lower three bars).

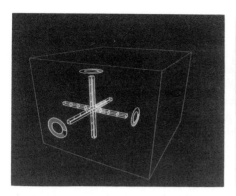 

Figure 2. A natural representation of the forces and torques may be proved in a pseudoperspective display (left panel) in which the bar graphs are aligned with unit vectors representing the direction of action of forces, and the roll, pitch, and yaw axes for torques. This format may be improved by coding torque as arc around each unit vector and viewing the graphic in perspective (right panel). Further improvements would involve maintaining alignment between the display axes and the control axes of the manipulator as seen by the human operator (Smith and Smith, 1962; Bernotat, 1970; Smith and Smith, 1987; Ellis et al., 1988, Chapter 16).

whether and to what extent the human is able to adapt to these distortions. His adaptability will be reflected in a better performance and shorter learning times.

Cunningham and Pavel (Chapter 17) study directional nonuniformities (deviation from the straightest path) in the performance of a 2D discrete aiming task, under transformed mappings between visual and motor space. This mapping could be a 2D rotation (realizable in 2D space) or a reflection (inverted polarity of one of the axes, non-realizable in 2D space). The nonuniformities under reflection were larger than under rotation, which indicates that the human representation of 2D space is constrained by physical realizability. An interesting finding is that the mechanisms responsible for visual motor control are sensitive to motor factors such as the number of joints involved in the motion of the human hand.

Welch and Cohen (Chapter 18) study the adaptation in eye-hand coordination, under conditions where they view their actively moving hand through a prism which displaces the visual field. For a fixed displacement this adaptation takes place in a matter of minutes. In this paper the adaptation to a continuously changing prismatic displacement is investigated. The adaptation in this case was found to be easily disturbed, which demonstrates the human flexibility to readjust quickly to different displacements or to normal vision.

Shebilske (Chapter 19) discusses discrepancies between two major functions of vision: perception, which provides a conscious representation of the environment, and visuomotor control, which guides movements that organisms make as they interact with their environment. He discusses mismatches between perception and visually guided motor responses, both in neurological patients and in healthy persons and demonstrates that perception and action do not share the same input operations. He suggests ecological theories based on the axiom that operations for encoding sensory information should approach optimal efficiency in the environment in which the species evolved. The Ecologically Insulated Event Input Operation (EIEIO) comprises an array of parallel, specialized input operations, which all mediate specific sensory-guided skills. They serve these skills optimally in a specific context, but do not generalize outside this context. The EIEIO hypothesis suggests as a guideline for training: train components which integrate visuomotor modules as opposed to separating them.

Bridgeman (Chapter 20) provides experimental evidence of the existence of two distinct representations of visual space, a visuo-cognitive one and a visuomotor one. The first is associated solely with visual perception; i.e. when subjects are asked verbally to report the apparent position of what they see, the information in the cognitive map is processed. The other representation drives visually guided behavior, both for eye and arm.

In the experiment reported in this chapter, a target was shown at five different positions on a wide-angle screen, surrounded by a rectangle, which could be centered or off-centered to both sides in order to produce a visual illusion called the "Roelofs" effect. In this phenomenon the target erroneously appears to be off-center in the direction opposite the frame. Subjects were asked to respond to the target in two ways: by physically turning a pointer, and by verbal judgement. He found that the Roelofs effect existed only for verbal judgements and not for pointing. In contrast to the cognitive map, the visuo-motor map has no memory and is not subject to illusions such as the Roelof's effect or induced motion. Thus, what he shows is that where you see isn't necessarily where you point.

Gregory (Chapter 21) points out that vision is an active process. Active exploration by touch is very important in vision. He presents a case study of a person, effectively blind from birth, who regained his sight in middle life. The person could see objects, earlier learned through touching. He was effectively blind to objects he did not know from his earlier touching experience. He was insensitive to spatial ambiguities, i.e. the Necker cube, and experienced it as a meaningless pattern of lines rather than a cube. He learned to conceive space not only by handling objects but also by walking. Horizontal distances were judged very accurately, whereas vertical distances were markedly misjudged, presumably due to a lack of prior experience with vertical excursions.

Orientation

This section discusses the interaction of the various sensory inputs in spatial orientation. This interaction is particularly important for virtual environments and head-mounted displays, in which the visual scene is generated independently from the inertially stable outside world. In these cases, the correspondence between visual and vestibular signals, as experienced in natural unconstrained viewing, does not necessarily exist. A subclass of these systems are head-mounted displays used in the space environment. Gravity, a vital cue in estimating the subjective vertical, is lacking and orientation in this case relies entirely on visual cues. The papers of Mittelstaedt and of Stoper and Cohen are here especially interesting, since they consider how the subjective vertical or eye level judgements are affected both by gravity and by visual cues.

Howard (Chapter 22) defines the reference frames necessary for understanding the mechanisms underlying spatial orientation and perceptual stability. These definitions are helpful in particular for head-mounted displays since the sensor, the retina, moves with respect to the head which moves with respect to the torso, which again moves with respect to the body, which again moves with respect to the outside world. In this chapter he discusses the oculogyral illusion, which is the apparent motion of a visual target as a result of stimulation of the semicircular canals, and vection, which is an illusion of self motion induced by looking at a large moving display. The magnitude of the vection depends on posture and axes of rotation. Finally, he discusses induced visual motion, which occurs when observing a stationary object against a moving background, i.e. the moon through the clouds.

Oman (Chapter 24) develops a theory of motion sickness on conflicts between sensory inputs or a mismatch with an expected motion pattern. He uses an optimal control framework for modelling and predicting motion sickness.

Mittelstaedt (Chapter 25) shows that spatial orientation influences the apparent form of an image, and form in turn affects apparent orientation. He formulates a theoretical model of trigonometric functions for determining the subjective vertical. He considers two main influences, which he models as "torques" that combine vectorially to produce the subjective vertical. His two influences are the gravito-idiotropic torque, resulting from an otolithic-sensed internal bias that influences the subjective vertical in the absence of visual cues, and a visual torque resulting from the orientation of the tilted image. He finds evidence that even in their influence on form perception, the gravito-idiotropic and visual effects combine vectorially rather than suppressing one another. His work is relevant when considering that objects on a head-mounted display might affect orientation judgements.

Stoper and Cohen (Chapter 26) study the various mechanisms active in judging the position of an object with respect to a viewer's eye level. The eye level can be referenced with respect to the gravity vector, to a visible surface or to a plane fixed to the head. Gravitational information originates from the otolithic sensors, surface information from purely visual cues, and head-referenced information from kinesthetic or proprioceptive measurements of the eye position. Their findings indicate that gravity is an important source of information for judging eye level. In the absence of gravity, head-referenced eye level judgements are very poor; the presence of a surface adds to the stability rather than precision of the judgement. This means that in the absence of gravity the judgement of the pitch angle of the observer's head relative to a surface is much less precise and subject to visual capture.

Zangermeister (Chapter 27) shows that with proper training, a subject is able voluntarily and continuously to eliminate the vestibular ocular reflex (VOR) response during a predictive gaze movement. This reflex normally serves to stabilize the visual image on the retina. He suggests that this stability is achieved by generating an image of the anticipated VOR and subtracting it from the actual one, so that the actual response is zero. His work is useful in particular for head-mounted displays, which are stationary with respect to the head (numerical, symbolical information) and thus move with respect to inertial space.

References

Bernotat, R.K. (1970). Rotation of visual reference systems and its influence on control quality. *IEEE Transactions on MMS*, **MMS-11**, 129–131

Ellis, S.R., Grunwald, A.J., Smith, S. and Tyler, M. (1988). Enhancement of man-machine communication: the human use of inhuman beings, *Proceedings of IEEE CompCon88*, February 29–March 4, 1988, San Francisco, Ca., 532–535.

Grunwald, A.J. and Merhav, S.J. (1976). Vehicular control by visual field cues—analytical model and experimental validation, *IEEE Transactions on Systems, Man and Cybernetics*, **SMC-6**, (12), 835–845.

Hannaford, B., Salganicoff, M. and Bejczy, A. (1987). Displays for telemanipulation, *Proceedings of the Spatial Displays and Spatial Instruments Conference*, Asilomar, Ca., Aug. 31–Sept. 3, 1987, *NASA CP* 10032, 32-1–32-7.

Ittelson, W.H. (1952). *The Ames Demonstrations in Perception*, Princeton: Princeton University Press.

Kim, W.S., Ellis, S.R., Hannaford, B. and Stark, L. (1987). A quantitative evaluation of perspective and stereoscopic displays in three axis manual tracking tasks. *IEEE Transactions on System Man and Cybernetics*, **SMC–17**, 61–71.

Kleinman, D.L., Baron, S. and Levison, W.H. (1970). An optimal control model of human response, Part I: Theory and validation. Part II: Prediction of human performance in a complex task, *Automatica*, **6**, 357–369.

McRuer, D.T. and Weir, D.H. (1969). Theory of manual vehicular control, *Ergonomics*, **12** (4), 599–633.

Smith, K.U. and Smith, W.M. (1962). *Perception and Motion: An Analysis of Space-Structured Behavior*. Philadelphia: W.B. Saunders.

Smith, T.J. and Smith, K.U. (1987). Feedback control mechanisms of human behavior. In Salvendy, G. (Ed.) *Handbook of Human Factors*. New York: Wiley.

Van De Graaff, R.C. and Wewerinke, P.H. (1983). Theoretical and experimental analysis of pilot failure detection. *Proceedings of the Nineteenth Annual Conference on Manual Control*, held at Massachusetts Institute of Technology, Cambridge, Massachusetts 02139, May 23–25, 1983, pp 352–362.

Van Lunteren, A. and Stassen, H.G. (1969). On-line parameter estimation of the human transfer in a man-bicycle system, *IFAC Congress*, Warsaw, p. 16.

Yajima, H., Sheridan, T.B. and Buharali, A. (1983). Human decision model in failure compensation, *19th Annual Conference on Manual Control*, May 23–25, 1983, MIT, 338–351.

Zacharias, G.L. and Caglayan, A.K. (1985). A visual cueing model for terrain following applications, *Journal of Guidance, Control and Dynamics*, **8**, No. 2, 201–207.

10

Spatial displays as a means to increase pilot situational awareness

Delmar M. Fadden, Rolf Braune and John Wiedemann

Boeing Commercial Airplane Company
Seattle, Washington

At least three elements influence the performance of an operator who must make a system achieve a desired goal: (1) the dynamics of the system itself, (2) the nature of the possible inputs, and (3) the means whereby the operator views the information concerning the desired and actual state of the system (e.g., Poulton, 1974; Wickens, 1984, 1987). In conventional airplanes manual control involves the coordination of "inner loop" controls. In this task the pilot is responsible for continuous manipulation of the controls to compensate for disturbances. Primary displays (Figure 1) provide the several essential flight parameters which the pilot is required to monitor, interpret, transform, and integrate.

It has long been recognized that intense concentration is necessary for a pilot to achieve high tracking performance using only "raw data." The underlying need for such concentration stems from the effort necessary to obtain timely error, error rate, and control input information in each of the three flight axes. Precision instrument approaches often have higher minimums if a suitable flight director or autopilot is not available and in use. Most pilots have come to depend on these aids. Some pilots express doubt about the precision of their own tracking ability any time they are unavailable.

Flight directors, which came into widespread airline use in the 1960s, aid the pilot in achieving improved performance by combining the error and error rate information; producing a control command appropriate to the situation. This command is then compared with the existing control input and the difference displayed as a steering command. The generation of the steering command entails automation of several logical and mathematical operations. Of course the pilot must set up the proper task for the flight director to perform and must follow the steering commands. In typical applications the automation is sufficiently complete that the pilot has no

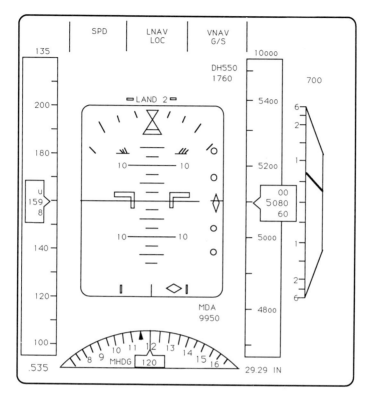

Figure 1. *Primary flight display.*

required intermediate data-interpretation role beyond that of recognizing and following the steering command.

While use of the flight director improves performance in precision tasks, it does not significantly reduce the continuous attention demands imposed on the pilot. Use of a path-following autopilot mode automates the process one step further by coupling the steering command to the control surfaces. Relieved of the continuous steering requirement, the pilot is able to devote more time to other tasks.

Both flight directors and autopilots achieve impressive performance gains. A side effect of these gains is a reduction in the necessity for the pilot to maintain a high level of awareness of the elements pertinent to the control task; namely the path error, error rate, and control input. To be sure, all modern aircraft present these parameters and most airline operating procedures dictate that the pilot monitor them while using either the flight director or autopilot. However, the monitoring task is fundamentally different from that of developing a control input given only "raw" data. In particular, the dynamic decision-making demands of the monitoring task are much lower than those of the control task.

Spatial displays, together with enhanced manual control, offer an opportunity to achieve the same high performance achieved with autopilots and flight directors while improving the pilot's overall situational awareness, particularly during flight tasks other than final approach. This is accomplished by revising the split of responsibilities between the pilot and the aircraft automation.

In transport operations, the need to alter the velocity (flightpath angle, track angle, or speed) is much less frequent than the need to compensate for wind effects, turbulence, configuration changes, and speed changes. In terminal area operations, the number of required velocity changes may be an order of magnitude or more lower than those attitude changes necessary to maintain a velocity. Furthermore, the needed velocity changes are typically separated by many seconds. By assigning the velocity-hold task to the basic flight-control system, the majority of the attitude adjustments can be made transparent to the pilot. This type of control frees the pilot from the continuous attention requirement of attitude steering while maintaining the pilot's direct involvement in airplane guidance.

Spatial displays make it possible for the pilot to be directly involved in developing the path error information and in selecting the specific tactic to be employed in correcting the error. To make this practical, current position and velocity information must be displayed in a consistent context. Operational displays based on work done at Boeing, NASA Langley, RAE-Weybridge, and other places have shown that a map display, with track angle and speed shown by means of predicted future positions, provides a suitable context.

The first generation of commercial airline spatial displays are in operation on the Boeing 757 and 767 and the Airbus A-310. These displays take the form of CRT maps with various types of integral predictors (Figure 2). The format consistency of these displays is quite high and pilot acceptance has been exceptionally good. The CRT maps are used for planning and assessing all types of lateral maneuvers. Direct manual aircraft control is still accomplished by reference to a separate attitude instrument, but virtually all of the decisions to maneuver laterally can be made looking at information contained in the map display.

The success of the map display and the potential for flightpath angle and track angle control to be used on the next generation of commercial aircraft encouraged us to consider expanding the role of spatial displays. Data from the NASA Aviation Safety Reporting System identifies altitude-related errors as the single largest category of reported problems (Reynard, Ames Research Center, 1987, personal communication). While the immediate causes of the reported errors are quite varied, we see a common thread emerging. The pilot's awareness of the vertical flight situation in most instances does not match the reality of the flight plan, the ATC clearance, or the equipment setup. A spatial display should be an ideal means of improving the pilot's vertical situation awareness (Baty, 1976).

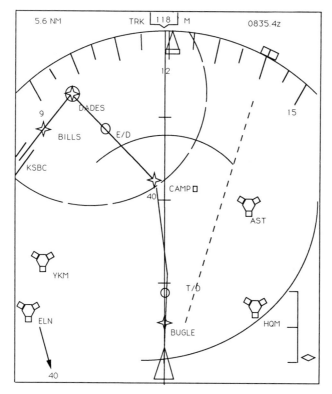

Figure 2. Map display.

For most transport flight operations except takeoff and landing, the tracking accuracy required of the pilot is at least an order of magnitude higher for the vertical task than for the lateral task. Typical tracking-performance goals as perceived by the pilot away from final approach are ±50 ft of altitude and ±0.5° of a VOR radial. At 40 miles from the VOR station, ±0.5° corresponds to over ±2000 ft. At this point the accuracy ratio is 40:1. Even on final approach the vertical accuracy requirements exceed the lateral by at least 2:1. If a conformal 3D display were used with sufficient resolution to satisfy the vertical task, the pilot would be overworked laterally. This concern, along with the difficulty of presenting future trend information in a forward-looking display, led us to concentrate on a separate 2D side-view display for the majority of vertical situation information (Grunwald et al., 1980; Filarsky and Hover, 1983).

Some past aircraft programs have referred to the attitude display as a vertical situation display. We prefer to use the more conventional terminology, ADI (attitude director indicator) or PFD (primary flight display) for the forward-looking display of attitude information and other fundamental flight data. We refer to a side-looking or profile display as a vertical-situation

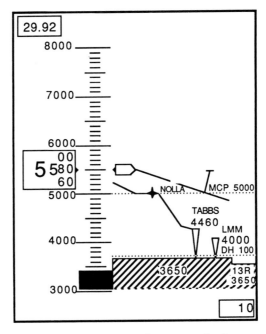

Figure 3. Vertical situation display.

display and expect that the pilot would obtain the majority of overall vertical situation awareness from this display (Figure 3).

Over the past two years we have been exploring ways of developing a useful and effective means to portray vertical-situation information. There are a number of practical problems which narrow the possible format options for vertical flight information. The remainder of this paper will outline the larger hurdles and indicate what progress has been made in solving them.

Three issues appear to be fundamental to the development of a successful vertical situation display:

(1) Handling of the large difference in resolution requirements between the longitudinal and vertical flight tasks.
(2) Determination of the appropriate level of control information to be contained in the instrument.
(3) Selection of a display context which will be intuitive to the pilot and provide useful assistance for on- and off-path vertical maneuvering.

Scaling issues

The disparity which exists between vertical and lateral resolution requirements applies as well to vertical and longitudinal information. In fact, since

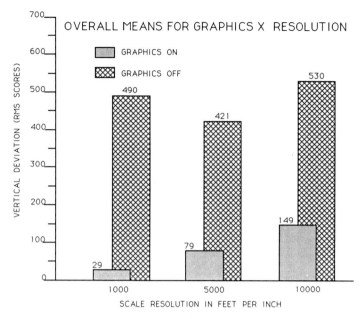

Figure 4. *Overall means for graphics × resolution.*

time constraints are seldom tighter than a minute or more, the difference in resolution requirements can be well in excess of two orders of magnitude. With this large a difference, equal vertical and horizontal display scaling is clearly impractical. By using a flightpath predictor we have been able to achieve a balance between vertical tracking performance and the desired path preview capability.

Initial test results indicate that when the vertical situation is presented spatially, a steady increase in mean deviation from an optimal descent occurs as scale resolution is decreased (Figure 4). However, even the largest deviation is significantly less than the lowest mean without the spatial graphics. This result could be attributed to the difference in the tactics the subject pilots employed to accomplish the task under the two presentations. Without the graphics the pilots had mentally to integrate various analog quantities according to their own individual rules of thumb. As can be seen in Figure 5, this results in an overall greater deviation from the optimal descent strategy and more variance among the individual pilot deviations. When given a spatial presentation of the situation, the subject pilots employed similar path-following tactics, resulting in greater tracking precision and a lower-rated workload level.

The fact of unequal scales causes the angle representations on the display to be exaggerated vertically. Through a preliminary test series we found that scale differences of as much as 20:1 do not have a negative influence on typical airline flying tasks. Obviously aircraft with significantly greater climb

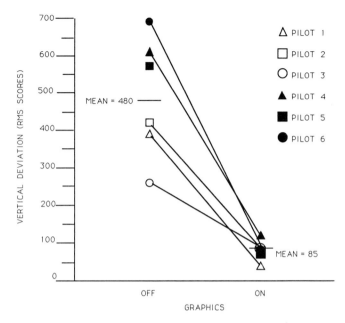

Figure 5. Overall means for graphics × pilots.

or descent capabilities than transports would encounter difficulty at lower-scale ratios. What appears more important to the pilots is that the longitudinal scaling of the side-view display and the map display be congruent so that the rate of movement between the two is compatible.

Another result from our initial investigations reveals that a digital readout of altitude takes on added importance as scale resolution is decreased (Figure 6). In seeking the proper balance between scale resolution for precision and scale range for preview, it was shown that a digital readout of altitude provides a good vernier indication while the graphics provides the necessary "big picture" overview. The graphic spatial information is effective in drawing the pilot's attention to the digital readout when precise control is needed.

Control issues

In all of today's transport aircraft, manual control is exercised using the attitude display with follow-up reference to the situational displays. This is the case for map-display-equipped aircraft as well. Laterally the track angle is two integrations removed from aircraft roll rate, over which the pilot has direct control. The resulting time delay between control input and map response is too long for track angle to provide primary inner-loop feedback to the pilot. Even when lateral acceleration is used to create a prediction of the dynamic path which will be flown, the pilot's primary control feedback comes from the bank indication on the attitude indicator.

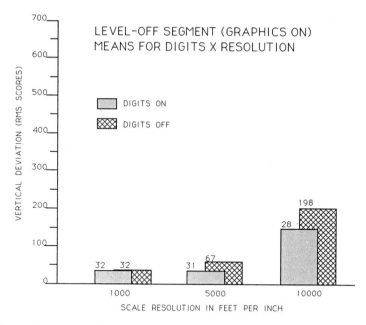

Figure 6. Level-off segment (graphics on) means for digits × resolution.

Vertically the conventional control parameter is pitch rate. This term is separated from flightpath angle by a single integration and some higher-order dynamics. For transports this places the flightpath response on the order of 1–2 sec behind the control input; long enough to be useless as the primary feedback term for most situations and short enough to interact negatively with pitch feedback. The primary dynamic term in the vertical situation is flightpath angle. Furthermore, flightpath angle, rather than pitch attitude, can be readily assessed in terms of the geometry or energy conditions of the vertical situation. If the response dynamics of flightpath angle were not so close to that of pitch attitude, the separation of control and situation assessment, which works very well in the lateral case, could be established for the vertical case as well.

Beginning with experimental work on the Boeing SST in the late 1960s and continuing through the early phases of the NASA TCV program, we became convinced that if flightpath angle, along with suitable situational reference information, is available to the flight crew, the crew will attempt to use it for control. Without good matching of the control and display dynamics, pilot workload may well increase.

If a flightpath-angle command-control system is in use, it is possible to display the flightpath which will be held. This term can be made as responsive as necessary to support the pilot's need for timely information. If a more conventional control system is used, a filter with appropriate lead compensation can be added to quicken the dynamics of the flightpath angle information (Bray and Scott, 1981).

The key situational element which makes control possible is the flightpath prediction based on flightpath angle. Remove the prediction and control reverts to conventional techniques. However, without the prediction, the usefulness of the display for enhancing current situational awareness is dramatically reduced. Even maintaining a constant altitude is difficult without the prediction. Thus the question about the desired level of control information is not an independent issue. If the display is to be useful, it must contain dynamic flightpath information. The presence of such information means that the display will be used for control. The real issue, then, is how to match the control and display dynamics to the information-processing capabilities of the pilot.

Display context

The third fundamental issue has to do with matching the frame of reference of the display to that of the pilot. The vertical component is straightforward. However, the options for the horizontal component are more complex. If information concerning the planned route of flight were always available and current, then distance along the route would be a good choice. However, the planned route is not always available. Furthermore, one of the more important uses of the display is during operations when the airplane is intentionally away from the planned path.

For these situations a narrow slice ahead of the airplane would be more useful. In either case close coordination between the vertical and horizontal situation displays is essential.

Development work aimed at clarifying the format orientation issue is now under way. We expect to have an understanding of the major tradeoffs late this year.

Conclusions

Our experience raises a number of concerns for future spatial-display developers. While the promise of spatial displays is great, the cost of their development will be correspondingly large. The cost goes well beyond time and materials. The knowledge and skills which must be coordinated to ensure successful results is unprecedented. From the viewpoint of the designer, basic knowledge of how human beings perceive and process complex displays appears fragmented and largely unquantified. Methodologies for display development require prototyping and testing with subject pilots for even small changes. Useful characterizations of the range of differences between individual users is nonexistent or at best poorly understood. The nature, significance, and frequency of interpretation errors associated with complex integrated displays is unexplored and undocumented territory.

Graphic displays have intuitive appeal and can achieve face validity much more readily than earlier symbolic displays. The risk of misleading the pilot is correspondingly greater. Thus while we in the research community are developing the tools and techniques necessary for effective spatial-display development, we must educate potential users about the issues so they can make informed choices. The scope of the task facing all of us is great. The task is challenging and the potential for meaningful contributions at all levels is high indeed.

References

Baty, D.L. (1976). Rationale and Description of a Coordinated Cockpit Display for Aircraft Flight Management, NASA Technical Memorandum 3457, November 1976, Ames Research Center, Moffett Field, CA.

Bray, R.S. and Scott, B.C. (1981). A head-up display for low visibility approach and landing. *Proceedings AIAA 19th Aerospace Sciences Meeting*, January 12–15, St. Louis, MO.

Filarsky, S.M. and Hover, G.W. (1983). The command flight path display. *Proceedings of the Society for Information Displays*, 196–197.

Grunwald, A.J., Robertson, J.B. and Hatfield, J.J. (1980). Evaluation of a Computer Generated Perspective Tunnel Display for Flight Path Following. NASA Langley Tech Report #1736.

Poulton, E.C. (1974). *Tracking Skills and Manual Control*. New York: Academic Press.

Wickens, C.D. (1984). *Engineering Psychology and Human Performance*. Columbus, Ohio: Charles Merrill.

Wickens, C.D. (1987). The effects of control dynamics on performance. In Boff, K. and Kaufman, L. (Eds) *Handbook of Perception and Performance*. Dayton, Ohio: University of Dayton Research Institute.

11

Experience and results in teleoperation of land vehicles[1]

Douglas E. McGovern

Sandia National Laboratories
Albuquerque, NM 87185

Abstract

Teleoperation of land vehicles allows the removal of the operator from the vehicle to a remote location. This can greatly increase operator safety and comfort in applications such as security patrol or military combat. The cost includes system complexity and reduced system performance. All feedback on vehicle performance and on environmental conditions must pass through sensors, a communications channel, and displays. In particular, this requires vision to be transmitted by closed-circuit television with a consequent degradation of information content. Vehicular teleoperation, as a result, places severe demands on the operator.

Teleoperated land vehicles have been built and tested by many organizations, including Sandia National Laboratories (SNL). The SNL fleet presently includes a number of vehicles of varying capability. These vehicles have been operated using different types of controls, displays, and visual systems. Experimentation studying the effects of vision-system characteristics on off-road, remote driving has been performed for conditions of fixed camera versus steering-coupled camera, color versus black and white video display, and reduced bandwidth video presentations. Additionally, much experience has been gained through system demonstrations and hardware development trials. This paper discusses experimental findings and the results of accumulated operational experience.

---

[1] This work performed at Sandia National Laboratories supported by the U.S. Department of Energy under contract number DE-AC0476DP00789.

Introduction

Remote control of land vehicles can be accomplished through provision of auxiliary sensory channels on-board the vehicle (inside-out control) or through observation of the vehicle in the world (outside-in control). Outside-in control is effective only over short visual ranges for vision with no obscuration such as smoke, fog, or obstacles. Inside-out control (referred to as teleoperation in the remainder of this paper) is generally applicable for activities such as security patrols or military combat in which any humans present will be at risk. The cost of such operation is increased complexity in the vehicle and control system, since all knowledge of the environment and the conditions of the vehicle have to be sensed, communicated to a control station, and displayed to the human operator. A further consequence of removing the operator from the vehicle is reduced capability for action, since the information content of the operator feedback is degraded by the intermediary channels.

Vehicles, control stations, and teleoperated systems have been built, tested, and demonstrated by a number of organizations. There is little definitive information, however, on the human factors involved in land vehicle teleoperation (McGovern, 1987b). Most information has taken the form of a description of vehicle design or proposed application, with only a few papers reporting actual experimental results. Most of the knowledge base is represented by personal experiences and unreported anecdotal evidence. This paper attempts to expand the data base through a presentation of some of the results of experimentation in teleoperation at Sandia National Laboratories and through discussion of the observations of SNL personnel gathered over several years of teleoperation experience.

Teleoperation systems

Sandia National Laboratories has been actively studying vehicular teleoperation. The major effort has entailed the development of a fleet of wheeled vehicles ranging in size from small, interior testbeds to large, on-road and off-road commercial and military vehicles (McGovern, 1987a). These vehicles (shown in Figure 1) are being used to conduct feasibility studies on the application of teleoperated vehicles to the physical security and military needs of the U.S. Government. In all of these vehicles, actuators operate the vehicle throttle, brakes, and steering. Control may be derived from manual input at a remote driving station or through some level of automatic control from a digital computer. Onboard processing may include simple vehicle control functions or may allow for unmanned, autonomous operation. Communication links are provided for digital communication between control computers, television transmission for vehicle vision, and voice for local control.

Figure 1. Teleoperated vehicles.

Control stations have been developed to support remote operation of the SNL vehicle fleet. Capabilities range from single television monitor stations with vehicle feedback limited to an audio channel (shown in Figure 2), through large, multiscreen, panoramic displays with computer-generated graphics representations of vehicle speed, pitch, roll, and heading (Figure 3). Vehicle camera mountings have included a single fixed camera, multiple fixed cameras, and cameras slaved to the vehicle steering gear. To date, SNL has not experimented with stereo vision or with head-slaved displays, although members of the staff have operated such equipment at other locations.

Experience

A large part of the knowledge base at SNL has been derived from operation of vehicles during hardware and software development and system demonstrations. Operators have ranged from well-trained, highly experienced personnel through people who had not previously driven a remotely controlled vehicle. The primary source of data has been the subjective comments of operators and observers.

The analysis of accidents involving teleoperated vehicles has provided additional information. Table 1 provides a listing of accidents experienced at SNL. Some of these accidents occurred while the operator was observing the vehicle directly (outside-in operation) and were predominantly depth-

Figure 2. Single monitor with audio feedback.

Table 1. Accident history

Vehicle incident		Cause
Outside-in operation		
Dune Buggy	Hit fence	Underestimated stopping distance
Dune Buggy	Hit tree	Depth perception
Dune Buggy	Hit fence	Underestimated stopping distance
Suzuki	Hit post	Depth perception
Suzuki	Hit car	Control reversal
Inside-out operation		
Suzuki	Roll-over	Loss of control on hill
Suzuki	Roll-over	Loss of control on hill
Suzuki	Roll-over	Hit traffic cone
Suzuki	Roll-over	Loss of control on hill
Suzuki	Roll-over	Loss of control on hill
Suzuki	Roll-over	Loss of control while backing
Suzuki	Roll-over	Loss of control, hit bump
Suzuki	Roll-over	Loss of control on hill
Suzuki	Roll-over	Loss of control on hill
Suzuki	Roll-over	Loss of control, hit bump

Figure 3. Panoramic display.

perception problems involving vehicle clearance or stopping distances. Control reversal caused one accident while operating the vehicle in the outside-in mode. In this accident, the vehicle was heading toward the operator. The operator wanted the vehicle to go toward the left of the operator (operator left). Since the vehicle was approaching the operator, this required the vehicle to turn to the right with respect to its direction of travel. The operator became disoriented and issued a left command. The vehicle responded by veering further to vehicle left (operator right) with a consequent collision with a parked car.

All of the accidents involving teleoperation (inside-out control) have been roll-overs. The particular vehicle involved is a small Suzuki LT50 four-wheel, all-terrain vehicle shown in Figure 4. The rear wheels are driven through a single-speed drive with a centrifugal clutch. The vehicle is capable of a 15 mph top speed on flat ground. Control inputs from the operator are through the control station illustrated in Figure 2. Figure 5 shows the view provided to the operator. In all but one incident, the vehicle was being operated off-road on a moto-cross track with steep slopes, high banked corners, and high berms at the edges of the track. The only exception was a roll-over caused by hitting a traffic cone while operating on a flat asphalt parking lot.

Figure 4. All terrain vehicle.

Figure 5. Operator's view via control station.

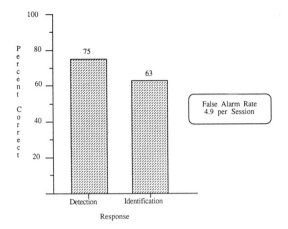

Figure 6. Search performance: obstacle detection and identification for a group of 35 obstacles on a remote driving course. Obstacles included such things as fences, oil drums, ditches, and trees. Driving conditions included remote control and video simulation with visual feedback provided by fixed black and white video, fixed color video, and steering slaved video. No significant differences were found among the six experimental conditions (McGovern and Miller, 1989).

Under the sponsorship of the U.S. Army Missile Command, through the Teleoperated Mobile Antiarmor Platform (TMAP) Project, SNL embarked on a major set of experiments to verify some of the observations regarding the "best" driving display (Miller, 1987, 1988; Miller and McGovern, 1988, 1989). In particular, the experimentation addressed the problems of detection and identification of obstacles in the path of the vehicle. Specific questions included the effect of color versus black and white, the utility of increasing the horizontal field-of-view through panning a camera in response to steering wheel movements (steering-slaved control), and the errors in operator interpretation of size and distance information as presented by the television system.

It was found that object and obstacle detection were not sensitive to camera differences. That is, the probability that an object would be seen was independent of the video system being used (illustrated in Figure 6). This did not support the initial hypothesis that color video and steering slaved camera control would result in superior operation. Objects were detected as well with a fixed, black and white camera.

Range data were sensitive to camera differences. When objects were detected, they were detected at a greater range when color was used, as shown in Figure 7. Color provided a fairly consistent 5 to 20 foot range advantage when compared to black and white video.

There was no experimental evidence that steering-slaved camera mounting provided benefit. This is in direct contradiction to previously reported subjective results. This may be explained by camera jitter in the experimental

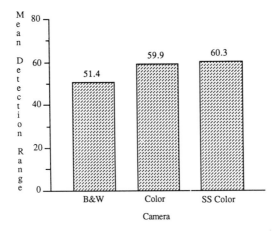

Figure 7. Detection range: obstacle detection range for 35 obstacles on a remote driving course. Two-factor analysis of variance of $\log_e$ (detection range) showed nearly significant camera differences. When data from both color conditions are combined, the black and white versus color difference becomes statistically significant ($F = 5.66$, $p = 0.02$), demonstrating an average advantage of about 9 feet for color cameras (McGovern and Miller, 1989).

set-up, specific steering (and camera pointing) ratios, and terrain and path marking. Also, object detection is just one part of the overall activity involved in remote driving, so a lack of demonstrated advantage in object detection does not establish that there are no other benefits.

Size and distance estimation produced inconsistent results. One study (Miller and McGovern, 1988) indicated that subjects using color video consistently judged the size and spacing of obstacles to be larger than subjects using black and white video. A follow-on study was conducted to investigate foreground texture effects on these size/distance judgment errors. In this study (Smith and Miller, 1989) color was not found to be an influential factor in estimating distance. Color did appear, however, to influence clearance judgment.

Teleoperation of land vehicles was studied to investigate the effects of video resolution on the ability of the operator to control the vehicle effectively (Schoeneman and McGovern, 1989). Experiments were constructed to allow evaluation of the lower limits of video quality necessary for degraded, yet effective, teleoperation. Subjects were asked to teleoperate a remote vehicle under varying conditions of reduced spatial and temporal resolution. A combination of video tape simulations and actual remote driving were used over a variety of terrains. Subjective assessments and teleoperation test data suggest the possibility of reducing video bandwidth to about 500 kHz. This seems to provide sufficient information for operation over relatively simple terrain. Reducing the bandwidth further results in loss of operator confidence and increased risk of getting lost or running into

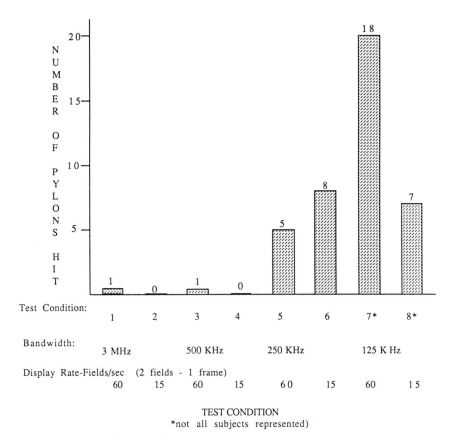

Figure 8. Pylons hit: total number of pylons struck while operating a small teleoperated vehicle under conditions of reduced spatial and temporal video feedback. Pairs of pylons established a figure-eight shaped course with 9 gates (18 total pylons). Not all operators could complete the course when spatial resolution was reduced to 125 kHz video bandwidth (Schoeneman and McGovern, 1989).

unrecognized obstacles. This is illustrated in Figure 8. Additional work is being done to validate these results and to investigate digitally processed imagery and variable resolution schemes.

Observations

A number of observations regarding important parameters, operational considerations, and system design features have been derived from SNL experiences. These are presented below strictly as indicators since, in the absence of hard experimental data, it is not clear that all are generally applicable. Likewise, not all system implementations are represented.

Field-of-view

It is very difficult to operate a vehicle in restricted space with a narrow field-of-view. Operation of a Jeep Cherokee on normal roads and parking lots was performed with a single camera, 40 degree field-of-view system. The operator was not comfortable turning corners. Installation of two additional cameras, to provide a total of 120 degrees field-of-view resulted in much "easier" operation. Tests have also been run using a steering-slaved camera both on the Jeep Cherokee and on the Suzuki all-terrain vehicle. Steering-slaved viewing provided sufficient effective field-of-view to allow turning tight corners and avoiding obstacles. Provision of a mechanism to allow the operator to force the camera further (an auxiliary pan control) was even more effective. As discussed above, however, obstacle recognition and detection is not necessarily improved through use of steering-slaved camera control.

Resolution

Camera resolution does not seem to be a factor in the ability to teleoperate a vehicle in the absence of obstacles. SNL has operated vehicles with malfunctioning communications links resulting in extremely poor resolution. Testing has also been performed with purposefully degraded video quality. As long as operations take place on well defined areas, an operator can successfully maneuver a vehicle from one point to another. High resolution does appear to be important when many sizes and types of obstacles are present and for operation off-road where identification of best path is important.

Color/black and white

Work with television surveillance systems has indicated that the increased resolution possible with black and white equipment is much more important than any additional information contained in the color signal. This does not necessarily appear true for teleoperation. Color provides additional cues leading to more accurate course planning. For example, the difference between dirt and asphalt is important for driving, but can not be determined from a black and white television picture. SNL has also found that orange traffic cones (with the color chosen for maximum visibility) tend to disappear on black and white television. These have been used to establish courses during demonstrations and experimentation. Using black and white television, it was found to be necessary to cover the cones with white paper so that they could be seen. Color is also rated as very highly desirable in all subjective preferences. Experimentation has not, however, supported a quantitative difference in obstacle detection between black and white and color video. More experimentation will be necessary to resolve these somewhat conflicting results.

Vehicle vibration

Vehicle vibration and bounce have not been observed significantly to degrade the displayed video scene. The small Suzuki has no suspension (springs or damping) other than its large, soft, off-road tyres. During operations which lead to the vehicle bouncing enough actually to leave the ground, the video remains relatively clear and usable. No operator has ever commented that vibration or bounce in the picture was bothersome.

Distance estimation

As seen from the accident reports and experimental results, distance estimation during outside-in driving is a problem. It also creates difficulties when using inside-out control. As reported by Spain (Spain, 1987) in a related set of experiments, operators using a head-mounted display consistently ran into pylons marking the end of a parking place. The feeling of being further from obstacles and landmarks than the actual position has also been reported by most operators of SNL vehicles. For all of the systems utilized in these observations, however, the display was smaller than geometric similarity resulting in a scene minification between 0.4 and 0.7. As discussed by Roscoe (Roscoe, 1979), it can be anticipated that size and distance judgment errors can be expected for these conditions. To achieve better results, scene magnification of approximately 25 per cent is required.

Negative obstacles

Terrain features such as ditches, holes, and drop-offs are extremely difficult to see using television. Negative obstacles such as these have contributed to many of the problems in teleoperating vehicles. In most cases, small ditches cannot be differentiated from variations in ground coloration until the vehicle has hit them. At that point, the horizon on the video scene changes, indicating that the vehicle just hit a ditch. It can be anticipated that stereo vision could help in this problem but no experimentation has been reported.

Tilt and roll

The large number of roll-overs reported establishes vehicle tilt and roll control as a major problem. In the Suzuki driving system, the only feedback is the video signal from the camera and an audio pickup providing engine sound. Vehicle attitude parameters are neither measured nor displayed. The typical accident scenario entails "launching" the vehicle from a ramp or attempting to traverse a side slope which is too steep for the vehicle to maintain stability. Most roll-overs have occurred at close to maximum vehicle speed (about 10–15 miles per hour) and have been a result of ground features representative of extremely challenging terrain. These have included hills with up to 45 degree slopes and highly banked corners on a moto-cross course. As the roll-over occurs, the operators express surprise. In debriefing, it appears that the operator had no indication that the vehicle was approaching a dangerous condition.

Over-control

A typical characteristic of novice operators is extreme steering over-control. The operator applies a small steering input to the vehicle but no result is immediately seen. The steering input is increased until a response is finally observed. The resulting turn is more than intended so the operator applies a small correction. Again, the response is not seen so more correction is applied, etc. The outcome is vehicle travel oscillating about the desired path. Operators report this to be a very stressful situation. Over-control has also contributed to several of the vehicle roll-over accidents. The operator applied excessive steering input, sending the vehicle over the edge of a berm. Observing novice drivers learning to control the vehicle, it is apparent that considerable internal control is being exercised as the operator adapts. After some minutes of operation, steering operation is considerably slower and at lower amplitude, resulting in smoother vehicle control. Spain (1987) reports similar findings.

Navigation

An associated problem in vehicle teleoperation is the difficulty of maintaining spatial orientation with respect to major landmarks, map features, or compass directions. It is not uncommon for operators to become lost on the moto-cross course. Even with landmarks and a map of the course, they have not been able to determine how to return to the starting location without assistance.

Summary and conclusions

Operational experience has been gathered at Sandia National Labs through development, test, and demonstration of a number of vehicles. A large experimental program in vision system requirements for teleoperation has also been completed. Through the knowledge gained in these programs, several key areas can be identified as critical to successful control of a teleoperated vehicle. The primary area is the quality and type of feedback provided to the operator. It has been shown that vehicles can be controlled in restricted environments with extremely poor conditions of viewing. As feedback capabilities improve (resolution, field-of-view, audio, instrument displays, etc.) better control can be expected.

Negative obstacles create difficulty in that operators can not distinguish them from other terrain features which do not affect vehicle travel. The result is hitting ditches, holes, or berms at excessive speed.

The interaction of the vehicle with the environment, as interpreted through the mediating effects of the television display system, can lead to poor control capabilities and hazardous operating conditions. Over-control of the vehicle steering, coupled with the operator's inability accurately to perceive vehicle attitude and terrain requirements has led to a number of accidents. This can be partially linked with the absence of kinesthetic feedback to the operator. Experimentation with vehicle simulators has shown a distinct lag in response to environmental inputs such as wind gusts when no kinesthetic feedback is present (Wierwille et al., 1983). With the addition of kinesthetic feedback to the operator (simulator platform motion) response time to sudden wind gusts dropped from an average of 0.56 sec to an average of 0.44 sec. Similar results have been reported for the addition of steering wheel torque feedback, thus providing "feel of the road" to the operator (Allen and Weir, 1984). The result of lack of kinesthetic feedback is similar to operating with a time delay in the control system. Additional lags are introduced by the communications systems and vehicle actuator and control systems.

Given the ability to maneuver a teleoperated vehicle in the real-world environment, the problem of navigation is encountered. Operators tend to get lost, disoriented, and confused when provided with visual input and maps. The effect of additional aids such as vehicle heading, and plotting of route traveled remains to be investigated.

References

Allen, R.W. and Weir, D.H. (1984). Analysis of Man-in-the-Loop Performance Measurement Technology for Crash Avoidance Research, STI TR-1200-1, Systems Technology, Inc., Hawthorne, California, March, 1984.

McGovern, D.E. (1987a). Advanced Technology Mobile Robotics Vehicle Fleet, SAND87-0033, Sandia National Laboratories, Albuquerque, NM, March 1987.

McGovern, D.E. (1987b). Technology Assessment for Command and Control of Teleoperated Vehicles, SAND87-1225, Sandia National Laboratories, Albuquerque, NM, May 1987.

McGovern, D.E. and Miller, D.P. (1989). Vision System Testing for Teleoperated Vehicles, SAND88-3123, Sandia National Laboratories, Albuquerque, NM, March 1989.

Miller, D.P. (1987). Evaluation of vision systems for teleoperated land vehicles. *IEEE Systems Man and Cybernetics Conference*, October 20–23, 1987, Alexandria, VA.

Miller, D.P. (1988). Evaluation of vision systems for teleoperated land vehicles. *IEEE Control Systems*, **8**, 37–41.

Miller, D.P. and McGovern, D.E. (1988). A laboratory-simulation approach to the evaluation of vision systems for teleoperated vehicles, prepared for presentation at the *International Symposium on Teleoperation and Control*, University of Bristol, England, 12–15 July, 1988.

Roscoe, S.N. (1979). When day is done and shadows fall, we miss the airport most of all. *Human Factors*, **21**, 721–731.

Schoeneman, J.L. and McGovern, D.E. (1989). Land Vehicle Teleoperation Under Conditions of Reduced Video Resolution, SAND89-1256, Sandia National Laboratories, Albuquerque, NM, August, 1989.

Smith, C.U. and Miller, D.P. (1989). The Effect of Color and Texture of Foreground on Size and Distance Perception Using Video Systems for Teleoperated Vehicles, SAND88-2596, Sandia National Laboratories, Albuquerque, NM, July, 1989.

Spain, E.H. (1987). Assessments of maneuverability with the teleoperated vehicle (TOV). *Fourteenth Annual Symposium of the Association for Unmanned Vehicle Systems*, July 19–21, 1987, Washington, D.C.

Wierwille, W.W., Casali, J.G. and Repa, B.P. (1983). Driver steering reaction time to abrupt-onset crosswinds, as measured in a moving-base driving simulator. *Human Factors*, **25**, 103–116.

12

A computer graphics system for visualizing spacecraft in orbit[1]

Don E. Eyles

Charles Stark Draper Laboratory
Cambridge, Massachusetts

Summary

To carry out unanticipated operations with resources already in space is part of the rationale for a permanently manned space station in Earth orbit. The astronauts aboard a space station will require an on-board, spatial display tool to assist the planning and rehearsal of upcoming operations. Such a tool can also help astronauts to monitor and control such operations as they occur, especially in cases where first-hand visibility is not possible. This paper describes a computer graphics "visualization system" designed for such an application and currently implemented as part of a ground-based simulation. The visualization system presents to the user the spatial information available in the spacecraft's computers by drawing a dynamic picture containing the planet Earth, the Sun, a star field, and up to two spacecraft. The point of view within the picture can be controlled by the user to obtain a number of specific visualization functions. The paper describes the elements of the display, the methods used to control the display's point of view, and some of the ways in which the system can be used.

Introduction

This paper describes a computer graphics display system designed to facilitate the visualization of spacecraft operations in Earth orbit.

The system was originally developed as a component of the Space Station Simulator project at the Charles Stark Draper Laboratory. The purpose of

---

[1] As part of the Draper Laboratory's involvement in navigation for manned spaceflight from the time of Apollo, the graphics display described below is in active use to assess advanced flight control concepts for the NASA Space Shuttle and Space Station Freedom.

this simulator is to assess the flying qualities of space station configurations, and to provide a software framework within which to develop control-system concepts applicable to space stations. Computer graphics were added to the simulator to provide qualitative information about the progress of the simulation, and to allow for a man-in-the-loop capability. As time went on it became evident that the displays required by engineers working on the ground might also be valuable to astronauts working aboard a space station.

To be able to carry out unanticipated tasks with resources already in Earth orbit is part of the purpose of a permanently manned space station. Operations will be required which cannot be rehearsed by the astronauts using ground-based simulators because the need for them will arise after the crew has been launched into space. On-board capabilities must exist to allow the crew to plan such orbital operations and to train themselves to execute them. In addition, the space station crew must perform a sort of air-traffic-control function in keeping track of other spacecraft operating nearby, and must control not only the space station itself and its movable appendages, but also free-flying spacecraft associated with the space station, including space-walking astronauts.

The display described in this paper, it attached to suitable mission and simulation software aboard a space station, can support both the on-board simulation capability and the real-time monitoring of operations. I shall speak of the visualization system as an on-board display instrument, with the understanding that its capabilities arose from, and are also applicable to, ground-based engineering simulation purposes.

The display is called a "visualization system" because it is a system designed to aid the user in visualizing a three-dimensional situation in space. In the terminology established for this conference, the visualization system fits in somewhere between a "spatial display" and a "spatial instrument." Like a spatial display, the system presents the user with an unembellished, undistorted image of a spatial situation. Like a spatial instrument, the system requires a degree of interaction with the user, who must control the point of view from which the image is drawn. Perhaps the best description of the visualization system is as a spatial instrument which can present a variety of spatial displays under the control of the user.

The existing implementation uses display equipment that produces "wire-frame" objects whose "hidden" parts are visible. Although in some cases wire frames may remain preferable, actual use aboard a space station will require flight-qualified display hardware capable of rendering solid objects with shading and shadowing.

I shall describe the elements of the scene created by the visualization system, discuss the means by which the point of view within the scene is controlled, and finally describe some of the specific ways in which the system can be used. A more detailed description of the visualization system is available in Eyles (1988). A short published description with color illustrations is available in Eyles (1986).

Elements

The principal elements of the display created by the visualization system are the planet Earth, the Sun, a field of stars, and one or two spacecraft. The planet Earth is drawn as a sphere made up of latitude and longitude grid lines and a map showing the outlines of major land masses and principal cities. Other Earth-fixed features such as circles indicating coverage from tracking sites can be added. The Earth is drawn from data expressed in a geodetic or Earth-fixed coordinate frame; that is, a frame of reference which moves with the Earth. The Sun is drawn, not to scale, as a yellow asterisk with 24 points. A star field of 123 stars is also drawn, and is valuable for two reasons. First, showing the stars in their correct astronomical positions provides a realistic star background for maneuvers being monitored or simulated, and allows maneuvers to be planned which may be dependent on the availability of specific navigational stars. Second, stars provide a motion cue when the point of view is rotating with respect to inertial space.

The visualization system also contains, in the present implementation, up to two spacecraft. One is often the space station. Each spacecraft may consist of a core and one or two movable appendages such as solar panels. For a spacecraft with thrusters, an exhaust plume is drawn when a jet is fired. Because the space station is not yet fully defined, and because other spacecraft may need to be represented, the visualization system allows spacecraft to be defined as an assemblage of simple cylindrical and plate elements. The visualization system also contains information such that a cylinder which is meant to represent an established type of module, for example a habitat module, can be given detail to make its appearance more realistic. In the case of an unusual or unknown spacecraft, simple cylinders and rectangular plates can be used to build up an image.

Additional minor elements of the visualization system include a gnomon, always drawn in the upper left corner of the square window occupied by the display, which indicates the orientation of the local-vertical, local-horizontal (LVLH), frame of reference pertaining to the principal spacecraft. The visualization system also has the capability of drawing a buoy, a yellow three-dimensional cross, which may be used to represent a spacecraft of unknown configuration, or to mark a spot in space, as, for example, a nominal position to be returned to after a maneuver. There are some other minor embellishments which apply to specific ways in which the visualization system can be used, and these are discussed later.

Information requirements

The Sun, stars, Earth, and spacecraft together form a sort of computerized orrery. The system is set into motion by computed transformations and positions which are used to locate each element in its proper relative position, either for the present time (for monitoring), or for some future time (for simulation). Besides initialization information specifying the configurations of the spacecraft that are to be drawn, the visualization system requires the following dynamic information from the simulation or mission software to which it is attached:

(1) Position of spacecraft center of mass.
(2) Position of spacecraft center of mass with respect to spacecraft structure.
(3) Position of spacecraft appendages.
(4) Attitude (orientation) of spacecraft.
(5) Jet firing information for spacecraft.
(6) Sun position.
(7) Transformation relating the Earth-fixed coordinate system to a reference inertial coordinate system.
(8) Transformation relating the spacecraft LVLH coordinate system, a frame which moves with the spacecraft and is defined in terms of its position and velocity, but not its attitude, to the reference system.
(9) Transformation relating the spacecraft "body" coordinate system, which is fixed with respect to the spacecraft's structure, to the LVLH frame.

New values for each quantity are required for each frame drawn by the visualization system.

For a given time, the relationships defined by this information form a scene which is representative of a real situation and not under the control of the user of the system. The point of view within the scene, however, can be controlled by the user to accomplish various specific visualization functions.

Control by the user

The point-of-view characteristics which are under the control of the user are the following:

(1) The coordinate system with respect to which the point of view will be defined.
(2) The origin, i.e. the object or point which is to occupy the center of the picture.
(3) The distance from the eye to the chosen origin.
(4) The line-of-sight vector, expressed in the chosen coordinate system, from the eye to the chosen origin.
(5) The angular field of view of the image presented.

In the present implementation all characteristics are dynamically under the control of the users as they use the display, with the exception of field of view, which is defined at initialization time. The user-controllable characteristics are input by means of an alphanumeric display and keystroke language based on the method used in the space shuttle. An analog dial and joystick may also be used in controlling point-of-view distance and direction. Although normally each characteristic is explicitly controlled, canned com-

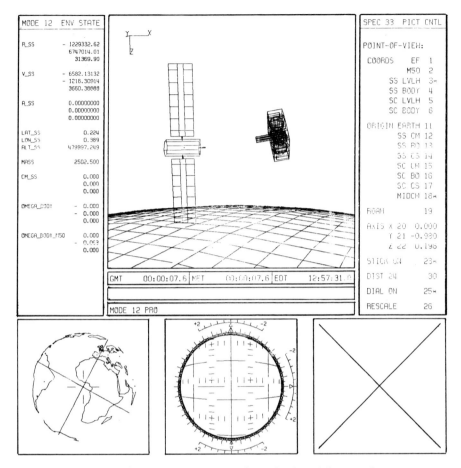

Figure 1. The complete space station simulator display. The visualization system forms the square window at upper center. The alphanumeric display page used to control the point of view is at upper right, simulation data are displayed at upper left, and special-purpose displays such as an orbit position indicator (OPI) and an attitude director indicator (ADI) are below. The visualization system shows an orbital maneuvering vehicle (OMV) nearing a satellite which it wishes to grapple, as it would be seen from an imaginary "chase-plane" flying beside them.

binations can be provided so that certain favorite set-ups can be obtained with a minimum of keystrokes. Figure 1 shows the alphanumeric display page used to control the visualization system point of view.

The point of view coordinate system may be chosen from among the usual frames of reference used in space applications. These include an inertial frame locked to the stars; an Earth-fixed frame which moves with the planet Earth; the LVLH frame which moves with the spacecraft, but is independent of its orientation; and a "body" frame which is locked to the spacecraft structure. When more than one spacecraft is included, the LVLH and body frames pertaining to each are available, although of course when the spacecraft are near each other their LVLH systems are not significantly different. All frames except the reference inertial are rotating coordinate systems. Additional coordinate systems can easily be added to the structure.

A second aspect of the point of view which is under the control of the user is the point upon which the display is centered. An early lesson in the design of the visualization system was that when the point of view is allowed to maneuver independently, it was easy to lose track of the object of interest. As a result, the point of view is normally centered on some chosen point. The choice of "origin" consists of the center of the Earth, the centers of mass, body coordinate system origin, or the crew station of either spacecraft, and the midpoint between the centers of mass of two spacecraft.

Having chosen an origin and a coordination system, the user must choose a line-of-sight vector. The line of sight is controlled by numerically specifying a unit vector expressed in terms of the chosen coordinate system, or by manipulating a joystick which is attached to the system. The line-of-sight distance, the distance between the "eye" and the chosen origin, may be controlled numerically or by means of an analog dial. Distances between 0 and 500,000 km (continuous) are permitted in the present implementation. A negative distance may be specified, but the usefulness is limited because that puts the chosen origin behind the eye.

The point of view may be thought of as looking inward from a spot on a sphere. The sphere is stationary with respect to the chosen coordinate system, it is centered on the chosen origin, and its radius is the chosen distance. The point of view's position on the sphere is specified by the line-of-sight vector.

The angular field-of-view of the display is also under the control of the user, although in the present implementation in the space station simulator, the field of view must be chosen ahead of time and is not subject to real-time modification. Fields of view between 10° and 90° are allowed. The most usual choice is 40°. While this angle does not correspond to the actual angle subtended by the display window when looked at from the usual viewing distance, it does roughly correspond to the field of view of the normal photograph taken with a medium length lens, and is satisfactory to most users.

Ways of using the visualization system

The visualization system is a general system which can present the scene resulting from any combination of the available coordinate systems, origins, and lines of sight. The following are some of the specific ways in which the system can be utilized:

Chase plane views
The view that would be available from an imaginary chase plane flying alongside can be obtained by choosing the LVLH framework, an origin

centered on the spacecraft of interest (or midway between two spacecraft of interest), and a line of sight and distance such as to achieve the desired view, whether from ahead, the side, behind, above, or below. Such a point of view can be useful when visualizing docking and berthing operations in which two spacecraft come together or separate. It can also be useful simply by presenting an "out of spacecraft" view of a single spacecraft, such that the spacecraft's location relative to the Earth in the background, its orientation, and the position of its movable appendages are simultaneously apparent. Figure 1 shows a chase plane view in which an OMV approaches a satellite to pick it up.

Pilot's-eye views
By selecting the body coordinate system of a given spacecraft, setting origin to "crew station," and choosing a distance of zero, the point of view can be placed in the driver's seat of any spacecraft, even an unmanned one for which an imaginary crew position is defined. Such views can serve a number of purposes, such as assessing what will be seen from the crew station window during a planned maneuver (including star availability and the problem of solar glare), presenting views that are not available in real life because there is no suitable window, and providing an on-board perspective for unmanned spacecraft which may be remotely controlled from the space station. If coupled to suitable simulation software, this point of view also allows the rehearsal of operations to be conducted by an astronaut using a Manned Maneuvering Unit, such as satellite capture. Figure 2 presents the view from a point behind the space station hatch to which the shuttle will dock. Such a view represents a "synthetic window" providing visibility in a case where spacecraft structure may preclude an actual window. (An illustration of the fact that the visualization system includes special cases equivalent to existing instruments is the fact that a pilot's eye view looking forward, perhaps with the horizon in view, corresponds to the stylized pattern presented by the attitudes reference instrument known as the 8-ball or artificial horizon.)

Whole-Earth views
The visualization system permits point-of-view distances large enough that the entire Earth is visible. Because at such distances the spacecraft appear as points of light, a capability called "rescale" is available which vastly expands each spacecraft (and shrinks the Earth), to produce a not-to-scale cartoon view in which both the position and orientation of the spacecraft are apparent. Such a point of view is useful for following a rendezvous operation in which the spacecraft may start out on opposite sides of the planet. For example, a whole-Earth view looking along the y-axis in LVLH coordinates shows the view normal to the orbital plane. z-axis views in the Earth-fixed or inertial framework show the Earth from its polar axis. In the inertial case the Earth will be seen to rotate during 24 hr. Figure 3 shows a scene in which two spacecraft are viewed from a polar axis point of view.

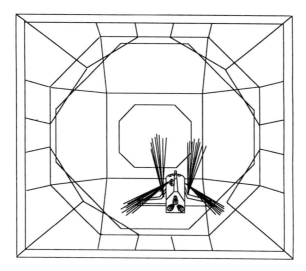

Figure 2. In this view the point of view has been locked to the body coordinate system of the space station and located just inside the shuttle docking hatch, looking in a forward direction. At a distance of approximately 150 m a space shuttle fires maneuvering jets to reach an attitude for docking. Such a point of view can be used to assess window visibility for upcoming operations, or, as in this case, to provide a synthetic window where none exists.

Isolating a factor

Another way of using the visualization system allows the effect on a spacecraft of some single factor to be isolated. Such a capability might come into play when a new control system is to be tested on-board before being given control of the space station. The spacecraft is drawn twice at the same location and time. One image represents the spacecraft as it actually appears in real time, the other represents a simulated version of the same spacecraft as if it were controlled by the new control system under test. Divergences between the two images will illustrate performance differences attributable to the new system.

Roam capability

In most cases it is desirable to center the point of view on the object of greatest interest. The "roam" capability can be selected to remove that constraint and allow the point of view to maneuver independently within the framework established by selected coordinate system and origin. During a roam, the point-of-view orientation is controlled by a joystick and a dial can be used to creep forward or backward. When the joystick is deflected a reticle is drawn at the center of the screen to facilitate pointing at the object of interest. The reticle disappears several seconds after the joystick is released to afford an unobstructed view. The roam capability can be used to mimic

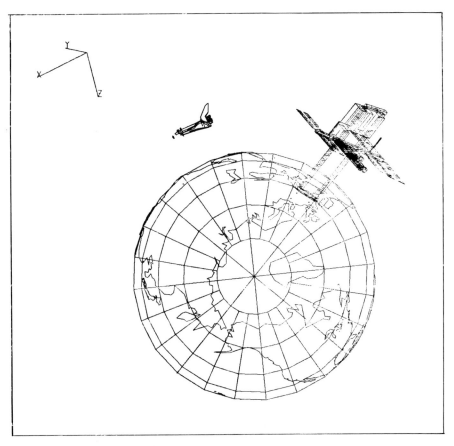

Figure 3. In this view the point of view has been located 30,000 miles from the center of the Earth and directly above the north pole. Two spacecraft are shown, the dual-keel space station and a space shuttle. The RESCALE option has been selected and therefore the spacecraft sizes are exaggerated. Such a point of view allows the positions and attitudes of multiple spacecraft to be simultaneously visualized, as might be desirable during a rendezvous maneuver.

a spacewalk, or EVA, by roaming within the spacecraft body coordinate frame. (However, control is geometric, and the orbital dynamics of an EVA are not simulated in this case.) The roam capability may be most important for inspecting the spacecraft's structure (as known to the computers) but, for example, the view from Boston or Los Angeles could be obtained by letting the point of view roam within the Earth-fixed frame.

The visualization system is not limited to the capabilities described. It can present any view that can be specified using the point-of-view variables under the control of the user. This can include points of view that are probably nonsensical. An example would be an Earth-centered view in the spacecraft body coordinate system. If the spacecraft is spun, the planet appears to gyrate in such a way that the spacecraft is kept in the same orientation.

Conclusion

The central strategies employed in designing the visualization system were, first, to use a picture to make available to the user the extensive information available in the space station's computer system; and second, rather than design a number of special-purpose instruments, to create a general display from which specific capabilities can be obtained by controlling the point of view in various ways.

It may be useful, in conclusion, to contrast the visualization system to concepts such as the "virtual cockpit" designed to assist the pilots of high-performance aircraft. While the virtual cockpit enhances the pilot's perceptual effectiveness, the "visualization system" enhances the crew's operational effectiveness.

The distinction follows from the dissimilar missions. The mission for which the virtual cockpit is designed may last only the few seconds it takes for a jet aircraft to carry out an attack. The pilot's success and survival depend on efficiency during this period. The virtual cockpit takes a single point of view and enhances its perceptions by introducing labels, speed posts, threat indicators, the terrain itself, and so forth. The attack pilots might appreciate a view of themselves as seen by the target, but the exigencies of the combat situation require instead that they stay within themselves.

The visualization system is also designed to enhance the pilot's effectiveness, but in this case the mission may last months and, despite the high absolute velocities, the relative speeds are often closer to sailboats than to jets. On the other hand, space is a place with no up or down, or rather a variety of ups and downs, depending on the particular situation. The visualization system responds by providing a tool that is suitable for the on-board planning and rehearsing that will be part of a long mission, and which offers a way of visualizing operations as they appear in several shifting frames of reference.

Aboard a space station, the pilot is sitting at a console which may face in an arbitrary direction and may be without a window. Split-second reactions are seldom necessary. There are no weather problems. What is necessary is the ability to plan and then to monitor spatial operations which may be hard to see and hard to visualize. For this case, the ability to assume a God's-eye view and follow the orbits leading to rendezvous, to fly alongside in a phantom chase plane, to take the vantage point of an imaginary window in your own spacecraft, or the viewpoint of another, perhaps unmanned satellite, may prove to be useful.

References

Eyles, D. (1986). Space station thrillers unfold at Draper Lab. *Aerospace America*, **38**.
Eyles, D. (1988). A computer graphics system for visualizing spacecraft in orbit, Charles Stark Draper Laboratory, R-2069.

13

Design and evaluation of a visual display aid for orbital maneuvering[1]

Arthur J. Grunwald[2] and Stephen R. Ellis[3]

Aero-Space Human Factors Research Division
Ames Research Center, Moffett Field, CA 94035

Summary

An interactive proximity operations planning system, which allows on-site planning of fuel-efficient, multi-burn maneuvers in a potential multi-spacecraft environment has been developed and tested. Though this display system most directly assists planning by providing visual feedback to aid visualization of the trajectories and constraints, its most significant features include (1) an "inverse dynamics" algorithm that removes control nonlinearities facing the operator and (2) a trajectory planning technique that reduces the order of control and creates, through a "geometric spread-sheet," the illusion of an inertially stable environment. This synthetic environment provides the user with control of relevant static and dynamic properties of way-points during small orbital changes allowing *independent* solutions to the normally coupled problems of orbital maneuvering. An experiment has been carried out in which experienced operators were required to plan a trajectory to retrieve an object accidentally separated from a dual-keel space station. The time required to plan these maneuvers was found to be predicted by the direction of the separation thrust and did not depend on the point of separation from the space station.

---

[1] An earlier version of this manuscript was presented at *Proceedings of the AGARD FMP Symposium Space Vehicle Flight Mechanics, 13–16 November 1989 Luxembourg, AGARD CP489*, pp. 29-1–29-13.
[2] Current address: Senior Lecturer, Department of Aeronautical Engineering, Technion Israel Institute of Technology, Haifa, Israel. The development of the proximity operations planning software described in this paper was part of Dr. Grunwald's NRC Senior Research Associateship project during his tenure at the Ames Research Center 1985–1987.
[3] Research Scientist at NASA Ames Research Center and Assistant Clinical Professor, School of Optometry, U.C. Berkeley, USA.

Introduction

Control of the hand

The insights of paleontology reveal the human hand to be a unique end product of millions of years of vertebrate and primate evolution. Indeed, it itself may have been an essential contributor to the development of the human capacity for abstract insight. The hand is a highly dextrous, general purpose manipulator capable of the fine touch needed to thread a needle and the more coarse control and force needed to lift an object heavier than the weight of its owner. However, like telerobotic effectors at the end of multilink robotic arms, control of the position and orientation of the hand in space can be computationally complex. The kinematics of the links that make up the arm complicate the relationship between the muscles which control each of them and the resulting position of the hand in space. Though cerebellar and other neuro-muscular diseases can reduce their victims to the necessity of conscious joint-angle control, in normal health our neurological control systems unburdened us of conscious control of the limb positions that determine our hand positions.

The unconscious ease of normal movement arises from the unique hierarchical control system that has evolved in association with the gross morphology of the hand. This system computationally separates lower order motor coordination functions from higher order commands concerning what to coordinate. One may think about some of the aspects of the lower order motor coordination as the inverse kinematics and dynamics that translate the higher order movement commands from egocentric coordinate space into a series of link movements in joint coordinate space (Atkeson, 1988). This transformation greatly simplifies the planning task confronting the higher order motor centers. It also reinforces the functional and spatial separation of task planning from muscle coordination and provides us at a conscious level with position control over our hands. We command a position and orientation and our hand effortlessly assumes it.

Kinematic complications of control: telerobotic arms

In light of the characteristics of the neurological control of hand position, it is not surprising that for generalized telerobotic manipulation tasks some form of resolved control of teleoperators is found to be easier for operators to control than control of the joint angles and positions of the various links of the arm (Pennington, 1983). This resolved control is achieved computationally by inverting the transformation matrix describing the arm's forward kinematics. However, due to computational singularities and physical constraints on joint motion an inverse motion may not be computable for some positions. Furthermore, if the manipulator has more degrees of freedom (links) than strictly necessary for the desired end effector motion, an infinite number of solutions exists. Accordingly, implementation of resolved control over an arm requires the addition of information. This information may take the form of arbitrary limits on the movement parameters but more usefully may be in the form of kinematic or dynamic optimization criteria such as time optimality, minimum energy, minimal path length, or smoothness of motion (Pellionisz, 1985; Stark, 1988). These criteria allow the resolution system to select one of the many possible patterns of joint movement that would result in the same movement or position.

Dynamic complications of control: order of control

In kinematically simple situations such as those that arise when a subject is engaged in a simple cross-and-square display tracking task using a two-dimensional joystick, factors other than kinematics determine the success of the tracking e.g., (1) the dynamic characteristics of the joystick itself, (2) the stimulus response compatibility of the control and display coordinate systems, (3) concurrent other tasks, and (4) the disturbance function and the dynamics of the controlled element. Large bodies of literature concerning the effects of these factors on tracking performance have been developed (Poulton, 1974; Wickens, 1986). Performance is generally best when the operator is provided with the lowest order of control possible subject to plant and disturbance characteristics.

The difficulties subjects encounter in higher order control environments, e.g., 2nd order or acceleration control, arise from the difficulties they encounter in estimating velocity and acceleration from position and the additional control movements needed for changing final position. Thus, Poulton's (1974, p. 360) recommendation for the design of a manual control system is to design an order of control as low as possible. This goal may be achieved within the control system itself, for example, by introducing an exponential lag that delays the full effect of the control input and reduces the likelihood that the operator of a higher order system will overshoot his target (Poulton, 1974). Higher order control situations also can be assisted by provision of displays using predictors that integrate the time derivatives of position and remove the need for the operator to perceive these rates directly (Palmer et al., 1980; Palmer, 1983). As will be discussed below, simple provision of a predictor is not, however, a sufficient display enhancement if control inputs interact in complex nonlinear ways.

Proximity operations planning display

The orbital environment

The proximate orbital environment of future spacecraft may include a variety of spacecraft co-orbiting in close vicinity. Most of these spacecraft will be "parked" in a stable location with respect to each other, i.e., they will be on the same circular orbit. However, some missions will require repositioning or transfers among them as in the case of the retrieval of an accidentally released object. In this case complex maneuvers are anticipated which involve a variety of spacecraft which are not necessarily located at stable locations and thus have relative motion between each other.

This multi-vehicle environment poses new requirements for control and display of their relative positions. Conventional scenarios involve proximity operations between two vehicles only. In these two-spacecraft missions, the maneuver may be optimized and precomputed in advance of the time of the actual mission. However, since the variety of possible scenarios in a multi-vehicle environment is large, a future spacecraft environment could require astronauts to execute maneuvers that may not have been precomputed. This demand will require an on-site planning tool which allows fast, interactive, informal creation of fuel-efficient maneuvers meeting all constraints set by safety rules.

The difficulties encountered in planning and executing orbital maneuvers originate from several causes (NASA, Lyndon B. Johnson Space Center, 1982, 1983; Allen, 1988). The first one is the counter-intuitive character of orbital motions as experienced in a relative reference frame. The orbital motions are expressed and tend to be perceived in a coordinate frame attached to a large proximate vehicle such as a space-station and, thus, represent relative rather than absolute motions. From experience in inertially fixed environments, it would be intuitively assumed that a thrust in the "forward" direction towards a target vehicle ahead but in the same orbit, i.e., in the direction of the orbital velocity vector, would result in a forward motion. However, after several minutes, orbital mechanics forces will dominate the motion pattern and move the chaser spacecraft "upwards", i.e., to a higher orbit. This will result in a backwards relative motion, since objects in a higher orbit have a slower orbital rate. Thus, a forward thrust ultimately has the opposite effect from that intended. The effect of this unexpected movement is compounded by the fact that a completed maneuver, which essentially is a timed orbital change, involves a potentially third order or higher order control process with departing, maneuvering and braking thrusts. Even without considering the counter-intuitive dynamics such a process is difficult to control! (Wickens, 1986). Furthermore, corrective thrusts produce significant nonlinear effects on spacecraft positions complicating iterative, manual efforts to drive a spacecraft to a desired stable position.

A second cause of the difficulty is the different and unusual way in which orbital maneuvering control forces are applied. In atmospheric flight control forces are applied continuously in a way to correct for randomly appearing atmospheric disturbances, or to compensate for atmospheric drag. In contrast, space-flight in the absence of atmospheric disturbances, has a near deterministic character. Therefore, space-flight is mainly unpowered and undamped along a section of an orbit with certain characteristics. By applying impulse-type maneuvering forces at a given way-point, the characteristics of the orbit are altered. After application of the maneuvering force the spacecraft will coast along on the revised orbit until reaching the next intermediate way-point along its planned trajectory. Once the spacecraft reaches the way-point, its relative motion can be stopped by an appropriate retro-burn. However, since the way-point will not generally be located at a stable relative position, the spacecraft will tend to drift under the influence of orbital mechanics unless corrective thrusts are continually applied.

Third, multi-vehicle orbital missions are subject to safety constraints, such as clearance from existing structures, allowable approach velocities, angles of departure and arrival and maneuvering burn restrictions due to plume impingement or payload characteristics. Design of a fuel-efficient trajectory which satisfies these constraints is a non-trivial task.

It is clear that visualization of the relative trajectories and control forces in an easily interpretable graphical format, will greatly improve the feel for orbital motions and control forces and will provide direct feedback of the operator's control actions. Furthermore, visualization of the constraints in a pictorial format will enable an interactive, graphical trajectory planning in which the design may be iteratively modified until all constraints are satisfied. Typical in-plane maneuvers are the R-bar burn, along the orbital radius vector and the V-bar burn, along the orbital velocity vector.

Consider a spacecraft located at the V-bar and thus at a stationary position relative to the space station. A small maneuvering burn outward along the direction of the R-bar will cause a small component **v** which will result in a small change in the direction of the orbital velocity vector. This will alter the parameters of the orbit. The orbit will become elliptical and after the burn the maneuvering spacecraft will be 90 degrees of orbital travel past the perigee of the new orbit. In Figure 1 the shape of the orbit and the corresponding relative motion trajectory is shown. The relative trajectory has a "closed" elliptical shape and after one orbit the spacecraft will return to its original location. The reason for this is that the radial burn did not significantly alter the magnitude of the velocity vector $\mathbf{v}_o$, and thus the total energy and mean motion, or orbital period, did not change. An outward R-bar burn will result in backward relative motion whereas an inward burn will cause forward relative motion.

In contrast to the R-burn, a maneuvering burn along the V-bar, will alter the magnitude of the vector $\mathbf{v}_o$ by the amount $|\mathbf{v}|$ and will therefore alter the total energy. Figures 2a,b show the shape of the orbit and the corresponding relative motion trajectory. For a burn in positive V-bar direction, the spacecraft will initially move forward but later on gain altitude and fall behind. The opposite is the case for a burn in negative V-bar direction. Here the spacecraft will initially move backwards but later on drop altitude and pull ahead. For a positive burn the spacecraft is initially at perigee and for a negative burn at apogee.

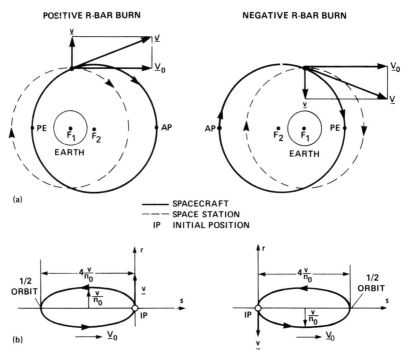

Figure 1. Orbital motion after R-bar burn. (a) Shape of orbit (b) Trajectory relative to the space station.

In general, a chasing vehicle's maneuvers in the orbital plane need not have solely V-bar or R-bar components but components of both. In addition it may also have out-of-plane components. Furthermore, its initial position may not be stable, i.e., offset with respect to its target's V-bar, and the desired flight time may be a fraction of an orbital period, i.e., 10–20 minutes. Under these circumstances the full effects of orbital dynamics are not given sufficient time to manifest themselves completely and are experienced as a kind of "variable orbital wind" blowing the controlled vehicle off a desired straight path. Figures 3 and 4 illustrate the kinds of deflections the "orbital wind" may produce for more generic maneuvers. In particular Figure 3 shows how the deflections caused by orbital dynamics can be partially overcome by using stronger thrusts, but this brute force technique can be very costly due to the fuel required both for departure and braking on arrival.

Limitations of present techniques

The present maneuvering techniques are well established and rely in most cases on visual contact and the use of a V-bar or R-bar reference in a

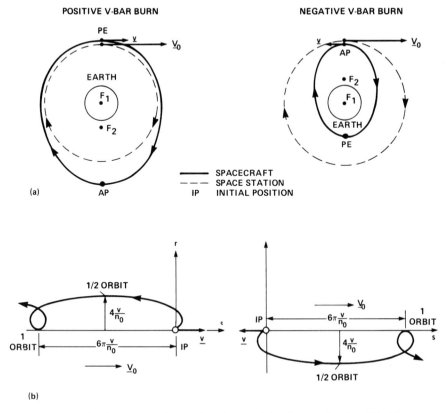

Figure 2. Orbital motion after V-bar burn. (a) Shape of orbit (b) Trajectory relative to the space station.

Crewman Optical Alignment Sight (COAS), (NASA, Lyndon B. Johnson Space Center, 1982, 1983; Kovalevsky, 1984). In a V-bar approach towards a target in positive V-bar direction, the initial burn is made in a direction slightly depressed downwards with respect to the V-bar. After a short while, the spacecraft will "ascend" again and cross the V-bar. At the V-bar crossing a small downward R-burn is initiated which again "depresses" the spacecraft below the V-bar. This process is repeated several times. The spacecraft thus proceeds along the V-bar in small "hops" until the target is reached. However, this technique is highly restricted, is not fuel-optimal, and may not conveniently satisfy other operational constraints of a multi-vehicle environment.

But it is clear from the previous examples that orbital motion can be complex, highly counter intuitive, and involve tightly interacting parameters. A burn towards the target might have an unintended opposite result. Relative motion is, in particular, difficult to visualize for a combined R-bar V-bar

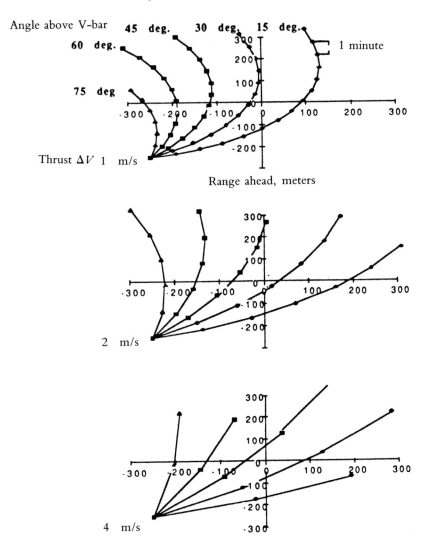

Figure 3. Relative orbital trajectories for different thrust magnitudes and angles for an insertion point below the space stations orbit and behind its center of mass. The space station is located with its center of mass at the origin. Note: that the effects of the orbital dynamics can be overpowered by increasing thrust.

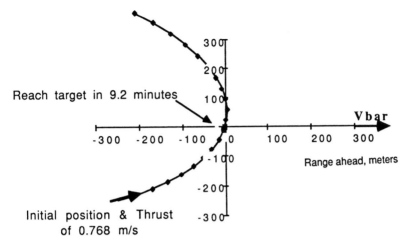

Figure 4. Rendezvous initiated by control of thrust and direction of a maneuvering burn, i.e. the forward method. Using a planning tool that provides a forward predictor of the effects of a planned maneuvering burn, a subject can find by trial and error a combination of thrust and insertion angle that will produce a trajectory to return to the space station from an offset position. Planning for a particular arrival time or selecting a fuel optimal maneuver is, however, manually very difficult with only a forward predictor to assist the operator.

burn at a non-stationary location. It is therefore very useful to visualize the relative motion trajectories graphically. Providing predictors on planning displays which foretell the final consequences of a maneuvering burn is, however, not sufficient symbolic enhancement to enable an operator to plan a timed maneuver. The nonlinear interaction between thrust magnitude and direction of thrust with time of arrival and final relative position preclude tractable manual control over the position and time of the predictor's endpoint.

Design of a pictorial orbital maneuvers planning system

The purpose of the interactive orbital planning system is to enable the operator to design an efficient complex multi-burn maneuver, subject to the stringent safety constraints of a future space-station traffic environment. The constraints include clearances from structures, relative velocities between spacecraft, angles of departure and arrival, approach velocity and plume impingement. The basic idea underlying the system is to present the maneuver as well as the relevant constraints in an easily interpretable pictorial

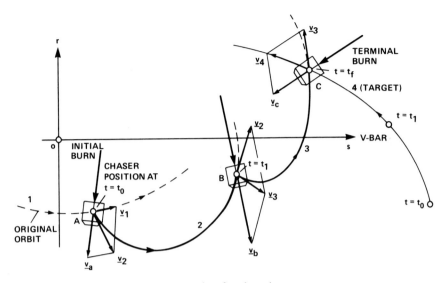

Figure 5. Example of a three-burn maneuver.

format. This format does not just provide the operator with immediate visual feedback on the results of his design actions to enable him to meet the constraints on his flight path, but goes beyond conventional approaches by introducing geometric, symbolic, and dynamic enhancements that bring the intellectual demands of the design process within normal human capacity (Palmer, 1983; NASA, Lyndon B. Johnson Space Center, 1982; McCoy and Frost, 1966; Brody, 1988; Eyles, 1988). The specific methods for enabling interactive trajectory design and visualization of constraints have been discussed in detail elsewhere and will not be repeated here (Grunwald and Ellis, 1988a, 1988b). Though the display also can handle planning out-of-plane maneuvers, the discussion will be limited to maneuvers in the orbital plane.

Example of a three-burn maneuver

An illustrative example of a three-burn maneuver is shown schematically in Figure 5. The trajectory originates from relative position A at time $t = t_o$, and is composed of two way-points B and C which specify the location in space station coordinates at which the chaser spacecraft will pass at a given time. At a way-point the orbital maneuvering system or other reaction control system can be activated, creating a thrust vector of given magnitude for a given duration in a given direction in or out of the orbital plane. The duration of the burn is considered to be very short in comparison with the total duration of the mission. In the orbital dynamics computations this means that a maneuvering burn can be considered as a velocity impulse which alters the direction and magnitude of the instantaneous orbital velocity vector of the spacecraft, inserting it into a new orbit.

Since the initial location A is not necessarily a stationary point, the magnitude and direction of the relative velocity of the chaser at point A are determined by the parameters of its orbit. If no maneuvering burn were initiated at $t = t_o$, the chaser would continue to follow the relative trajectory 1, subject to the parameters of its original orbit, see dotted line in Figure 5. However, a maneuvering burn at $t = t_o$ will alter the original orbit so that the chaser will follow the relative trajectory 2, subject to the parameters of this new orbit.

In Figure 5 $\mathbf{v}_1$ and $\mathbf{v}_2$ indicate the relative velocity vector of the chaser just before and after the maneuvering burn, respectively, where $\mathbf{v}_1$ and $\mathbf{v}_2$ are tangential to the relative trajectories 1 and 2, respectively. The vector difference between $\mathbf{v}_1$ and $\mathbf{v}_2$, $\mathbf{v}_a$, is the velocity vector change initiated by the burn, and corresponds with the direction and magnitude or duration at which the orbital maneuvering system is activated. Likewise, at way-point B the burn $\mathbf{v}_b$ alters the orbit to orbit 3.

Location C is the terminal way-point and is in this case the location where the target will arrive at $t = t_f$. Since the target has an orbit of its own, orbit 4, it will have a terminal relative velocity vector $\mathbf{v}_4$ at $t = t_f$. The relative velocity between target and chaser is the vector difference between $\mathbf{v}_3$ and $\mathbf{v}_4$, $\mathbf{v}_c$. This vector determines the retro-burn that is needed at the target location, in order to bring the relative velocity between chaser and target to the minimum required for the docking operation.

Inverse method of solving orbital motion

Interactive trajectory design demands that the operator be given free control over the positioning of way-points. However, the usual input variables of the equations of orbital motion are the magnitude and direction of the burn, rather than the time and relative position of way-points. Therefore an "inverse method" is required to compute the values of a burn necessary to arrive at a given way-point positioned by the operator. This method is outlined hereafter.

The equations of orbital motion can be computed from its momentary position and velocities, relative to a reference spacecraft with a known circular orbit (Kovalevsky, 1984; Grunwald and Ellis, 1988a, 1988b; Taff, 1985; Thomson, 1986). Thus, for a given initial relative position A with $x(t_0)$ and an initial relative velocity $v(t_0)$, the relative position and velocities of a way-point at time $t = t_1$ can be computed. However, a maneuvering burn at $t = t_0$ will cause a change in the direction and magnitude of the relative velocity vector $\mathbf{v}(t_0)$. As a result of this maneuvering burn, the position of the way-point at time t_1, will change as well.

Consider v_a and α_a to be the magnitude and direction of the velocity change due to the maneuvering burn. Then the relative position and velocity at $t = t_1$, $x(t_1)$, will be a complex non-linear function of v_a and α_a. (Grunwald and Ellis, 1988a, 1988b) Consider now that the operator is given direct control over v_a and α_a by slaving these variables respectively to the x and y motions of a controller such as a joystick or a mouse. A displacement of the mouse in either x or y direction will result in a complex non-linear motion pattern of $x(t_1)$ (see Appendix). Furthermore, this motion pattern will change with the initial conditions. This arrangement is highly undesirable in an interactive trajectory design process, in which the operator must have direct and unconstrained control over the positioning of way-points.

It is therefore essential to give the operator direct control over the position and relative time of way-points rather than over the magnitude and direction of the burn. The inverse method by which this is accomplished computes the magnitude and direction of the burn required to bring the spacecraft from initial location $x(t_0)$ to the way-point $x(t_1)$ at $t = t_1$. This inverse technique contrasts with conventional display aids for proximity operations which are generally forward looking and provide a predictor (McCoy and Frost, 1966; Brody, 1988). While forward looking displays are probably well suited as flying aids for real-time, out-the-window control, a planning system need not conform to this style of aiding.

The active way-point concept

Although a trajectory may be composed of several way-points, only one way-point at a time, the active way-point, is controlled by the operator. While the position and time-of-arrival of the active way-point can be varied, the position and time-of-arrival of all other way-points remains unchanged. However, variations in the active way-point will cause changes in the trajectory sections and way-point maneuvering burns just preceding and just following the active way-point. The on-line solution of the inverse algorithm enables these changes to be visualized almost instantaneously and provides the operator with on-line feedback on his design actions.

Way-point editing

The trajectory design process involves changes in existing way-points, addition of new points or deletion of existing undesired points. An illustrative example of this way-point editing process is shown in Figure 6. In the program the way-points are managed by a way-point stack, which includes an up-to-date sequential list of the position x, the time-of-arrival t and the relative velocity v just after initiating the burn, of all way-points.

Figure 6a shows two way-points, the initial point x_0 and the terminal point x_1. The initial way-point is defined by the initial conditions of the situation and cannot be activated or changed by the operator. The terminal way-point x_1 is thus the active way-point which can be changed and placed at a required location. The corresponding way-point stack is shown on the right. The

Visual display aid for orbital maneuvering

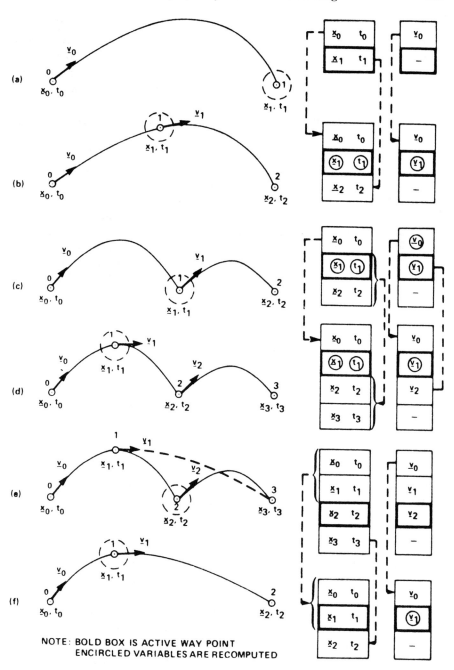

Figure 6. Editing of way-points.

active way-point box is drawn in bold. The relative velocity stack shows only the velocity v_0, which is the relative velocity just after the burn at way-point 0, computed by the inverse algorithm, and required to reach point x_1 at time t_1.

Figure 6b shows the addition of a new way-point. Though its time of occurrence may be manually adjusted later, the new way-point is added half-way in time on the trajectory section just preceding the active way-point. Thus its time-of-arrival is chosen to be $t = 0.5\ (t_i + t_{i-1})$, where i is in this case 1 and relates to the stack before modification. The new position x_1 and relative velocity v_1 are computed by a conventional "forward" method, by computing the orbital position at the new time t, using the existing orbital parameters previously computed with x_0, v_0 and t_0. The newly computed way-point position, time and relative velocity are inserted between points 0 and 1 of the stack before modification and the new way-point is chosen to be the active one. The dotted lines in Figure 6 indicate variables which are transferred without modification and the encircled variables are the newly computed ones. It is important to note that since the relative velocity vectors $\mathbf{v}_0$ and $\mathbf{v}_1$ are matched to the required way-points x_1 and x_2, respectively, the inverse algorithm does not need to make any adjustments.

Figure 6c shows the results of changes in the newly created way-point on the way-point stack. Since x_1 and t_1 are varied, the relative velocity at way-point 0, v_0 will be readjusted by the inverse algorithm and likewise the relative velocity v_1.

Figure 6d shows the creation of an additional new way-point. Since the active way-point prior to the addition was point 1, the new point is added half-way between point 0 and 1 and its position and relative velocity are computed with the forward method. The new values are inserted between points 0 and 1 of the stack before modification and the new way-point is again set to be the active one.

In Figure 6e way-point 2 is activated. Apart from the shift in active way-point, the stack remains unchanged. The dotted line shows the direct-path section between point 1 and point 3 without the intermediate burn at point 2. Deletion of the way-point 2 will remove this point from the stack and after that close the gap (see Figure 6f). However v_1 has to be readjusted to fit the new direct-path section. This adjustment is made on-line by the inverse algorithm.

The repetitive use of the inverse algorithm to calculate the trajectories linking each pair of way-points presents the planner with a kind of "geometric spread-sheet" that preserves certain relationships between points in space, namely that they are connected by fuel-minimum trajectories for their particular separation in time, while their other properties, namely their relative positions in space, may be freely varied. To our knowledge this application of inverse dynamics to this kind of display problem is new and has some very helpful side-effects. The constant background computation to preserve the relative position and time of each way-point creates an illusion of an inertially stable space that assists planning of relative movements about a target spacecraft. Additionally, this technique assists planning by allowing separable solutions to the plume impingement, velocity limit, and traffic conflict problems. Once a way-point has been positioned to bring an aspect of a maneuver within prescribed limits, e.g., relative velocity, the adjustment can be isolated from the effects of earlier adjustment such as to satisfy a plume impingement constraint. This isolation of the solutions of the separate problems is an essential characteristic since without it, the solution to one maneuver problem would undo a solution to another.

Operational constraints

The multi-spacecraft environment will require strict safety rules regarding the clearance from existing structures. Thus, spatial "envelopes" through which the spacecraft is not allowed to pass can be visualized on the display.

Restrictions on angles of departure and arrival may originate from structural constraints at the departure gate or the orientation of the docking gate or grapple device at the target craft. Limits for the allowable angles of departure or arrival can be visualized as cones on the display. In addition the terminal approach velocity at the target might be limited by the target characteristics. Limits for the allowable start and end velocities can be visualized as limit arcs associated with the approach or departure cones (see Figure 7). The limit arc symbols on the display graphically show allowable ranges of magnitude and direction for thrusts and relative velocities.

Way-point maneuvering burns are subject to plume impingement constraints. Hot exhaust gasses of the orbital maneuvering systems may damage the reflecting surfaces of sensitive optical equipment such as telescopes or infra-red sensors. Even cold nitrogen jets might disturb the attitude of the target satellite. Maneuvering burns towards this equipment are restricted in direction and magnitude, where limits for the allowable direction and magnitude are a function of the distance to the equipment and plume characteristics. These limits can be visualized as limit arcs on the display.

Flight safety requires that the relative velocity between spacecraft is subject to approach velocity limits. In conventional docking procedures this limit was proportional to the range (NASA, Lyndon B. Johnson Space Center, 1982, 1983). A commonly used rule-of-thumb is to limit the relative approach velocity to 0.1 per cent of the range. This conventional rule is quite conservative and originates from visual procedures in which large safety margins are taken into account to correct for human or system errors. Although the future traffic environment will be more complex, and will therefore demand larger safety margins, more advanced and reliable measurement and control systems may somewhat relax these demands.

In this display the relative approach velocity is defined as the component of the relative approach velocity vector between the two spacecraft along their mutual line-of-sight. The limit on this relative approach velocity is a function

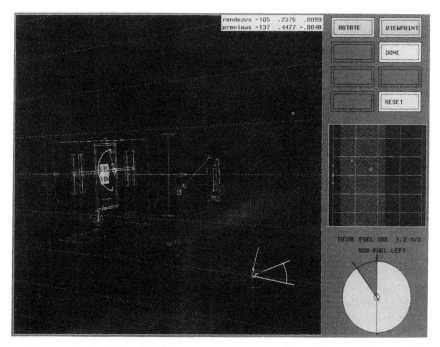

Figure 7. Screen image of the proximity operations planning tool. The upper right window shows the soft buttons for the viewpoint control mode. The out-of-plane display is shown in the middle right window. The lower right window shows the fuel display. The main viewport shows an incompletely planned mission for which three burns have been selected. The velocity vector or +V-bar is shown by the arrow pointing to the right on the central grid line. Note that the relative velocity vector on arrival, shown by the arrow in the lower right of the viewport, is outside the entry cone indicating the acceptable range of relative velocity on arrival with the target craft.

of the range between the spacecraft. This function will depend on the environment, the task and the reliability of measurement and control equipment and cannot be determined at this stage. For this display a simple proportional relation has been chosen. The approach velocity limit is visualized on the display as a circle drawn around the chaser indicating the minimum range between the two spacecraft allowed for the present approach velocity. If the target craft appears within this circle, the approach limit has been violated.

General comments

The proposed interactive orbital planning system should be seen as a step in determining a display format which may be useful in a future dense spacecraft traffic environment. The examples shown here deal with the most general situation, which involves departures from or arrival at non-stationary locations. Such missions with spacecraft at non-stationary positions and substantial out-of-plane motion may represent worst-case situations but these are the ones most likely to require customized maneuvering.

It is hard to predict whether the constraints used here will be relevant and realistic in the future spacecraft environment. However, they encompass in a broad sense the general type of restrictions which are expected in the multi-vehicle environment, e.g., limitations on approach rates, plume impingement and clearance from structures.

A final restriction of the present display relates to the way the orbital maneuvering system is activated. Only pure impulse maneuvering burns are considered, in which the duration of the burn is negligible with respect to the duration of the mission and in which these burns cause major changes in the relative trajectories. Station keeping or fly-by missions however, require a more sustained type of activation, such as periodic small burns with several-second intervals over a time-span of several minutes. A more "distributed" way of activating the orbital maneuvering system could be introduced in which the operator has control over the frequency and time-span of the activation.

A last comment relates to the way the spatial trajectory is visualized. The perspective main view shows the projection of the actual trajectory on the orbital plane, rather than the trajectory itself. The reason for this is two-fold. The orbital trajectory, with its typical cycloidal shape, when shown without lines projecting onto the orbital reference plane, is ambiguous and might seem to come out of the orbital plane. This illusion may result from the viewer's familiarity with common objects such as a coil spring and has been first reported earlier (Ellis and Grunwald, 1988a, 1988b, 1987). Therefore, the trajectory cannot be clearly shown without its projection on the orbital plane. However, since the symbolic enhancements and burn vectors relate to the in-plane motion and match with the trajectory projection on the orbital plane, both the trajectory and its projection should be visualized. However, for most views of a 3D display, both the trajectory and its projection on the orbital plane will show up as separate curves, a fact that may cause confusion. Therefore, a compromise has been sought, in which the projection is shown together with "pedestals" placed at the way-points orthogonal to the orbital plane, which mark the actual trajectory at the way-points. In spite of these restrictions the proposed display illustrates the usefulness of interactive graphical trajectory design.

Experimental study

An experiment has been conducted with the above planning tool to determine the time required to plan a variety of rendezvous missions for which the target's orbital insertion parameters were systematically varied. In particular, we attempt to develop a regression predictive of the time required to plan a rendezvous with a vehicle simulating an inadvertently released object from a variety of positions along the main structures of a dual-keel space station configuration. The space-station is modeled in a 480 km circular orbit inclined 28.5 degrees with respect to the equator. This corresponds to a v_o orbital velocity of 7623 m/sec or an orbital angular rate 0.0011 rad/sec. The chasing vehicle for the maneuver departed from a +V-bar location on the station and may be thought of as a craft attempting to recover an astronaut or small object such as a wrench accidentally released with either 0 or moderate (1.0 m/sec) delta v and which is drifting away under the influence of orbital mechanics forces. Out-of-plane insertion velocity components of the target were randomly selected to be ±0.25 or ±0.5 m/sec. The direction of the added delta v at insertion was systematically varied in eight equal directions about the +V-bar. The ten orbital insertion points for the targets were distributed along the port keel of the space station from 200 m above the center of mass to 150 m below it and randomly selected to produce ninety different recovery scenarios. The planned one way flight time was twenty minutes and the maneuver took place during orbital daylight. The scenarios simulate the rendezvous phase of a rendezvous and retrieval mission.

Task

The subject's task was expeditiously to plan a feasible in- and out-of-plane trajectory from a space station +V-bar departure port to rendezvous with the target, subject to plume impingement constraints on the station, avoidance of the station's structure, and alignment of the relative velocity vector on rendezvous to fit within the 30 degree entrance cone. All subjects were told to complete their planning task quickly, much as they would wish to walk across a room without wasting time and not to worry about minimizing overall fuel use, though they were limited to 12 m/sec delta v of maneuvering fuel. Figure 9 illustrates a three-burn partial solution to one of the experimental scenarios in which the relative velocity on rendezvous has not yet been adjusted to fit within the approach constraint shown by the approach cone.

Subjects

Three highly practiced subjects very familiar with the operation of the planning tool planned rendezvous for the set of initial conditions which were generally novel to them.

Procedure and design

Before beginning the experiment the subjects reviewed a training manual describing the display's controls and practiced their operation on a sample rendezvous. Thereafter, subjects were automatically presented through a UNIX C-shell script with the ninety rendezvous problems in four approximately equal groups of randomly ordered conditions. Data collection took

Effect of Insertion Angle on Planning Time

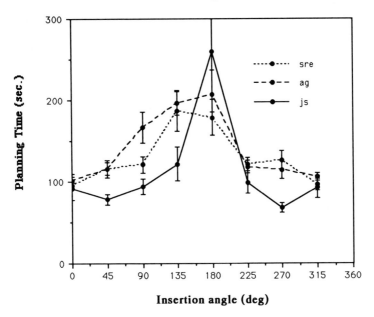

Figure 8. The mean planning time for each of three subjects shows a marked peak for target insertion angles around 200 deg. The error bars are ±1 standard error.

about six hours and was generally spread across two days. Descriptive statistics were collected automatically by the IRIS computer to summarize planning time, fuel use, number of way-points used, and a large number of other detailed characteristics of the mission planning process.

Results

The planning time required for the selected missions was highly dependent upon the angle of the insertion of delta v (Figure 8). Analysis of variance on time ($F = 7.968$, $df = 7,14$, $p < .001$) showed that the effects of insertion angle were large, statistically reliable and dominated effects due to the point of insertion. Planning time was not significantly affected by the selection of an insertion point ($F = 0.584$, $df = 9,18$, ns) and there was no statistical interaction between the insertion angle and the insertion point ($F = 0.870$, $df = 63,126$, ns.).

Discussion

In earlier experiments differences in planning time appeared mainly due to the misalignment of the approach cone axis with the relative velocity resulting from a two burn fuel-minimum intercept trajectory. These differences were reduced by practice, but maintained a rough proportionality (Ellis and Grunwald, 1988). Thus, a rendezvous that took twice as long as another approximately maintained this proportion as practice reduced the times for both of them. Thus, the differences observed in the present experiment with highly practiced subjects probably reflect genuine differences in planning difficulty arising from the interaction of orbital dynamics and the mission constraints. For the particular conditions used, insertion angles near 135 degrees are particularly hard. Accordingly, potential chase vehicles should be positioned at several ports, V-bar and R-bar, to facilitate capture of inadvertently separated objects.

The results of the present and previous experiments make clear that experienced human operators can manually, quickly plan complex orbital maneuvers when their planning tool is adapted to their capabilities. It is, none the less, also clear that properly programmed automatic systems could also plan these maneuvers. These results can help set performance criteria for these automatic systems since they should at least be capable of producing feasible plans in less than two minutes to beat a manually determined plan. Incorporation of all the mission constraints, however, can greatly complicate and lengthen an automatic search since these constraints may be arbitrarily placed in space and in some cases may be discrete. The development of usable search algorithms is a current direction of research and certainly constrained random search strategies could be adopted if more efficient analytic methods do not work well (Soller *et al.*, 1990).

But is it also clear that however the maneuver is planned, any astronaut who would be flying the mission would want to foresee what the system has planned for him and be able to visualize his trajectory, if for no other reason than to monitor its unfolding as it is flown. Automatically generated trajectories will only be as good as the designer's hindsight in selecting optimization criteria and mission constraints. Unique mission features or failures may arise that require the custom-tailoring of a trajectory. Significantly, the mission planning interface described in this paper also can serve as an interface to an mission "editor" that would allow an astronaut to visualize the automatically planned trajectories and edit them if necessary to suit his special requirements.

References

Allen, J.P. (1988). Physics at the edge of the earth. In French, A.P. (Ed.), *Physics in a Technological World*. New York: American Institute of Physics.

Atkeson, C.G. (1988). Learning arm kinematics and dynamics. *Annual Review of Neuroscience*, **12**, 157–183.

Brody, A.R. (1988). A forward-looking interactive orbital trajectory plotting tool for use with proximity operations. *NASA CR 177490*, June, 1988.

Ellis, S.R. and Grunwald, A. (1988). A visualization tool for planning orbital maneuvers for proximity operations: design philosophy and preliminary results.

Proceedings of the 23rd Annual Conference on Manual Control. June 22–24, 1988, BBN; Cambridge, Massachusetts.
Ellis, S.R. and Grunwald, A.J. (1987). A New Illusion of Projected Three-Dimensional Space, *NASA TM 100006*, July, 1987.
Eyles, D. (1988). A computer graphics system for visualizing spacecraft in orbit. In Ellis, S.R. Kaiser, M., and Grunwald, A. (Eds). *Spatial Displays and Spatial Instruments*, NASA-CP 10032.
Grunwald, A.J. and Ellis, S.R. (1988a). Interactive orbital proximity operations planning system. *Proceedings of the 1988 IEEE International Conference on Systems, Man, and Cybernetics*. Peking, China., IEEE CAT 88CH2556-9, 1305-13-12.
Grunwald, A.J. and Ellis, S.R. (1988b). Interactive orbital proximity operations planning system. *NASA TP 2839*.
Kovalevsky, J. (1984). *Introduction to Celestial Mechanics* Translated by Express Translation Service, New York: Springer-Verlag New York Inc.
McCoy, W.K. Jr., and Frost G.G. (1966). Predictor Display Techniques for On-Board Trajectory Optimization of Rendezvous Maneuvers, Aerospace Medical Research Laboratories AMRL-TR-66-60, May, 1966.
NASA, Lyndon B. Johnson Space Center (1983). *Rendezvous/Proximity Operations Workbook* RNDZ 2102, 1983.
NASA, Lyndon B. Johnson Space Center (1982). *Flight Procedures Handbook*, JSC-10566, November, 1982.
Palmer, E. (1983). Conflict resolution maneuvers during near miss encounters with cockpit displays of traffic information. *Proceedings of the Human Factors Society 27th Annual Meeting*, Human Factors Society, Santa Monica, CA. 757–761.
Palmer, E., Jago, S., Baty, D.L. and O'Conner, S. (1980). Perception of horizontal aircraft separation on a cockpit display of traffic information. *Human Factors*, **22**, 605–620.
Pellionisz, A.J. (1985). *Cerebellar functions*, 201–229. Berlin: Springer.
Pennington, J.K. (1983). A rate-controlled teleoperator task with simulated transport delay. *NASA TM 85653*, September, 1983.
Poulton, E.C. (1974). *Tracking Skill and Manual Control*. New York: Academic Press.
Soller, J., Grunwald, A.J. and Ellis, S.R. (1990). Optimization of multiburn orbital maneuvers in proximity operations. *NASA TM* in preparation.
Stark, L. (1988). Biological redundancy. *Proceedings of the NATO Workshop on Telerobotics* (Abstract 303).
Taff, L.G. (1985). *Celestial Mechanics, A Computational Guide for the Practitioner*. New York: John Wiley and Sons.
Thomson, W.T. (1986). *Introduction to Space Dynamics*. New York: Dover Publications, Inc.
Wickens, C.D. (1986). The effects of control dynamics on performance. In Boff, K.R., Kaufman, L. and Thomas, J.P. (Eds) *Handbook of Perception and Human Performance*. New York: Wiley.

Appendix

Relative motion in space-station coordinates

The relative position and velocity of a co-orbiting spacecraft in space station coordinates are $x^0 = \{x^0, y^0, z^0\}$ and $\dot{x}^0 = \{\dot{x}^0, \dot{y}^0, \dot{z}^0\}$ respectively, and are obtained from space station based measurement equipment where: $\dot{x}^0 = dx^0(t)/dt$ etc. denotes the

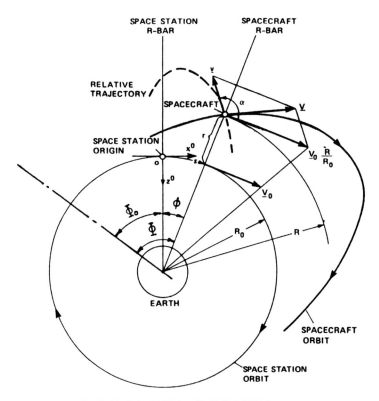

Figure 9. In-plane orbital situation showing the difference in phase angles Φ_0 and Φ for the space station and another spacecraft. Note that the bold solid line represents the absolute spacecraft orbit and the dotted line represents its trajectory relative to the space station. The ratio of r to R_0 has been greatly exaggerated.

time derivative. x^0 points in the direction of the orbital velocity vector, y^0 is perpendicular to the orbital plane and points to the right, and z^0 points toward the center of the earth. The $x^0 o z^0$ plane is the orbital plane.

Since the displacements and velocities in the out-of-orbital-plane direction y^0 are usually much smaller than the ones in the orbital plane, in x^0 and z^0 direction, and since out-of-plane maneuvering burn does not significantly alter the total orbital energy, the out-of-plane motion can be decoupled from the in-plane motion. Hence, the in-plane motion will be analysed first.

At the initial time $t = t_0$ the radius of the space station orbit R_0 is given by: $R_0 = R_E + h$, where $R_E = 6,378,140$ m and is the equatorial earth radius and $h = 480,000$ m and is the altitude of the space-station orbit above the earth surface. The absolute orbital velocity of the space-station is then given by: $V_0 = (GM/R_0)^{1/2}$, where $GM = 3.986005 \times 10^{14}$ m^3/sec^2 and is the geocentric gravitational constant. For simplicity the curvature of the V-bar is assumed to be negligible, so that:

$$s \sim x^0$$
$$r \sim -z^0 \quad\quad (1a,b,c)$$
$$v \sim \sqrt{\dot{s}^2 + \dot{r}^2}$$

$$\alpha = \tan^{-1}(\dot{r}/\dot{s}) \quad\quad \dot{s} \geq 0$$
$$\alpha = \tan^{-1}(\dot{r}/\dot{s}) + 180° \quad \dot{s} < 0 \quad\quad (2a,b)$$

where s is the distance measured along the V-bar between the space station and the spacecraft's R-bar, r the distance of the spacecraft above the V-bar, measured along the R-bar, v the magnitude of the relative velocity and α its direction, measured from the V-bar in upwards direction (positive rotation in the right-hand system). The relative velocity reflects the situation just after the activation of a maneuvering burn.

Although the general solution for orbital motion is fairly complex and non-linear, the relative motion between two co-orbiting spacecraft in close proximity can be simplified by a first order approximation. This is known as the Euler-Hill solution which allows both the "forward" solution and the "inverse" solution. The forward solution computes at a given final time t_f and for a given relative initial position and initial velocity vector, the final relative position and velocity. On the other hand, the inverse solution computes for given t_f and given initial location, the required initial velocity vector to reach a given desired final location. The linearized Euler-Hill solution is derived as follows.

The in-orbital-plane motion of a spacecraft orbiting the earth is described by two degrees of freedom: the radius R and the orbital angle, Φ, with respect to an arbitrary fixed reference (see Figure 9). Making use of Newton's first law and his law of gravity, the equations of motion of the spacecraft in the absence of external forces are given by:

$$\ddot{\Phi}(t) R(t) + 2\dot{\Phi}(t) \dot{R}(t) = 0$$
$$\ddot{R}(t) - \{\dot{\Phi}(t)\}^2 R(t) + \frac{GM}{R(t)^2} = 0 \quad\quad (3a,b)$$

The motion of the space station is described by $R_0(t)$ and $\Phi_0(t)$. $R(t)$ and $\Phi(t)$ can now be expressed as:

$$R(t) = R_0 + r(t)$$
$$\Phi(t) = \Phi_0 + \phi(t) \quad\quad (4a,b)$$

Since the spacecraft is in close proximity to the space station r and ϕ are relatively very small in comparison to R_0 and Φ_0. Since the space station is in circular orbit, it follows that

$$R_0(t) = R_0 = \text{constant and}$$
$$\dot{\Phi}_0(t) = n_0 = \text{constant} \quad\quad (5a,b)$$

where n_0 is the mean motion or angular orbital rate of the space station in radians/sec. Differentiating Equations (4a,b) and using Equations (5a,b) yields:

$$\dot{R}(t) = \dot{r}(t); \; \ddot{R}(t) = \ddot{r}(t)$$
$$\dot{\Phi}(t) = n_0 + \dot{\phi}; \; \ddot{\Phi} = \ddot{\phi} \quad\quad (6a-d)$$

Substituting Equations (6a–d) in Equations (3a,b), neglecting products of r and ϕ and of their derivatives, using Kepler's third law:

$$n_0 = \sqrt{\frac{GM}{R_0^3}} \qquad (7)$$

and using the definitions:

$$\dot{s}(t) \equiv R_0\dot{\phi}(t); \quad \ddot{s}(t) \equiv R_0\ddot{\phi}(t) \qquad (8a,b)$$

yields the linearized equations for the relative motion of a spacecraft with respect to the space station (Euler-Hill equations).

$$\ddot{s}(t) + 2n_0\dot{r}(t) = 0$$
$$\ddot{r}(t) - 3n_0^2 r(t) - 2n_0\dot{s}(t) = 0 \qquad (9a,b)$$

For given initial conditions $\dot{r}_0, r_0, \dot{s}_0, s_0$ where $r_0 = r(t_0)$ etc., the solution for $r(t)$ and $s(t)$ written in matrix form is:

$$\begin{bmatrix} r(t) \\ s(t) \end{bmatrix} = \begin{bmatrix} a_{1,1} & a_{1,2} \\ a_{2,1} & a_{2,2} \end{bmatrix} \begin{bmatrix} \dot{r}_0 \\ \dot{s}_0 \end{bmatrix} + \begin{bmatrix} b_{1,1} & 0 \\ b_{2,1} & 1 \end{bmatrix} \begin{bmatrix} r_0 \\ s_0 \end{bmatrix} \qquad (10)$$

where

$$a_{1,1} = \frac{1}{n_0}\sin(n_0 t)$$

$$a_{1,2} = \frac{2}{n_0}[1 - \cos(n_0 t)]$$

$$a_{2,1} = -a_{1,2}$$

$$a_{2,2} = \frac{1}{n_0}[4\sin(n_0 t) - 3n_0 t] \qquad (11a\text{–}f)$$

$$b_{1,1} = 4 - 3\cos(n_0 t)$$

$$b_{2,1} = 6[\sin(n_0 t) - n_0 t]$$

and the solution for the velocity vector components $\dot{r}(t)$ and $\dot{s}(t)$ is:

$$\begin{bmatrix} \dot{r}(t) \\ \dot{s}(t) \end{bmatrix} = \begin{bmatrix} d_{1,1} & d_{1,2} \\ d_{2,1} & d_{2,2} \end{bmatrix} \begin{bmatrix} \dot{r}_0 \\ \dot{s}_0 \end{bmatrix} + \begin{bmatrix} e_{1,1} & 0 \\ e_{2,1} & 0 \end{bmatrix} \begin{bmatrix} r_0 \\ s_0 \end{bmatrix} \qquad (12)$$

where

$$d_{1,1} = \cos(n_0 t)$$
$$d_{1,2} = 2\sin(n_0 t)$$
$$d_{2,1} = -d_{1,2}$$
$$d_{2,2} = [4\cos(n_0 t) - 3] \qquad (13a\text{–}f)$$
$$e_{1,1} = 3n_0 \sin(n_0 t)$$
$$e_{2,1} = 6n_0[\cos(n_0 t) - 1]$$

Equations (10–13) constitute the "forward" solution. For a given final time or time-of-arrival $t = t_f$ first the coefficients of Equations (11) and (13) are computed and after that with Equations (10) and (12) the final position r_f, s_f and final velocity $\dot{r}_f$, $\dot{s}_f$, are found, where $r_f = r(t_f)$. With the "inverse" solution, the initial velocities $\dot{r}_0$ and $\dot{s}_0$ have to be computed for given time-of-arrival t_f and given initial position r_0, s_0 and final position r_f, s_f. These velocities are easily obtained from Equation (10) according to:

$$\begin{bmatrix} \dot{r}_0 \\ \dot{s}_0 \end{bmatrix} = \begin{bmatrix} a_{1,1} a_{1,2} \\ a_{2,1} a_{2,2} \end{bmatrix}^{-1} \begin{bmatrix} r_f - b_{1,1} r_0 \\ s_f - b_{2,1} r_0 - s_0 \end{bmatrix} \tag{14}$$

where

$$\begin{bmatrix} a_{1,1} a_{1,2} \\ a_{2,1} a_{2,2} \end{bmatrix}^{-1} = \frac{1}{(a_{1,1} a_{2,2} + a_{1,2}^2)} \begin{bmatrix} a_{2,2} & -a_{1,2} \\ -a_{2,1} & a_{1,1} \end{bmatrix} \tag{15}$$

and then may be used to compute the angle and magnitude of relative velocity, α and v from Equation (1) and Equation (2).

14

Telepresence, time delay and adaptation

Richard Held

Massachusetts Institute of Technology, Cambridge, Massachusetts

Nathaniel Durlach

Massachusetts Institute of Technology, Cambridge, Massachusetts
and
Boston University, Boston, Massachusetts

Introduction

Displays, which are the subject of this conference, are now being used extensively throughout our society. More and more of our time is spent watching television, movies, computer screens, etc. Furthermore, in an increasing number of cases, the observer interacts with the display and plays the role of operator as well as observer. To a large extent, our normal behavior in our normal environment can also be thought of in these same terms. Taking liberties with Shakespeare, we might say that "all the world's a display and all the individuals in it are operators in and on the display".

Within this general context of interactive display systems, we begin our discussion with a conceptual overview of a particular class of such systems, namely, teleoperator systems. We then consider the notion of telepresence and the factors that limit telepresence, including decorrelation between the (1) motor output of the teleoperator as sensed directly via the kinesthetic/ tactual system, and (2) the motor output of the teleoperator as sensed indirectly via feedback from the slave robot, i.e., via a visual display of the motor actions of the slave robot. Finally, we focus on the deleterious effect of time delay (a particular source of decorrelation) on sensori-motor adaptation (an important phenomenon related to telepresence).

Teleoperator systems

A schematic outline of a highly simplified teleoperator system is presented in Figure 1. As pictured, the major components of a teleoperator system are a

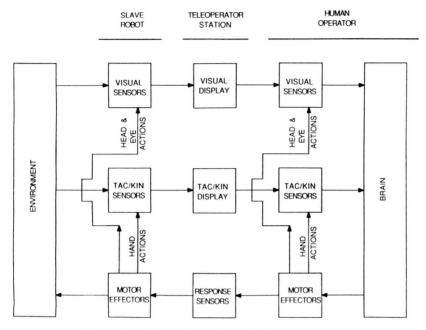

Figure 1. Schematic outline of teleoperator system.

human operator, a teleoperator station (or "suit"), a slave robot, and an environment which is sensed and acted upon by the slave robot. As indicated by the arrows flowing from left to right, sensors on the slave robot are stimulated by interaction with the environment, the outputs of these sensors are displayed in the teleoperator station to the sensors of the human operator, and the received information is then transmitted to higher centers (brain) within the human operator for central processing. As indicated by the arrows flowing from right to left, the central processing results in motor responses by the human operator which are detected in the teleoperator station and used to control motor actions by the slave robot. The upward flowing arrows depict the role played by the motor system (at both the slave robot and human operator levels) in controlling the sensors and therefore the flow of information from environment to brain.

The normal situation in which the human interacts directly with the environment can be pictured as a special case of the teleoperator situation by ignoring the teleoperator station and identifying the slave robot's sensors and effectors with those of the human operator. Similarly, imaginary or virtual environments can be pictured in terms of the teleoperator situation by retaining the human operator and teleoperator station, but replacing the real environment and slave robot by a computer simulation. Finally, robotic systems can be realized by replacing the human operator and teleoperator station by an automatic central processor, and interpolations between teleoperator systems and robotic systems can be realized by assigning lower-level control functions to automatic processing and higher-level control functions (supervisory control) to the human operator.

Note also that the sensor and effector channels need not be restricted in the manner illustrated in Figure 1. Not only are there many cases in which the visual channel pictured would be paralleled by an auditory channel, but for certain purposes the slave robot might also include sensors for which the human has no counterpart (e.g., to sense infrared energy or magnetic fields). Furthermore, on the response side, the teleoperator station might detect and exploit responses other than simple motor actions. For example, it might be useful for certain purposes to measure changes in skin conductivity, pupil size, or blood pressure.

In general, the purpose of a teleoperator system is to augment the sensorimotor system of the human operator. The structure of the teleoperator system will depend on the specific augmentation envisioned, as well as on the technological limitations. A continuum that relates directly to the issue of telepresence considered below concerns the extent to which the structure of the slave robot is the same as that of the teleoperator. At one extreme are systems meant simply to transport the operator to a different place. In the ideal version of such a system, the slave robot would be isomorphic to the operator and the various sensor and effector channels would be designed to realize this isomorphism. In a closely related set of systems, the basic anthropomorphism is preserved, but the slave robot is scaled to achieve, for example, a reduction of size or magnification of strength. At the opposite extreme are systems involving radical structural transformations and highly non-anthropomorphic slave robots. In these systems, there is no simple correspondence between slave robot and human operator, and the design and organization of the sensor and effector channels generally becomes very complex and difficult to optimize, even at the abstract conceptual level. General reviews of teleoperation and teleoperator systems can be found in Johnsen and Corliss (1971) and Vertut and Coiffet (1986).

Telepresence

Although the term "telepresence" is often used in discussions of teleoperation, it never has been adequately defined. According to Akin *et al.* (1983), telepresence occurs when the following conditions are satisfied:

> 'At the worksite, the manipulators have the dexterity to allow the operator to perform normal human functions. At the control station, the operator receives sufficient quantity and quality of sensory feedback to provide a feeling of actual presence at the worksite.'

A major limitation of this definition is that it is not sufficiently operational or quantitative. It does not specify how to measure the degree of telepresence. Also, as indicated by the phrase "perform normal human functions" in the first sentence, it fails to address the issue of telepresence for systems that are designed to transform as well as to transport and perform abnormal human functions.

Independent of the precise definition of telepresence, why should one care about telepresence? What is it good for? Certainly, there is no theorem which states that an increase in telepresence necessarily leads to improved performance. In our opinion, a high degree of telepresence is desirable in a teleoperator system primarily in situations when the tasks are wide-ranging, complex, and uncertain, i.e., when the system must function as a general-purpose system. In such situations, a high degree of telepresence is desirable because the best general-purpose system known to us (as engineers) is us (as operators). In a passage that is relevant both to this issue and to the definition of telepresence, Pepper and Hightower (1984) state the following:

> 'We feel that anthropomorphically-designed teleoperators offer the best means of transmitting man's remarkably adaptive problem solving and manipulative skills into the ocean's depths and other inhospitable environments. The anthropomorphic approach calls for development of teleoperator subsystems which sense highly detailed patterns of visual, auditory, and tactile information in the remote environment and display the non-harmful, task-relevant components of this information to an operator in a way that very closely replicates the pattern of stimulation available to an on-site observer. Such a system would permit the operator to extend his sensory-motor functions and problem solving skills to remote or hazardous sites as if he were actually there.'

In addition to the value of telepresence in a general-purpose teleoperator system, it is likely to be useful in a variety of other applications. More specifically, it should enhance performance in applications (referred to briefly in the first section) where the operator interacts with synthetic worlds created by computer simulation. The most obvious cases in this category are those associated with training people to perform certain motor functions (e.g., flying an airplane) or with entertaining people (i.e., providing imaginary worlds for fun). Less obvious, but equally important, are cases in which the system is used as a research tool to study human sensorimotor performance and cases in which it is used as an interactive display for data presentation (e.g., Bolt, 1984).

An important obstacle at present to scientific use of the telepresence concept is the lack of a well-defined means for measuring telepresence. It should not only be possible to develop subjective scales of telepresence (using standardized scale-construction techniques), but also to develop tests, both psychological and physiological, to measure telepresence objectively. For example, some test based on the "startle response" might prove useful. Certainly, such a test could distinguish reliably between different degrees of realism in the area of cinematic projection. Also, of course, given both some subjective scales and some objective tests, it would be important to study the relations among the two types of measures.

Beyond questions related to the definition and measurement of telepresence, the core issue is how one achieves telepresence. In other words, what are the factors that contribute to a sense of telepresence? In fact, what are the essential elements of just plain "presence?" Or alternately, looking at the other side of the coin, how can the ordinary sense of presence be destroyed (short of damaging the brain)?

Given the vague and qualitative character of definitions and estimates of telepresence, it is not surprising that there is no scientific body of data and/or theory delineating the factors that underlie telepresence. Our remarks on this topic thus make substantial use of intuition and speculation, as well as extrapolation from results in other areas.

Sensory factors that contribute to telepresence include high resolution and large field of view. Obviously, reduction of input information either by degraded resolution or restricted field of view will interfere with the extent to which the display system is transparent to the operator. Perhaps these two variables are tradable in the sense that the effective parameter in determining the degree of telepresence is the number of resolvable elements in the field, or, equivalently, for fields with uniform resolution over the field, Area of Field/Area of Resolvable Element. Also important, of course, is the consistency of information across modalities: the information received through all channels should describe the same objective world (i.e., should be consistent with what has been learned through these channels about the normal world during the normal development process). In addition, the devices used for displaying the information to the operator's senses in the teleoperator station should, to the extent possible, be free from the production of artifactual stimuli that signal the existence of the display. Thus, for example, the visual display should be sufficiently large and close enough to the eyes to prevent the operator from seeing the edges of the display (or anything else in the teleoperator station, including the operator's own hands and body). At the same time, the display should not be head-mounted in such a way that the operator is aware of the mounting via the sense of touch. Clearly, attempting to satisfy both of these constraints simultaneously is a very challenging task.

Motor factors necessary for high telepresence involve similar issues. Perhaps the most crucial requirement is to provide for a wide range of sensorimotor interactions. One important category of such interactions concerns movements of the sensory organs. It must be possible for the operator to sweep the direction of gaze by rotating the head and/or eyeballs and to have the visual input to the retinas change appropriately. This requires using a robot with a rotating head, the position of which is slaved to the position of the operator's head. The desired result can then be achieved in two ways, depending upon whether the system is designed to have the position of the robot's eyeballs (1) fixed relative to the the robot's head (e.g., pointing straight ahead) or (2) slaved to the position of the operator's eyeballs in the operator's head. In the first case, appropriate results can be obtained using binocular images that remain fixed relative to the operator's head position during eyeball scanning. In the second case, the positions of the projected images must be slaved to the position of the operator's eyeballs. If they were instead held fixed, then whenever the operator's eyeballs were rotated, the projected images would rotate. For example, if the operator's eyeballs were rotated to look at an object whose images were on the right side of the projection screens, the slave robot's eyeballs would rotate to the right, the images of the object in question would move to the center of the two screens, and these images would then be sensed to the left of the foveal region. In order to eliminate this problem, the projected images would also have to be rotated to the right. In other words, if the position of the robot eyeballs is slaved, the position of the projected images must also be slaved. To the best of our knowledge, no such system has yet been developed (although monitoring of operator eyeball position is being used to capitalize on reduced resolution requirements in the peripheral field in the pursuit of reduced bandwidth).

Another category of sensorimotor interactions that is essential for high telepresence concerns movements of viewed effectors. It must be possible for the operator simultaneously to move his/her hands (receiving the internal kinesthetic sensations associated with these movements) and see the slave robot hands move accordingly. Also, as with the sensory display, the devices used in the teleoperator station to detect and monitor the operators movements should, to the extent possible, be undetectable to the operator. The more the operator is aware of these devices, the harder it will be to achieve a high degree of telepresence. An amusing picture that is addressed to the issue of viewing one's own effectors, or more generally, one's own body parts, and that is of some historical interest, is shown in Figure 2 (Mach, 1914).

The most crucial factor in creating high telepresence is, perhaps, high correlation between (1) the movements of the operator sensed directly via the internal proprioceptive/kinesthetic senses of the operator and (2) the actions of the slave robot sensed via the sensors on the slave robot and the displays in the teleoperator station. Clearly, the destruction of such correlation in the normal human situation (in which the slave robot is identified with the operator's own body) would destroy the sense of presence.

In general, correlation will be reduced by time delays, internally generated noises, or non-invertible distortions that occur between the actions of the operator and the sensed actions of the slave robot. How these variables interact, combine, and trade in limiting telepresence and teleoperator performance is a crucial topic for research. In the next section we look more closely at the effects of one of these variables, namely, time delay.

Note also that telepresence will generally tend to increase with an increase in the extent to which the operator can identify his or her own body with the slave robot. Many of the factors mentioned above (in particular, the correlation between movements of the body and movements of the robot) obviously play a major role in such identification. Additional factors, however, may also be important. For example, it seems plausible that identification, and therefore telepresence, would be increased by a similarity in the visual appearance of the operator and the slave robot.

Figure 2.   Mach observing visible parts of his own body and the surroundings.

Finally, it is important to consider the extent to which telepresence can increase with operator familiarization. Even if the system is designed merely to transport rather than to transform, it will necessarily involve a variety of transformations that initially limit the sense of telepresence. A fundamental topic for research concerns the extent to which such limitations can be overcome by appropriate exposure to the system and development of appropriate models of the transformed world, task, self, etc. (through adaptation, training, learning, etc.). Figure 3 illustrates schematically how the internal dynamics of the operator are originally established and may be altered over time when interaction with the world is transformed. The representation (in brain) of the operator's interaction with the world is an important factor in the sense of presence. The operator identifies his or her own actions as such in accord with the concomitant sensory changes. Loss of such concomitance may reduce the sense of presence. But an updating of the internal model may promote the recovery of a lost sense of presence within that world. The figure shows how the motor command originating in the

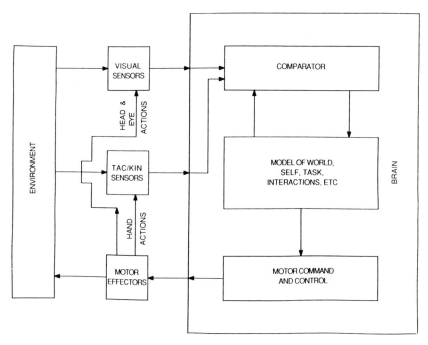

Figure 3. Information flow and feedback loops involved in actions of the operator in the environment.

central nervous system (CNS) activates the musculature which in turn causes sensory changes which feed back to the CNS. The comparator is designed to receive a feedforward signal from the internal model, which derives from past experience and anticipates the consequences of activity based upon that previous experience. That signal is then compared with the contemporary consequences of action. Any transform in the feedback loop will alter the expected feedback and be discrepant with the feedforward signal. In that event, the discrepant signal may be used to update the world model and lead to more accurate anticipations of action and an improved sense of presence.

Time delays and adaptation

Time delays between action of the teleoperator and the consequences of these actions as realized on the displays in the teleoperator station can arise from a variety of sources, including the transmission time for communication between the teleoperator station and the worksite and the processing time required for elaborate signal-processing tasks. Independent of the causes, it is clear that such feedback delays degrade both telepresence and performance. Research on the effects of time delays on manual tracking and remote manipulation and on methods for mitigating these effects are discussed in a variety of sources (e.g., Adams, 1962; Arnold and Braisted, 1963; Black, 1970; Ferrell, 1965, 1966; Johnsen and Corliss, 1971; Kalmus et al., 1960; Leslie, 1966; Leslie et al., 1966; Levison et al., 1979; Pennington, 1983; Pew et al., 1967; Poulton, 1974; Sheridan, 1984; Sheridan and Ferrell, 1963, 1967, 1974; Sheridan and Verplank, 1978; Starr, 1980; Wallach, 1961; Wickens, 1986). Of particular interest has been the development of systems that combat the effects of time delay through judicious supplementation of human teleoperation by automatic processing (involving predictive models and use of the human operator for supervisory control).

The particular effect of time delay on which we shall focus in the remainder of this paper is the effect on sensori-motor adaptation. As suggested at the end of the last section, the degree of telepresence that can be achieved with a given system depends ultimately on the extent to which the operator can adapt to the system.

Basic demonstration of adaptation was discussed by Helmholtz in his *Physiological Optics* (Helmholtz, 1962). In the typical experiment, the subject wears prism spectacles over his or her eyes which optically shift the apparent location of objects seen through them. When the subject reaches for a seen target without correction (open loop), the termination of his or her reach will obviously be in error by an amount approximating the apparent displacement of the target produced by the prism. Correction of a reach can be prevented in one of two ways. If the subject (S) is required to make a rapid ballistic movement of his or her hand to the target, the duration of hand travel is too short to allow correction. However, if both target and hand are visible at the termination of the reach, the error may be noted by S and subsequent reaches corrected. Alternatively, the target may be presented in a location where the hand may reach but not be seen. Following the initial measurements of reaching accuracy, the subject views either his or her hand or a surrogate for it through the prisms for a period of time called the exposure period. During that period he may or may not receive visual information concerning the error of the reaching. Following the exposure period a second measure is obtained of the accuracy of open-loop reaching for visible targets. The result is generally a decrease of error from that of the initial localizations in a direction which indicates correction for the presence of the prism displacement. Further open-loop measurements may be made with the prisms removed, in which case the error of reaching for a target *increases*. This increased error shows that the shift in localization is not dependent upon the presence of the prisms, but is a more generalized change in eye-hand coordination adaptive for the presence of the prisms.

Some sort of adaptive process occurs during the exposure period which compensates for the error introduced by the prism. Information available during the exposure period produces an update of the internal model of the visuospatial coordinates which are anticipated as the goal of reaching for the target. The nature of the necessary and sufficient information required for adaptation, and of the subsystems that actually adapt, has been the subject of much debate and experimentation (Welsh, 1978, 1986). It appears that any of a number of sources of information about the transformed relation between the seen position of the hand and its location as known through other information may serve to produce adaptation. One such source of information is the error seen when reaching for targets. When the reaching subject can see the error, he or she is bound to correct for it by a process of which he or she is usually quite conscious. Among other cognitive factors, knowledge of the optical effects of the prism may enhance adaptive responses. Active movement of the arm which produces visual feedback enhances adaptation, perhaps by sharpening the sense of position of bodily parts. More interesting from several points of view is the adaptive process which occurs during exposure when visible error feedback appears to be absent. For example, subjects adapt while looking through the prism at only a luminous spot fixed to the hand in an otherwise dark field. The spot moves with the hand, but when no other targets or even visible landmarks are present, there can be no explicit visible error. There may, however, be a discrepancy with the expectations based upon the concomitance of visual location of the hand with its non-visually sensed position. But this condition raises a further question. If the subject sees only a luminous spot on the hand as it moves, how does the nervous system identify this spot with the sensed positions of the hand? Aside from cognitive factors, we must hypothesize that the movements of the visible spot concomitant with the sensed movements of the hand allow this identification. The problem then becomes one of correlation between signals. Moreover, we recognize that this form of identification may well be a basis for establishing presence itself. This realization led to the following experiment.

The experiment concerns the effect of time delay on adaptation of eye-hand coordination to prism displacement. Changes in the seen position of the hand are delayed during a period of exposure between test and retest. For a given exposure, the delay is fixed, but over a series of exposures, the delay is varied. The question we asked was: what are the effects of delaying feedback by various amounts on the adaptive process that takes place during exposure with continuous monitoring by the subject of his or her hand movements in a frontal plane? In other words, how much is the effective correlation of identifying signals degraded by delay of visual feedback of varying intervals? In an earlier experiment (Held et al., 1966), we found that delays as small as 300 msec eliminated adaptation to prism displacement. Consequently, the following experiment incorporated delays of smaller magnitude.

As shown in Figure 4, the subject (S) stood at the apparatus. He positioned his head in a holder mounted on top of a light-proof box and looked down through an aperture into a mirror. The mirror reflected the image of a luminous spot, formed on a ground glass screen, which appeared on an otherwise dimly illuminated background. The image originated on an oscilloscope face and was focused on the screen. S's right hand grasped a

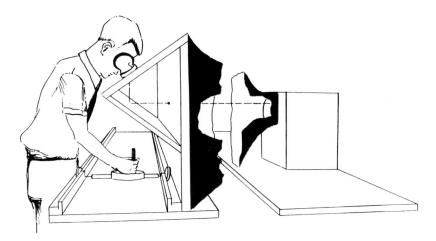

Figure 4. Experimental setup for studying adaptation to visual displacement and delay.

handle consisting of a short vertical rod located at arm's length beneath the box. The rod was attached to a lightweight roller-bearing arrangement which minimized inertia and friction but restricted hand movements to a region in the horizontal plane. When the hand moved the cursor, sliding contacts were driven along two linear potentiometers aligned at right angles to each other. This movement varied DC signals corresponding to the coordinates of the cursor on the horizontal surface. These signals were applied to the vertical and horizontal channels of the oscilloscope, thereby producing a single spot on the screen, the position and motion of which corresponded to that of the cursor. The optical system (lens and mirror) caused the spot to appear superimposed on the handle of the cursor when neither positional displacements nor temporal delays were introduced. The apparatus could be set to displace the spot 1.5 inch laterally to either the right or the left side. Temporal delays ranging from 20 to 1000 msec could be introduced in either the lateral or the vertical dimension, or both.

In addition to driving the trace by movements of the cursor, the loop could be opened and the trace spot set to display, one a a time, five stationary visible targets. The target coordinates were determined by applying paired X and Y voltages to the oscilloscope under the experimenter's control. Ss were instructed to set the handle of the cursor so that the top of the vertical rod felt superimposed on the visible target. Ss pressed a switch when they felt that the cursor was correctly positioned and the position was recorded.

Ss were 12 right-handed male college undergraduates with adequate vision and were naive as to the purpose of the experiments. Each S performed six runs separated by rest periods. Each run consisted of six steps:

(1) *Practice.* S was instructed to track the luminous spot with his eyes as he moved the cursor back and forth across the horizontal surface and to change the left-right direction of his hand movement with the beat of a metronome. This beat varied in a 60-sec cycle from 50 to 90 beats/min. Practice lasted a minute or two during which the subject traced the limits of movement of the cursor. He was instructed to avoid hitting the limiting stops during subsequent exposure and target localization, thus eliminating one potential source of information regarding the position of his hand on the surface.
(2) *Pre-Exposure Localization.* S was instructed to look at and localize the apparent positions of each of the five visual targets presented four times in a pseudo-random sequence. The moveable spot was extinguished prior to target presentations and the subject was instructed to move the cursor randomly about the surface before and between target presentations.
(3) *First Exposure.* S performed for 2 min as he did during the practice period. Both positional displacement and delayed visual feedback were introduced. One of six delay conditions, 0, 120, 150, 210, 330, and 570 msecs, was presented during each run. The six delays were presented to each S in a different order; half of the Ss were exposed to the spot laterally displaced in one direction (right or left) during this exposure and half with the same order of delayed conditions, but with the direction of displacement in the opposite direction.
(4) *First Post-Exposure Localization.* Identical to the pre-exposure localization.
(5) *Second Exposure.* Identical to the initial exposure, but with lateral displacement in the opposite direction.
(6) *Second Post-Exposure Localization.* Same as pre-exposure localization.

The results were analysed by taking the differences between the first and second post-exposure localizations as the primary measure of compensatory shift. These differences tend to be larger and more reliable than those between pre-exposure and post-exposure localizations (Hardt *et al.*, 1971).

Four experiments were performed. They were identical except for variations in the exposure procedure. In the first experiment, S tracked the hand-driven spot with his eyes as described above. In the second, S's eyes fixated a dim cross during exposure, thereby precluding tracking of the spot with the eyes. In the third, each S was trained to relax his arm while grasping the cursor and the experimenter moved the cursor in the manner discussed above (passive condition). The fourth experiment was identical to the second except that two shorter time delays were used, namely, 30 and 60 msec.

The S's mean compensatory shifts at various time delays are shown in Figure 5. The overall effect of delay in the first experiment (no fixation) is significant. All of the mean shifts are different from zero and all the shifts under delay are significantly less than the shift at zero delay. The results of

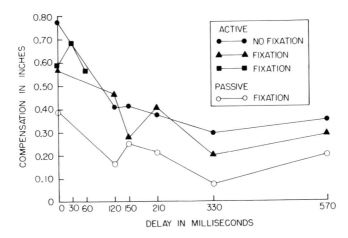

Figure 5. Results of experiments.

the second experiment (fixation) did not differ significantly from those of the first, showing that tracking the hand-driven target with the eyes was not a factor in promoting adaptation. While the passive condition of the third experiment reduced the overall level of adaptation, significant adaptation still occurred, and the overall shape of the curve with delay was similar to that of the active conditions. Finally, the effects of very short delays in the fourth experiment did not differ significantly from zero delay, although delays of 120 msec clearly do reduce adaptation. We conclude that delays must exceed 60 msec if they are to be sufficient to reduce adaptation significantly under the conditions of the experiment. For reasons we do not understand, the curves appear to asymptote at 30 to 40 per cent of compensation under zero delay.

It should also be noted that subjective impressions varied strongly with the delay. At the shorter delays (not too far above threshold), the viewed hand seems to be suffering simply a minor lag, as if it were being dragged through a viscous medium. At delays beyond a couple of hundred msec, however, the image seen becomes more and more dissociated from the real hand (i.e., identification, and therefore presence, breaks down).

In general, it is obvious that some degree of identification is necessary in order for adaptation to occur. Moreover, when adaptation occurs, it is obvious that identification increases. Thus, adaptation and identification (and therefore telepresence) must be very closely related. Note, however, that adaptation will fail to occur when either (1) no identification is possible or (2) identification is complete. Thus, tests of adaptation cannot, by themselves, be used to measure identification; other kinds of tests must also be included. Clearly, a precise characterization of the relations between adaptation, identification, and telepresence (or presence) requires further study.

Acknowledgements

The authors thank Hubert Dolezal for his help in performing the experiment reported.

References

Adams, J.L. (1962). An investigation of the effects of time lag due to long transmission distances upon remote control, Phase II, NASA TN-1351.
Akin, D.L., Minsky, M.L., Thiel, E.D. and Kurtzman, C.R. (1983). Space applications of automation, robotics, and machine intelligence systems, (ARAMIS)-Phase II, Vol. 3: Executive Summary, M.I.T., Contract NAS 8-34381, NASA Marshall Space Flight Center.
Arnold, J.E. and Braisted, P.W. (1963). Design and evaluation of a predictor for remote control systems with signal transmission delays, NASA TN D-2229.
Black, J.H., Jr. (1970). 'Factorial study of remote manipulation with transmission time delay', M.S. Thesis, Dept. Mechanical Engineering, M.I.T., Cambridge, MA.
Bolt, R.A. (1984). *The Human Interface*. Belmont, California: M.I.T. Lifetime Learning Publications (a division of Wadsworth, London).
Ferrell, W.R. (1965). Remote manipulation with transmission delay, M.I.T., NASA TN D-2665.
Ferrell, W.R. (1966). Delayed force feedback, *Human Factors*, **8**, 449–455.
Hardt, M.E., Held, R. and Steinbach, M.J. (1971). Adaptation to displaced vision: a change in the central control of sensorimotor coordination, *Journal of Experimental Psychology*, **89**, 229–239.
Held, R., Efstathiou, A. and Greene, (1966). Adaptation to displaced and delayed visual feedback from the hand, *Journal of Experimental Psychology*, **72**, 887–891.
Helmholtz, H. Von (1962). *Treatise on Physiological Optics (Vol. III)*, translated from the 3rd German (Ed., J.P.C. Southall), Dover, New York.
Johnsen, E.G. and Corliss, W.R. (1971). *Human Factors: Applications in Teleoperator Design and Operation*. New York: Wiley Interscience.
Kalmus, H., Fry, D.B. and Denes, P. (1960). Effects of delayed visual control on writing, drawing, and tracing, *Language Speech*, **3**, 96–108.
Leslie, J.M. (1966). Effects of time delay in the visual feedback loop of a man-machine system, NASA CR-560.
Leslie, J.M., Bennigson, L.A. and Kahn, M.E. (1966). Predictor aided tracking in a system with time delay, performance involving flat surface, roll, and pitch conditions, NASA CR-75389.
Levison, W.H., Lancraft, R.E. and Junker, A.M. (1979). Effects of simulator delays on performance and learning in a roll-axis tracking task, AFFDL-TR-79-3134, *Proceedings of 15th Annual Conference on Manual Control, Dayton, OH*.
Mach, Ernst (1914). *The Analysis of Sensations*. Chicago, Illinois: Open Court Publishing Company.
Pennington, J.E. (1983). A rate-controlled teleoperator task with simulated transport delays, NASA TM-85653.
Pepper, R.L. and Hightower, J.D. (1984). Research issues in teleoperator systems, *28th Annual Human Factors Society Meeting*, San Antonio, TX.
Pew, R.W., Duffenbach, J.C. and Fensch, J.C. (1967). Sine wave tracking revisited, *Transactions of the IEEE*, HFE-8, 130–134.

Poulton, E.C. (1974). *Tracking Skill and Manual Control*. New York: Academic Press.
Sheridan, T.B. (1984). Supervisory control of remote manipulators, vehicles, and dynamic processes: experiments in command and display aiding. In Rouse, W.B. (Ed.), *Advances in Man-Machine Systems Research*, (Vol. I, pp. 49–137). Greenwich, Conneticut: J.A.I. Press.
Sheridan, T.B. and Ferrell, W.R. (1963). Remote manipulation control with transmission delay, *Transactions of the IEEE*, Vol. HFE-4, 25–28.
Sheridan, T.B. and Ferrell, W.R. (1967). Supervisory control of manipulation, NASA SP-144, 315–323.
Sheridan, T.B. and Ferrell, W.R. (1974). *Man-Machine Systems*. Cambridge, Massachusetts: M.I.T. Press.
Sheridan, T.B. and Verplank, W.L. (1978). Human and computer control of undersea teleoperators, Department of Mechanical Engineering, M.I.T., Technical Report, Engineering Psychology Program, ONR.
Starr, G.P. (1980). A comparison of control modes for time-delayed remote manipulation, *Proceedings 16th Annual Conference on Manual Control*, M.I.T., Cambridge, MA.
Vertut, J. and Coiffet, P. (1986). *Robot Technology, Teleoperations and Robotics: Evolution and Development* (Vol. 3A) and *Applications and Technology* (Vol. 3B). Englewood Cliffs, New Jersey: Prentice-Hall.
Wallach, H.C. (1961). Performance of a pursuit tracking task with different time delay inserted between the control mechanism and display cursor, TM 12-61, U.S. Ordinance Human Engineering Laboratory.
Welch, R.B. (1986). Adaptation of space perception. In Boff, K.R., Kaufman, L., Thomas, J.P. (Eds), *Handbook of Perception and Human Performance*, Vol. I, Ch. 24. New York: John Wiley and Sons.
Welch, R.B. (1978). *Perceptual Modification, Adaptions to Altered Sensory Environments*. New York: Academic Press.
Wickens, C.D. (1986). The effects of control dynamics on performance. In Boff, K.R., Kaufman, L., Thomas, J.P. (Eds), *Handbook of Perception and Human Performance*, Vol. II, Ch. 39. New York: John Wiley and Sons.

15

Multi-axis control in telemanipulation and vehicle guidance

G.M. McKinnon[1] and R.V. Kruk[2]

CAE Electronics Ltd.
Montréal, Canada

This paper describes the development of a family of multi-axis hand controllers for use in telemanipulator systems and vehicle guidance. Experience in the control of the Shuttle Remote Manipulator System (SRMS) arm is reviewed together with subsequent tests involving a number of simulators and configurations, including use as a side-arm flight control for helicopters. The factors affecting operator acceptability are reviewed.

The success of in-orbit operations depends on the use of autonomous and semiautonomous devices to perform construction, maintenance and operational tasks. While there are merits to both fully autonomous and man-in-the-loop (or teleoperated) systems, as well as for pure extravehicular activity (EVA), it is clear that for many tasks, at least in early stages of development, teleoperated systems will be required.

The design of control input devices for complex multi-axis tasks includes many considerations. A primary and overriding concern is that the configuration and orientation of the input device must coincide in a logical and consistent way with those of the task. Any feedback signals, to be effective, must be easily interpreted, should respect a single, logical, coordinate system, and should be in phase with the controlled system within the limits of human perception. While these principles appear simple and obvious, in reality they are often compromised.

A typical robotic task involves six degrees of freedom: three translations and three rotations. Control can be achieved either by end point control, essentially piloting the end point of the manipulator, or through joint by joint control. End point control, of course, implies sufficient intelligence in the control electronics to convert control inputs into joint commands and to provide feedback loops to ensure that the joint commands generated result in

---

[1] Director, R&D. (Dr. McKinnon died in autumn of 1988.)
[2] Group Leader, Human Factors.

desired movements and orientations of the end effector. The robot may include redundancy so that the number of joints controlled exceeds the degrees of freedom of the manipulator. Redundancy of this sort provides the advantage of alternate solutions for a single end point location and orientation so that collision avoidance is possible. End point control is analogous to the stability augmentation interfaces used to improve the handling qualities of aircraft controls. If end point control is used to control manipulators with redundant degrees of freedom, control algorithms will be required based on avoidance of contacts, energy optimization or other concepts. These should be transparent to the operator.

To employ a robot in a telemanipulator mode requires the design of a man-machine interface with primary control achieved, conventionally, through the use of hand controls. An initial step in configuring the man-machine interface is to define the relationship between the input device and the task. Numerous possibilities can be conceived but typical solutions are generally variants of the following:

(1) Joint by joint control with one control input for each joint.
(2) Multi-axis joysticks, one for each hand, used to control the manipulator end point. One control may be used for three axis position control and the second to control orientation.
(3) A single multi-axis control to control both position and attitude of the end effector.
(4) A master/slave system where a replica, or master, with similar geometry is used to control a manipulator or slave. The replica may be scaled depending on physical constraints. The term telepresence implies a master/slave system with a scale factor of unity. Exoskeletal controllers are a variation of this approach.

In addition to the number and function of control input devices, several possibilities exist for mode of operation. In some applications alternate modes can be selected, possibly using a mode switch or selector mounted on the input device. Commonly used modes include direct position control, a standard mode for master/slave systems often augmented by an indexing arrangement to avoid sustained uncomfortable positions, resolved rate control, and on-off or bang-bang control. Resolved rate or bang-bang control can be implemented using an isometric or force stick or an input device which includes compliance or displacement. If displacement is used, the handling characteristics in each axis must be comfortable for the operator and suitable for the task. Handling characteristics include break out, gradient, travel, damping and definition of hard and soft stops in each axis. Isometric control provides a solution which is mechanically elegant. However, the loss of kinaesthetic feedback results in a tendency to over control, particularly in stressful situations.

To facilitate control, various forms of feedback may be provided by a direct view, often with a limited viewing angle from a fixed position, through fixed cameras, or through cameras mounted on the manipulator. Cameras may have slew or zoom capability. Three dimensional displays have been proposed for some applications. Feedback may also be provided by force cues either by direct force feedback in a master/slave configuration or by tactile, aural or visual indicators (Hannaford, 1987).

A fundamental issue in the design of the man-machine interface is the use of visual feedback. Visual feedback is normally the primary source of feedback information. Ideally, an unobstructed direct view of the workspace should be provided but even if this is available the question arises as to the ideal eye point. In an anthropomorphic system the ideal viewpoint may be considered to be near the head location of the pseudo-person. For vernier tasks or due to obstructions other views may be required. For dextrous tasks, depth perception is required. This can be achieved by stereoscopic displays, by multiple orthogonal displays, to a certain extent by two dimensional displays which include parallel information or, in some cases, through the use of visual aids or markers for specific tasks. In space applications in particular, the problem is often complicated by limited available transmission bandwidths and by inherent transmission delays.

In the authors' experience, the fundamental principle for visual feedback is that the coordinates of the display must correspond in a meaningful way to the principal axes of the control input device and to the motions of the manipulator end effector. In the case of dextrous tasks involving multiple manipulators the selection of viewpoints requires careful consideration. To switch from a fixed viewpoint to a camera mounted on the manipulator requires mental gymnastics from the operator. Interpretation of two different viewpoints simultaneously may substantially increase operator workload. Systems have been constructed with as many as five simultaneous feedback displays.

One option available to the designer of a control system is to resort to mode control. This technique has many valid applications and is accepted, of course, in advanced flight control systems. In the case of robotic systems, mode changes may be desirable in many applications, however, some basic concerns must be addressed. In general, the requirement for mode changes depends on the specific task. A typical requirement in space is to break tasks into a gross positioning mode followed by a vernier or dextrous mode. In some cases the manipulator may in fact be a small dextrous manipulator mounted on a larger positioning system. In the gross positioning mode, the operator may prefer to use a "God's eye view", or a view from the end of the manipulator so that he can "fly" the manipulator or, perhaps, a view from the eventual workspace so that he can guide the manipulator toward it. Regardless of the viewpoint, effective control depends on a clear understanding of the geometry of the situation by the operator and a natural relationship between the axes displayed and the principal axes of the controller. The negative effects of axis misalignment in such cases have been documented (Lippay et al., 1981).

At the transition from gross positioning to dextrous or vernier mode, it is essential that the operator be clearly aware of changes in coordinate systems or viewpoints. Again, the coordinate system definition of the control input should be clearly related to that of the task.

In some robotic systems, the use of force feedback has proven useful in providing task related cues. Force feedback is usually associated with master/slave devices operating in position control mode, and permits the operator to sense force limits, contact with rigid objects, loads on the end effector, etc. The implementation of effective force feedback is difficult from a mechanical point of view since typical robotic systems involve abrupt non-linearities, hysteresis, dead zones, friction, as well as flexibility which impose severe constraints on the fidelity of force feel which can be achieved. Even fairly crude force feedback can provide essential cues for some tasks such as mating connectors or sensing interferences. Effective implementation is also hampered in space applications, where power is at a premium and inadequate to resist an attempted input by the operator. In many space applications significant lags due to data transmission severely limit the effectiveness of force feedback. With lags in excess of 100 msec, force feedback becomes a deterrent to effective control on a continuous basis although an operator may be trained to use delayed information to detect limits or contact. The delays of multiple seconds anticipated for remote control of space based robotics systems clearly limit the applicability of force feedback, particularly where the delay period varies randomly (Martin-Marietta Denver Aerospace, 1987).

One solution to the transportation lag problem is to provide the operator with a responsive model of the remote system, that is a model which replicates system dynamics but eliminates delays due to communication of data (Bejczy and Handlykon, 1981). The operator controls displays of the model to perform required tasks and the resulting control signals are transmitted to the remote manipulator which will perform the same motions after the transmission delay. In this case force feedback can be implemented in the predictive model to the extent that reliable task based information is accessible. Such a system can improve performance in dextrous tasks but requires careful design.

A further key design objective in the implementation of human-machine interfaces for space is that of standardization. Astronauts should naturally and comfortably interpret their input motions in terms of motions of the manipulator or task. This "transparency" is achieved by careful design to ensure that task coordinates and views are always presented in a clear, unambiguous and logical way, and by ensuring that standardized input devices are used in standardized modes. If conventions are established and systematic modes of control are respected, training time is reduced and effectiveness and performance are improved. The end objective in the design of displays and controls for telemanipulators is to establish a "remote presence" for the operator.

The SRMS system

A number of manual control input devices have been used in space over the years. For the most part these devices were designed as flight controls for the various satellites and modules which have flown. The first truly robotic control device was that used on the SRMS or CANADARM system of the Space Shuttle. The control interface in this case consisted of two three-degree-of-freedom devices used in conjunction with a displays and controls panel, CCTV visual feedback from cargo bay and arm-mounted cameras, augmented by limited direct viewing. A Translational Hand Control (THC) allowed the astronaut to control the end point of the arm in the three rectilinear degrees of freedom with the left hand, and a Rotational Hand Control (RHC) was used in the right hand to control rotational degrees of freedom (Lippay, 1977).

The THC was designed specifically for the SRMS application by CAE Electronics, while the RHC was a modified version of the Shuttle flight control produced by Honeywell. The geometry and overall configuration of the RHC was thus predetermined and was not matched to the task. The device does not have the single centre of rotation advantageous in generalized manipulator control. The RHC differed from the flight control version in several ways:

(1) The forces and travels were modified to reflect task requirements.
(2) Auxiliary switches and functions were changed to comply with task requirements. In fact all auxiliary switches were located on the RHC-COARSE/VERNIER, RATE HOLD and CAPTURE RELEASE.
(3) A switch guard was added to CAPTURE/RELEASE to prevent inadvertent release of a payload.
(4) Redundant electronics were eliminated in view of the reduced level of criticality.

The THC differed from the RHC in that it incorporated rate-dependent damping through the use of eddy current dampers driven by planetary gears. A hand index ring was added to the THC after initial evaluations of prototype units. The ring provided a reference for position and led to the use of the device as a fingertip control, whereas the RHC with its larger hand grip was clearly a hand control. Force levels and gradients on the THC were low, and the rate dependent damping enhanced the smooth feel of the device. The x and y inputs of the THC were not true translations, but an effort was made to optimize a linkage in the available space to reduce the curvature due to a displaced pivot point.

The SRMS system has proven to be operable but is less than optimal. Required tasks must be carefully programmed and even an experienced operator requires training and practice for any task requiring coordinated or dextrous motions. Several modes can be selected; vernier, rate hold, etc. and views can be selected from various camera locations. In general, a resolved

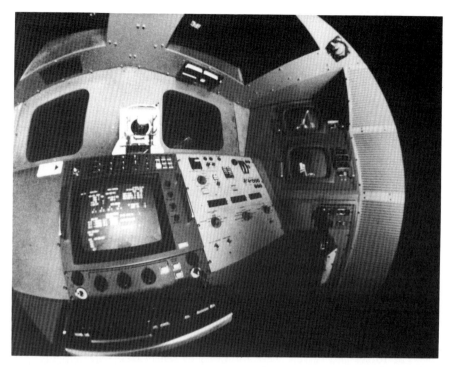

Figure 1. Fish-eye view of SIMFAC, the control console in the aft of the Space Shuttle cockpit where the translational (THC, left center next to the right window) and rotational hand controllers (RHC, below the small CRT displays on the rignt) for the robot arm are mounted.

rate mode of control is used with a consistent correlation between controller axes and task coordinates. The resolved rate coordinate system is selectable, and in general, a shuttle axis-based coordinate system is used for "out the window" operation, whilst the arm-wrist axes are used in conjunction with the wrist camera, thereby retaining a direct correlation between command and observed coordinates. Some automated sequences are available and joint control can be effected as a fallback mode in the event of a system failure.

With training, astronauts can become proficient in performing required tasks. In general, however, the tasks must be carefully programmed and significant training and practice is required before an astronaut feels comfortable with the system. Even with training, the skill of the astronaut is still a limiting factor on system capability. Tasks requiring coordinated or dextrous motions are difficult to achieve.

While there is no hard data to compare alternatives, the shortcomings of the SRMS design in part can be attributed to the limitations of the RHC and THC described above, but mainly to the unfortunate location of the two hand controls and lack of direct correspondence between the axes of the controls and those of the visual displays.

The SRMS system incorporated no force-reflective feedback aside from indications of motor parameters from each joint. Positional feedback of the end point is strictly visual—either direct viewing or through CCTV. The axes of the presented display depend on the view selected: direct, cargo bay or arm-mounted camera. Control is in the resolved rate mode. In the case of a largescale arm such as the SRMS, a master slave or indexed position mode is not suitable because of scaling problems.

Figure 1 shows a simulation of the SRMS Displays and Controls System in SIMFAC. The RHC is located to the lower right of the D&C panel and a breadboard model of the THC to the upper left. The CCTV displays are to the right and the direct viewing ports are overhead and immediately above the D&C panel.

Multi-axis study

Following the design of the SRMS system, a study of multi-axis controls was conducted (McKinnon and Lippay, 1981; Lippay et al., 1981). The purpose of the study was to determine the feasibility of controlling six degrees of freedom with a single hand control. According to the guidelines laid down for the study, mode changes were to be avoided so that coordinated control was required simultaneously in all axes. No specific application was defined; however, the controller was to be usable either to fly a spacecraft or to "fly" the end point of a manipulator.

The study included a review of the literature, observation of available multi-axis controllers, and discussions with experts. Although a prototype device was not required by the contract, one was assembled. Interestingly, the consensus of opinion at the time amongst the knowledgeable community was that coordinated control in six axes was desirable, but probably not feasible.

A number of six-degree-of-freedom controls were reviewed. The most notable were devices with force feedback operated in the indexed position mode. A prototype laboratory version was developed by R. Skidmore at Martin Marietta and evaluated in various dynamic and graphic simulations. A similar design and evaluation was done at Jet Propulsion Laboratories by A. Bejczy (Bejczy and Handlykon, 1981). These devices were both unsuitable in design for implementation in a mature control system, but permitted laboratory evaluation of force characteristics, displacements, and interactions with visual feedback. Another approach was developed by D. Whitney at the Draper Laboratory. This was elegantly designed from the mechanical viewpoint, but difficult to use due to the absence of tactile feedback.

This study uncovered no mature or workable concept for a six-degree-of-freedom controller and a lot of skepticism amongst practitioners as to the feasibility of implementing more than four degrees of freedom. A more recent study of hand controls was done by Brooks and Bejczy (1985).

Development process

At the conclusion of the study, in spite of the climate of skepticism, it was felt that there was no reason why a well-coordinated, six-degree-of-freedom controller could not be designed. Experiments with a variable-geometry test rig demonstrated that the only way to avoid inherent cross-coupling between axes, achieve the ability to make discrete inputs where required, and still have a direct correlation between control inputs and resulting action was to center all axes at a single point positioned at the geometric center of the cupped hand. In this way, control of the end point related to hand motions. Alignment of controller axes in a logical way to the axes of visual displays was also considered essential.

One initial concern was the issue of isometric (purely force) versus displacement control. An isometric controller is rugged and easily constructed from a mechanical standpoint. Unfortunately, the concept leads to overcontrol, particularly in stressful situations, because of the lack of proprioceptive indication of input commands. In some situations operators tend to saturate the controller to the extent that they quickly suffer fatigue. While there may be tasks in which isometric control is adequate and acceptable, in general the addition of displacement with suitable breakout gradients and hard-stop positions improves performance. For this reason, most manual controls designed on the isometric principle have been modified to include compliance.

Our initial designs were based on the use of force transducers to generate input signals. The controls were designed to allow for the inclusion of compliance and adjustable force characteristics, although the device could also be configured for isometric operation in all axes. It was quickly established that some compliance was advantageous. Since there was always significant displacement, the force transducers were replaced by position transducers, thus permitting the use of rugged, compact, noncontact, optical position sensors and eliminating the tendency to generate noise signals due to vibration or shock. In addition, a purely position system made it easier to eliminate cross-coupling between axes when pure motions in a single axis were required.

An intermediate step of isometric translational axes and displacement in rotation, a so-called "point and push" approach, was unsuccessful because of the problems described above in the isometric axes.

In the final analysis, a prototype design was constructed which included significant displacement in all six axes. The prototype unit is shown in Figure 2.

Prototype design

The design concept was to ensure that all six axes pass through a single point. The mechanical components and transducers for the rotational axes

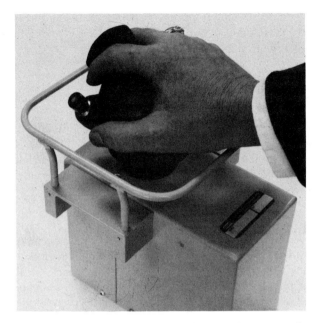

Figure 2. Prototype six-degree-of-freedom hand controller.

were mounted within a ball. The ball in turn was mounted on a stick which was free to translate in three mutually orthogonal axes. All axes had appropriate breakout forces, gradients and stop-force characteristics generated by passive components. The output of the device was a position signal sensed by optical transducers. No additional rate-dependent damping was included. While rate-dependent damping does enhance the "feel" of the controller, the additional mechanical complexity may not be justified.

The relationship between breakout forces and gradients is task-dependent. In general, the breakouts should be sufficient that pure inputs can be generated easily in a single axis; however, breakouts do have a negative impact on controllability for small simultaneous coordinated movements in multiple axes.

Various handgrip shapes were investigated, but with the emergence of the coincident axis concept as previously described, there was a fundamental need to provide a face perpendicular to the direction of commanded motion. The other prime requirement was a shape which ensured the correct positioning of the hand relative to the geometric center of the system. The natural solution was a sphere. As development of the mechanism and sensing systems progressed, the ball size was reduced to its present configuration. This approximates to the size of a baseball, and has shown to be comfortable for bare-handed, gloved, and pressure-suited operation.

Several derivatives of the basic design evolved for special applications. A bang-bang device was configured for tests on the MMU simulator. A four-axis (three rotations on a vertical purely rate-dependent damped linear axis) model was evaluated for flight control in helicopters. In some configurations a protuberance was added to provide a tactile cue for orientation. Auxiliary switches were added on this protuberance.

Test and evaluation

To date a number of tests have been carried out. It is difficult to compare data between tests since different tasks and performance metrics were used. In general, though, subjective ratings and measures of performance were consistent and some basic design principles were established. The tests performed were as follows:

Johnson Space Flight Center

Initial tests were performed using the controller to control computer graphic representations of docking tasks.

Subsequent tests were also made using the full-scale mockup of the SRMS arm (MDF). Comparisons were made between the conventional SRMS (two three-degree-of-freedom controllers) configuration and the single six-axis device. NASA human factors personnel, technicians and astronauts participated in the tests.

The most striking feature of many of these tests was achievement of adequate levels of control within minutes with a single six-axis device whereas in some cases equivalent levels with 2 × 3 axis devices requires weeks of training.

Martin Marietta

The controller was evaluated with computer graphics representations of docking maneuvers.

Astronaut evaluations of a bang-bang configuration were done on the MMU simulator. Tests were performed for operation in pressurized space suits, as shown in Figure 3.

Unfortunately, we do not have access to hard data on those trials.

Marshall Space Flight Center

A six-axis controller was used to control a six-jointed arm shown in Figure 4. The system has been operated from 1985 with a variety of operators and tests (e.g., Martin-Marietta Denver Aerospace, 1987).

Grumman

Tests were carried out using two six-degree-of-freedom controllers using resolved rate algorithms to control two six-degree-of-freedom dextrous

Figure 3. MMU tests at Martin-Marietta.

Figure 4. Control of robot at Marshall Space Flight Center.

manipulators as shown in Figure 5. Comparisons were done with master/slave control in the same environment. The visual was direct view in all conditions and the tasks included disassembly of a nut and bolt and maneuvering of a ring along a track. Conditions were master/slave with force feedback (MSFF), master/slave with no force feedback (MS-FF) and six degree-of-freedom handcontroller with no force feedback (HC). In the disassembly task, mean time to complete the task was significantly faster with MSFF (106.2 sec) than either MS-FF (160.2 sec) or HC (190.1 sec). In the

Figure 5. Simultaneous control of two arms.

track maneuvering task completion of the task, as shown in Figure 6, was significantly faster with the master/slave controller versus the handcontrollers, but, as shown in Figure 7, errors with the master/slave controllers were significantly greater than with the handcontrollers (McKinnon et al., 1987). The Grumman report of the study (Grumman Space Systems, 1987) indicates that master/slave controllers are much faster, but that handcontrollers are better for control of fine motions (i.e., more accurate). As well, operation of handcontrollers was reported as considerably less fatiguing than master/slave systems. However, the Grumman report also expresses some concern about possible problems of lack of force feedback in occasions whereby two arm tasks are conducted in coordination in constrained conditions. It was felt that without force feedback large forces could be developed on constrained objects, unbeknownst to the operator—this could result in damage if the constraints were suddenly released. The issue of the importance of force feedback in teleoperation tasks remains controversial.

National Aeronautical Establishment—Flight Research Laboratory

Four-axis versions of the design have been developed, installed and flown in a variable stability helicopter as shown in Figure 8. Over a period of three years, three formal studies were conducted. In addition to system data, Cooper-Harper ratings for task performance were recorded. The primary tasks in all three studies were precision hover tasks, including landing, take–off, off–level landings, hover turns, acceleration and quickstops. In the third study, NOE (NAP of the EARTH) and cruise flight were added. Table

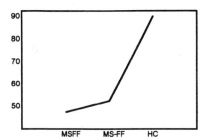

Figure 6. Track maneuvering time, in seconds.

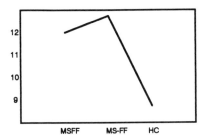

Figure 7. Track maneuvering errors.

Figure 8. Four-degree-of-freedom controller installed in a helicopter.

1 shows a summary of Cooper-Harper ratings across the ranges encountered in the trials.

In the first study (Lippay et al., 1985) the control configuration of the aircraft (a variable stability BELL 205A helicopter) was primitive (minimal augmentation to remove cross-coupling) with rate damping. The intent was to compare performance with conventional controls (cyclic and collective) to a 4 DOF isometric controller; to a CAE 4 DOF displacement controller with a ball (teleoperation) grip; and to a CAE 4 DOF displacement controller with a vertical (stick) grip. Table 2 shows the results. Taking the conventional control result as baseline it can be seen that the force controller and CAE controller with ball grip significantly degraded performance, whereas the CAE controller with stick grip produced results similar to conventional controls.

In the second study (Kruk et al., 1986), precision hover tasks were conducted with conventional controls and a further developed version of the CAE controller with vertical grip. In this study as well, more advanced control system augmentation was used (roll and pitch axes) while the heave (collective) axis was direct drive with variations in damping. Table 3 shows the results. Our conclusions from these results were that a four-axis displacement controller could produce performance equivalent to conventional controls across a variety of levels of augmentation and that fully satisfactory performance could be achieved.

In the third study, four U.S. Army test pilots evaluated the CAE four-axis controller in a variety of augmentation conditions: (1) unaugmented, undamped, with the aircraft axes fully coupled (e.g., power increases would affect all axes), (2) rate damped (similar to study 1), (3) rate command (4) attitude command. The aircraft was flown in NOE as well as cruise and hover flight. Table 4 shows the results. In this study it was demonstrated that although operational tasks could not be achieved with zero augmentation, the aircraft could be flown safely to landing without augmentation in all axes (albeit with extremely high pilot workload). With control augmentation, the controller provided satisfactory performance, although different levels of augmentation (not necessarily the highest) appeared to be optimal for specific tasks. In sum, the results of a series of studies evaluating the CAE controllers at the Canadian National Research Council's Flight Research Laboratory indicate that a displacement sidearm controller could replace all current conventional controls in helicopters.

It should be noted that, in the case of the four-axis version, the use of a relatively conventional handgrip superimposed on the ball was possible while respecting the principle of a single centre. The addition of another translation axis with a similar handgrip would likely introduce cross coupling.

Canadian space program (space station)

The Canadian component of Space Station Freedom is the mobile servicing Center (MSS). Concerned with the weight, space and power consumption

Table 1. Cooper-Harper ratings

Number	Rating	Level of pilot compensation required to achieve desired performance	
2	Good–Desirable	Not a factor	Satisfactory without improvement
3	Fair–Some mildly unpleasant characteristics	Minimal	
4	Minor but annoying shortcomings	Moderate	Shortcomings warrant improvement
5	Moderately objectionable shortcomings	Considerable	
6	Very objectionable shortcomings	Extensive	
7	Major deficiencies	Not attainable	Deficiencies require improvement

Table 2. Mean Cooper-Harper ratings, study 1

Conventional Controls	3.3
Isometric (force) Controller	4.7
CAE Controller—Ball Grip	4.1
CAE Controller—Stick Grip	3.6

Table 3. Cooper-Harper ratings, study 2

Control System	Heave Axis Damping	Conventional Controls	CAE 4 axis Controller
Rate command	underdamped	4.5	4.0
Rate command	moderately damped	3.0	3.0
Attitude command	zero damping	3.0	4.0
Attitude command	highly damped	2.0	2.0

Table 4. Mean Cooper-Harper ratings, study 3

Flight Regimes	Control Systems			
	Unaugmented	Rate Damped	Rate Command	Attitude Command
Precision hover	6.3	3.3	3.1	3.4
NoE	not flown	3.0	3.3	4.0
Cruise	5.5	3.0	2.5	3.0

penalties associated with master-slave control systems, (e.g., Martin-Marietta Denver Aerospace, 1987) the approach for manipulator control in this system is to evaluate the effectiveness of a variety of handcontroller configurations. Current trials at SPAR Aerospace of Toronto (prime Canadian space station contractor) involve comparisons of performance in manipulation of a high fidelity model of the SRMS in SIMFAC (Figure 1) between the two 3 DOF handcontrollers for the SRMS and a current version of the CAE 6 DOF handcontroller. The participants in these trials include NASA astronauts and the tasks include track and capture of free flying objects, payload berthing, and payload unberthing.

Discussion

Tests conducted to date have demonstrated that multi-axis control using a single hand is not only feasible, but, providing certain design guidelines are respected, preferable to approaches in which axes are distributed amongst separate controllers. Statements to the effect that six degrees of freedom is too much for one hand ignore the fact that the humans have the ability to make complex multi-axis movements with one hand using only "end point" conscious control. The coordinate transformations required are mastered at an early age and the inverse kinematics are resolved with no conscious effort on the part of the operator to separate translational and rotational components, which would increase his or her work load. The operator requires considerable training and practice with a 2 × 3 axis system before achieving the same level of control as is immediately possible with the single six-axis device. NASA experience has shown that the weeks of training necessary for the former can become less than 30 sec for the latter. While the guidelines have been verified only in specific environments for specific tasks, the authors feel confident in making the following statements:

(1) A proportional displacement controller will provide improved performance and in many cases more relaxed control than an isometric device. Performance with isometric devices varies more between individual subjects than that with displacement control.
(2) Force gradients and characteristics should be correlated to the task being performed. There may be a justification for standardizing force characteristics and controller configurations for all space-related equipment to ensure commonality and to reduce training requirements.
(3) An obvious and consistent orientation between controller axes and those of visual feedback displays is essential. This is an area where standardization between tasks and systems is a key element. A single controller design would be suitable for all applications, provided that basic axis orientation and control mode standards are maintained.
(4) The use of force-reflecting feedback has not been evaluated by the

authors, although a program is under way to investigate some novel approaches. In general, direct force feedback is useful only in a system with high mechanical fidelity. In the presence of abrupt nonlinearities such as stiction or backlash and particularly transport lag in excess of 100 msec, force feedback can in fact be detrimental.

(5) For some tasks with some manipulators, a master/slave system can provide equal or superior performance to that of a manual control in resolved-rate mode. Resolved rate is, however, universally applicable and can provide a standardized approach for virtually all manipulator or flight-control tasks.

(6) There are specific problems for the operator in the zero-g environent. Sufficient work has been done to demonstrate the importance of adequate reference support, such as armrests and index rings similar to that used on the translational controller of the Shuttle SRMS. The more delicate the task, the more important this consideration becomes.

(7) In tasks in which lag exceeds 1 sec, it may be assumed that real-time interactive control in the strict sense is not feasible. Providing physical relationships are stable or static, a reconstructive mode using generated graphics for a "rehearsal" of manipulator movement may be used, stored in memory, then activated. When lags are 100 msec or less, resolved-rate control may be used directly to control the end-effector (position control is inadequate when any substantial excursion may be required). The lag regime between 100 msec and 1 sec causes difficulty because there is a tendency to compensate for delay or system instability (e.g., arm-flexing modes) with more complex drive and "prediction" algorithms. Our experience thus far is that the simplest control algorithm which permits stable response generally provides the best performance.

In a variety of applications, multiple-degree-of-freedom controllers can be used naturally and effectively, to control tasks requiring high levels of dexterity and coordination.

References

Bejczy, A.K. and Handlykon, M. (1981). Experimental results with a six degree of freedom force reflecting hand controller, *Proceedings of the Seventeenth Annual Conference on Manual Control*, Los Angeles, June.

Brooks, T.L. and Bejczy, A.K. (1985). *Hand Controls for Teleoperation*, 85-11, March. California Institute of Technology, Pasedena: VPL Publication.

Grumman Space Systems (1987). *Telerobotic Work System*, Report No. SA-TWS-87-R0002 in response to NAS9-17229.

Hannaford, B. (1987). Task level testing of the JPL-OMV smart end effector, *Proceedings of the Workshop on Space Telerobotics*, Pasadena.

Kruk, R.V., Runnings, D.W., King, M., Lippay, A.L. and McKinnon, G.M. (1986). Development and evaluation of a proportional displacement sidearm controller for helicopters, *Proceedings of the Human Factors Society 30th Annual Meeting*, 865–869. Dayton, Ohio.

Lippay, A.L. (1977). Multi-axis hand controller for the shuttle remote manipulator system. *Proceedings of the Thirteenth Annual Conference on Manual Control*, Cambridge, Mass.

Lippay, A.L., King, M. and Kruk, R.V. (1985). Helicopter flight control with one hand, *Journal of the Canadian Aeronautics and Space Institute*, **31** (4), 333–345.

Lippay, A.L., McKinnon, G.M. and King, M.L. (1981). Multi-axis hand controllers, a state of the art report, *Proceedings of the Seventeenth Annual Conference on Manual Control*, Los Angeles, June.

Martin-Marietta Denver Aerospace (1987). *Telepresence Work Station Definition Study*, Report No. MCR-86-528 in response to NAS9-17230.

McKinnon, G.M. and Lippay, A. (1981). *Bibliography on Manual Controllers and Human Factors*, CAE Electronics Ltd.

McKinnon, G.M., King, M.L. and Runnings, D.W. (1987). Co-ordinated control of multi-axis tasks. *Annual IEEE Conference on Robotics and Automation*. Raleigh, North Carolina.

16

Visual enhancements in pick-and-place tasks: human operators controlling a simulated[1] cylindrical manipulator

Won S. Kim, Frank Tendick and Lawrence Stark

University of California
Berkeley, California

Summary

A visual display system serves as an important human/machine interface for efficient teleoperations. However, careful consideration is necessary to display three-dimensional information on a two-dimensional screen effectively. A teleoperation simulator is constructed with a vector-display system, joysticks, and a simulated cylindrical manipulator in order to evaluate various display conditions quantitatively. Pick-and-place tasks are performed, and mean completion times are used as a performance measure. Two experiments are performed. First, effects of variation of perspective parameters on a human operator's pick-and-place performance with monoscopic perspective display are investigated. Then, visual enhancements of monoscopic perspective display by adding a grid and reference lines are investigated and compared with visual enhancements of stereoscopic display. The results indicate that stereoscopic display does generally permit superior pick-and-place performance, while monoscopic display can allow equivalent performance when it is defined with appropriate perspective parameter values and provided with adequate visual enhancements. Mean-completion-time results of pick-and-place experiments for various display conditions shown in this paper are observed to be quite similar to normalized root-mean-square error results of manual tracking experiments reported previously.

---

[1] This work was supported in part by the Jet Propulsion Laboratory, Contract No. 956873 (Dr. A.T. Bejczy, Technical Monitor), and by the NASA Ames Research Center, NCC 2-86 Cooperative Agreement (Dr. S.R. Ellis, Technical Monitor).

Introduction

Visual display systems serve as an important human/machine interface for efficient teleoperations in space, underwater, and in radioactive environments (Johnsen and Corliss, 1967, 1971; Heer, 1973; Vertut and Coiffet, 1986). Closed-circuit television systems, presenting two-dimensional (2D) images captured by remote video cameras, have been commonly used for these visual displays. As technology evolves from manually controlled teleoperations to sensor/computer-aided advanced teleoperations (Bejczy, 1980; Vertut et al., 1984) or telerobotics, (Ferrell and Sheridan, 1967; Bronez et al., 1986; Montemerlo, 1986; Clarke and Bronez, 1986) graphics displays have been drawing attention as a means to provide an enhanced human/machine interface. A graphic display can present an abstract portrayal of the working environment or state of the control system based on sensor signals and a data base (Johnsen and Corliss, 1971; Leinemann and Schlechtendahl, 1984). A force-torque display (Bejczy et al., 1982) and a "smart" display (Bejczy et al., 1980) are examples of graphic displays developed for efficient teleoperations.

There are two types of visual displays: monoscopic and stereoscopic. The stereoscopic display provides two slightly different perspective views for the human operator's right and left eyes. A stereoscopic view enables the human to perceive depth by providing a distinct binocular depth cue called stereo disparity. Some earlier studies with television displays showed that stereoscopic displays, as compared to monoscopic displays, did not provide significant advantage in performing some telemanipulation tasks (Kama and DuMars, 1964; Rupert and Wurzburg, 1973; Freedman et al., 1977). Careful recent studies (Pepper et al., 1981; Pepper et al., 1983), however, indicated that stereo performance was superior to mono under most conditions tested, while the amount of improvement varied with visibility, task, and learning factors. These results showed that the advantage of the stereoscopic television display became pronounced with increased scene complexity and decreased object visibility.

Monoscopic and stereoscopic graphic displays were recently compared by employing three-axis manual tracking tasks (Kim et al., 1985; Ellis et al., 1985; Kim et al., 1987). Root-mean-square (rms) tracking error was used as a performance measure for quantitative evaluation. Results were consistent with previous television display results, indicating that stereoscopic graphic displays did generally permit superior tracking performance, while monoscopic displays allowed equivalent performance when they were defined with appropriate perspective parameters and provided with adequate visual-enhancement depth cues such as reference lines.

The purpose of our present study is to examine generality or consistency of the above results. A three-axis pick-and-place task, instead of the three-axis manual tracking task, is employed in our present study as a realistic teleoperations task. Two experiments similar to those in Kim et al. (1987) are performed. In the first experiment, we quantitatively evaluate monoscopic

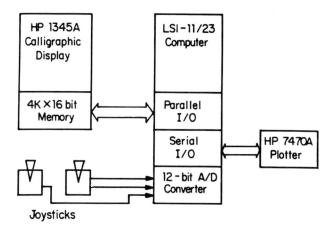

Figure 1. The experimental setup.

perspective display by investigating individual effects of perspective parameters. Perspective projection alone, however, does not provide sufficient three-dimensional (3D) depth information for monoscopic display. Thus, a 5-line-by-5-line horizontal grid representing a base plane and a vertical reference line representing vertical separation from the base plane are introduced as two visual-enhancement depth cues. In the second experiment, we investigate effects of these two visual-enhancement depth cues on pick-and-place performance for both monoscopic and stereoscopic displays.

Methods

In order to evaluate various display conditions quantitatively, a teleoperations simulator is constructed with a vector-display system, joysticks, and a simulated cylindrical manipulator. Figure 1 shows a schematic diagram of the experimental setup, with which three-axis pick-and-place tasks are performed.

Real-time simulation of the manipulator

The Hewlett-Packard 1345A vector-display module is used for real-time dynamic display. It has high resolution (2048 × 2048 addressable data points), and high vector-drawing speed (8194 cm of vectors at 60-Hz refresh rate). It also has a fast vector-updating speed (approximately 10 μsec/vector), communicating with a host computer through a 16-bit parallel I/O port. Two isotonic (displacement) joysticks are employed for the Cartesian position control of the manipulator gripper. An LSI-11/23 computer with the

RT-11 operating system is used as a host computer. It performs computations for the simulated manipulator motion and perspective or stereoscopic display, and measures task completion time.

The human operator indicates the desired gripper position of the manipulator in robot base Cartesian coordinates by using three axes of the two joysticks. The computer senses the joystick displacements through 12-bit A/D converters. The joystick gain for each axis is chosen to be 1 so that the full range of the joystick displacement for each axis corresponds to the full movement range of the gripper position for the corresponding axis. The computer transforms the desired gripper position in Cartesian coordinates to the desired joint angle (θ_1 for the revolute joint 1) and joint slidings (d_2 and d_3 for the prismatic joints 2 and 3) by employing the inverse kinematic position transformation. The next two sections describe how to present 3D information of the manipulator on the 2D display screen.

Monoscopic perspective display

A monoscopic perspective display can be constructed by a perspective projection of an object onto the view plane (projection plane) followed by a mapping of the view plane onto the screen. There are two approaches to obtaining the perspective projection of an object. One is to leave the object stationary and choose a desired viewpoint and a projection plane, called the viewpoint-transformation method. The other approach is to fix the viewpoint and transform the object, called the object-transformation method. These two approaches are mathematically equivalent (Kim et al., 1987; Foley and Van Dam, 1983; Carlbom and Paciorek, 1978). The latter will be described here.

In order to derive the perspective display formulas based on the object-transformation method, a right-handed XYZ world coordinate system is established. The viewpoint is fixed at the origin (0, 0, 0) and the view plane at the $z = -d$ plane. Perspective projection can be obtained by three transforms: rotation R, translation T, and perspective transform P.

Initially, an object is located so the view reference point of the object is at the origin. Then the object is appropriately rotated and translated to achieve the desired viewing angles and distance. In general, an arbitrary orientation of an object can be described by successive principal-axis rotations about the Y, X, and Z axes.

$$R = \text{Rot}(Y, -\theta_1) \, \text{Rot}(X, \theta_2) \, \text{Rot}(Z, \theta_3) \tag{1}$$

where the yaw, pitch and roll angles are $-\theta_1$, $-\theta_2$, and θ_3, respectively. It can be shown that the yaw and pitch angles used in the object transformation approach are equivalent to the azimuth and elevation angles in the viewpoint-transformation approach (Kim et al., 1987).

For simplicity, 4-space homogeneous coordinate transformations are used.

The rotation of a point at position (x, y, z) to a new position (x', y', z') can be described by

$$(x', y', z', 1) = (x, y, z, 1)\mathbf{R} \qquad (2)$$

where

$$\mathbf{R} = \begin{bmatrix} R_{11} & R_{12} & R_{13} & 0 \\ R_{21} & R_{22} & R_{23} & 0 \\ R_{31} & R_{32} & R_{33} & 0 \\ 0 & 0 & 0 & 1 \end{bmatrix}. \qquad (3)$$

From equation (1), each element of the 4×4 matrix $\mathbf{R}$ can be calculated as $R_{11} = C_1C_3 - S_1S_2S_3$, $R_{12} = -C_1S_3 - S_1S_2C_3$, $R_{13} = S_1C_2$, $R_{21} = C_2S_3$, $R_{22} = C_2C_3$, $R_{23} = S_2$, $R_{31} = -S_1C_3 - C_1S_2S_3$, $R_{32} = S_1S_3 - C_1S_2C_3$, $R_{33} = C_1C_2$. S_i and C_i denote $\sin \theta_i$ and $\cos \theta_i$, respectively.

After the rotation, the object is translated by D along the negative Z axis.

$$\mathbf{T} = \text{Trans}(0, 0, -D) \qquad (4)$$

$$= \begin{bmatrix} 1 & 0 & 0 & 0 \\ 0 & 1 & 0 & 0 \\ 0 & 0 & 1 & 0 \\ 0 & 0 & -D & 1 \end{bmatrix}. \qquad (5)$$

The length D represents the distance from the viewpoint to the view reference point, called the object distance.

The UV coordinate system is embedded in the view plane. Perspective transformation of a point $Q(x, y, z)$ in the world coordinate to its projection $Qp(u, v)$ on the view plane can be described by

$$(x', y', z', w) = (x, y, z, 1)\mathbf{P} \qquad (6)$$

$$(u, v) = (x'/w, y'/w) \qquad (7)$$

where

$$\mathbf{P} = \begin{bmatrix} 1 & 0 & 0 & 0 \\ 0 & 1 & 0 & 0 \\ 0 & 0 & 0 & -1/d \\ 0 & 0 & 0 & 0 \end{bmatrix}. \qquad (8)$$

The symbol d denotes the view plane distance from the viewpoint. Increase of the view plane distance results in uniform magnification of the perspective projection. Thus, d can be specified in terms of the zoom or magnification factor, which can be defined as $M = d/D$. Distance d can also be specified in terms of field-of-view (fov) angle, which is the angle at the viewpoint subtended by the view-plane window. If the view-plane window is specified as a square region $(u_{min}, u_{max}, v_{min}, v_{max}) = (-1, 1, -1, 1)$, then the fov angle is related to the view-plane distance by $d = \cot(\text{fov}/2)$. The perspective projection obtained with a wide fov angle is similar to the picture taken by a wide-angle camera lens, and a narrow fov angle is similar to one taken by a telephoto lens.

After the object is projected onto the view plane, mapping of the view plane onto the physical display screen is performed. Mapping of a point from (u, v) in the UV coordinate to (x_s, y_s) in the screen coordinate can be achieved by appropriate translations and scalings:

$$x_s = \text{VSX } u + \text{VCX} \qquad (9)$$

$$y_s = \text{VSY } v + \text{VCY} \qquad (10)$$

where VSX and VSY are scaling factors, and VCX and VCY are translation factors.

Stereoscopic display

The monoscopic display does not give true depth perception. The human brain merely interprets the 2D monoscopic picture as 3D space. The stereoscopic display presents two views of an object on the display: one for the right eye, and the other for the left. This pair of pictures is called a stereo pair or a stereogram. The human operator views a stereogram through a stereoscope. Most people can fuse the stereo pair into one 3D image, perceiving relative depth by the human steroscopic vision ability. The stereoscope is composed of two converging lenses and a supporting frame (septum) separating right and left views. As illustrated in Figure 2, two converging lenses form the image of the stereo pair onto the image plane behind the actual display screen, which can provide fairly correct accommodation and convergence conditions for the human eyes, if the geometrical and optical conditions are appropriately arranged.

In order to obtain the formulas for the stereoscopic display, an XYZ coordinate system is established with its origin in the middle of the two optical centers for the right and left eyes, as depicted in Figure 2. The display screen, on which a stereogram is presented, is located at the picture plane (view plane, projection plane) $z = -d$. The two converging lenses of the stereoscope form the virtual image of the stereogram on the image plane $z = -D$. By denoting the focal length of the binocular lens as F, the converging lens formula yields

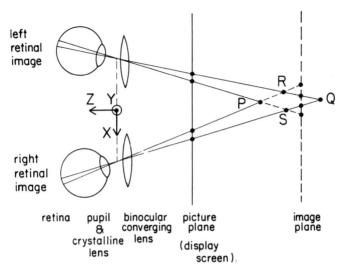

Figure 2. Stereoscopic display.

$$\frac{1}{d} - \frac{1}{D} = \frac{1}{F} \qquad (11)$$

When D is infinity, $d = F$. When $D = 40$ cm and $F = 20$ cm, $d = 13.3$ cm.

As in the object-transformation approach used previously for the monoscopic perspective display, the object is initially located so the view-reference point of the object is at the origin. Then the object is appropriately rotated and translated using equations (3) and (5) to achieve the desired viewing angles and distance.

Denoting the interocular distance (IOD) (approximately 5.5 to 6.5 cm), we can express the positions of the two optical centers by $(x_{or}, 0, 0)$ for the right eye and $(x_{ol}, 0, 0)$ for the left eye, where $x_{or} = \text{IOD}/2$, and $x_{ol} = -\text{IOD}/2$. The projection of a point $P(x, y, z)$ onto the view plane for each eye is formed at the intersection of the projection line with the view plane. By representing the right and left projection points by $P_r(x_r, y_r)$ and $P_l(x_l, y_l)$, respectively, the following equations can be obtained:

$$x_r = x_{or} + (x - x_{or})(-d/z) \qquad (12)$$

$$x_l = x_{ol} + (x - x_{ol})(-d/z) \qquad (13)$$

$$y_r = y_l = y(-d/z) \qquad (14)$$

Finally, these projection points on the projection plane can be mapped onto the physical screen coordinates by appropriate translations and scalings.

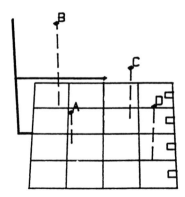

Figure 3. A monoscopic perspective presentation using nominal perspective parameters.

Experimental procedures

Two sets of experiments were performed, varying perspective parameters and visual enhancement conditions. In both experiments, subjects were seated in front of the display (on which the manipulator, the objects to pick up, and the boxes to place them in were presented) (Figure 3), and the subjects were asked to perform three-axis pick-and-place tasks. The subjects controlled the manipulator using two joysticks to pick up each object with the manipulator gripper and place it in the corresponding box. One hand, using two axes (forward-backward and right-left) of one joystick, controlled the gripper position for the two axes parallel to the horizontal base plane. The other hand, using one axis (forward-backward) of the other joystick, controlled the vertical axis.

Each of the four objects (point targets A, B, C, D) was positioned randomly within the manipulator reach space. Each object position was marked by a tiny diamond and a letter. Picking up an object was accomplished when the manipulator gripper touched the object within the boundary of the error tolerance, defined by a hypothetical cube. The size of the cube was set so that the picking process was neither too easy nor too hard within the range of experimental variation. Accomplishment of picking up an object was indicated by doubling the object letter. Thereafter, the object moved together with the gripper until it was placed in the right box. Placing an object was accomplished by touching the correct box with the gripper, similar to the picking process. After the touch, the object symbol letter became single again, and the object remained in the box, while the gripper was free to move for the next operation.

One run of the pick-and-place task consisted of five sessions of four pick-and-place operations in order from object A to D, totaling twenty pick-and-place operations.

Perspective parameter experiment

In this experiment, we investigated the effects of different perspective parameters on the human operator's pick-and-place performance with monoscopic perspective display. The five perspective parameters, azimuth, elevation, roll, fov angle, and object distance were independently varied, keeping the other variables fixed at their nominal values. The nominal perspective parameter values were chosen as elevation = $-45°$, azimuth = $0°$, roll = $0°$, fov angle = $12°$, and object distance = 40 cm.

Experimental variables were varied as follows: (1) seven elevation angles: $0°$, $-15°$, $-30°$, $-45°$, $-60°$, $-75°$, and $-90°$; (2) eight azimuth angles: $-135°$, $-90°$, $-45°$, $0°$, $45°$, $90°$, $135°$, and $180°$; (3) eight roll angles: $-135°$, $-90°$, $-45°$, $0°$, $45°$, $90°$, $135°$, and $180°$; (4) five fov angles: $8°$, $12°$, $24°$, $48°$, and $64°$, (5) four object distances: 30, 40, 80, and 160 cm.

The monoscopic perspective presentation with the nominal perspective parameters is shown in Figure 3. Some examples of variations in perspective parameter values used in this experiment are shown in Figure 4. In this experiment, a 5-line-by-5-line horizontal grid and vertical reference lines were always presented. The experiment was run with each of the 32 experimental conditions presented in random order. There were two runs of 20 pick-and-place operations per condition for each subject. For the monoscopic conditions, the subjects were seated 40 cm in front of the display screen.

Visual enhancement experiment

In this experiment, effects of visual enhancements on the human operator's pick-and-place performance were investigated. The visual-enhancement depth cues used for both monoscopic and stereoscopic displays were a grid and reference lines. Three-axis pick-and-place tasks were performed for four visual-enhancement conditions at each of five different perspective parameter conditions with both monoscopic and stereoscopic displays. The four visual-enhancement conditions were: GL (presence of both grid and reference line), L (reference line only), G (grid only), and O (neither). The five perspective parameter conditions used were: (1) $0°$ in elevation, (2) $-90°$ in elevation, (3) nominal perspective parameter values, (4) $45°$ in azimuth, and (5) 80 cm in object distance.

Monoscopic presentations for the four visual-enhancement conditions with the nominal perspective parameters (condition III) are shown in Figure 5. Monoscopic presentations for the five perspective parameter conditions, when both grid and reference lines are presented (condition GL), are shown above the mean completion time plot in Figure 8. A stereoscopic presentation with the nominal perspective parameters, when both grid and reference lines are presented, is shown in Figure 6. The experiment was run first with each of the 20 monoscopic display conditions presented in random order, then with each of the 20 stereoscopic display conditions presented in random

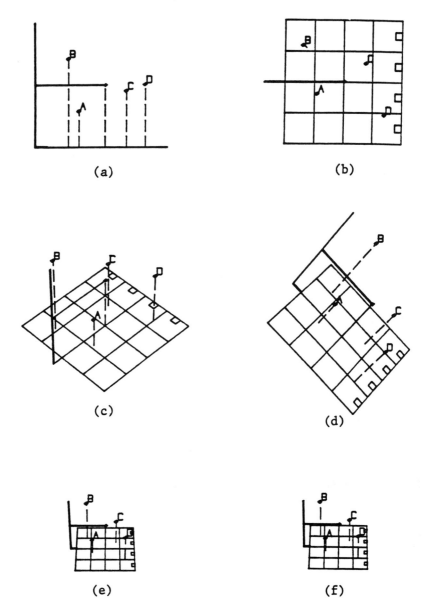

Figure 4. Examples of various monoscopic perspective presentations with (a) an extreme 0° elevation angle, (b) the other extreme −90° elevation angle, (c) 45° azimuth angle, (d) 45° roll angle, (e) fov angle doubled to 24°, and (f) object distance doubled to 80 cm. A 5-line-by-5-line horizontal grid and vertical reference lines are presented.

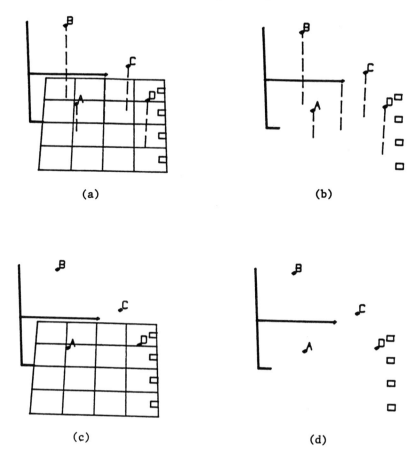

Figure 5. Monoscopic presentations under four visual-enhancement conditions: (a) GL (presence of both grid and reference line), (b) L (reference line only), (c) G (grid only), and (d) O (neither).

order. There were two runs of 20 pick-and-place operations per condition for each subject.

In the monoscopic display conditions, subjects were seated 40 cm in front of the screen. In the stereoscopic display conditions, subjects were seated 13.3 cm in front of the screen, viewing the stereogram through the stereoscope. The focal length of the converging lens of the stereoscope was 20 cm, and thus the virtual image of the stereogram was formed at 40 cm from the lens (Equation (11)).

Subjects

Two young adult male subjects with normal stereo vision participated in each of the two experiments. Each subject was trained for at least 5 hours

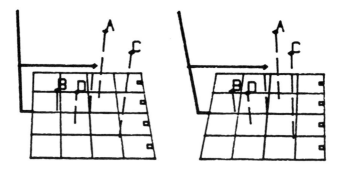

Figure 6. An example of a stereoscopic presentation.

before the experiments to saturate the "learning" effect. During the training period, mean completion times were regularly checked to see whether the subject reached an asymptotic, steady-state, pick-and-place performance. However, during the actual experiment, mean completion times were not checked until all the experimental runs were completed. Each subject repeated the experiment once more in order to examine intra-subject variation as well as inter-subject variation.

Experimental results

Mean completion time was used as the performance measure in our pick-and-place tasks. Each of the mean completion time data points in Figures 7 and 8 is the average obtained from one run of 20 pick-and-place operations.

The experimental results for two subjects with two runs each plotted in Figure 7 with mean completion time as the ordinate and perspective parameter values as the abscissa. The effects of elevation, azimuth, roll, fov angle, and object distance are plotted in Figure 7(a), (b), (c), (d), and (e), respectively.

The experimental results for two subjects with two runs each are shown in Figure 8. Mean completion time (ordinate) is plotted for the various display conditions (abscissa). The monoscopic display data are marked by squares and dashed lines, and the stereoscopic display data are marked by filled diamonds and solid lines. The five separate columns represent five different perspective parameter settings, conditions 1–5. Each column has four different visual-enhancement conditions, GL, L, G, and O.

Figure 7. Perspective parameter experiment. Three-axis pick-and-place performance ▶ with various monoscopic perspective displays: (a) mean completion time as a function of elevation, (b) mean completion time as a function of azimuth, (c) mean completion time as a function of roll, (d) mean completion time as a function of fov angle, (e) mean completion time as a function of object distance.

Visual enhancements in pick-and-place tasks

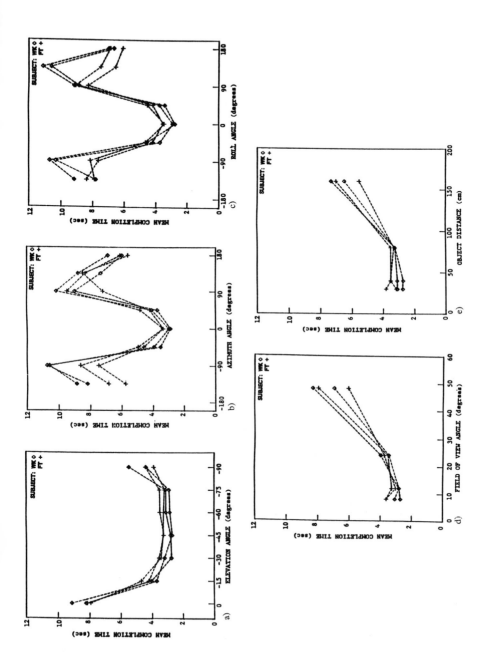

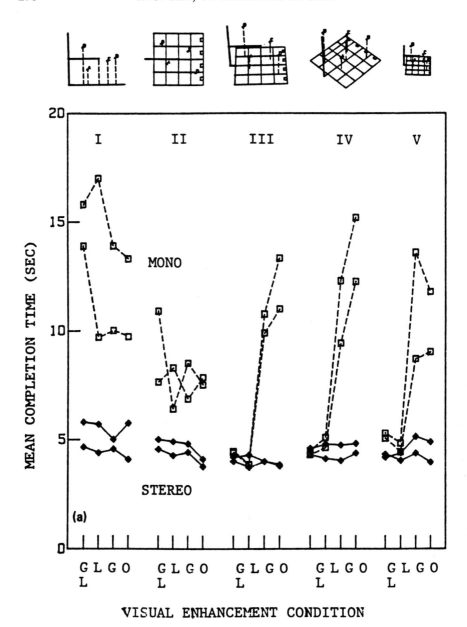

Figure 8. Visual-enhancement experiment. Three-axis pick-and-place performance for four visual-enhancement conditions at each of five different perspective parameter conditions with both monoscopic display and stereoscopic display. The monoscopic presentations for the five perspective parameter conditions are shown above the plot. Four visual-enhancement conditions are GL (presence of both grid and reference line), L (reference line only), G (grid only), and O (neither).

Discussion

Effects of perspective parameters

The mean-completion-time plots of Figure 7 show the effects of variation of perspective parameters on pick-and-place performance. Plot (a) shows that as the elevation angle approaches 0° or −90°, mean completion time increases. This is due to the loss of one axis's position information. Performance at −90° elevation was better than at the 0° extreme because the perspective view at −90° elevation made it possible to see some of the height of the reference line if it was not near the center of the projected image. Thus, there was a partial view of the "lost" axis. Plot (b) shows that as the azimuth angle exceeds the range of −45° to +45°, the mean completion time increases markedly. An azimuth angle other than 0° implies rotation of the display reference frame relative to the joystick control axes, thus making the joystick control more difficult compared to the 0° azimuth angle. When the azimuth angle is beyond −45° to +45°, it is difficult for the human operator to compensate. Performance is especially poor when the azimuth angle is about −90° or +90°, even worse than the case when azimuth angle is 180°. At 180° azimuth angle, the human operator uses inversion rather than rotation. Plot (c) shows that change in roll angle produces an effect similar to changing the azimuth angle, because of analogous disorientation. Plots (d) and (e) show that as the fov angle or the object distance increases, and the displayed object picture becomes smaller, task performance degrades.

Effects of visual enhancements

The results of the visual-enhancement experiment appear in Figure 8. Monoscopic display results in columns I and II show that when the elevation angle is 0° or −90°, the mean completion times are very long, even with grid or reference line enhancements. This is because position information for one axis is lacking, and the subject must sweep the gripper along that axis until it touches the correct position. At −90° elevation, the reference lines almost disappear. At 0° elevation, the grid appears as a single line. Monoscopic display results in columns III, IV, and V show that by choosing adequate elevation angles, mean completion times can be shortened, and fast pick-and-place performance can be attained with monoscopic perspective display, if reference lines are provided (GL, L). However, the grid alone without the reference line (G) does not appear to shorten completion time.

The stereoscopic display results in Figure 8 show that mean completion times with stereoscopic display are short over all visual conditions, regardless of the presence of a grid or reference lines. Especially, stereoscopic display data in columns I and II show that stereoscopic displays maintain fast performance even with extreme elevation angles. Comparable mean completion times between monoscopic and stereoscopic displays in columns III, IV, and

V demonstrate that pick-and-place performance with monoscopic perspective displays, if reference lines are provided and suitable perspective parameters are chosen, can be as good as that with stereoscopic displays.

Comparison with three-axis manual tracking tasks

It is observed that the mean-completion-time plots obtained from pick-and-place experiments in this paper are quite similar to the normalized rms tracking error plots obtained from three-axis manual tracking experiments. This strong similarity suggests that the results obtained in this paper are not task-specific, but may be applicable to other tasks.

Choice of display

There are many kinds of depth cues that a display can provide. Monoscopic display can provide monocular depth cues such as interposition (occlusion), brightness (light and shade), perspective projection (size), and monocular motion parallax. The human operator's knowledge and learning can also provide strong depth information pertaining to a 3D model of a working environment. Stereoscopic display also provides a distinct binocular depth cue, called stereo disparity or binocular parallax. Consideration of these cues basically explains the experimental results of Pepper *et al.*, (1981). Their results indicated that stereoscopic display performance was superior to monoscopic display performance under most conditions tested, although the amount of improvement varied with task, visibility, and learning factors. For some simple telemanipulation tasks, monocular depth cues and cognitive depth cues from knowledge and learning may be enough for successful and reliable performance, and there will be no advantage in using stereoscopic display (Kama and DuMars, 1964). However, for some complex tasks, monocular and cognitive depth cues may be insufficient or unavailable for successful performance with monoscopic display, and the use of stereoscopic display could significantly enhance performance. In our experiments, monocular depth cues were minimized, and target positions were randomly arranged to minimize learning effect. Consequently, our experimental results showed that pick-and-place performance with stereoscopic display was superior to monoscopic display when visual-enhancement depth cues were not presented.

Our results also showed that when reference lines were presented for visual enhancement, monoscopic display performance with adequate perspective parameters was equivalent to stereoscopic display performance. In order to present reference lines on the monoscopic display, 3D position information of the displayed objects must be available. In a graphic display of current manipulator and camera positions, 3D position information is normally available via joint position sensors, and reference lines can be easily provided. In a television image display of the working environment, only camera views are normally available for 3D position information. Under current technology, a machine vision system that extracts 3D position information of each pixel in real time from a stereo camera view is too difficult to construct, (Drumheller and Poggio, 1986) although the human visual system can easily produce a 3D image from a stereoscopic view. However, a special-purpose, machine-vision system that extracts 3D position information of only some salient points in real time from a stereo camera view can be built. Then, reference lines for these points can be presented or superimposed on the monoscopic television display for enhanced teleoperation.

Conclusion

Results of the perspective parameter experiments indicate that in order to attain good performance with a monoscopic perspective display, adequate parameter values should be chosen. For example, extreme elevation angles or excessive azimuth angles result in very long mean completion times. Results of the visual-enhancement experiment indicate that the horizontal grid does not appear to improve pick-and-place performance in our task. The vertical reference line, however, was significant in improving performance with monoscopic perspective display. When the monoscopic display was defined with appropriate perspective parameters and provided with adequate visual-enhancement depth cues such as reference lines, the monoscopic display allowed pick-and-place performance equivalent to that of the stereoscopic display. Stereoscopic display showed short mean completion times over all visual display conditions regardless of the presence of the grid or the reference lines.

Strong similarities were observed between the mean completion time results of the three-axis pick-and-place experiments for various display conditions and the normalized rms error results of the three-axis manual tracking experiments reported previously. This demonstrates that the effects seen are robust and not task-dependent.

References

Bejczy, A.K. (1980). Sensors, controls, and man-machine interface for advanced teleoperation, *Science*, **208**, 4450, 1327–1335.

Bejczy, A.K., Brown, J.W. and Lewis, J.L. (1980). Evaluation of smart sensor displays for multidimensional precision control of space shuttle remote manipulator, *16th Annual Controllers Conference on Manual Control*, 607–627.

Bejczy, A.K., Dotson, R.S., Brown, J.W. and Lewis, J.L. (1982). Manual control of manipulator forces and torques using graphic display. *IEEE Proceedings of the International Conference on Cybernetics and Society*, 691–698.

Bronez, M.A., Clarke, M.M. and Quinn, A. (1986). Requirements development for a free-flying robot—The Robin. *Proceedings of IEEE Robotics and Automation*, **1**, 667–672.

Carlbom, I. and Paciorek, J. (1978). Planar geometric projections and viewing transformation, *Computing Survey*, **1**, 4, 465–502.

Clarke, M.M. and Bronez, M.A. (1986). Telerobotics for the Space Station, *Mechanical Engineer*, **108**, 2, 66–72.

Drumheller, M. and Poggio, T. (1986). On parallel stereo, *Proceedings of IEEE International Conference on Robotics and Automation*, **3**, 1439–1448.

Ellis, S.R., Kim, W.S., Tyler, M., McGreevy, M.W. and Stark, L. (1985). Visual enhancement for telerobotics: perspective parameters, *IEEE 1985 Proceedings of the International Conference on Cybernetics and Society*, 815–818.

Ferrell, W.R. and Sheridan, T.B. (1967). Supervisory control of remote manipulation, *IEEE Spectrum*, **4**, 10, 81–88.

Foley, J.D. and Van Dam, A. (1983). *Fundamentals of Interactive Computer Graphics*. Addison-Wesley.

Freedman, L.A., Crooks, W.H. and Coan, P.P. (1977). TV requirements for manipulation in space, *Mechanism and Machine Theory*, **12**, 5, 425–438.

Heer, E. (1973). *Remotely Manned Systems: Exploration and Operation in Space*, California: California Institute of Technology.

Jenkins, L.M. (1986). Telerobotic work system—Space robotics application, *Proceedings of IEEE Robotics and Automation*, **2**, 804–806.

Johnsen, E.G. and Corliss, W.R. (1967). Teleoperators and Human Augmentation, *NASA SP-5047*.

Johnsen, E.G. and Corliss, W.R. (1971). *Human Factors Applications in Teleoperator Design and Operation*, New York: Wiley-Interscience.

Kama, W.N. and DuMars, R.C. (1964). Remote Viewing: A Comparison of Direct Viewing, 2-D and 3-D television, AMRL-TDR-64-15, Wright-Patterson Air Force Base.

Kim, W.S., Ellis, S R., Tyler, M. and Stark, L. (1985). Visual enhancement for telerobotics, *IEEE 1985 Proceedings of the International Conference on Cybernetics and Society*, 819–823 .

Kim, W.S., Ellis, S.R., Hannaford, B., Tyler, M. and Stark, L. (1987). A quantitative evaluation of perspective and stereoscopic displays in three-axis manual tracking tasks. *IEEE Transportation Systems, Man, Cybernetics*, **SMC-17**, 1, 61–72.

Leinemann, K. and Schlechtendahl, E.G. (1984). Computer graphics support for remote handling simulation and operation. *Transportation and the American Nuclear Society*, **46**, 771–772.

Montemerlo, M.D. (1986). NASA's automation and robotics technology development program. *Proceedings of IEEE Robotics and Automation*, **2**, 977–986.

Pepper, R.L., Smith, D.C. and Cole, R.E. (1981). Stereo TV improves operator performance under degraded visibility conditions, *Optics Engineering*, **20**, 4, 579–585.

Pepper, R.L., Cole, R.E., Spain, E.H. and Sigurdson, J.E. (1983). Research issues involved in applying stereoscopic television to remotely operated vehicles, *Proceedings of SPIE*, **402**—Three-Dimensional Imaging, 170–173.

Pischel, E.F. and Pearson, J.J. (1985). Image processing and display in three dimensions, *Proceedings of SPIE*, **528**—Digital Image Processing, 23–28.

Rupert, P. and Wurzburg, D. (1973). Illumination and television considerations in teleoperator systems. In Heer, E. (Ed.), *Remotely Manned Systems*, 219–228, California: California Institute of Technology.

Vertut, J. and Coiffet, P. (1986). Teleoperations and robotics: evolution and development. *Robot Technology*, **3A**. Englewood Cliffs: Prentice-Hall.

Vertut, J., Fournier, R., Espiau, B. and Andre, G. (1984). Sensor-aided and/or computer-aided bilateral teleoperator system (SCATS), RO.MAN.SY *Proceedings of the 5th Symposium* (also in Morecki, A., Bianchi, G. and Kedzior, K. (Eds), *Theory and Practice of Robots and Manipulators*, pp. 281–292).

17

Target axis effects under transformed visual-motor mappings

H.A. Cunningham and M. Pavel

NASA Ames Research Center and Stanford University

Summary

Performance of a discrete 2D aiming task under transformed visual-motor mappings (rotations of 45 deg, 90 deg, 135 deg, and 180 deg, and reflections about the horizontal, vertical, and oblique midlines) exhibits target axis effects. That is, the magnitude of aiming error varies with the axis in 2D space on which the aiming target lies. In our experiments, eight aiming targets corresponded to four target axes: horizontal (right, left), vertical (up, down), right oblique (up-right, down-left), and left oblique (up-left, down-right). Observed target axis effects are of two varieties. The first is higher aiming error along the left oblique axis than along the right oblique, observed previously under non-transformed mappings and tentatively attributed to biomechanical factors (Keele, 1968). Our transformation paradigm allows examination of biomechanical and other "motor" factors independently of vision, cognition, and other "non-motor" factors, by decoupling the motor (manual input device) and non-motor (display feedback) representations of the axes. The second variety of target axis effects are reflection-specific effects observed here for the first time. We characterize reflection-specific target axis effects in terms of a vector sum process (Cunningham and Vardi, in press) that interacts with the formal properties of reflections so as to produce qualitatively different behavior along different target axes.

Introduction

This research investigated spatial aiming error in a two-dimensional discrete aiming task performed under rotations and reflections of the normal spatial mapping between visual space (a CRT display) and manual space (a planar digitizing tablet). In this computerized task, hand movements on the horizontal tablet produced cursor movements on the vertical screen; the goal was to capture a target with the cursor by moving in as straight a path as possible from a starting position to the target. Transformations were rotations of 45 deg, 90 deg, 135 deg, and 180 deg, and reflections about horizontal, vertical, and oblique bisecting axes. Straight-path movement to a target under these conditions requires a remapping of the familiar correspondence between directions in visual and manual spaces into a new correspondence. The normal, or familiar, mapping is such that right and left on the tablet map to right and left on the screen, and forward and back on the tablet map to up and down on the screen. Eight aiming targets, corresponding to four axes of movement: horizontal (right, left), vertical (up, down), right oblique (up-right, down-left), and left oblique (up-left, down-right), were used. Figure 1a shows the four target axes under the familiar visual-motor mapping. Figures 1b and 1c show the right oblique axis as it is transformed by rotation (1b) and reflection (1c).

Rotations and reflections are linear transformations. They preserve line length, line parallelism, and angle size, but they do not preserve movement direction (with the single exception of the axis of reflection, under reflection transformations). Rotations, but not reflections, preserve the sign or direction of angles. That is, a right turn in manual space remains a right turn in display space, whereas under reflections a right turn in manual space becomes a left turn in display space. Certain general-purpose transformation algorithms (such as matrix multiplication) are insensitive to these differences and will compute all linear transformations in a qualitatively similar way. If the human visual-motor control system were such a general-purpose calculator, then the errors produced during visual-motor remapping would not vary according to target axis. Therefore the finding of target axis effects suggests that the mechanisms of human visual-motor control are not general purpose but rather they have constraints on their processing which are sensitive to certain formal properties of transformations.

Spatial error was defined as deviation of movement paths from straight lines connecting their beginning and ending points, analysed separately for the four target axes. The target axis provides a good basis for summarizing directional aiming error, although aiming trajectories often exhibit multiple directions of movement. Aiming error tends to group by axis, with targets that share an axis having similar error magnitudes. Moreover, the target axis determines the initial state of the system which in turn strongly influences subsequent correction processes, in this task. This point will be taken up again in the discussion.

In a 1968 review of the motor literature, Keele reported results of studies by Brogden and colleagues (Brogden, 1953; Briggs and Brogden, 1953; Briggs et al., 1954) showing that aiming error for movements on a horizontal surface is higher for targets lying on a left oblique axis than for targets on a right oblique axis, for right-handed subjects. We have replicated this result in our own task, using the familiar mapping. Figure 2 is a polar plot of aiming error for the four target-defined movement axes. The two plotted points for

FAMILIAR MAPPING

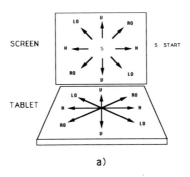

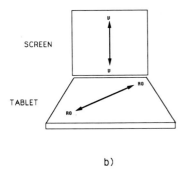
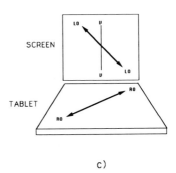

Figure 1. Display screen and digitizing tablet used in the task: (a) eight target positions corresponding to four target axes; (b) under a 45 deg rotation, tablet right-oblique corresponds to screen vertical; (c) under a vertical axis reflection, tablet right-oblique corresponds to screen left-oblique.

each axis represent the average error for the axis. Keele (1968) suggested that biomechanical differences between movement from the shoulder and movement from the elbow could account for low error on the right oblique and high error on the left oblique.

The transformation paradigm offers a new way of testing independent contributions of motor versus non-motor processing factors to target axis effects. Under the well- learned familiar mapping, directions of movement in motor, or manual, space are equivalent to directions of movement in non-motor, or display, space. The familiar mapping, though essentially arbitrary, is well established through years of visually-guided manual activity. Transformation of the visual-motor mapping decouples visual and motor directions of movement (e.g., Figure 1), and makes it possible to separate motor and non-motor influences on aiming error. That is, motor "right" may now

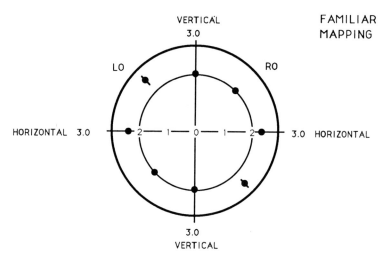

Figure 2. Polar plot of RMS aiming error, by axis, under the familiar mapping where display and tablet axes are aligned. Average of six subjects. Light circle corresponds to error for the right-oblique axis. Error bars on left-oblique data points reflect +/− one standard error of the mean.

map to visual "up", and so forth. This method is superior to at least one possible alternative, comparing left-handed and right-handed subjects, because handedness cannot be assumed to be a strictly motor phenomenon.

This paper reports on two different target axis effects. First, we use the transformation paradigm to examine the previously reported right oblique/ left oblique effect. Second, we examine target axis effects under qualitatively different transformations (i.e., reflections versus rotations). Under rotations, all directions of movement are transformed by the same angular amount and in the same angular direction (e.g., clockwise). Under reflections, the amount and direction of angular transformation depends on the angular relation between the movement axis in question and the axis of reflection. Along the axis of reflection, direction is preserved. With increasing rotation of an axis away from the axis of reflection, the angular difference between visual and motor directions increases, up to a maximum of 180 deg along the axis perpendicular to the axis of reflection.

Rotations, but not reflections, preserve the local sign of angles. To illustrate this point, Figure 3a shows visual and motor representations of a movement path under a 45 deg rotation. Note that a righthand turn in visual space is produced by a righthand turn in motor space, with the orientations of the pre-turn and post-turn movement directions transformed by the same rotational amount. Figure 3b shows the same turn under a vertical axis reflection. Note that to produce a righthand turn in visual space, the hand must execute a lefthand turn in motor space.

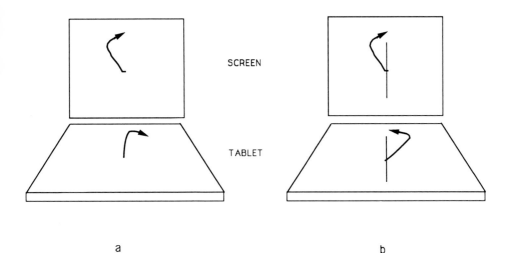

Figure 3. Rotations preserve local sign or direction of angles whereas reflections do not: (a) righthand turn under a 45 deg rotation; (b) righthand turn under a vertical axis reflection.

Method

Six right-handed subjects performed a discrete aiming task using a vertical CRT screen for visual display of target and cursor, and a hand-held stylus on a horizontal digitizing tablet for producing movements of the cursor. Eight possible target positions were arranged at 45 deg intervals around the center. Each aiming trial followed this sequence: (1) subject aligns cursor with a marker at the center of the display screen; (2) a cueing tone sounds; (3) after a variable foreperiod [250 to 750 milliseconds] a target appears in one of the eight target positions; (4) subject captures target by moving the cursor into alignment with it; (5) target extinguishes. Subjects were instructed to move in as straight paths as possible, and to emphasize this accuracy goal over speed, within a 3-second time limit imposed by the experimental software.

Each experimental session consisted of 32 baseline trials under the familiar mapping, followed by 128 trials under one of the six transformed mappings: rotation of 45 deg, 90 deg, 135 deg or 180 deg, or reflection about the vertical or horizontal midline. Transformations of the motor space relative to the visual space were obtained using a combination of software and physical rotation of the digitizing tablet. Root mean squared (RMS) error measured the deviation of a trajectory from a straight line connecting the starting and ending points of the trajectory:

$$\text{RMS} = \sqrt{\frac{\sum_{i}^{N} d_i^2}{N}}$$

where d is the euclidean distance between the ith x,y trajectory coordinate and a straight line connecting the trajectory's endpoints, and N is the number of points in the trajectory.

Results and discussion

There was an effect of trialblock, with error decreasing over practice, but trialblock did not interact with target axis for rotations ($F(21,105) = 1.02$; $p = 0.45$) or for reflections ($F(21,105) = 1.13$; $p = 0.33$). Therefore target axis effects are independent of learning or practice effects, and we will report average aiming error over trialblocks.

Figures 4a–d show polar plots of average RMS error for the four target axes (horizontal, vertical, right oblique, and left oblique), separately for the four rotation conditions. Axes in the polar plots correspond to motor axes (axes of movement on the digitizing tablet, irrespective of display axis). Plotted points for each axis are the average of the two component target positions. In each plot, the inner circle represents the magnitude of right oblique error. Error bars on the left oblique points represent +/− one standard error of the mean. The distance scales are the same for the 45 deg and 180 deg plots, and at half scale for the 135 deg and 90 deg plots, so that absolute error magnitude can be compared among the rotation conditions. The expected right oblique "advantage" was found for the 45 deg, the 90 deg and the 135 deg rotations. (Note that standard error for the 45 deg rotation is smaller than the width of a plotting point.) The 180 deg rotation does not exhibit the right oblique advantage. A planned statistical comparison between right and left oblique for the four rotations gave a statistically significant main effect ($F = 12.76$; $df = 1,5$; $p = 0.016$). Statistical analysis using visual rather than motor directions yielded the opposite effect of axis ($F(3,15) = 3.48$; $p = 0.04$), with left oblique error significantly lower than right oblique error. This effect was carried entirely by the 90 deg rotation, in which visual and motor axes are exchanged. This interpretation is supported by the significant Rotation × Target Axis interaction ($F(9,45) = 3.16$; $p = 0.005$).

Figures 5a and 5b show RMS error for the four motor axes under the two reflections. The right oblique advantage is exhibited by the horizontal axis reflection, but not the vertical axis reflection. The planned comparison computed on both reflections was not statistically reliable ($F(1,5) = 4.69$; $p = 0.08$). However, there is a striking target axis effect in these plots that was not seen under rotations: the axis perpendicular to the axis of reflection has

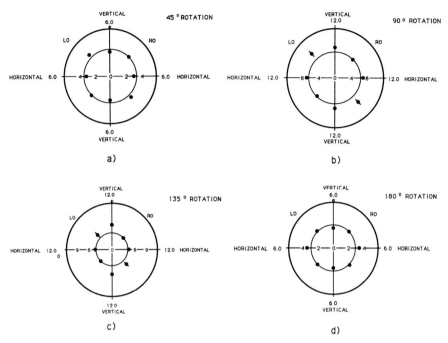

Figure 4. Polar plots of RMS aiming error for the four rotations where display and tablet axes are not aligned. Axes shown are tablet axes. Averages of six subjects. Light circles correspond to error for the right-oblique axis. Error bars reflect +/− one standard error of the mean.

significantly lower error than do the other three axes. The oblique axes have relatively high error, as does the axis of reflection. Analysis of variance gives a significant Target Axis × Reflection interaction ($F(3,15) = 11.16$; $p < 0.001$).

To summarize the results of Experiment 1, directional nonuniformities were found under both rotations and reflections. Under rotations, a previously observed difference between right and left obliques was found to hold for motor axes but not for visual axes. The effect was not found in all transformations, however, and so gives only partial support to the suggestion of Keele (1968) that the right oblique advantage is due to motor factors such as biomechanical properties of the hand/arm system.

Both reflections produced reflection-specific patterns of aiming error, with the axis of reflection having higher error than the axis perpendicular to it. That this pattern is seen only in reflections, and is specific to the type of reflection, indicates that it is related to the orientation of axes with respect to the reflection transformation itself, not with respect to objective spatial coordinates. To test this interpretation, Experiment 2 examined the effects of

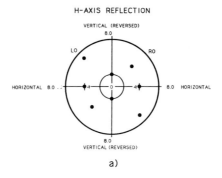

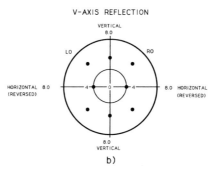

Figure 5. Polar plots of RMS aiming error for the horizontal-axis and vertical-axis reflections. Axes are tablet axes: (a) light circle corresponds to error for the reversed vertical axis, error bars as before; (b) light circle corresponds to error for the reversed horizontal axis, error bars as before.

reflection about a new axis: the right oblique. Two subjects from the first experiment (the authors) plus two new, naive, subjects performed eight blocks of 16 trials under the ro-axis reflection. Figure 6a shows RMS error for four target axes, separately for the two authors. Figure 6b shows RMS error for four target axes, separately for the two naive subjects. The expected pattern is confirmed: the left oblique (reversed) axis has significantly lower error than the axis of reflection (right oblique) and the horizontal and vertical axes.

General discussion

The fact that the human visual-motor control system does not respond uniformly with respect to different target axes suggests that the mechanisms of visual-motor control are not equivalent to a general-purpose calculator,

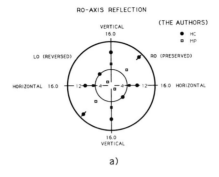

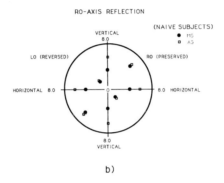

Figure 6. Polar plots of RMS aiming error for the right- oblique-axis reflection, tablet axes: (a) authors HC and MP, light circle corresponds to error for the reversed left-oblique axis of subject HC; (b) naive subjects MS and AG, light circle corresponds to subjects' mean error for the reversed left-oblique axis. Error bars as before.

but rather have constraints on their processing which are sensitive to particular properties of transformations. These constraints may be primarily biomechanical under normal mappings and rotations. Under reflections, however, something more is involved. To what properties of reflections might visual-motor mechanisms be sensitive?

Rigidity and physical realizability

In our task, hand movements were constrained to lie in the 2D space defined by the digitizing tablet. Rotation is a physically realizable transformation in 2D space. For example, lines drawn on the surface of a piece of paper can be rotated in 2D space simply by rotating the piece of paper. However, reflection is not physically realizable in 2D space, in this sense. There is no rigid transformation of a piece of paper such that lines drawn on its surface can be reflected about the midline of the paper. Under reflection, each individual line or axis of movement is transformed by a different rotational distance in 2D space. If the mechanisms of visual-motor control seek a physically realizable solution to a transformation, then under reflections they might respond to the rotational equivalent of each axis. This would produce axis-wise ordering of aiming error that is identical to the rotation-wise ordering seen under rotations. The low error for the reversed axis under reflection is consistent with the fact that the 180 deg rotation produces less error than other rotations. However, the high error for the non-transformed axis of reflection is not consistent with the fact that the 0 deg rotation (i.e., the familiar mapping) produces the least error of all. Therefore some other mechanism must be involved.

Vector–sum correction process

A previous report (Cunningham and Vardi, in press) advances a vector sum account of rapid, automatic error correction under visual-motor transformations. Two input vectors determine instantaneous direction of movement: (1) a target-directed vector, **T**, which represents the direction in which the system intends to go; and (2) a transformation vector, **R**, which represents the influence, in visual space, of the transformation. Instantaneous direction of movement is determined by their vector sum. Under rotations, the angle between **T** and **R** is the same at all positions in 2D space; but under reflections, the angle varies from 0 deg to 180 deg depending on the position of the cursor with respect to the axis of reflection.

Figures 7a–c depict movement paths predicted by the vector sum account under a vertical-axis reflection, as they would be seen in visual space. In each, two possible initial conditions are shown. Initial conditions arise from random directional error added to the first movement. The movement paths are given by heavy black lines with sequential positions marked along them. At each marked position, a horizontal-vertical coordinate system is given in lighter lines, along with the short target vector **T** and the longer transformation vector **R**. The transformation vector **R** is longer than the target vector **T** early in practice, before adaptation has occurred. Note that **R** is always reflected about the vertical axis with respect to **T**. Each subsequent position is produced by their vector sum. Figure 7a shows the results for a target lying on the vertical axis of reflection. Figure 7b shows the results for a target lying on the axis perpendicular to the axis of reflection. Figure 7c shows the results for a target on one of the remaining major axes (right oblique). Along the vertical axis of reflection and the right oblique axis, each new vector-sum correction causes the movement path to veer farther from the original target axis. Along the horizontal axis perpendicular to the axis of reflection, however, each new correction causes the movement path to move toward the original target axis. Once the cursor has returned to the original target axis, a reversal mechanism (see Cunningham, 1989) can be used to move again

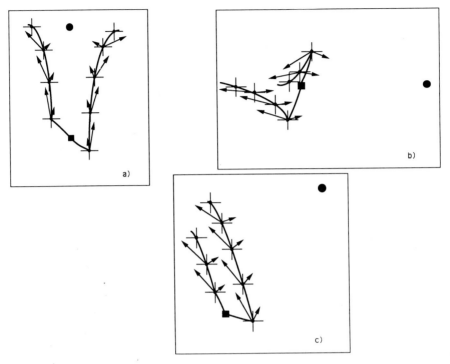

Figure 7. Movement paths produced by vector sum correction process operating under vertical-axis reflection, as seen in visual space. Filled square is starting position; filled circle is target: (a) target lies on preserved vertical axis; (b) target lies on reversed horizontal axis; (c) target lies on right oblique axis. Each shows the results of two possible initial states near the starting position. Reflection is more directionally nonuniform than performance under rotations.

along the desired path to the target. RMS aiming error for this case will be lower than for the cases shown in Figures 7a and 7b.

Conclusions and applications

Target axis effects have important implications for our understanding of human visual–motor control, for they imply nonuniformity with respect to movement direction in the operation of the underlying mechanisms. Sensitivity to properties of spatial transformations, reported here, suggests constraints on the mechanisms involved in visual–motor adaptation and control. Target axis effects are also of practical importance. They can lead to biases in an operator's input to a system, biases which may not be detected when evaluating average or overall performance of a task, because they affect only a subset of the inputs. Understanding these biases allows the development of systems that minimize biases or correct for them during operation.

We have found some evidence for biomechanical effects on aiming performance when the hand-arm system is used. These effects can cause operator-induced error, and they are probably not amenable to change through learning or practice. However, controllers can be designed so as to minimize them. We have also found that performance under reflections is more directionally nonuniform than performance under rotations. Reflections thus present a greater challenge both to the operator who must adjust his or her inputs according to instantaneous position of the cursor, and also to the experimenter or performance evaluator who must take account of target axis effects in interpreting performance data.

References

Briggs, G.E. and Brogden, W.J. (1953). Bilateral aspects of the trigonometric relationship of precision and angle of linear pursuit movements. *American Journal of Psychology*, **66**, 472–478.

Brogden, W.J. (1953). The trigonometric relationship of precision and angle of linear pursuit movements as a function of amount of practice. *American Journal of Psychology*, **66**, 45–56.

Cunningham, H.A. (1989). Aiming error under transformed spatial mappings reveals spatial structure of visual-motor maps. *Journal of Experimental Psychology: Human Perception and Performance*, **15**, 493–506.

Cunningham, H.A. and Vardi, I. (In press). A vector-sum process produces curved aiming paths under rotated visual-motor mappings. *Biological Cybernetics*.

Harris, C.S. (1965a). Perceptual adaptation to inverted, reversed, and displaced vision. *Psychological Review*, **72**, 419–444.

Held, R. and Hein, A. (1958). Adaptation of disarranged hand-eye coordination contingent upon re-afferent stimulation. *Perceptual and Motor Skills*, **8**, 87–90.

Held, R., Efstathiou, A. and Greene, M. (1966). Adaptation to displaced and delayed visual feedback from the hand. *Journal of Experimental Psychology*, **72**, 887–891.

Howard, I.P. (1982). *Human Visual Orientation*. Toronto: John Wiley and Sons.

Keele, S.W. (1968). Movement control in skilled motor performance. *Psychological Bulletin*, **70**, 387–403.

Kim, W.S., Ellis, S.R., Tyler, M.E., Hannaford, B. and Stark, L.W. (1987). Quantitative evaluation of perspective and stereoscopic displays in three-axis manual tracking tasks. *IEEE Transactions on Systems, Man, and Cybernetics*, **17**, 61–72.

Thompson, R.F., Voss, J.F. and Brogden, W.J. The effect of target velocity upon the trigonometric relationship of precision and angle of linear pursuit movements. *American Journal of Psychology*, **69**, 258–263.

18

Adapting to variable prismatic displacement[1]

Robert B. Welch and Malcolm M. Cohen

NASA Ames Research Center
Moffett Field, California

Summary

In each of two studies, subjects were exposed to a continuously changing prismatic displacement with a mean value of 19 prism diopters ("variable displacement") and to a fixed 19-diopter displacement ("fixed displacement"). In Experiment 1, we found significant adaptation (post-pre shifts in hand-eye coordination) for fixed, but not for variable, displacement. Experiment 2 demonstrated that adaptation can be obtained for variable displacement, but that it is very fragile and will be lost if the measures of adaptation are preceded by even a very brief exposure of the hand to normal or near-normal vision. Contrary to the results of some previous studies, we did not observe an increase in within-S dispersion of target-pointing responses as a result of exposure to variable displacement.

Introduction

Human observers who are allowed to view their actively moving hands through an optical medium that displaces, inverts, right-left reverses, or otherwise rearranges the visual field reveal significant adaptive changes in hand-eye coordination (Welch, 1978). For example, the initial errors made when one looks through a wedge prism and attempts to touch a target are typically corrected within a minute. Depending on the nature of the exposure conditions, this prism-adaptive shift in hand-eye coordination can be based on changes in (1) the felt position of the limb (e.g., Harris, 1965); (2) visual

---

[1] The authors wish to thank Arnold Stoper for his valuable comments on a preliminary draft of this paper and Michael Comstock for creating the computer program used for data acquisition.

localization (e.g., Craske, 1967); or (3) the algebraic sum of both of these events (e.g., Wilkinson, 1971).

An alternative to prismatic displacement of constant strength (which may be referred to as "fixed displacement") is displacement that varies continuously in both magnitude and direction ("variable displacement"). It has been shown by Cohen and Held (1960) that active exposure to a variable displacement in the lateral dimension with a mean value of zero fails to produce an adaptive shift in the *average* location of the subject's repeated target-pointing attempts, although it does appear to increase the *variability* of these responses around the mean value. The latter observation has been interpreted as a degradation in the precision of hand-eye coordination.

The absence of adaptation to this form of variable displacement should not come as a surprise, since, over the course of the prism exposure period, there is no net prismatic displacement to which adaptation can occur. What remains to be determined, however, is whether it is possible to adapt to a situation of variable displacement in which the mean value is significantly different from zero, since in this case the occurrence of adaption is at least plausible. The aim of the present investigation was to answer this question and, in addition, to compare the magnitude of such adaptation with that produced by comparable fixed prismatic displacement.

Method

General design

Two experiments were carried out. In both, subjects were used as their own control under conditions of fixed and variable prism exposure to the same average displacement (19 prism diopters). This comparison is seen in Figure 1. Experiment 1 also included the between-groups factor of direction (up *vs.* down) of the optical displacement of the hand that was present during exposure. Prism adaptation was indexed by the difference between pre- and postexposure target-pointing accuracy without visual feedback (visual open-loop).[2] Also obtained were post-pre differences in the within-S variability (standard deviation) of target-pointing over the 10 pre- and 10 postexposure trials. Finally, potential intermanual transfer of the prism-adaptive shifts in

---

[2] An attempt was also made to obtain measures of prism-adaptive shifts in felt-limb position. During the pre- and postexposure periods, subjects (with eyes shut) were to try to place the right and left index finger (alternately) at a position on the touch pad that they felt to be directly in a horizontal line with an imaginary point in the center of the bridge of their nose. Unfortunately, many subjects reported that they approached this task as if it were merely another form of target-pointing. Furthermore, their responses were erratic and the data were difficult to interpret. For these reasons, the results from these measures have been omitted from this report.

target-pointing was examined by testing both exposed and nonexposed hands.

General procedure and apparatus

At the outset of the testing period, subjects sat at a table with face pressed into the frame of a pair of prismless (normal-vision) goggles built into a box. Looking into this box, they viewed the reflection of a back-illuminated 1 × 1 inch cross, the apparent position of which was straight ahead at approximately eye level and at a distance of 48 cm. At the same distance was a vertically positioned 12 × 12 inch touch pad. For the preexposure (and later the postexposure) measures of target-pointing accuracy, subjects pointed alternately with the right and left index fingers (ten responses each), attempting to contact the touch pad at a place coincident with the apparent center of the cross. The inter-response interval was approximately 3 sec. View of the pointing hand was blocked by the mirror thereby precluding error-corrective visual feedback. When subjects touched the pad, the X and Y coordinates of the finger's position were immediately signaled and written to a floppy disk, using a program supported by an Apple II Plus computer.

During the prism-exposure period, the prismless goggles were replaced by binocular prisms (variable or fixed) and the mirror was moved out of the way, allowing subjects to see the touch pad as well as the hand when it was brought into view. In addition, a vertical rod serving as a hand-movement guide was situated parallel to and approximately 9 cm away from the surface of the pad.

The exposure period consisted of a series of 55-sec cycles. During the first half of each cycle, subjects, who were looking through the (upward- or downward-displacing) prisms, actively moved the preferred hand up and down along the guide rod, fixating the limb at all times. They grasped the rod with the thumb hooked around the rod and the palm of the hand facing them. Hand movements were made to the beat of a 1-Hz electronic metronome; the limb was moved up on the first beat, down on the next beat, and so forth, for exactly 27.5 sec. Then for the next 27.5 sec the subjects rested the hand on the table and fixated the cross while looking through the goggles, now set to produce displacement in the opposite direction. This was followed by 27.5 sec of observed hand movement, with the direction of prismatic displacement returned to its original state. Subjects alternated between these two displacements for a total of nineteen 55-sec cycles (17:25 min). Finally, postexposure measures of target-pointing accuracy were obtained in the same manner as the preexposure measures.

The conditions of fixed downward and fixed upward displacement were achieved by means of paired base-up and base-down wedge prisms, respectively. The prisms were attached to a sliding panel that moved them to a position directly in front of the goggle eyepieces. Variable displacement in the vertical dimension was produced by a pair of binocular, motor-driven Risley prisms which rotated in opposite directions; the net result was a binocular optical displacement that continuously changed in the vertical dimension over a range of ±30 diopters (±17.1°).

Measures of potential prism-adaptive shifts in target-pointing accuracy in the vertical dimension were obtained by subtracting (for each hand separately) the mean of the ten preexposure responses from the mean of the ten postexposure responses. Potential prism-induced changes in within-S variability of target pointing were determined by subtracting the standard deviation of a given subject's ten preexposure measures (for a particular hand) from the standard deviation of the corresponding ten postexposure measures.

Experiment 1

Design

Twelve subjects (8 males and 4 females, ages 19–33) were randomly divided into two six-subject groups. For one group the visual field was displaced upward during that half of each cycle in which the subject viewed the actively moving hand; for the other, the field was displaced downward. Subjects were tested individually in two conditions—variable displacement and fixed displacement—occurring 48 hr apart. The order of the two conditions was counterbalanced across subjects.

Procedure

Following the preexposure measures of open-loop target pointing, the mirror was removed and subjects looked through prismless (i.e., nondisplacing) goggles while undergoing the nineteen 55-sec cycles. On each cycle the hand was viewed for 27.5 sec, followed by 27.5 sec of viewing the target cross while the hand was resting on the table out of view. The purpose of this long period of normal vision was to establish an accurate and reliable baseline measure of each subject's perception of the hand's location under nondistorted visual circumstances before introducing the prismatic displacement. After a short rest break, subjects repeated the procedure, but this time they viewed the moving hand through prisms that were set either for fixed or for variable displacement. In order to reduce the possibility of significant loss of adaptation through spontaneous decay, the postexposure measures were obtained immediately after the subjects had viewed the prismatically displaced hand, which necessitated terminating the last cycle after the first 27.5 sec. It is important to note that because of this procedural decision the last view of the hand for the fixed displacement condition was one of 19 diopters

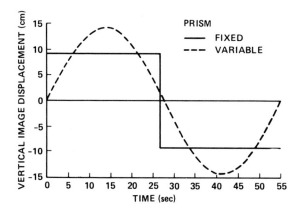

Figure 1. Prismatic exposure conditions: Fixed and variable prism displacements.

displacement, while for the variable-displacement condition it entailed little or no displacement (see Figure 1).

Results

As shown in Figure 2, prism-adaptive shifts in target-pointing accuracy for the exposed hand were obtained in the fixed, but not in the variable, displacement condition for both the upward and downward displacement groups. The finding of adaptive post-pre shifts for both directions of displacement confirms that these changes represent adaptation to the prisms *per se*, rather than some form of "drift" of pointing accuracy over time due to arm fatigue or other factors unrelated to the prismatic displacement. Analysis of variance revealed main effects for Direction (up/down), $F(1,4) = 14.49$, $p = 0.22$, and Displacement (variable/fixed), $F(1,4) = 30.01$, $p < 0.01$, and for the Direction/Displacement interaction, $F(1,4) = 82.14$, $p < 0.001$. Figure 2 indicates that the difference between the variable and fixed displacement conditions was greater for the upward displacement group. There was no main effect for order, nor was this factor involved in any interactions. Adaptation for the nonexposed hand (due to intermanual transfer) was obtained only for the fixed/upward displacement condition.

No statistically significant post-pre shifts in the dispersion (standard deviations) of target pointing were obtained for either hand in any condition.

Finally, for none of the conditions was there evidence of any decay of adaptation over the ten postexposure trials for either hand.

Discussion

Since adaptation occurred for fixed but not variable displacement, the answer to the original experimental question would seem to be that human observers

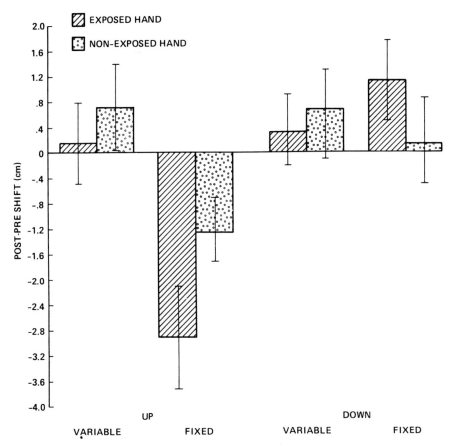

Figure 2. Experiment 1: Post-pre shifts (cm) in target-pointing accuracy.

are not capable of adapting to nonzero variable displacement, at least with exposure periods of the length used here. There is, however, an alternative possibility, raised by the fact that for subjects in the variable-displacement condition, the last experience during the prism exposure period was of normal or near-normal vision (Figure 1). It may be suggested that the adaptation produced in this experiment (or perhaps specifically in the variable-displacement condition) is quite *fragile* and therefore easily destroyed by subsequent exposure to normal vision. If so, then one might speculate that adaptation was actually produced in both conditions, but eliminated for the variable-displacement condition because of the "unlearning" that occurred at the very end of the exposure period. Experiment 2 attempted to examine this possibility by asking the following question: does the difference in adaptation in favor of fixed displacement that was obtained in Experiment 1 remain when the exposure period for the variable-displacement condition is caused to end on maximum displacement, rather than on no displacement?

Experiment 2

Design

Six subjects (2 males and 4 females, ages 21–39) were used as their own control in conditions of variable and fixed displacement in the upward direction only. The two conditions were separated by 48 hr and their order of occurrence counterbalanced across subjects.

Procedure

During the prism-exposure period, subjects viewed the preferred hand in the same manner as in Experiment 1, with the addition of one extra half-cycle. The latter ended after only 13.75 sec, which meant that the prismatic displacement for the variable condition was at its maximum of 30 diopters while the displacement for the fixed condition remained at its constant level of 19 diopters (see Figure 1).

Pre- and postexposure measures of target-pointing accuracy for both hands were taken in the same manner as in Experiment 1.

Results

As may be seen in Figure 3, prism-adaptive shifts in target-pointing accuracy were found for both variable and fixed-displacement conditions and both exposed and nonexposed hands. All of the post-pre shifts were significantly different from zero, but there were no main effects for the factors of Hand (exposed/non-exposed) or Displacement (variable/fixed), nor any interactions. Once again, no prism-induced changes in target-pointing precision (within-S standard deviations) or postexposure decay of adaptation were observed.

Discussion

The results of Experiment 2 are consistent with the "fragility hypothesis", since when the most recent visual experience in the variable-displacement condition was of maximum displacement, adaptation was substantial and, in fact, as great as that produced by fixed displacement. An interesting secondary finding was the large amount (i.e., 100 per cent) of intermanual transfer produced.

Conclusions

The present study has demonstrated that human subjects are capable of adapting their hand-eye coordination to nonzero variable displacement,

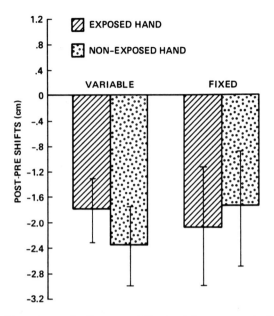

Figure 3. Experiment 2: Post-pre shifts (cm) in target-pointing accuracy.

although this adaptation is quite easily destroyed. It is possible, of course, that this fragility is unique to the current situation in which the prism-exposure task did not involve visual error-corrective feedback and exposure periods were repeatedly interrupted by rest periods. Furthermore, the present design does not allow us to exclude the possibility that the adaptation produced in the fixed-displacement condition was also fragile and would therefore have been quickly eliminated by exposure to normal vision.

A surprisingly large amount of adaptation was observed for the non-exposed hand, especially in Experiment 2. This may have been due to the use of alternating exposure and rest periods, since "distribution of practice" has been demonstrated to facilitate intermanual transfer of prism adaptation (e.g., Cohen, 1973). Such intermanual transfer has frequently been used as evidence that prism-adaptive changes in *vision* have occurred. Evidence against this interpretation of the present observations, however, comes from studies (e.g., Uhlarik and Canon, 1971) showing that prism exposure not involving target-pointing, is generally ineffective in producing this kind of adaptation. An alternative interpretation of intermanual transfer of prism adaptation is that it represents a central change in motor programming that is usable, at least partially, by the nonexposed hand.

Contrary to the results of Cohen and Held (1960), neither of the present experiments revealed an increase in the dispersion of target pointing as a result of exposure to variable displacement. Two explanations for this failure to replicate may be proposed. First, it is possible that the presence of only one target for the pre- and postexposure trials (in contrast to the four used by Cohen and Held, 1960) was conducive to a "stereotyping" of target-pointing responses. Such a potential constraint on trial-to-trial variability would be likely to counteract any disruptive effects that variable displacement might have on the within-subject dispersion of responses. Second, the present exposure period was relatively brief in comparison to that used in the Cohen-Held experiment. Indeed, in the latter, no increase in dispersion was obtained until after 30 min of variable displacement. In the present experiment, actual exposure to the hand (excluding the 27.5-sec "rest" periods) amounted to only a little over 8 min.

It is of interest to speculate why variable prismatic displacement should produce adaptation that is so easily destroyed (assuming that future research supports this conclusion). One possibility is that exposure to variable displacement causes the adaptive system to be quite labile and therefore easily changed, even by very brief exposures to new visual displacements or to normal vision. This interpretation fits with the finding by Cohen and Held (1960) of degraded hand-eye precision after exposure to variable displacement, but is weakened by the present failure to replicate the Cohen-Held observation.

A second possibility is that subjects exposed to variable displacement experience only "visual capture," a nearly instantaneous shift in felt limb position when viewing the prismatically displaced hand (Welch and Warren, 1980). Since visual capture is extremely fragile, it will be destroyed by even a brief exposure to normal vision and will also rapidly decay when view of the hand is precluded. The quick decay of visual capture, however, conflicts with the absence of postexposure decay in either of the present experiments, rendering this interpretation questionable.

The most likely explanation of the present results is that when human observers are actively exposed to a systematically changing prismatic displacement, they acquire the ability to adapt (or readapt) nearly instantaneously, as required. Such presumptive adaptive flexibility would represent a clear advance over the situation with fixed displacement since the latter involves relatively slow acquisition of adaptation and the presence of substantial after effects upon return to normal vision. In short, it is possible that prolonged exposure to variable displacement provides the observer with the ability to shift from one set of visuomotor relationships to another with a minimum of disruption. There is evidence to suggest that humans are capable of acquiring and maintaining multiple adaptations to disparate sensory environments as the result of repeated alternation between them. For example, it has been demonstrated that professional deep sea divers experience much less perceptual distortion when first viewing the underwater world through a face mask than do novice divers and likewise undergo minimal after effects upon emerging (Kinney *et al.*, 1970; Luria and Kinney, 1970; Luria *et al.*, 1967). Furthermore, subjects who are used as their own control in prism adaptation studies frequently reveal an "instantaneous," albeit partial, adaptive shift when returned to the testing room in which they had previously adapted (e.g., Hein, 1972; McGonigle and Flook, 1978). The possibility that multiple adaptations are involved in the present experiments is currently being evaluated.

References

Cohen, M.M. (1973). Visual feedback, distribution of practice, and intermanual transfer of prism aftereffects. *Perception in Motor Skills*, **37**, 599–609.

Cohen, M.M. and Held, R. (1960). Degrading visual-motor coordination by exposure to disordered re-afferent stimulation. Paper presented at meetings of *Eastern Psychological Association*, New York City.

Craske, B. (1967). Adaptation to prisms: Change in internally registered eye-position. *British Journal of Psychology*, **58**, 329–335.

Harris, C.S. (1965). Perceptual adaptation to inverted, reversed, and displaced vision. *Psychology Review*, **72**, 419–444.

Hein, A. (1972). Acquiring components of visually guided behavior. In Pick, A.D. (Ed.), *Minnesota Symposia on Child Psychology*. Minneapolis: University of Minneapolis Press.

Kinney, J.A.S., Luria, S.M., Weitzman, D.O. and Markowitz, H. (1970). *Effects of Diving Experience on Visual Perception Under Water*. NSMRL Report No. 612, U.S. Naval Submarine Medical Center, Groton, CN.

Luria, S.M. and Kinney, J.A.S. (1970). Underwater vision. *Science*, **167**, 1454–1461.

Luria, S.M., Kinney, J.A.S. and Weissman, S. (1967). Estimates of size and distance underwater. *American Journal of Psychology*, **80**, 282–286.

McGonigle, B.O. and Flook, J. (1978). Long-term retention of single and multistate prism adaptation in humans. *Nature*, **272**, 364–366.

Uhlarik, J.J. and Canon, L.K. (1971). Influence of concurrent and terminal exposure conditions on the nature of perceptual adaptation. *Journal Experimental Psychology*, **91**, 233–239.

Welch, R.B. (1978). *Perceptual Modification: Adapting to Altered Sensory Environments*. New York: Academic Press.

Welch, R.B. and Warren, D.H. (1980). Immediate perceptual response to intersensory discrepancy. *Psychology Bulletin*, **88**, 638–667.

Wilkinson, D.A. (1971). Visual-motor control loop: A linear system? *Journal of Experimental Psychology*, **89**, 250–257.

19

Visuomotor modularity, ontogeny and training high-performance skills with spatial instruments

Wayne L. Shebilske

Department of Psychology
Texas A&M University
College Station, Texas

Summary

Spatial display instruments convey information about the identity and the location of objects in order to assist surgeons, astronauts, pilots, blind individuals, and others in identification, remote manipulation, navigation, and obstacle avoidance (Bach-Y-Rita, 1972; Committee on Vision, 1986). Computer generated spatial displays are revolutionizing modern technology in automated systems such as aircraft collision-avoidance systems and virtual reality (VR) systems in which human operators interact with a computer generated image. Ironically, the net effect of automatization has been to increase the need for training human operators who have inherited critical roles that are difficult to learn and constantly changing (Rasmussen and Rouse, 1981). In fact, many automated systems operators must learn high-performance skills, which are defined by three characteristics: (a) more than 100 h are required to reach proficiency, (b) more than 20 per cent of highly motivated trainees fail to reach proficiency, and (c) substantial qualitative differences in performance exist between a novice and an expert (Schneider, 1985).

The purpose of the present paper is to discuss potential discrepancies between perception and sensory guided performance with a focus on implications for training. The empirical foundation of known perception-action incongruities is reviewed as is the theoretical foundation for understanding them in terms of visuomotor modularity. Inadequacies of these foundations with respect to training are suggested along with recommendations for steps toward understanding, predicting, and controlling stimulus information, perceptual impressions, and visuomotor performance during training of high-performance skills with spatial display instruments.

Discrepancies between two functions of vision

Two major functions of vision are (a) to provide a conscious representation of the environment (perception), and (b) to guide movements that organisms make as they interact with their environment (visuomotor control). An assumption common to many diverse philosophies and scientific theories is that visuomotor control is based on perception (Goodale, 1988). On one hand, there is already enough evidence to call this assumption into question in contexts that are relevant to spatial display technology. On the other hand, there is not enough evidence to support alternative principles for training perception-action skills.

Dramatic evidence of perception-action discord appeared in neurological studies. For example, Weiskrantz *et al.* (1974) reported that patients with striate lesions can point accurately to targets that they cannot "see". This phenomenon, called "blind-sight", has been replicated in several laboratories (e.g., Perenin and Jeannorod, 1975; Bridgeman and Staggs, 1982). The opposite discordance between perception and action has also been reported. Perenin and Vighetto (1983) described neurological patients who could verbally describe the spatial locations and relative positions of objects in space, but could not reach accurately toward the objects.

Psychophysical studies have also revealed mismatches between perception and visually guided motor control in healthy subjects. Lee and Thomson (1982), for instance, reported one that occurs when subjects stand on a stable floor surrounded by walls that oscillate back and forth. The oscillating visual stimulus influences a motor response of swaying forward and backward in phase with the stimulus. However, subjects remain unaware of the oscillations of either the stimulus or themselves.

Mismatches between perception and visually guided motor responses have also been observed during simple reaching tasks. For example, Goodale (1988) observed the performance of subjects who reached with their unseen hand for targets that appeared in peripheral vision. Subjects initiated eye movements and hand movements at about the same time. A first saccadic eye movement usually terminated short of the target and a second saccade brought the eye on target before the hand reached the target. On half of the trials the target remained in its original position. On the other trials, the target jumped to a new position 10 per cent further away than the original target. The second jump occurred during the first saccade, and subjects reported that they did not see it. (A control experiment showed that subjects performed at a chance level when they were asked to detect whether or not the second jump occurred.) Yet, subjects corrected the trajectory of their limb to accommodate the second displacement. They did so without increasing the duration of the limb movement that would have been required for a target in the same final position and without introducing a "breakpoint" in the velocity and acceleration profiles. Bridgeman (this volume) reports similar incongruities between conscious perception of locations and visually guided behavior.

A simple demonstration might give you first hand experience with perception-action discordance in a situation where you will probably expect concordance. Stand next to table and look down at a can with a small ball on top. Tilt your head forward so that it is a directly over the top of the can. Compare the apparent distance of the ball during binocular and monocular viewing. Then try hitting the ball during binocular and monocular viewing by swinging a ruler in one rapid ballistic swing parallel to the can surface at the level of the ball. Most people see no difference in apparent distance or they see the ball about 1cm farther away during monocular vision, which agrees with results reported by Roscoe (this volume). In contrast, most people hit the ball during binocular viewing but swing 3 cm or more above the ball during monocular viewing. If you are among the many people who are surprised to see themselves swing well above the ball, you will be in a better position to appreciate the main point of the demonstration, which is that we must go beyond our intuitions in training perception-action skills.

Our intuitions predict concordance between perception and action unless cognitive variables intervene or unless the motor control system is overloaded. In the ball hitting example, no cognitive variables come to mind for those subjects who see no apparent change in distance and the motor task seems simple since it is done easily during binocular vision. Accordingly, perceptual impressions and action responses should be interchangeable in training people to hit a ball during monocular vision. This prediction would be correct if perception and visually guided action shared the same input operations. But evidence is becoming increasingly clear that they do not.

Current theories of separate visual input operations

When comparing theories that can account for separate input operations, it is useful to note that they all begin with the conversion of distal information, which is in the environment, into proximal information, which interfaces the sensory system with the environment. Theories differ, however, in how proximal patterns relate to perception and visually guided action.

The "New Look" school emphasizes top-down influences. For example, Gogel (1977) stated that objects can be cognitively judged to be in a different location than they appear and that performance can reflect these cognitive judgments. Many information processing models come out of this tradition (e.g., Foley, this volume), and are characterized in Figure la as perception plus transformation models.

The "New Look" emphasis on top-down processing has been challenged by Fodor (1983) in his book, *The Modularity of Mind*. He argues that input operations are insulated from other information and beliefs that an observer has about the environment. They are also insulated from each other and operate as independent and parallel "smart systems" (Pylyshyn, 1984). Such modular models characterize most computational approaches to vision, including the pioneering work of Marr (1982). Many cognitive scientists, however, remain committed to the old tradition of identifying vision with visual perception. In this respect, such a modular approach is no different than the "New Look" school. Models from both approaches can be characterized as perception plus transformation models (Figure 1a). Consider, for example, the model by Marr (1982). He suggested that a single 2½D sketch (a viewer centered abstract symbolic representation of spatial layout) is constructed in order to serve all sensory guided systems (including those that use 3D object centered frames of reference) and that different systems utilize this representation according to different rules to suit different purposes. On this account, modularity comes into play after perceptual processes have structured an initial representation of spatial layout.

In contrast, Figure 1b illustrates a two modes of visual representation model according to which separate input modules diverge already at the initial sampling of the proximal pattern before abstraction processes begin. Examples of such multiple modes of visual representation models include those proposed by Bridgeman (this volume) and by Goodale (1988). Such models were motivated in part by neurophysiological studies into the organization of the visual system. Bridgeman (this volume), for example, cites evidence that the visual world is represented by several topographic maps in the cortex (Van Essen *et al.*, 1982). Goodale (1988) cites evidence of independent visuomotor modules that are separate from retinal projections right through to motor nuclei (e.g., Ingle, 1982). In other words, these multiple modes of visual representation models postulate separate hard-wired visual modules.

At the same time, it is recognized that such modules alone cannot account for the integration and coordination of action in a stimulus rich environment. Goodale (1988), for instance, calls upon researchers to work out ways that visuomotor modules are integrated in behaving animals. The remainder of this paper argues that ecological considerations must guide these efforts and their application to training in spatial display technology.

Toward an ecological theory of separate visual representations

Ecological theories are based on the axiom that operations for encoding sensory information should approach optimal efficiency in the environment in which a species evolved (Gibson, 1979; Shebilske and Fisher, 1984; Shebilske *et al.*, 1984; Turvey, 1979; Turvey and Solomon, 1984). The dominant ecological theories have adhered to a direct perception approach, which assumes a one-to-one correspondence between stimulus information, perception, and action (e.g., Gibson, 1979; Turvey and Solomon, 1984). Accordingly, this approach allows no room for multiple visual representations. In contrast, Shebilske (1984; 1987a; 1987b) proposed an ecological efference mediation theory according to which efference-based information (e.g., from the oculomotor system) interacts with higher order light-based information (e.g., from optical flow patterns) to determine performance during natural events. Thus, fallibility in both visual and efference-based information functions synergistically to shape both the phylogeny and ontogeny of the visual system (Shebilske et al., 1984). This theory adds the important dimension of ontological development to analyses of multiple visual representations.

The ecological considerations that motivated the theories that we considered earlier are limited to phylogeny. For example, Goodale (1988, p. 263) notes that "vision evolved in vertebrates and other organisms to control movements.... In 'designing' these control systems, natural selection appears to have favored the development of a visual system that consists of a network of relatively independent sensorimotor channels, each of which supports a particular kind of visually guided behavior. ...In a now classic series of studies of vision in the frog, *Rana pipiens*, David Ingle (1973, 1982) was able to show that visually elicited feeding, visually elicited escape, and visually guided avoidance of barriers are mediated by entirely separate pathways from the retina to the motor nuclei."

In addition to studying such phylogenetic development, the ecological efference mediation theory demands consideration of ontological development throughout the life-span. The question is whether ecological interactions between an organism and its environment can lead to the development of multiple visual representations. Recent neurophysiological theories suggest that such ontological development might be possible. Pribram (1988), for example, proposed that distributed memorial representations affect the structure of dendritic networks throughout the visual pathways. Each synapse is a processing unit and networks of synapses enable us to extract spatial-temporal relationships from our environment. Past experience can modify the structure of dendritic networks and thereby modify the way that they extract information from the environment (e.g., Kandel and Schwartz, 1982). Pribram argues that such distributed memory representations can affect the earliest stages of sensory information processing. Accordingly, interactions with the environment could yield multiple visual representations.

Within cognitive science, such theorizing has characterized parallel distributed processing (PDP) computational models (e.g., Feldman and Ballard, 1982: Hinton and Anderson, 1981; Pribram, 1988; Rumelhart and McClelland, 1986; Smolensky, 1988). Such models combined with the ecological efference mediation theory provide a heuristic framework for pursuing empirical evidence that ecological interactions during natural events can produce separate visual representations.

Specifically, Figure 1c illustrates a hypothesis about Ecologically Insulated Event Input Operations (EIEIOs). This hypothesis postulates parallel input

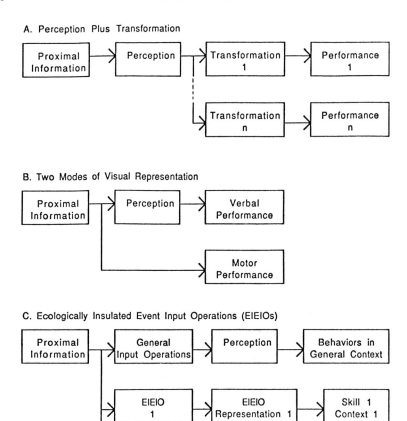

Figure 1. Hypotheses proposing framework within which to investigate the many-to-one relationship existing between stimulus information, perceptual impressions and performance.

modules that are molded by interactions of an organism with its environment in an attempt to achieve maximally efficient performance of sensory guided skills within the prevailing internal and external states in which the skill is performed. Each module structures its own representation of spatial-temporal relationships according to its own set of rules in order to fulfill its unique function.

The EIEIO hypothesis is designed to account for human adaptability to the wide range of conditions in which perception and action systems must operate. Perception guides our behavior day and night in highly structured environments such as forests and in relatively low structured environments such as snow covered fields. In contrast, specific skills are often performed within a narrow range of this total variability. Night landing on an aircraft carrier, for instance, includes only the low-information end of the total range, and gymnastics usually includes only the high-information end. It would be efficient, therefore, to develop separate representations for visual perception and visuomotor coordination in specific skills. Whereas input operations corresponding to conscious representations (e.g., apparent target location) must be designed to be maximally efficient over the entire range of contextual variability to which an organism is exposed, EIEIOs need only to be maximally efficient within a narrower range of contextual variability within which a particular skill (e.g., moving an unseen hand quickly and accurately to a target) is performed. It is tentatively proposed further that an EIEIO representation is only utilized when a skill is performed in the environment in which it was learned. Specifically, interaction between an organism and its environment directly determines the formation and selection of modules without conscious decisions. Following the lead of Pribram (1988), I call the bases of the selection process *ecoholonomic matches* (eco = ecological; holo = holistic; nomic = lawful). These matches are assumed to occur between stimulus information and distributed memorial information in parallel channels of dendritic networks. Ecological experiences during ontogeny modify the dendritic networks so that parallel channels in the visual system are adapted to resonate to the appropriate internal and external conditions. Through coordinated cooperative processing these modules afford an adaptive advantage relative to all the conditions to which humans are exposed (Shebilske, 1987a; 1987b).

This ontological development is hypothesized to occur when an organism has the opportunity to perform a skill repeatedly in an environment that (a) has contextual variability over a narrower range than the one for the perceptual system, and (b) provides an opportunity to learn that the conscious representation is inefficient. Such conditions could be incorporated in training high-performance skills with spatial display instruments.

Separate visual representations, training and spatial displays

An important theme for training is that practice can change the way sensory information is used to guide performance. Schmidt (1987) reviewed the history of thought on this theme starting with William James' observation (1890) that practice of skills seems to lead to more automatic, less mentally taxing behavior. This observation spawned considerable research leading to evidence for three separate process level changes that seem to contribute to this practice effect as follows: (a) tasks that are slow and guided shift from dependence on exproprioceptive information to dependence on proprioceptive information (e.g., Adams and Goetz, 1973); (b) tasks that have predictable parameters, such as predictable target locations in pointing tasks, shift to open-loop control (e.g., Schmidt and McCabe, 1976); and (c) tasks that have unpredictable parameters shift to fast, automatic, and parallel processing of the information needed to make decisions (e.g., Schneider and Shiffrin, 1977). The EIEIO hypothesis is proposed as a fourth kind of practice effect in which training produces separate visuomotor modules with their own visual representations.

This fourth possibility incorporates numerous implications of two relatively recent theoretical concepts: (a) separate visual representations with the demand for analysing visuomotor, not visual, modularity (Goodale, 1988), and (b) PDP systems with many suggestions such as memorial mediation without increased reaction time (Pribram, 1988).

These ideas must merge with research that is amassing on cognitive issues in human-machine interactions including capacity limitations, attention, decision processes, automatic and controlled processes, movement control, adaptive control, and other cognitive issues related to training (e.g., Frese et al., 1987; Hancock, 1987; Hancock and Chignell, 1989; Helander, 1988; Rasmussen, 1986). Important training principles were summarized by Schneider (1985). Four principles were: train components of a skill, promote consistent processing, vary aspects of the task that vary in the operational situation, and come as close to the functional mission context as possible without overloading the trainee. The EIEIO hypothesis provides a rationale for these guidelines and suggests another: train components that integrate visuomotor modules as opposed to separate visual and motor components.

Consider this additional guideline in the context of training with VR displays, which are so realistic that observers get an impression of actually being at the scene. Held (this volume) outlined a framework for analysing determinants of compellingness of these impressions. Compellingness is surely desirable in VR displays for many reasons, but when it comes to training, separate visuomotor modules imply that variables, such as time lags, might have different effects on perceptual compellingness and performance proficiency. Accordingly, training programs should integrate visuomotor modules and evaluate progress toward acceptable performance accuracy. Similarly, scientists might find that factors influence perceptual impressions and performance differently in training pilots to respond appropriately to relative vertical separation in collision-avoidance situations (Sherry Chappell, personal communication, September 1, 1987). Scientists studying stereopsis (e.g., Enright, this volume; Foley, this volume; Schor, this volume; and Stevens, this volume) might also find different training factors affecting impressions of depth and performance with 3D displays.

Finally, the principle of visuomotor modularity during training is relevant to research on questions raised by performance in unique contexts. For instance, Hart and Brickner (this volume) summarized unusual stimulus characteristics of monocular helmet mounted displays and reported location errors, which are attributed to a displaced sensor. Other factors might also be involved. The display could cause localization errors during performance even when locations are seen accurately. Recent evidence suggests further that localization errors might be related to lateral heterophoria, the tendency to fixate in front of or beyond a target when vision in the two eyes is dissociated (Park and Shebilske, 1990; Shebilske, 1990). Current research is investigating whether these errors can be overcome through training that repeatedly demands integrated visuomotor responses in the new context. The EIEIO hypothesis provides a heuristic for this and other research aimed at understanding capacities for perceptual and performance adaptability in unique contexts.

References

Adams J.A. and Goetz, E.T. (1973). Feedback and practice of variables in error detection and correction. *Journal of Motor Behavior*, **5**, 217–224.
Bach-Y-Rita, P. (1972). *Brain Mechanisms in Sensory Substitution*. New York: Academic Press.
Bridgeman, B. and Staggs, D. (1982). Placticity in human blindsight. *Vision Research*, **22**, 1199–1203.
Committee on Vision, National Research Council. (1986). *Electronic Travel Aids: New Directions for Research*. Washington, D.C.: National Academy Press.
Feldman, J.A. and Ballard, D.H. (1982). Connectionist models and their properties. *Cognitive Psychology*, **6**, 205–254.
Fodor, J. (1983). *The Modularity of Mind*. Cambridge, MA: MIT Press.
Frese, M., Ulich, E. and Dzida, W. (Eds) (1987). *Psychological Issues of Human-Computer Interaction in the Work Place*. New York: Elsevier.
Gibson, J.J. (1979). *The Ecological Approach to Visual Perception*. Boston: Houghton-Mifflin.
Gogel, W.C. (1977). The metric of visual space. In Epstein, W. (Ed.), *Stability and Constancy in Visual Perception: Mechanisms and Processes*. New York: Wiley.
Goodale, M.A. (1988). Modularity in visuomotor control: from input to output. In Pylyshyn, Z.W. (Ed.) *Computational Processes in Human Vision: An Interdisciplinary Perspective*. Norwood, N.J.: Ablex Publishing Co.
Hancock, P.A. (Ed.) (1987). *Human Factors Psychology*. New York: Elsevier.
Hancock, P.A. and Chignell, M.H. (Eds) (1989). *Intelligent Interfaces*. New York: Elsevier.
Hinton, G.E. and Anderson, J.A. (1981). *Parallel Models of Associative Memory*. Hillsdale, NJ: Erlbaum.
Helander, M. (Ed.) (1988). *Handbook of Human-Computer Interaction*. New York: Elsevier.
Ingle, D.J. (1973). Two visual systems in the frog. *Science*, **181**, 1053–1055.
Ingle, D.J. (1982). Organization of visuomotor behaviors in vertebrates. In Ingle, D.J., Goodale, M.A. and Mansfield., R.J.W. (Eds), *Analysis of Visual Behavior*. Cambridge, MA: MIT Press.
James, W. (1890). *The Principles of Psychology*. New York: Holt.
Kandel, E.R. and Schwartz, J.H. (1982). Molecular biology of learning: Modulation of transmitter release. *Science*, **218**, 433–443.
Lee, D.N. and Thomson, J.A. (1982). Vision in action: the control of locomotion. In Ingle, D.J., Goodale, M.A. and Mansfield, R.J.W. (Eds), *Analysis of Visual Behavior*. Cambridge, MA: MIT Press.
Marr, D. (1982). *Vision*. New York: Freeman.

Park, K. and Shebilske, W.L. (1990). Monocular perception of egocentric direction and Hering's laws of visual direction. *Journal of Experimental Psychology: Perception and Performance*, in press.

Perenin, M.T. and Jeannerod, M. (1975). Residual vision in cortically blind hemifields. *Neuropsychologica*, 13, 1–7.

Perenin, M.T. and Vighetto, A. (1983). Optic ataxia: a specific disorder in visuomotor coordination. In Hein, A. and Jeannerod, M. (Eds), *Spatially Oriented Behavior*. New York: Springer-Verlag.

Pew, R.W. (1986). Human performance issues in the design of future Air Force systems. *Aviation. Space, and Environmental Medicine*, 57, 78–82.

Pribram, K.H. (1988). *A Holonomic Brain Theory: Cooperativity and Reciprocity in Processing the Configural and Cognitive Aspects of Perception*. Hillsdale, NJ: Erlbaum.

Pylyshyn, Z.W. (1984). *Cognition and Computation: Toward a Foundation for Cognitive Science*. Cambridge, MA: MIT Press.

Rasmussen, J. (1986). *Information Processing and Human-Machine Interaction*. New York: Elsevier.

Rasmussen, J. and Rouse, W.B. (Eds) (1981). *Human Detection and Diagnosis of System Failures*. New York: Plenum Press.

Rumelhart, D.E. and McClelland, J.L. (Eds) (1986). *Parallel Distributed Processing I and II*. Cambridge, MA: MIT Press.

Schmidt, R.A. (1987). The acquisition of skill: some modifications to the perception-action relationship through practice. In Heuer, H. and Sanders, A.F. (Eds), *Perspectives on Perception and Action*. New Jersey: Erlbaum.

Schmidt, R.A. and McCabe, J.F. (1976). Motor program utilization over extended practice. *Journal of Human Movement Studies*, 2, 239–247.

Schneider, W. (1985). Training high performance skills: Fallacies and guidelines. Special Issue: Training. *Human Factors*, 27, 205–300.

Schneider, W. and Shriffin, R.M. (1977). Controlled and automatic human information processing: I. Detection, search, and attention. *Psychological Review*, 84, 1–66.

Shebilske, W.L. (1984). Efferent Factors in Cognition and Perception. In Prinz, W. and Sanders, A.F. (Eds), *Cognition and Motor Processes*. New York: Plenum.

Shebilske, W.L. (1987a). An ecological efference mediation theory of natural event perception. In Prinz, W. and Sanders, A.F. (Eds), *Perspectives on Perception and Action*. Hillsdale, NJ: Erlbaum.

Shebilske, W.L. (1987b). Baseball batters support an ecological efference mediation theory of natural event perception. In Bouwhuis, D.G., Bridgeman, B., Owens, D.A., Shebilske, W.L. and Wolff, P. (Eds), *Sensorimotor Interactions in Space Perception and Action*. Amsterdam: North-Holland.

Shebilske, W.L. (1990). Lateral heterophoria and dart throwing errors. Paper presented at the meeting of the *Association for Research in Vision and Ophthalmology*, May.

Shebilske, W.L. and Fisher, S.K. (1984). Ubiquity of efferent factors in space perception. In Semmlow, J.L. and Welkowitz, W. (Eds) *Frontiers of engineering and computing in health care*. New York: IEEE Publishing.

Shebilske, W.L. and Proffitt, D.R. (1983). Paradoxical retinal motion during head movements: apparent motion without equivalent apparent displacement. *Perception and Psychophysics*, 34, 467–481.

Shebilske, W.L., Proffitt, D.R. and Fisher, S.K. (1984). Efferent factors in natural event perception can be rationalized and verified: a reply to Turvey and Solomon. *Journal of Experimental Psychology: Human Perception and Performance*, 10, 455–460.

Smolensky, P. (1988). On the proper treatment of connectionism. *Behavioral and Brain Science*, 11(11), No. 1, 1–74.

Turvey, M.T. (1979). The thesis of the efference-mediation of vision cannot be rationalized. *Behavioral and Brain Sciences*, **2**, 59–94.

Turvey, M.T. and Solomon, J. (1983). Visually perceiving distance: a comment on Shebilske, Karmiohl, and Proffitt. *Journal of Experimental Psychology: Human Perception and Performance*, **10**, 449–454.

Van Essen, D.C., Newsome, W.T. and Bixby, J.L. (1982). The pattern of interhemispheric connections and its relationship to extrastriate visual areas in the macaque monkey. *Journal of Neuroscience*, **2**, 265–283.

Weiskrantz, L., Warrington, E., Sanders, M.D. and Marshall. J. (1974). Visual capacity in the hemianopic field following restricted occipital ablation. *Brain*, **97**, 709–729.

20

Separate visual representations for perception and for visually guided behavior

Bruce Bridgeman

Department of Psychology
University of California
Santa Cruz, California

Summary

Converging evidence from several sources indicates that two distinct representations of visual space mediate perception and visually guided behavior, respectively. The two maps of visual space follow different rules; spatial values in either one can be biased without affecting the other. Ordinarily the two maps give equivalent responses because both are veridically in register with the world; special techniques are required to pull them apart. One such technique is saccadic suppression: small target displacements during saccadic eye movements are not perceived, though the displacements can change eye movements or pointing to the target.

A second way to separate cognitive and motor-oriented maps is with induced motion: a slowly moving frame will make a fixed target appear to drift in the opposite direction, while motor behavior toward the target is unchanged. The same result occurs with stroboscopic induced motion, where the frame jumps abruptly and the target seems to jump in the opposite direction.

A third method of separating cognitive and motor maps, requiring no motion of target, background or eye, is the "Roelofs effect": a target surrounded by an off-center rectangular frame will appear to be off-center in the direction opposite the frame. Again the effect influences perception, but in half of our subjects it does not influence pointing to the target. This experience also reveals more characteristics of the maps and their interactions with one another—the motor map apparently has little or no memory, and must be fed from the biased cognitive map if an enforced delay occurs between stimulus presentation and motor response.

In designing spatial displays, the results mean that "what you see isn't necessarily what you get". Displays must be designed with either perception or visually guided behavior in mind.

The visual world is represented by several topographic maps in the cortex (Van Essen et al., 1982). This characteristic of the visual system raises a fundamental question for visual physiology: do all of these maps work together in a single visual representation, or are they functionally distinct? And if they are distinct, how many functional maps are there and how do they communicate with one another? Because these questions concern visual function in intact organisms, they can be answered only with psychophysical techniques. This paper presents evidence that there are at least two functionally distinct representations of the visual world in normal humans; under some conditions, the two representations can simultaneously hold different spatial values. Further, we are beginning to understand some of the ways in which the representations communicate with one another.

Experiments in several laboratories have revealed that subjects are unaware of sizeable displacements of the visual world if they occur during saccadic eye movements, implying that information about spatial location is degraded during saccades (Ditchburn, 1955; Wallach and Lewis, 1965; Brune and Lücking, 1969; Mack, 1970; Bridgeman et al., 1975). Yet people do not become disoriented after saccades, implying that spatial information is maintained. Experimental evidence supports this conclusion. For instance, the eyes can saccade accurately to a target that is flashed (and mislocalized) during an earlier saccade (Hallett and Lightstone, 1976), and hand-eye coordination remains fairly accurate following saccades (Festinger and Cannon, 1965). How can the loss of perceptual information and the maintenance of visually guided behavior exist side by side?

To begin a resolution of this paradox, we noted that the two kinds of conflicting observations use different response measures. The saccadic suppression of displacement experiments requires a nonspatial verbal report or button press, both symbolic responses. Successful orienting of the eye or hand, in contrast, requires quantitative spatial information. The conflict might be resolved if the two types of report, which can be labeled as cognitive and motor, could be combined in a single experiment. If two pathways in the visual system process different kinds of information, spatially oriented motor activities might have access to accurate position information even when that information is unavailable at a cognitive level that mediates symbolic decisions such as button pressing or verbal response. The saccadic suppression of displacement experiments cited above address only the cognitive system.

In our first experiment on this problem (Bridgeman et al., 1979), the two conflicting observations (saccadic suppression on one hand and accurate motor behavior on the other) were combined by asking subjects to point to the position of a target that had been displaced and then extinguished. Subjects were also asked whether the target had been displaced or not. Extinguishing the target, and preventing the subjects from viewing their hands (open-loop pointing), guaranteed that only internally stored spatial information could be used for pointing. On some trials, the displacement was detected, while on others it went undetected, but pointing accuracy was similar whether the displacement was detected or not.

This result implied that quantitative control of motor activity was unaffected by the perceptual detectability of target position. But it is also possible (if a bit strained) to interpret the result in terms of signal detection theory as a high response criterion for the report of displacement. The first control for this possibility was a two-alternative, forced-choice measure of saccadic suppression of displacement, with the result that even this criterion-free measure showed no information about displacement to be available to the cognitive system under the conditions where pointing was affected (Bridgeman and Stark, 1979).

A more rigorous way to separate cognitive and motor systems was to put a signal only into the motor system in one condition and only into the cognitive system in another. We know that induced motion affects the cognitive system, because we experience the effect and subjects can make verbal judgments of it. But the above experiments implied that the information used for pointing might come from sources unavailable to perception. We inserted a signal selectivity into the cognitive system with stroboscopic induced motion (Bridgeman et al., 1981). A surrounding frame was displaced, creating the illusion that the target had jumped, although it remained fixed relative to the subject. Target and frame were then extinguished, and the subject pointed open-loop to the last position of the target. Trials where the target had seemed to be on the left were compared with trials where it had seemed to be on the right. Pointing was not significantly different in the two kinds of trials, showing that the induced-motion illusion did not affect pointing.

Information was inserted selectively into the motor system by asking each subject to adjust a real motion of the target, jumped in phase with the frame, until the target was stationary. Thus the cognitive system specified a stable target. Nevertheless, subjects pointed in significantly different directions when the target was extinguished in the left or the right positions, showing that the difference in real target positions was still available to the motor system. The visual system must have picked up the target displacement, but not reported it to the cognitive system, or the cognitive system could have ascribed the visually specified displacement to an artifact of frame movement. Thus a double dissociation occurred: in one condition the target displacement affected only the cognitive system, and in the other it affected only the motor behavior.

Dissociation of cognitive and motor function has also been demonstrated for the oculomotor system by creating conditions in which cognitive and motor systems receive opposite signals at the same time. Again the experiment involved stroboscopic-induced motion; a target jumped in the same direction as a frame, but not far enough to cancel the induced motion. The spot still appeared to jump in the direction opposite the frame, while it actually jumped in the same direction. Saccadic eye movements followed the veridical direction even though subjects perceived stroboscopic motion in the opposite direction (Wong and Mack, 1981). If a delay in responding was required, however, eye movements followed the perceptual illusion, implying that the motor system has no memory and must rely on information from the cognitive system under these conditions.

All of these experiments involve motion or displacement, leaving open the possibility that the dissociations are associated in some way with motion systems rather than with representation of visual space *per se*. A new series of experiments in my laboratory, however, has demonstrated dissociations of cognitive and motor function without any motion of the eye or the stimuli at any time. The dissociation is based on the Roelofs effect (Roelofs, 1935), a tendency to misperceive target position, in the presence of a surrounding frame presented asymmetrically, in the direction opposite the offset of the frame. The effect is similar to a stroboscopic induced motion in which only the final positions of the target and frame are presented (Bridgeman and Klassen, 1983).

Method

Subjects

The subjects were nine undergraduate volunteers and the author. Six of the subjects were naive with respect to the purposes of the experiment; the others assisted with the experiments, as well as serving as subjects.

Apparatus

Subjects sat with stabilized heads before a hemicylindrical screen that provided a clear field of view 180° wide × 50° high. A rectangular frame 21° wide × 8.5° high × 1° in width was projected, via a galvanic mirror, either centered on the subject's midline, 5° left, or 5° right of center. Inside the frame, an "x" 0.35° in diameter could be projected via a second galvanic mirror in one of five positions, 2° apart, with the middle "x" on the subject's midline (Figure 1). A pointer with its axis attached to a potentiometer mounted near the center of curvature of the screen and its tip near the screen gave a voltage proportional to the tip's position, with a simple analog circuit. The voltage was fed into an A/D converter of a laboratory computer that controlled trial presentation and data collection. Perceived target position was recorded from a detachable computer keyboard placed in front of the subject. All keys except the five keys corresponding to the five target positions were masked off.

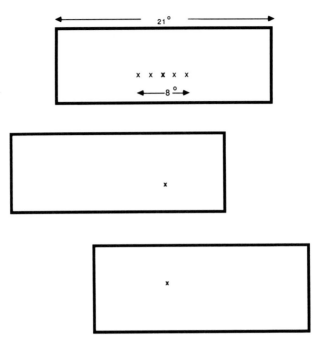

Figure 1. Stimulus array used in pointing/judging experiments. The frame could be centered (top), biased 5° left (middle), or biased 5° right (bottom). A target appeared in one of the five positions indicated in the top frame. Other frames show the position of the center target.

Procedure

Training

Subjects were first shown the five possible positions of the target in sequence on an otherwise blank screen. Then they saw targets exposed for 1 sec and estimated their positions with the five response keys ("judging trials"), until they were correct in five consecutive trials. Next, they were trained on pointing, with the same stimuli ("pointing trials"), until they spontaneously returned the pointer to its rightmost position (as initially instructed) for five consecutive trials. In both conditions, subjects were instructed to wait until the offset of the stimulus before responding. Presentation of the target alone forced the subjects to use an egocentric judgment, and the long display time reduced the possibility of target onset eliciting a spurious motion signal that might affect responses.

No delay condition

The 30 types of judging and pointing trials were mixed in a pseudorandom order. Each trial type was repeated 5 times, for a total of 150 trials/block.

Trial order was restricted so that pointing trials and judging trials with the same target and frame positions would alternate in the series. At stimulus offset, subjects heard a short "beep" tone to indicate a judging trial or a longer "squawk" tone to indicate a pointing trial. There was a rest period after each 50 trials.

Trials were collated by the computer and a separate two-way ANOVA was run for each response type (assessing target main effect, frame main effect, and interaction).

Delay condition

Procedures were the same except that a 4 sec interval was interposed between stimulus offset and the tone that indicated the type of response.

Results

No delay condition

For all subjects, there was a significant main effect of target position in both trial types and a significant main effect of frame position for judging trials. Thus, all subjects showed a Roelofs effect (Figure 2).

The main effect of frame position in pointing trials showed a sharp division of the subjects into two groups: 5 of the 10 subjects showed a highly significant Roelofs effect ($p < 0.005$), while the other 5 showed no sign of an effect ($p > 0.18$). Thus, responses to pointing and judging trials were qualitatively different for half of the subjects, showing a Roelofs effect only for judging.

Four of the five subjects who showed a Roelofs effect in pointing were females. Thus, a sex effect is possible in this condition, with females more likely to code the target position in a symbolic form. The number of subjects, however, is too small to draw firm conclusions on this issue.

Delay condition

With a 4 sec delay interposed between display offset and tone, 9 of the 10 subjects showed a significant Roelofs effect for the judging task ($p < 0.01$) and 8 of the 10 showed a significant effect for the pointing task. One of the two remaining subjects showed no significant effect of frame position for either task. The other subject whose pointing behavior still showed no effect of the frame (Figure 3) was retested with an 8 sec delay between display offset and tone. A Roelofs effect was found for both pointing and judging trials ($p < 0.001$) (Figure 4).

In summary, interposing a long enough delay before the response cue forces all subjects to use pointing information that is vulnerable to bias from

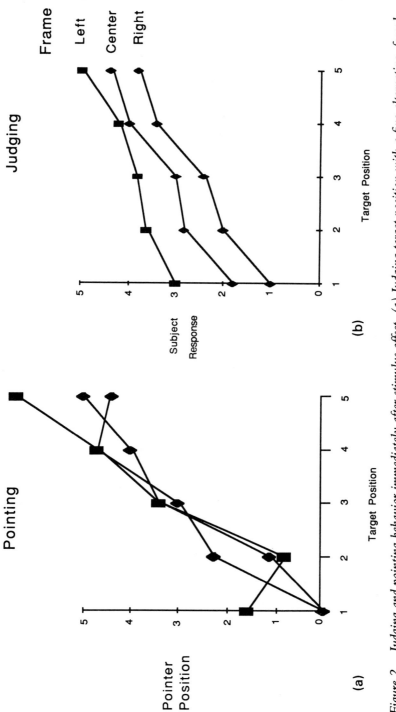

Figure 2. Judging and pointing behavior immediately after stimulus offset. (a) Judging target position with a five-alternative, forced-choice procedure. The separation of the three curves corresponding to the three frame positions is due to the Roelofs effect. (b) Pointing to targets under the same perceptual conditions, in trials intermingled with the judging trials. Overlap of the three curves indicates lack of influence of frame position on pointing behavior. Data are from one subject.

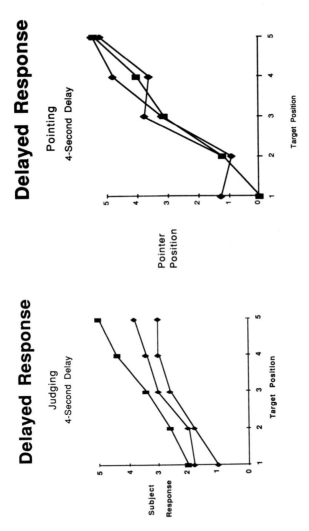

Figure 3. *Judging and pointing after a 4 sec delay. In this subject, no Roelofs effect is evident for pointing; the other subjects showed an effect at this delay.*

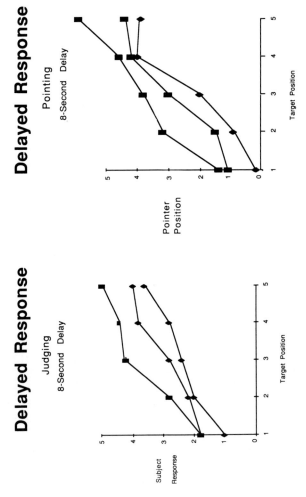

Figure 4. Judging and pointing after an 8 sec delay. A Roelofs effect for pointing has appeared.

the frame position, even though half of the subjects were not vulnerable to this bias when responding immediately.

Discussion

These experiments show that perception of a Roelofs effect is robust, being seen by nearly all subjects under all delays. The Roelofs effect in visually guided behavior, though, depends much more strongly on the subjects and conditions. Half of the subjects showed an effect of a surrounding frame on pointing behavior. The remainder showed the effect only when a long enough delay was interposed between target presentation and response.

The appearance of the Roelofs effect with a delay between stimulus and motor response is reminiscent of the results of Wong and Mack (1981): saccadic eye movements followed a veridical motion with a short delay, but followed a perceived motion in the opposite direction after a longer delay. If eye movements and visually guided behavior of the arm were controlled by a single motor-oriented internal map of the visual world, then we would expect the effects of delay to influence eye and arm similarly, and the Wong and Mack results and our results could be explained in the same way.

There is now some evidence that oculomotor and skeletal motor systems do indeed share one map of visual space (Nemire and Bridgeman, 1987). Normally, eye and hand behavior are not correlated (Prablanc *et al.*, 1979), in our interpretation because eye and hand motor systems read their information from the same visual map through separate, independent noise sources. To show the identity of visual information driving these two systems, we disturbed the normally veridical mapping process by having subjects make repeated saccades in darkness. This resulted in saccade undershoot, but equally great undershoot of manual pointing.

Our conclusion is that the normal human possesses two maps of visual space. One of them holds information used in perception: if subjects are asked what they see, the information in this "cognitive" map is accessed. The other map drives visually guided behavior, for both eye and arm. The "motor" map is not subject to illusions such as induced motion and the Roelofs effect. In this sense it is more robust, but as a result it is less sensitive to small motions or fine-grained spatial relationships. It also has no memory, being concerned only with the here-and-now correspondence between visual information and motor behavior. If a subject must make motor responses to stimuli no longer present, this system must take its spatial information from the cognitive representation, and brings any cognitively based illusions along with it.

An alternative explanation of the results has been suggested (Ian Howard, personal communication, Sept. 2, 1987); presentation of an off-center frame might bias the subject's subjective straight-ahead in the same direction as the frame's offset. Judging of point position would then be biased in the opposite direction because the subject bases his or her judgments on an offset straight ahead direction. Pointing, however, would remain the same because the subject has not in fact moved, and arm position must be egocentric. This alternative can be tested empirically by having subjects point to the center of the apparatus when the frame is presented in center, left, or right position. Preliminary data from three subjects indicate that frame position has no effect on pointing straight ahead.

Finally, we can apply this conception of two maps of visual space to design of spatial displays. Any display where perception is the primary goal, such as displays of the status of instruments, is subject to induced-motion illusions, Roelofs effects, and other cognitive biases. The designer can take advantage of these effects in designing such displays, but must beware that they do not distort the data displayed.

Displays which guide real-time behavior, on the other hand, are not subject to such illusions. The designer need not worry, for instance, about background motions affecting visually guided behavior toward a target (Bridgeman et al., 1981). But information must be available continuously, for the internal map guiding these behaviors has no significant memory.

References

Bridgeman, B. and Klassen, H. (1983). On the origin of stroboscopic induced motion. *Perception and Psychophysics*, **34**, 149–154.
Bridgeman, B. and Stark, L. (1979). Omnidirectional increase in threshold for image shifts during saccadic eye movements. *Perception and Psychophysics*, **25**, 241–243.
Bridgeman, B., Hendry, D. and Stark, L. (1975). Failure to detect displacement of the visual world during saccadic eye movements. *Vision Research*, **15**, 719–722.
Bridgeman, B., Kirch, M. and Sperling, A. (1981). Segregation of cognitive and motor aspects of visual function using induced motion. *Perception and Psychophysics*, **29**, 336–342.
Bridgeman, B., Lewis, S., Heit, G. and Nagle, M. (1979). Relation between cognitive and motor-oriented systems of visual position perception. *Journal of Experimental Psychology: Human Perception and Performance*, **5**, 692–700.
Brune, F. and Lücking, C.H. (1969). Oculomotorik, Bewegungswahrnehmung und Raumkonstanz der Sehdinge. *Der Nervenarzt*, **40**, 692–700.
Ditchburn, R. (1955). Eye-movements in relation to retinal action. *Optica Acta*, **1**, 171–176.
Festinger, L. and Canon, L.K. (1965). Information about spatial location based on knowledge about efference. *Psychology Review*, **72**, 373–384.
Hallett, P.E. and Lightstone, A.D. (1976). Saccadic eye movements towards stimuli triggered during prior saccades. *Vision Research*, **16**, 99–106.
Mack, A. (1970). An investigation of the relationship between eye and retinal image movement in the perception of movement. *Perception and Psychophysics*, **8**, 291–298.
Nemire, K. and Bridgeman, B. (1987). Oculomotor and skeletal motor systems share one map of visual space. *Vision Research*, **27**, 393–400.
Prablanc, C., Echallier, J.F., Komilis, E. and Jeannerod, M. (1979). Optimal response of eye and hand motor systems in pointing. *Biological Cybernetics*, **35**, 113–124.

Roelofs, C. (1935). Optische localisation. *Archives für Augenheilkunde*, **109**, 395–415.
Van Essen, D.C., Newsome, W.T. and Bixby, J.L. (1982). The pattern of interhemispheric connections and its relationship to extrastriate visual areas in the macaque monkey. *Journal of Neuroscience*, **2**, 265–283.
Wallach, H. and Lewis, C. (1965). The effect of abnormal displacement of the retinal image during eye movements. *Perception and Psychology*, **1**, 25–29.
Wong, E. and Mack, A. (1981). Saccadic programming and perceived location. *Acta Psychology*, **48**, 123–131.

21

Seeing by exploring

Richard L. Gregory

Department of Psychology
University of Bristol
Bristol, England

The classical notion of how we see things is that perception is passive—that the eyes are windows, and in floods reality. This was how the Greeks saw perception, and it is the basis of the accounts of the seventeenth and eighteenth century Empiricist philosophers. But physiological work of the nineteenth century cast doubt on this view that perception is passive acceptance of reality. The doubt arose from discoveries of elaborate neural mechanisms, of the delay of signals, and of the time required to process the signals and then make decisions. The doubt was fueled by interest in phenomena of visual and other illusions; for how could passively accepted truth be illusory? It was clear to Hermann von Helmholtz and others a hundred years ago that illusions suggest active processes of perception, which do not always work quite correctly or appropriately. This discovery, and surely this was an important discovery, was not at all popular with philosophers—for perception as the principal basis for true statements became suspect. Worse, evidently perception needed scientific backup (and indeed, what was discovered with instruments did not always agree with how things seem to the senses), so philosophers lost out to scientists as the discoverers and arbiters of truth. Fortunately for them, scientists often disagree on their observations, and how they should be interpreted, so philosophy gradually took on other roles, especially advising scientists what to do.

Perhaps curiously, perception is not at the present time a popular topic for philosophers. This must be partly because scientific accounts of perception have now gone a long way away from appearances. They depend on physiological and psycho-physical experiments (as well as curious phenomena including various kinds of illusions) which require technical investigation and do not fall within traditional concepts of philosophy. For example, it has become clear over the last 20 years or so that visual perception works by selecting various features from the environment, by specialized information channels of the eye and brain. This is an extension of the nineteenth century physiological concept of the Specific Energies of nerves, suggested by the founder of modern physiology Johannes Muller (1801–58). His notion that there are many special receptors and neural pathways, each giving its own distinct sensation, has recently been confirmed and extended for touch, hot and cold, and tickle (Iggo, 1982). In vision, various features (such as the position and orientation of edges, direction and velocity of movement, stereoscopic depth, brightness, and colors) are signaled by dedicated channels having special characteristics for transmitting and analysing significant features of the world. There are also "spatial frequency" channels, tuned to separations of features, which suggest that spectral analysis plays some part in pattern recognition. All this implies that a great deal of parallel processing goes on in the visual system—leading to integrated pattern vision in which many sources of information, sensory and stored from the past, come together—to give powerfully predictive hypotheses, which are our reality of the object world. It seems appropriate and useful to think of perceptions as "hypotheses" (Perceptual Hypotheses) by analogy with the hypotheses of science which make effective use of limited data for control and prediction (Gregory, 1974, 1981).

We may go on to ask further what, perceptually, is an object? What is accepted or seen as an object depends greatly on use—on what is handled, or what behaves, as a unit. It seems that we map the world into individual objects in infancy, by exploring with our hands and discovering what can be pushed or pulled as units, and generally how things behave to us and to each other. Thus when we read a book, each page is an object, as we turn them separately; but on the shelf each book is an object, as they are selected and picked as a unit. And on a printed page letters, words, sentences, or paragraphs may be units, according to how we read. Perceptual units are set up early in life, but it is an interesting possibility that new structuring might be continued throughout adult life—by continuing to explore the world with our hands and eyes. Then we might continue the remarkable perceptual and intellectual development of childhood throughout life. This is the hope (one might almost say religion) of interactive "hands-on" science centers, including the *Exploratorium* founded by Frank Oppenheimer in San Francisco, and the *Exploratory* we have started in Bristol (Gregory, 1986). They allow people of all ages to discover the world of objects (and something of science and technology, as well as their own perceptions) by active exploration.

The importance of experience through interaction with objects was impressed upon me 25 years ago when my colleague Jean Wallace and I studied the rare case of someone (S.B.) who, after being effectively blind from infancy, received corneal grafts in middle life. This is the situation envisaged by John Locke, following a letter he received from his friend Samuel Molyneux who asked, "Suppose a man born blind, and now adult, and taught by his touch to distinguish between a cube and a sphere of the same metal.... Could he distinguish and tell which was the globe, which the cube?" Locke (1690, Bk. II, Chapt. 9, Sect. 8) was of the opinion that "the blind man, at first, would not be able with certainty to say which was the globe, which the cube." And later, George Berkeley (1707) said similarly that we should expect such a man not to know whether anything was "high or low, errect or inverted ... for the objects to which he had hitherto used to apply the terms up and down, high and low, were such only as affected or were in some perceived by touch; but the proper objects of vision make a new set of ideas, perfectly distinct and different from the former, and which can in no sort make themselves perceived by touch". Berkeley goes on to say that it would take a long time to associate the two. But, contrary to the expectations of the philosophers, we found that directly after the first operation, S.B. could see things immediately that he knew from his earlier touch experience; although for many months, and indeed years, he remained effectively blind for things he had not been able to explore by touch. So Berkeley's assumption that vision and touch are essentially separate is not correct; knowledge based on touch is very important for vision. Most dramatically, S.B. could immediately tell the time by sight from a wall clock on the hospital ward; as he had read time by touch from the hands of his pocket watch, from which the glass had been removed so that he could feel its hands. Even more surprising: following the operation he could immediately read uppercase, though not lowercase letters. It turned out that he had learned uppercase, though not lowercase, letters by touch as a boy at the Blind School from uppercase letters engraved on wooden blocks. The blind children were given only uppercase letters, as lowercase was not used at that time for street signs or brass plates, which it would be useful to read by touching. So the blind school had inadvertently provided the needed controlled experiment, which suggested that active exploration is vitally important for the development of meaningful seeing in children.

Most moving, and most informative, was S.B.'s response to seeing a lathe (which he knew from descriptions) for the first time. Shortly after leaving the hospital, we showed him a simple lathe in a closed glass case at the science museum. Though excited by interest, he made nothing of it. Then, with the cooperation of the Museum staff, we opened the case to let S.B. touch the lathe. As reported at the time (Gregory, 1974):

> 'We led him to the glass case, which was closed, and asked him to tell us what was in it. He was quite unable to say anything about it, except that he thought the nearest part was a handle. (He pointed to the handle of the transverse feed.) He complained that he could not see the cutting edge, or the metal being worked, or anything else about it, and appeared rather agitated. We then asked a Museum Attendant for the case to be opened, and S.B. was allowed to touch the lathe. The result was startling; he ran his hands deftly over the machine, touching first the transverse feed handle and confidently naming it as a "handle," and then on to the saddle, the bed and the head-stock of the lathe. He ran his hands eagerly over the lathe, with his eyes shut. Then he stood back a little and opened his eyes and said: "Now I've felt it, I can see."'

S.B.'s effective blindness to objects he did not know was remarkably similar to clinical agnosia, and to Ludwig Wittgenstein's (1953) notion of "aspect blindness". In our own experience (or rather lack or it) of ambiguous figures, such as Jastrow's Duck-Rabbit—while it is accepted as a rabbit, the duck features are scarcely seen, disappearing into aspect blindness. This is also dramatic in Rubin's Face-Vases, which disappear in turn, sinking into the ground of the invisibility of aspect blindness, to emerge from nothing as materializing figures. Thus Wittgenstein (1953, p. 213) asks of an imaginary aspect-blind person, presented with the reversing-skeleton Necker Cube figure:

> 'Ought he to be unable to see the schematic cube as a cube? For him it would not jump from one aspect to another. The aspect-blind will have altogether different relationship to pictures from ours.'

We found that S.B. did not experience reversals of these (to us) ambiguous figures. For him they were meaningless patterns of lines, and, in general, pictures were hardly seen as representing objects. From this, I suggest (Gregory, 1981) that perceptual phenomena of ambiguity should be highly useful for investigating meaning and understanding.

There was evidence that he learned to conceive and perceive space, not only by handling objects but also by walking. In the hospital ward he was able to judge distances of objects such as chairs with remarkable accuracy. But looking down from the window—which was some forty or more feet high—he described the distance of the ground as about his own body height. He said that if he hung from the windowsill with his fingers, he feet would just touch the ground. Blind people avoid jumping down for they do not know what (if anything!) is below them; they feel carefully with their feet first. So he would have had little or no experience of distances below his feet, except for stairs and occasionally ladders. We may conclude that experience of walking was necessary for seeing distance. This is borne out by our, normal, loss of size scaling looking down from a high building, when cars and people and so on look like toys, though for the same horizontal distance they look almost their "correct" sizes.

All this is evidence that perception depends neurally on reading or interpreting sensory signals in terms of experience and knowledge, or by assumptions (which may, however, be wrong and misleading to produce illusions (Gregory, 1968, 1980)) of the object world. The *Exploratory* aim is to amplify and extend first-hand experience to enrich perception and understanding for children and throughout adult life. The effectiveness of the hands-on approach for teaching has been questioned. But in any case, surely capturing interest is the first essential for more formal methods to be effective. It is hard to believe that learning has to be serious; it is far more likely that play is vitally important for primates to learn how to exist in the world in which they find themselves. It is fascinating to watch children and adults in this play-experiment situation of individual discovery. Although research is needed to be sure, they certainly give every indication of thinking and learning by doing.

It seems that children do not approach questions or experiments from a vacuum; they generally have performed ideas, which may not be appropriate or coherent, but may be held robustly. They may be discovered (both by their parent or teacher) by setting up predictions. Thus in the *Exploratory*, experiments with gyroscopes, or the Bernoulli effect, are highly surprising and so reveal erroneous conceptions. Assumptions may of course also be discovered through questioning, and spontaneous questions may reveal how children or adults see, or think they see. According to Jean Piaget and several other authorities, young children hold magical notions of cause, not distinguishing between their own responses and the behavior of inanimate objects, and they tend to hold Aristotelean notions of physics of motion and forces. In 1929, Piaget described children as believing that all objects capable of movement—such as bicycles, and the sun and moon—are alive. And Piaget reported many investigations on perception of conservation (or lack of conservation) of matter, finding that most children before the age of nine, when given various shapes of a lump of clay, do not appreciate conservation of substance. Presumably hands-on experience tends to correct such errors; but how good are adults? A marketing trick is to use odd-shaped bottles to make the contents look larger, which fools most people.

Do children, if implicitly, apply the scientific method to generate their understanding of the world? This was the view of Jean Piaget (1896–1980), the greatest name in the field. Piaget came to favour of an outright empiricism, where logic itself is learned. In *The Child and Reality* (1972), Piaget proposes the following hypothesis (p. 94):

'(a) That at every level (including perception and learning), the acquisition of knowledge supposes the beginning of the subject's (child's) activities in forms which, at various degrees, prepare logical structures; and (b) therefore that the logical structures already are due to the coordination of the actions themselves and hence are outlined the moment the functioning of the elementary instruments are used to form knowledge'.

Piaget offers experiments to show effects of inferences during perceptual development in children, showing that perceptions change as inferences change. For example (*The Child and Reality*, p. 95): "A young child is shown briefly two parallel rows of four coins, one being spaced out more than the other: The subject will then have the impression that the longer row has the more coins". Piaget goes on to say that joining the corresponding coins of each row by lines, or joining them in other ways, has different effects for different ages or stages of perceptual development. So Piaget suggests that different inferences about the lines are made, each making the rows of coins appear somewhat different. He also cites an experiment from his laboratory in which the numbers 1 and 7 are shown with their tops hidden, and at different orientations. When the 1 is tilted to the slope of the 7, it is still read as a 1 when ending a sequence likely to be a 1, but otherwise it is seen as a 7. So probability affects perception in children.

Older children's notions are reported in *Children's Ideas in Science*, edited by Rosalind Driver, Edith Guesne, and Andree Tiberghien (1985). This starts with an account by Rosalind Driver of two 11-year-old boys in a practical class measuring the length of a suspended spring, as equal weights are added to a scale pan. In the middle of the experiment one of the boys unlocked the clamp and moved the top of the spring up the retort stand. He explains:

> 'This is farther up and gravity is pulling it down harder the farther away. The higher it gets the more effect gravity will have on it because if you just stood over there and someone dropped a pebble on him, it would just sting him, it wouldn't hurt him. But if I dropped it from an airplane it would be accelerating faster and faster and when it hit someone on the head it would kill him.'

This reveals the boy's view of gravity, which is not quite ours.

Whether young children ask abstract or philosophical questions has been asked by an American teacher of philosophy, Gareth Matthews, in *Philosophy and the Young Child* (1980). As an example, a boy who had often seen airplanes take off, disappearing in the distance, flew for the first time at the age of 4 years. After takeoff, he turned to his father and said in a puzzled voice: "Things don't really get smaller up here".

How do children come to derive reality from appearances? Is a single dramatic experience such as flying for the first time—or discovering that patterns of spectral lines from glowing gases correspond to light from the stars—sufficient for a paradigm change of view or understanding in children? Can adults go back to the drawing board to see the world afresh?

For looking at the details of how perception works, it is convenient to consider somewhat separately the early stages of how patterns and colors are signaled by the retina and analysed by the initial stages of the brain's perceptual systems, and then the cognitive (knowledge-based) processes of selecting and testing perceptual hypotheses of the objects and situations that we have to deal with to survive. A particular question that concerns us—and we have no clear answer—is how the various signaled features finally come together, without obvious discrepancies. For example, given that color and brightness are signaled by different parallel systems, why don't they lose their registration to separate and produce spurious edges at borders of objects?

Curiously, our mammalian ancestors did not have effective color vision before the primates, including ourselves at the top of the evolutionary tree. So it might be expected that for us brightness contrast is more significant than color contrast for recognizing objects, and this is generally so. The importance of brightness rather than color contrast is clear from the effectiveness of black and white photography. Switching out the color of a TV set does little to impair our perception (apart from watching snooker) except in rather special, though sometimes biologically important, situations. From this simple experiment we can see that color is useful for spotting red berries in green foliage, seeing through camouflage, remotely sensing the edibility of fruit and meat, which could be a major reason why color vision developed in primate evolution. It had already developed, in various forms, in insects, fishes, and birds, but curiously it was lost for mammals, to be reinvested in our immediate primate ancestors.

In some of our experiments, we do the converse of switching out the color of a TV set: we remove brightness differences while preserving color contrast. This gives "isoluminant" displays, which can be seen only by color vision because there are no brightness differences. We have developed several techniques for producing color-without-brightness contrast, usually for a pair of colors, such as red and green. It is important to ensure that they are set to equal brightness for each observer, for there are individual differences of color sensitivities which, when extreme, are color-blindness (or better, "color anomaly") which is usually reduced sensitivity to (so-called) red or green light. For these experiments it is important that neighboring color regions do not overlap, or have gaps, because such registration errors would produce brightness differences at the color borders. So producing truly isoluminant displays presents some technical problems (and it rarely occurs in nature), but some of the phenomena can be seen in formal color printing when the print has the same brightness as its different-color background. When the print and background have the same brightness, it is difficult to read and the edges of the letters appear "jazzy". The print is unstable, moving around disconcertingly. In spite of the loss of stability, and uncertainty of just where the edges are, there is hardly any loss of visual acuity as measured with a grating test, although letters are more difficult to read. The fact that letter acuity though not grating acuity is impaired suggests that precise *position* of edges (called "phase" information) is lost at isoluminance, though *separations* between nearby features are signaled almost normally. Reading is particularly difficult when letters are closely spaced. They can also lose their individual identities, breaking up into unfamiliar units.

Losses may also be of neurally higher-level brain processes. Most striking is the appearance (or rather, disappearance) of an isoluminant face. This can be shown best with a matrix of red and green dots as in coarse screen printing: when the two colors are set to isoluminance, the face immediately loses all expression and looks flat, with meaningless holes where the eyes and mouth should be. It no longer looks like a face: it becomes meaningless shapes. Although this is a "subjective" observation, it is unmistakable. It is very strong evidence of drastic perceptual loss when only color is available, for almost anything is normally accepted as a face. This, indeed, makes the cartoonist's work possible because just a few lines can evoke an expressive face; so it is remarkable that face perception is so completely lost with isoluminant color contrast. It is important to note that this loss does not occur when a normal brightness-contrast picture is blurred, for example by being projected out of focus, so this loss of face seems to be a central perceptual phenomenon.

The kinds of losses that occur with normal observers at isoluminance are strikingly like the clinical symptoms of amblyopia, or a lazy eye. This "artificial amblyopia" of isoluminance is convenient for experiments because it can be switched on and off and compared with the normal vision in the same individual. Also, we can see what happens and compare our experience with the reports of people who suffer from amblyopia, which is a help for at least intuitive understanding.

A further and dramatic loss is of a certain kind of stereoscopic depth. The American psychologist Bela Julesz discovered, over twenty years ago, that when slightly different random dot patterns are presented, one to each eye, in a stereoscope, regions of dots which are shifted sideways for one eye are seen as lying at a different distance from the rest of the dots which are not displaced. This shows that the brain can compare meaningless dot patterns presented to the eyes and compute depth from small horizontal shifts—which normally occurs for different distances, as the eyes receive slightly different views as they are horizontally separated by a few centimeters. But when the dots are, for example, green on a red background of the same brightness, this stereoscopic depth is lost. We are now comparing this dramatic loss of stereoscopic depth for meaningless dot patterns (which, however, is perhaps never quite complete) with what happens when there are lines and meaningful objects presented in stereoscopic depth to the two eyes. There is some evidence that edges activate different neural mechanisms from the random dots, because a few people have "line" but not "random dot" stereo vision. Perhaps also the meaning, or object-significance, of what is presented may be important in how the brain compares features for perceiving depth.

There is a corresponding phenomenon for movement. When a pair of such random dot figures is alternated, about 10 times/sec, and viewed with one or both eyes, the shifted dot region separates from the rest of the dots and moves right and left. We find that when the dots are set to isoluminance, the displaced dots are lost among the others and no movement is seen (Ramachandran and Gregory, 1978). This is remarkable, because the dots can be quite large, and clearly visible individually, and yet this kind of stereo depth and movement are lost without brightness information.

Visual channels may be isolated in various ways, including selective adaptation to colors (giving colored afterimages); to prolonged viewing of tilted lines (making vertical lines look tilted in the opposite direction); to movement (as in the "movement after effect," which was known to Aristotle). We have recently found that continuous *real* movement is signaled by the same neural channel as discontinuous *apparent* (or phi) movement, which may be seen when stationary lights are switched on and off in sequence—provided the gaps in space and time of the apparent movement are not too great (Gregory and Harris, 1984). When the gaps are large (greater than about 10 min arc subtended angle), movement can still be seen, but now it is signaled by a different neural channel, or cortical analysing system. This we have found by showing that real movement can cancel opposite-direction apparent movement. This is done by illuminating a readily rotating sector disk with stroboscopic short flashes of light set to make it appear to rotate backwards from its true motion, and also with a variable-intensity continuous light. This produces, say, real clockwise movement and, at the same time, apparent anticlockwise movement of the disc. These movements can be set to cancel, or null, by adjusting the relative intensities of the strobe and continuous lights. At the null point there is only a random jitter, with no systematic movement. The null point is not affected by the disturbing effect of adapting to prolonged viewing of movement. The movement after effect affects the real and apparent movement equally, which is strong evidence that they are sharing a common channel. The nulling of real against short-range, apparent movement occurs even though the strobe and the continuous lights have different colors, so the eye's three color channels share a common movement system.

There is, however, an interesting limit to the real/apparent-movement shared channel. When the strobe's flash rate is set to give large jumps of the rotating sectors, nulling no longer occurs. The two movements are now seen passing through each other, simultaneously. These observations indicate a shared channel for real- and short-range apparent movement, but a separate channel for long-range movement. It is well known to cartoon film animators that the long-range movement of large jumps between frames has cognitive characteristics, such as being affected by which features are parts of the same object, or are likely to move separately.

An intriguing question is how the various sources of information from different parallel neural channels combine to give unified perceptions of objects. Although neural channels have different characteristics, and in spite of selective adaptations (which affect some channels but not others), and in spite of distortions (which may be dramatic), we do not experience spurious multiple edges. This surely requires some explanation. We suggest that misregistrations are avoided by a process of "border-locking," such that luminance borders pull nearby color edges to meet them (Gregory and Heard, 1979). So spatial registration discrepancies are prevented, although at the cost of some distortions, which may be very evident. Presumably, some visual distortion of size and curvature is not important in nature, although multiple edges, where there should be but one, would be seriously confusing. So, we suggest, registration is maintained by border-locking (where color is slave to luminance) at the cost of some distortion.

It turns out that the classical perspective distortion illusions (such as the Muller-Lyer and the Poggendorf illusions) remain essentially unchanged when presented with their lines having color contrast to their backgrounds, and set to isoluminance (Gregory, 1976). But some illusions, notably the Cafe Wall illusion (Gregory and Heard, 1979), which has no perspective-depth features, appear undistorted when isoluminant. It seems that early sensory processing is affected by isoluminance (as in the parallel lines of the Cafe Wall illusion), but the cognitive reading (or misreading) of perspective depth from converging lines, which can give spatial distortions (Gregory,

1974), is unaffected by isoluminance—it does not matter how the information arrives for cognition.

Recently, David Hubel and Margaret Livingstone (1987) have found strong evidence for separate cortical systems for representing and analysing luminance and color information. It now seems that color is primarily analysed by blobs in the third layer of the striate cortex, while orientations, etc., signalled by luminance differences are analysed by interblob cells at this early stage of visual processing. On a matter of detail, we disagree with one of Hubel and Livingstone's observations, for, as mentioned above, we find that the perspective depth distortion illusions remain at isoluminance; but they claim that these and all perspective depth disappear. This is not our experience, but no doubt this discrepancy will soon be resolved.

References

Berkeley, G. (1707). A New Theory of Vision (Everyman (Ed.), London, 1910).
Driver, R., Guesne, R.D. and Tiberghien, A. (1985). *Children's Ideas in Science.* Milton Keynes: Open University Press.
Gregory, R.L. (1968). Perceptual illusions and brain models, *Proceedings of the Royal Socitey*, B, **171**, 278–296.
Gregory, R.L. (1976). Visions with isoluminant color contrast: 1. A projection technique and observations. *Perception*, **6** (1), 113–119.
Gregory, R.L. (1980). Perceptions as hypotheses, *Philosophical Transactions of the Royal Society*, B, **290**, 181–197.
Gregory, R.L. (1981). *Mind in Science.* London: Weidenfeld and Nicolson.
Gregory, R.L. (1986). *Hands-On Science: An Introduction to the Bristol Exploratory.* London: Duckworth.
Gregory, R.L. and Harris, J.P. (1984). Real and apparent movement nulled. *Nature*, **307**, 729–730.
Gregory, R.L. and Heard, P.F. (1979). Border locking and the cafe wall illusion. *Perception*, **8**, 4.
Gregory, R.L. and Wallace, J.G. Recovery from Early Blindness: A Case Study, *Exp. Psychol. Soc. Mongr.* II, Cambridge: Heffers. Reprinted in Gregory, R.L. (1974) *Concepts and Mechanisms of Perception*, London: Methuen, 65–129.
Iggo, A. (1982). Cutaneous sensory mechanisms. In Barlow, H.B. and Mollon, J.D. (Eds), *The Senses*, 369–406. Cambridge: CUP.
Livingstone, M.S. and Hubel, D.H. (1987). Psychophysical evidence for separate channels for the perception of form, color, movement and depth. *Journal Neuroscience*, **7**(11), 3416–3468.
Locke, J. (1690). In Niddich, P.H. (Ed., 1975), *Essay Concerning Human Understanding.* Oxford: OUP.
Matthews, G.B. (1980). *Philosophy and the Young Child.* Cambridge, Massachusetts: HUP.
Piaget, J. (1972). *The Child and Reality: Problems of Genetic Psychology.* Translated by Arnold Rosin. London: Frederick Muller.
Ramachandran, V.S. and Gregory, R.L. (1978). Does color provide an input to human motion perception? *Nature*, **275**, 55–56.
Wittgenstein, L. (1953). *Philosophical Investigations*, Translated by G.E.M. Anscombe. Oxford: Blackwell.

22

Spatial vision within egocentric and exocentric frames of reference

Ian P. Howard

Human Performance Laboratory, Institute for Space and Terrestrial Science, York University, Toronto, Ontario.

Introduction

It is remarkable that we are able to perceive a stable visual world and judge the directions, orientations and movements of visual objects given that images move on the retina, the eyes move in the head, the head moves on the body and the body moves in space. An understanding of the mechanisms underlying perceptual stability and spatial judgements requires precise definitions of relevant coordinate systems. An *egocentric frame* of reference is defined with respect to some part of the observer. There are four principal egocentric frames of reference, *a station-point frame* associated with the nodal point of the eye, a *retinocentric* frame associated with the retina, a *headcentric frame* associated with the head, and a *bodycentric frame* (torsocentric) associated with the torso. Additional egocentric frames can be defined with respect to any segment of the body. An *egocentric task* is one in which the position, orientation or motion of an object is judged with respect to an egocentric frame of reference. A *proprioceptive task* is a special kind of egocentric task in which the object being judged is also part of the body. An example of a proprioceptive task is that of directing the gaze toward the seen or unseen toe. An *exocentric frame* of reference is external to the observer. Geographical coordinates and the direction of gravity are examples of exocentric frames of reference. These various frames of reference are listed in Table 1, together with examples of judgements of each type.

The station-point frame

We start with an illuminated three-dimensional scene of fixed objects, *the visual world*. A *station point* is defined with respect to some arbitrary coordinate system anchored in the world. Any optical system has two nodal points which have the geometrical property that all light rays passing through the

Table 1. *Frames of reference for visual spatial judgements.*

Frame of reference	Sensory components	Examples of tasks
PROPRIOCEPTIVE O and RF internal		
Non-visual	Sense of position of body parts	Points to the unseen toe
Purely visual	Locations of images of body parts	Align two seen parts of the body
Intersensory	Location of image plus proprioception	Point unseen finger to seen toe
EGOCENTRIC O external, RF internal		
STATION POINT	Abstract or inferred	Specify objects visible from a vantage point
RETINOCENTRIC	Retinal local sign plus retinal landmark	Fixate an object. Place line on retinal meridian
HEADCENTRIC	Eye position + retinal local sign	Place an object in the median plane of the head
BODYCENTRIC		
Purely visual	Relative retinal location	Align a stick to the seen toe
Intersensory	Neck + eye position + retinal local sign	Point stick to the unseen toe. Place an object to left of the body
SEMI-EXOCENTRIC O internal, RF external		
Purely visual	Relative retinal location	Align self with two objects
Intersensory	Seen part of body and gravity senses	Point upwards
EXOCENTRIC O and RF external		
Absolute	Vision with appropriate reference frame	Judge geographical directions
Relative	Relative retinal location with appropriate constancies.	Align three objects. Judge the shape of an object
Intersensory	Visual and non-visual stimuli compared	Set a line vertical. Point line to unseen sound

O signifies the object, the position or orientation of which is being judged or set. RF signifies the reference frame with respect to which the object is being judged or set.

first emerge from the second without having changed direction. The nodal points of the human eye are close together and can be regarded as one *nodal point* situated near the centre of the eye. The nodal point is a geometrical abstraction, light rays do not necessarily pass through it. The nodal point of the eye is the *visual station point*.

The *visual surroundings* or *ambient array* is the set of light sources and reflecting surfaces which surround the station point and from which light rays can reach the station point. The spherical array of light rays that reach the station point constitutes the station-point frame of reference. Within this frame of reference the distance of any point in the ambient array from the station point and the angle subtended at the nodal point by any pair of points in the ambient array can be specified. The station-point frame of reference itself contains no natural fiducial lines for specifying orientation or direction, since a point has no defined orientation. It is therefore meaningless to talk about the effects of rotating the station point or rotating the ambient array round the station point. Every linear motion of the station point changes the distances of points and the angular subtense of pairs of points in the ambient array. The ambient array is sometimes thought of as the projection of the ambient array onto a fixed surface, usually a spherical surface centred on nodal point. This simply means that distances to points in the visual surroundings are not directly specified in this form of the ambient array, although the ambient array may contain enough information to allow distances to be recovered.

Visual attributes which may be defined in terms of the station-point frame of reference include: (1) what is in view from a given place, (2) the relative directions (angular subtense) of two or more objects, (3) the distance of an object, (4) the relative angular velocities of moving objects, (5) velocity flow fields created by linear motion of the station point, expressed as a set of differential directed angular velocities and (6) the set of objects which define the locus of zero parallax (heading direction) in a three-dimensional array of objects as the station point is moved along a linear path. Judgements of these attributes are station-point judgements.

The retinocentric frame

We now add a pupil, lens, retina and associated structures of an eye. For a given position and orientation of the eye, the bundle of light rays which enter the pupil is the *optic array* and the portion of the ambient array from which these light rays originate is the *distal visual stimulus*, or field of view. For most purposes we can assume that the optic array projects onto a spherical retina centred on the nodal point. A *visual line* is any line which passes through the pupil and nodal point from a point in the distal stimulus to its image on the retina. The *visual axis* is the visual line through the fixation point and the centre of the fovea. A three-dimensional polar coordinate system centred on the nodal point can be used to specify the retinocentric distance, position, and direction of any object. An object's distance is its distance from the nodal point. Its eccentricity is the angle between its visual line and the visual axis. Its meridional direction is the angle between the plane containing the visual line of the object and the visual axis and the plane containing the visual axis and the retinal meridian which is vertical when the head is in a normal upright posture. These three-dimensional coordinates project onto the surface of the retina as two-dimensional polar coordinates, with the fovea as origin for eccentricity and the normally vertical meridian as the fiducial line for meridional direction. This is the *retinocentric frame of reference*. Note that the linear velocity of an image is proportional to the angular velocity of the object relative to the eye. This retinal coordinate system may be projected through the nodal point onto the concave surface of a perimeter or onto a tangent screen, which allows one to specify the oculocentric eccentricity and meridional angle of a stimulus on a chart. For certain purposes it may be more convenient to specify retinocentric positions in terms of elevation and azimuth or longitude and latitude. For instance, longitude and latitude are useful when describing the retinal flow field created by linear motion over a flat surface because the flow vectors conform to lines of longitude. The visual axis provides a natural reference which allows one to specify the direction of gaze with respect to selected landmarks in the ambient array.

Visual attributes which may be defined in terms of the retinocentric frame of reference include: (1) the eccentricity and meridional direction of an object (its visual direction relative to the fovea and the prime retinal meridian), and (2) the orientation of a line, relative to prime retinal meridia. These are absolute retinocentric visual features. Relative retinocentric features involve only the specification of the relative positions, orientations or motions of images on the retina. Examples are: (1) the shape of a retinal image, (2) the retinal velocity of an image and hence the angular velocity of an object moving with respect to the eye, (3) the angular velocity of the eye with respect to a stationary object, and (4) retinal flow fields created by translations of the eye with respect to an ambient array. For a perfectly spherical retina relative retinocentric features are geometrically equivalent to those defined in terms of the ambient array projected onto a spherical surface.

All absolute retinocentric attributes change when the eye rotates with respect to a fixed nodal point and distal display. However, absolute retinocentric features are not necessarily affected by all types of eye rotation. For instance, the retinocentric direction of an object is invariant when the eye rotates about the visual line of the object. The eyes rotate as if about an axis at right angles to the meridian along which the gaze moves (Listing's law). An interesting consequence of this fact is that the retinocentric orientation of a line is invariant when the gaze moves along the line (Howard, 1982, p. 185). For a spherical retina and distortion-free optical system, relative oculocentric attributes, such as the shape of the retinal image, are not affected by any rotations of the eye. If the retina were not spherical this would not be true and the task of shape perception would be more complex.

The station-point and retinocentric frames of reference are both *oculocentric frames of reference*.

The headcentric frame

We now add a head. The orientation of an eye in the head about each of three axes may be specified objectively in terms of either the Fick (latitude and longitude), the Helmholtz (elevation and azimuth) or the Listing (polar) coordinate system (see Howard, 1982 for details). The headcentric position of a visual object may be specified in terms of angles of elevation relative to a transverse plane through the eyes and angles of azimuth relative to the median plane of the head. The headcentric orientation of an object is usually specified with respect to the the normally vertical axis of the head. The head is defined as being vertical when the line from the ear hole to the angle of the eye socket and the line joining the two pupils are both horizontal. Particular headcentric spatial features of objects may be defined in terms of the types of head motion that leave them unchanged. If we assume that the centre of rotation of the eye is the same as the nodal point then the headcentric position of an object is the vector sum of its retinocentric position and the position of the eye in the head. For instance, if an object is 10° to the left of the fixation point and the eye is elevated 10° then the headcentric position of the object is about 14.1° along the upper left diagonal with respect to the eye socket. Similar arguments apply to the headcentric orientation and motion of an object. Of course the coordinate systems used for specifying retinocentric position and eye position must correspond. Visual attributes that may be defined in headcentric terms include: (1) the direction of an approaching object relative to the head, (2) the direction of gaze in the head, (3) an object's inclination to the mid-head axis and (4) a shape defined by the path an eye follows when pursuing a light spot.

The bodycentric frame

We now add a body. The bodycentric (torsocentric) position, orientation or movement of an object may be specified with reference to any of the three principal axes or planes of the body. The defining characteristic of bodycentric attributes is that they are affected by specific types of body motion.

If no part of the body is in view, bodycentric judgments require the observer to take account of oculocentric information, eye-in-head information and information from the neck joints and muscles regarding the position of the head on the body. Thus the oculocentric, headcentric and bodycentric reference systems form a hierarchical, or nested set, as indicated in the second column of Table 1. But this is not all. For certain types of bodycentric judgement the observer must appreciate the lengths of body parts, in addition to their angular positions. For instance, a person can place the finger tip of the hidden hand on a visual target only if the length of the arm is taken into account. Conscious knowledge is not involved, but rather the implicit knowledge of the body that is denoted by the term body schema. If the body as well as the object being judged is in view, bodycentric judgments are much simpler since they can be done on a purely visual basis without the need to know the positions of the eyes or head.

Examples of bodycentric attributes include: (1) the direction of an object relative to a part of the body. This would need to be appreciated by a person who wished to direct the hidden hand towards an object, (2) motions of an object with respect to a part of the body and (3) the inclination of an object relative to the mid-body axis.

The exocentric frame

Finally, the exocentric position, orientation or movement of an object are specified with respect to coordinates external to the body. The defining characteristic of exocentric spatial attributes is that they are not affected by changes in the position or orientation of the observer or any part of the observer. Exocentric attributes may be absolute or relative. Absolute exocentric attributes are defined with respect to a coordinate system which is assumed to be fixed in inertial space. Examples of extrinsic coordinate systems are the one-dimensional gravitational coordinate, the two-dimensional geographical coordinates and a set of three-dimensional Cartesian coordinates. Absolute exocentric attributes include: (1) the gravitational orientation of a line, (2) the compass direction of an arrow and (3) the movements of an object within a defined space.

Relative exocentric attributes are defined in terms of the position, orientation or motion of one object relative to another or of parts of an object relative to other parts. The reference frame is now intrinsic to the object or set of objects being judged. The distinction is analogous to that between extrinsic and intrinsic geometries. Relative exocentric attributes include: (1) the shape of an object (the relative dispositions of parts), (2) rotation of an object relative to an intrinsic axis. For instance, the rotation of an aircraft about the yaw, roll or pitch axis and (3) the motion of one object relative to another.

Exocentric judgements about an isolated visual object can be with respect to a frame of reference provided by memory, as when we relate the position of a light to the remembered positions of the contents of a room. Otherwise, the exocentric position of an isolated visual object can be specified with respect to a frame of reference supplied by a second sense organ. Thus we can judge the position of a light in relation to a frame of reference provided by sounds or by things we touch or we can judge the orientation of a line in terms of stimulation registered by the vestibular organs. These are all intersensory tasks.

In theory, the variance of performance on an *intersensory task* should equal the sum of the variances of directional tasks that involve the separate component senses. A *multisensory task* is one in which the position, orientation or movement of an object is detected by more than one sense organ at the same time. For instance, we perform a multisensory task when we determine the headcentric direction of an object both by sight and by the sound that it makes. Given that the observer believes that the seen and heard object is one, the variance of performance on a multisensory task should, theoretically, be less than the variance of performance on tasks using only one or other of the component senses (see Howard, 1982, Chapter, 11 for more details on the distinction between intersensory and multisensory tasks).

Finally there are cases where the frame of reference is external but the object is the self. I shall refer to them as *semi-exocentric frames of reference*. Examples of semi-exocentric attributes include: (1) the position of an observer on a map, (2) the compass direction of an observer with respect to an object, and (3) the position of an observer with respect to being under or over something. Note that, unlike purely exocentric attributes, semi-exocentric attributes vary with changes in the location of the observer.

In what follows I shall discuss the extent to which perceptual judgements within egocentric and exocentric frames of reference are subject to illusory disturbances and long-term modifications. I shall argue that well-known spatial illusions, such as the oculogyral illusion and induced visual motion have usually been discussed without proper attention being paid to the frame of reference within which they occur, and that this has lead to the construction of inadequate theories and inappropriate procedures for testing them.

Perceptual judgements within the oculocentric frame

The subjective registration of the station-point or retinocentric features of an object depends on the local sign mechanism of the visual system. This is the mechanism whereby, for a given position of the eye, each region of the visual field has a unique and stable mapping onto the retina and visual cortex.

Any misperception of the oculocentric position or movement of a visual object can arise only as a result of some disturbance of the retinal local sign-system or of the oculocentric motion-detecting system. In a geometrical illusion, lines are apparently distorted or displaced when seen in the context of a larger pattern. In a figural after effect a visual test object seen in the neighborhood of a previously seen inspection object appears displaced away from the position of the inspection object. Such effects operate only over distances of about one degree of visual angle and the apparent displacement rarely exceeds a visual angle of a few minutes of arc (Kohler and Wallach, 1944). We must conclude that the local-sign system is relatively immutable. This is not surprising, since the system depends basically on the anatomy of the visual pathways. Several claims have been made that oculocentric distortions of visual space can be induced by pointing with hidden hand to visual targets seen through displacing prisms (Cohen, 1966; Held and Rekosh, 1963). Others have claimed that these effects were artifactual and we are left with no convincing evidence that oculocentric shifts can be induced in this way (see Howard, 1982, p. 501, for a more detailed discussion of this subject).

The movement after effect is a well known example of what is almost certainly an oculocentric disturbance of the perception of motion. I will not discuss this topic here.

Perceptual judgements within the headcentric frame

A person making headcentric visual judgements must take account of both oculocentric and eye-in-head information. The question of how and to what extent people make accurate use of eye-in-head information when making headcentric judgements is a complex one. One complication arises because the two eyes are in different positions. The visual system must construct a headcentric frame of reference that is common to both eyes. It can be shown that people judge the headcentric directions of an object as if the eyes were superimposed in the median plane of the head, somewhere between the actual positions of the two eyes. This is known as the cyclopean eye, or *visual egocentre* (see Howard, 1982, for a fuller discussion of all these issues).

A misjudgment of the headcentric direction or motion of a visual object can arise from a misregistration of the position or motion of either the retinal image or the eyes. In this section I shall consider only phenomena due to misregistration of the position or movement of the eyes.

Illusory shifts of headcentric visual direction

Deviations of the apparent straight ahead due to misregistered eye position are easy to demonstrate. If the eyes are held in an eccentric position a visual target must be displaced several degrees in the direction of the eccentric gaze to be perceived as straight ahead. When the observer attempts to look straight ahead after holding the eyes off to one side, the gaze is displaced several degrees in the direction of the previous eye deviation. Attempts to point to visual targets with unseen hand are displaced in the opposite direction. The magnitude of these deviations has been shown to depend on the duration of eye deviation and to be a linear function of the eccentricity of gaze (Hill, 1972; Morgan, 1978; Paap and Ebenholtz, 1976). Similar deviations of bodycentric visual direction occur during and after holding the head in an eccentric posture (Howard and Anstis, 1974). It has never been settled whether these effects are due to changes in afference or to changes in efference associated with holding the eyes in a given posture. Whatever the cause of these effects, it is evident that the headcentric system is more labile than the oculocentric system. This is what one would expect because headcentric tasks require the neural integration of information from more than one sense organ.

The oculogyral illusion

The oculogyral illusion may be defined as the apparent movement of a visual object induced by stimulation of the semicircular canals of the vestibular system (Graybiel and Hupp, 1946). The best visual object is a small point of light in dark surroundings and fixed with respect to the head. When the vestibular organs are stimulated, as for instance by accelerating the body about the mid-body axis, the point of light appears to race in the direction of body rotation. The oculogyral illusion also occurs when the body is stationary but the vestibular organs signal that it is turning. This happens, for instance, in the 20 or 30 seconds after the body has been brought to rest after being rotated. It is not surprising that a point of light attached to the body should appear to move in space when the observer feels that the body is rotating. I shall refer to this perceived motion of the light with the body as the exocentric component of the oculogyral illusion. The exocentric component is not very interesting because it is difficult to see how a rotating person could do other than perceive a light which is attached to the body as moving in space. But even casual observation of the oculogyral illusion reveals that the light appears to move with respect to the head in the direction of body acceleration. This is the headcentric component of the oculogyral illusion.

Whiteside et al., (1965) proposed that the headcentric component of the oculogyral illusion is due to the effects of unregistered efference associated with the vestibulo-ocular response (VOR). The idea is that when the subject fixates the point of light, VOR engendered by body acceleration is inhibited by voluntary innervation. The voluntary innervation is fully registered by the perceptual system but the VOR efference is not, and this asymmetry in registered efference causes the subject to perceive the eyes as moving in the direction of body rotation. This misperception of the movement of the eyes is interpreted by the subject as a headcentric movement of the fixated light. To support this theory we need evidence that the efference associated with VOR is not fully registered by the perceptual system responsible for making judgments about the headcentric movement of visual objects.

For frequencies of sinusoidal head rotation up to about 0.5 Hz, the vestibulo-ocular reflex (VOR) is almost totally inhibited if the attention is directed to a visual object fixed with respect to the head (Benson and Barnes, 1978). The most obvious theory is that VOR suppression by a stationary object is due to cancellation of the VOR by an equal and opposite smooth pursuit generated by the retinal slip signal arising from the stationary light. This cannot be the whole story because Barr et al. (1976) reported that the gain of VOR produced by sinusoidal body rotations decreased to about 0.4 when subjects imagined that they were looking at an object rotating with them. It looks as though VOR efference can be at least partially cancelled or switched off even without the aid of visual error signals (McKinley and Peterson, 1985; Melvill Jones et al., 1984). Tomlinson and Robinson (1981) were concerned to account for how an imaginary object can inhibit VOR but for our present purposes, the more important point is that VOR is not totally inhibited.

Perhaps an imagined object is not a satisfactory stimulus for revealing the extent of voluntary control over VOR. We wondered whether an afterimage might be a better stimulus because it relieves subjects of the task of imagining an object and requires them only to imagine that it is stationary with respect to the head. We had already found optokinetic nystagmus to be totally inhibited by an afterimage even though it was not inhibited by an imaginary object. The results of all these experiments are reported by Howard *et al.* (1988).

Subjects in total darkness were subjected to a rotary acceleration of the whole body of $14°/s^2$ to a terminal velocity of $70°/s$, which was maintained for 60 s. In one condition subjects were asked to carry out mental arithmetic. In a second condition they were asked to imagine an object rotating with the body, and in a third condition an afterimage was impressed on both eyes just before the trial began and the subject was asked to imagine that it was moving with the body. The same set of conditions was repeated but with lights on so that the stationary OKN display filled the visual field. Under these conditions both VOR and OKN are evoked at the same time.

In all conditions the velocity of the slow phase of each nystagmic beat was plotted as a function of time from the instant that the body reached its steady-state velocity. For none of the subjects was VOR totally inhibited at any time during any of the trial periods. For the OKN plus VOR condition subjects could see a moving display, but they could totally inhibit the response only after about 30 s, when the VOR signal had subsided.

We propose that VOR is not completely inhibited by an afterimage seen in the dark because the mechanism used to assess the headcentric motion of visual objects does not have full access to efference associated with VOR. Thus the system has no way of knowing when the eyes are stationary. The component of the VOR which cannot be inhibited by attending to an afterimage gives an estimate of the extent to which VOR efference is unregistered by the system responsible for generating voluntary eye movements and for giving rise to the headcentric component of the oculogyral illusion.

Perceptual judgements within the exocentric frame

Information about the position, orientation and movement of the body in inertial space is provided by the normally stationary visual surroundings, by proprioception and by the otolith organs and semicircular canals of the vestibular system. The otolith organs respond to the pitch and roll of the head with respect to gravity but provide no information about the rotation or position of the head around the vertical axis. The otolith organs also respond to linear acceleration of the body along each of three orthogonal axes but cannot distinguish between head tilt and linear acceleration. The semicircular canals provide information about body rotation in inertial space about each of three orthogonal axes. But if rotation is continued at a constant angular velocity the input from the canals soon ceases. The integral of the signal from the canals can provide information about the position of the body but only with respect to a remembered initial position.

Vection

Vection is an illusion of self motion induced by looking at a large moving display. For instance, illusory self rotation, or circularvection, is induced when an upright subject observes the inside of a large vertical cylinder rotating about the mid-body axis (yaw axis). For much of the time the cylinder seems to be stationary in exocentric space and the body feels as if it moving in a direction opposite to that of the visual display. Similar illusions of self motion may be induced by visual displays rotating about the visual axis (roll axis) or about an axis passing through the two ears (pitch axis). Judgements about the motion of the self with respect to an external frame of reference are semi-exocentric judgements since they involve an external frame and a reference to the self. Rotation of a natural scene with respect to the head is normally due to head rotation and the vestibular system is an unreliable indicator of self rotation except during and just after acceleration. Therefore it is not surprising that scene rotation is interpreted as self rotation, even when the body is not rotating. There is a conjunction of visual and vestibular inputs into the vestibular nuclei (Waespe and Henn, 1978) and the parietal cortex (Fredrickson and Schwarz, 1977) which probably explains why visual inputs can so closely mimic the effects of vestibular inputs.

Vection for different postures and axes of rotation

If the vection axis is vertical, the sensation of self rotation is continuous and usually at the full velocity of the stimulus motion. If the vection axis is horizontal, the illusory motion of the body is restrained by the absence of utricular inputs that would arise if the body were actually rotating. Under these circumstances a weakened but still continuous sensation of body rotation is accompanied by a paradoxical sensation that the body has tilted only through a certain angle (Held *et al.*, 1975). There are three vection axes with respect to the body (yaw, roll and pitch) and in each case the vection axis can be either vertical or horizontal. Of these six stimulus conditions only three had been investigated. We decided to measure vection and illusory body tilt under all six conditions (Howard *et al.*, 1987). The subject was suspended in various postures within a large sphere that could be rotated about a vertical or horizontal axis. The magnitude of vection and of illusory body tilt were measured for yaw, pitch and roll vection for both vertical and horizontal orientations of each axis (see Figure 1).

For body rotation about both vertical and horizontal axes, yaw vection was stronger than pitch vection which was stronger than roll vection. When

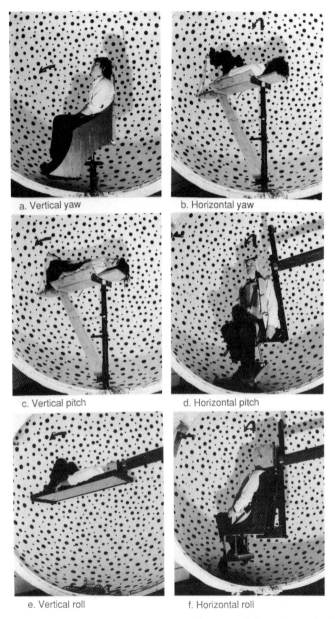

Figure 1. The set of postures and vection axes used by Howard, Cheung and Landolt (1987) to study vection and illusory body tilt. The subject is seen through the open door of the 3 m diameter sphere which could be rotated about either the vertical or horizontal axis. The subject was supported in different postures by air cushions and straps (not shown) so as to produce the six possible combinations of vection axis (yaw, pitch and roll) and gravitational orientation of the axis.

the vection axis was vertical, sensations of body motion were continuous and usually at, or close to the full velocity of the rotating visual field. When the vection axis was horizontal, the sensations of body motion were still continuous but were reduced in magnitude. Also for vection about horizontal axes, sensations of continuous body motion were accompanied by sensations of illusory yaw, roll or pitch of the body away from the vertical posture. The mean illusory body tilt was about 20° but the body was often reported to have tilted by as much as 90°. Two subjects in a second experiment reported sensations of having rotated full circle. Held *et al.* (1975) reported a mean illusory body tilt of 14°. We obtained larger degrees of body tilt probably because our display filled the entire visual field and subjects were primed to expect that their bodies might really tilt. In most subjects, illusory backwards tilt accompanied by pitch vection was much stronger than illusory forward tilt. Only two of our 16 subjects showed the opposite asymmetry, that was also reported by Young *et al.* (1975).

Vection and the relative distances of competing displays

The more distant parts of a natural scene are less likely to rotate with a person than are nearer parts of a scene, so that the headcentric motion of more distant parts provides a more reliable indicator of self rotation than does motion of nearer objects. It follows that circularvection should be related to the motion of the more distant of two superimposed displays. In line with this expectation Brandt *et al.* (1975) found that vection was not affected by stationary objects in front of the moving display but was reduced when the objects were seen beyond the display. Depth was created by binocular disparity in this experiment and there is some doubt whether depth was the crucial factor as opposed to the perceived foreground-background relationships of the competing stimuli. Furthermore, the two elements of the display differed in size as well as distance.

Ohmi *et al.* (1987) conducted an experiment using a background cylindrical display of randomly placed dots which rotated around the subject, and a similar stationary display mounted on a transparent cylinder which could be set at various distances between the subject and the moving display. The absence of binocular cues to depth allowed the perceived depth order of the two displays to reverse spontaneously, even when they were well separated in depth. Subjects were asked to focus alternately on the near display and the far display while reporting the onset or offset of vection. They were also asked to report any apparent reversal of the depth order of the two displays, which was easy to notice because of a slight difference in their appearance.

In all cases vection was experienced whenever the display that was perceived as the more distant was moving and was never experienced whenever the display perceived as more distant was stationary. Thus circular vection is totally under the control of whichever of two similar displays is perceived as background. This dominance of the background display does not depend on depth cues, because circularvection is dominated by a display that appears more distant, even when it is nearer. We think that perceived distance is not the crucial property of that part of the scene interpreted as background. When subjects focused on the moving display, optokinetic pursuit movements of the eyes occurred, and when they focused on the stationary display, the eyes were stationary. But such a change in the plane of focus had no effect on whether or not vection was experienced, as long as the apparent depth order of the two displays did not change.

Thus sensations of self rotation are induced by those motion signals most reliably associated with actual body rotation, namely, signals arising from that part of the scene perceived as background. Vection sensations are not tied to depth cues, which makes sense because depth cues can be ambiguous. Furthermore, vection sensations are not tied to whether the eyes pursue one part of the scene or another, which also makes sense because it is headcentric visual motion that indicates self motion, and this is detected just as well by retinal image motion as by motion of the eyes.

Vection and the central-peripheral and near-far placement of stimuli

It has been reported that circularvection is much more effectively induced by a moving scene confined to the peripheral retina than by one confined to the central retina (Brandt et al., 1973). In these studies, the central retina was occluded by a dark disc which may have predisposed subjects to see the peripheral display as background and it may have been this rather than its peripheral position which caused it to induce strong vection. Similarly, when the stimulus was confined to the central retina subjects may have been predisposed to see it as a figure against a ground, which may have accounted for the weak vection evoked by it.

Howard and Heckmann (1989) conducted an experiment to test this idea. The apparatus is depicted in Figure 2. The subject sat at the center of a vertical cylinder covered with randomly arranged black opaque dots. A 54° × 44° square display of dots above the subject's head was reflected by a sheet of transparent plastic onto a matching black occluder in the center of the large display. The central display could be moved so that it appeared to be suspended 15 cm in front of or 15 cm beyond the peripheral display. In the latter position it appeared as if seen through a square hole. In some conditions, one of the displays moved from right to left or from left to right at 30°/s while the other was occluded. In other conditions both displays were visible but only one moved and in still other conditions, both displays moved, either in the same direction or in opposite directions. In each condition subjects looked at the center of the display and rated the direction and strength of circularvection.

The results are shown in Figure 3. They reveal that vection was driven better by the peripheral stimulus acting alone than by the central stimulus acting alone. Indeed it was driven just as well by a moving peripheral display

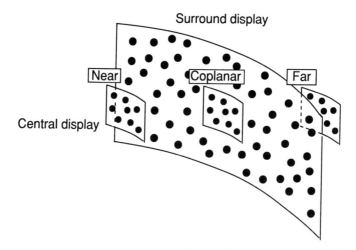

Figure 2. Diagramatic representation of the displays used by Howard et al. (1987) to study the interaction between central-peripheral and far-near placement of two displays in generating circularvection. The two displays could be moved in the same or in opposite directions, or one of them could be stationary or blacked out.

with the center black or visible and stationary as by a full-field display. However, vection was reduced when the central display moved in a direction opposite to that of the peripheral display. When the peripheral display was visible but stationary the direction of vection was determined by the central display but only when it was farther away than the surround. This result is understandable when we realize that this sort of stimulation is produced, for example, when an observer looks out of the window of a moving vehicle. The moving field seen through the window indicates that the viewer is carried along with the part of the scene surrounding the window on the inside. When the moving central display was nearer than the stationary surround, a small amount of vection was evident in the same direction as the motion of the central display. We believe that the motion of the center induced apparent motion in the stationary surround, which in turn caused vection. We call this induced-motion vection. These experiments are a confirmation and extension of experiments conducted by Howard *et al.*, (1987).

Induced visual motion, an oculocentric, headcentric and exocentric phenomenon

Induced visual motion occurs when one observes a small stationary object against a larger moving background and was first described in detail by Duncker (1929). For instance, the moon appears to move when seen through moving clouds. In a commonly studied form of induced motion the station-

Spatial vision

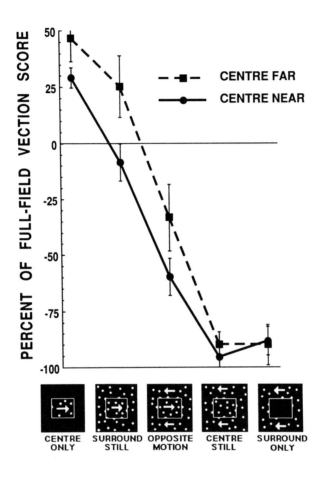

Figure 3. The magnitude of circular vection (expressed as a percentage of the magnitude obtained in the full-field condition) for each Distance × Display condition. Positive vection scores indicate vection in a direction consistent with (opposite) the motion of the central display and negative scores indicate vection in a direction appropriate to the motion (or assumed contrast-motion) of the surround. Whether each display was moving, stationary or blacked out is indicated in the chart below each pair of data points. The dashed line is for when the central display was seen beyond the surround display and the solid line is for when the central display was seen nearer than the surround display. The error bars are standard errors of the mean. Each data point is the mean for eight subjects. Although points on the X-axis are categories, data points have been linked by lines to reveal important trends in the data. (Reprinted from Howard and Heckmann, Perception, **18**, 657–667, Pion Ltd., 1989).

ary object is seen within a frame which moves from side to side. In this stimulus configuration the moving frame changes in eccentricity and this may be responsible for some of the illusory motion of the stationary object. In order to study the effects of relative motion alone it is best to present the stationary object on a large moving background that either fills the visual field or remains within the confines of a stationary boundary.

We have evidence that induced visual motion occurs within the oculocentric, the headcentric and the exocentric system and that the mechanisms in the three cases are very different. As an oculocentric effect, it could be due to contrast between oculocentric motion-detectors. As a headcentric effect, it could be due to misregistration of eye movements. This could occur in the following way. Optokinetic nystagmus (OKN) induced by the moving background is inhibited by voluntary fixation on the stationary object. If the efference associated with OKN were not fully available to the perceptual system, but the efference associated with voluntary fixation were available, this should create an illusion of movement in a direction opposite to that of the background motion. This explanation, which I proposed in Howard (1982, p. 303) is analogous to that proposed by Whiteside et al. (1965) to account for the oculogyral illusion. It has been championed more recently by Post and Leibowitz (1985), Post (1986) and Post and Heckmann (1986).

Induced visual motion can also be an exocentric illusion. It has been explained above that inspection of a large moving background induces an illusion of self motion accompanied by an impression that the background is not moving. A small object fixed with respect to the observer should appear to move with the observer and therefore to move with respect to the exocentric frame provided by the perceptually stationary background. This possibility was mentioned by Duncker.

We have recently devised psychophysical tests which can be used to dissociate the oculocentric, headcentric and exocentric forms of induced visual motion. These tests will now be described.

The key to measuring the oculocentric component of induced visual motion is to have two inducing displays moving in opposite directions, with a stationary test object on or near each. Nakayama and Tyler (1978) reported that a pair of parallel lines pulsing in and out in opposite directions induced an apparent pulsation of a pair of stationary lines placed between them. The apparent velocity of this induced motion was only about $0.1°/s$. This display is not ideal for measuring oculocentric induced visual motion since the outward and inward motion of the two induction lines mimics visual looming produced by forward body motion. An outwardly expanding textured surface is known to induce forward linear vection (Andersen and Braunstein, 1985; Ohmi and Howard, 1988).

A better stimulus for measuring oculocentric induced visual motion is that shown in Figure 4a. The two inducing stimuli move in a shearing fashion which does not mimic visual looming. If the gaze is directed at the boundary between the two moving displays, neither optokinetic nystagmus nor vection

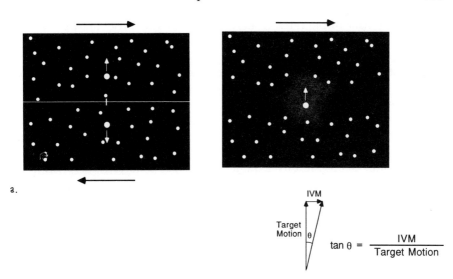

Figure 4. Stimuli used by Heckmann and Howard (1989) to dissociate oculocentric and headcentric components of induced visual motion. Left: Two displays moving in opposite directions each have a test spot which moves vertically at 2°/s with quick returns. The apparent slant of the path of motion of each test spot is the resultant of its actual vertical motion and its horizontal induced motion. The relative slant of the paths of the two spots, as the subject fixates between them, gives a measure of oculocentric induced motion. Right: The subject visually pursues a test spot as it moves vertically over a moving display. The apparent slant of its path of motion gives a measure of headcentric induced motion.

should occur. Any perceived relative motion between the two test spots must reflect oculocentric induced motion since headcentric or exocentric induced motion would affect the two objects in the same way. The task of judging the relative velocity of the test spots is simplified by using a procedure described by Wallach et al. (1978). The two test spots were moved vertically at a velocity of 2°/s with periodic fast returns and subjects estimated the apparent inclination of the path of motion of one spot relative to that of the other spot. The apparent direction of motion of each spot is the resultant of its actual vertical motion and its apparent horizontal motion. With this display we have found the velocity of oculocentric induced motion to be about the same as that reported by Nakayama and Tyler.

The next step is to isolate the headcentric component of induced visual motion. Since the oculocentric component is confined to the region of the inducing stimulus, placing the test dot on a black band, as shown in Figure 4b, ensures that this form of induced motion will not occur. Again subjects judged the apparent slant of the path of a vertically moving spot, but this time pursuing it with the eyes. In a series of experiments we have shown that the apparent slant of the track is determined by headcentric induced motion and is not influenced by exocentric induced motion. This is probably because the frame of reference for judging the vertical is carried with the illusory motion of the body. The magnitude of headcentric induced motion was found to be about 2°/s, which is considerably larger than oculocentric induced motion (Heckmann and Howard, 1989; Post and Heckmann, 1986).

Finally we measured exocentric induced visual motion by having subjects estimate the velocity of illusory self motion induced by the motion of a large moving display. By definition this is a measure of the exocentric induced visual motion. People readily experience 100 per cent vection at stimulus velocities of up to 60°/s and stationary visual objects appear to move in space at the same velocity as the apparent movement of the body. Thus exocentric induced visual motion can be many times larger than headcentric induced motion which in turn is several times larger than oculocentric visual motion.

The task of distinguishing between oculocentric, headcentric and exocentric components of any perceptual phenomenon and the task of discovering which sensory or cognitive processes may be responsible for a given phenomenon, require tests and procedures specifically designed for each case.

Acknowledgements

The experiments on vection and induced motion described in this chapter were part of DCIEM Contract 97711-4-7936/8SE.

References

Andersen, G.J. and Braunstein, M.L. (1985). Induced self-motion in central vision. *Journal of Experimental Psychology, Human Perception and Performance*, **11**, 122–132.
Barr, C.C., Schulthies, L.W. and Robinson, D.A. (1976). Voluntary, non-visual control of the human vestibulo-ocular reflex. *Acta Otolaryngology*, **81**, 365–375.
Benson, A.J. and Barnes, G.R. (1978). Vision during angular oscillation: The dynamic interaction of visual and vestibular mechanism. *Avaition Space and Environmental Medicine*, **49**, 340–345.
Brandt, T., Dichgans, T. and Koenig, E. (1973). Differential effects of central versus peripheral vision on egocentric and exocentric motion perception. *Experimental Brain Research*, **16**, 476–491.
Brandt, T., Wist, E.R. and Dichgans, J. (1975). Foreground and background in dynamic spatial orientation. *Perception and Psychophysics*, **17**, 497–503.
Cohen, H.B. (1966). Some critical factors in prism adaptation. *American Journal of Psychology*, **79**, 285–290.
Duncker, K. (1929). Uber induzierte Bewegung. *Psychol Forsch.*, **22**, 180–259.
Fredrickson, J.M. and Schwarz, D.W.F. (1977). Vestibulo-cortical projection. In Naunton, R.F. (Ed.), *The Vestibular System*, 203–210. New York: Academic Press.
Graybiel, A. and Hupp, D.I. (1946). The oculo-gyral illusion. *Journal of Avaition Medicine*, **17**, 3–27.

Heckmann, T. and Howard, I.P. (1989). Induced visual motion: dissociation of exocentric and headcentric components. *Investigative Ophthalmology in Visual Science*, **30** (ARVO Abs.), 323.

Held, R. and Rekosh, J. (1963). Motor-sensory feedback and geometry of visual space. *Science*, **141**, 722–723.

Held, R., Dichgans, J. and Bauer, J. (1975). Characteristics of moving visual areas influencing spatial orientation. *Vision Research*, **15**, 357–365.

Hill, A.L. (1972). Direction constancy. *Perception and Psychophysics*, **11**, 175–178.

Howard, I.P. (1982). *Human Visual Orientation*, 388–397. Chichester: Wiley.

Howard, I.P. and Anstis, T. (1974). Muscular and joint-receptor components in postural persistence. *Journal of Experimental Psychology*, **103**, 167–170.

Howard, I.P. and Heckmann, T. (1989). Circular vection as a function of the relative sizes, distances and positions of two competing visual displays. *Perception*, in press.

Howard, I.P., Cheung, B. and Landolt, J. (1977). Influence of vection axis and body posture on visually-induced self rotation and tilt. *Proceedings of the AGARD Conference on Motion Cues in Flight Simulation and Simulator Sickness*. Brussels.

Howard, I.P., Giaschi, D. and Murasugi, C.M. (1988). Inhibition of optokinetic and vestibularnystagmus by head-fixed stimuli. *Experimental Brain Research*, Submitted for publication.

Howard, I.P., Ohmi, M., Simpson, W. and Landolt, J. (1987). Vection and the spatial disposition of competing moving displays. *Proceedings of the AGARD Conference on Motion Cues in Flight Simulation and Simulator Sickness*. Brussels.

Kohler, W. and Wallach, H. (1944). Figural aftereffects: An investigation of visual processes. *Proceedings of the American Philosophy Society*, **88**, 269–357.

McKinley, P.A. and Peterson, B.W. (1985). Voluntary modulation of the vestibuloocular reflex in humans and its relation to smooth pursuit. *Experimental Brain Research*, **60**, 454–464.

Melvill Jones, G., Berthoz, A. and Segal, B. (1984). Adaptive modification of the vestibulo-ocular reflex by mental effort in darkness. *Brain Research*, **56**, 149–153.

Morgan, C.L. (1978). Constancy of egocentric visual direction. *Perception and Psychophysics*, **23**, 61–68.

Nakayama, K. and Tyler, C.W. (1978). Relative motion induced between stationary lines. *Vision Research*, **18**, 1663–1668.

Ohmi, M. and Howard, I.P. (1988). Effect of stationary objects on illusory forward self motion induced by a looming display. *Perception*, **17**, 5–12.

Ohmi, M., Howard, I.P. and Landolt, J.P. (1987). Circular vection as a function of foreground-background relationships. *Perception*, **16**, 17–22.

Paap, K.R. and Ebenholtz, S.M. (1976). Perceptual consequences of potentiation in the extraocular muscles: An alternative explanation for adaptation to wedge prisms. *Journal of Experimental Psychology*, **2**, 457–468.

Post, R.B. (1986). Induced motion considered as a visually induced oculogyral illusion. *Perception*, **15**, 131–138.

Post, R.B. and Heckmann, T. (1986). Induced motion and apparent straight ahead during prolonged stimulation. *Perception and Psychophysics*, **40**, 263–270.

Post, R.B. and Leibowitz, H.W. (1985). A revised analysis of the role of efference in motion perception. *Perception*, **14**, 631–643.

Tomlinson, R.D. and Robinson, D.A. (1981). Is the vestibulo-ocular reflex cancelled by smooth pursuit? In Fuchs, A. and Becker, J. (Eds) *Progress in Oculomotor Research*. North Holland: Elsevier.

Waespe, W. and Henn, V. (1978). Conflicting visual-vestibular stimulation and vestibular nucleus activity in alert monkeys. *Experimental Brain Research*, **33**, 203–211.

Wallach, H., Bacon, J. and Schulman, P. (1978). Adaptation in motion perception: alteration of induced motion. *Perception and Psychophysics*, **24**, 509–514.

Whiteside, T.C.D., Graybiel, A. and Niven, J.I. (1965). Visual illusions of movement. *Brain*, **88**, 193–210.

Young, L.R., Oman, C.M. and Dichgans, J.M. (1975). Influence of head orientation on visually induced pitch and roll sensations. *Aviation Space and Environment Medicine*, **46**, 246–269.

23

Comments on "Spatial vision within egocentric and exocentric frames of reference"

Thomas Heckmann

Human Performance Laboratory
Institute for Space and Terrestrial Science
York University, North York, Ontario, Canada M3J 1P3

and

Robert B. Post

Department of Psychology
University of California, Davis, CA 95616

Induced visual motion is the name assigned to a group of phenomena which can be described with more or less the same words: "illusory motion of stationary contours opposite the direction of moving ones". As Dr. Howard has pointed out, it is possible that oculocentric, headcentric and exocentric mechanisms generate experiences which may be described by the words "induced visual motion". We have found Dr. Howard's framework very helpful in organizing our thoughts about the multiple sources of these apparently similar phenomena. We also accept that some forms of induced visual motion may depend on vection and cannot be explained by suppression of nystagmus (e.g., phenomenal tilt of a stationary stimulus during roll vection induced by a contoured disc rotating in a frontal plane). We are less certain than Dr. Howard, however, that there is only one mechanism for induced visual motion.

In Dr. Howard's study, phenomenal motion of a stationary display which was positioned in front of a moving display occurred only when there was vection. We have reliably obtained induced visual motion of small fixation targets in the complete absence of vection (Post and Heckmann, 1987; Post and Chaderjian, 1988; Heckmann and Post, 1988). Dr. Howard would likely explain this finding with his statement that "... visual consequences of vestibular stimulation have a lower threshold than sensations of bodily motion." We agree wholeheartedly: optokinetic afternystagmus (OKAN), which is a good indicator of the vestibular effects of visual stimulation, has been found at moving-contour velocities too low to elicit vection (Koenig et al., 1982). We have also reliably obtained OKAN after exposure to a moving-contour stimulus which elicits no vection (Heckmann and Post, 1988). In fact, induced visual motion may be elicited by a single moving dot stimulus (Post and Chaderjian, 1988) which is not capable of producing vection.

If induced visual motion occurs because a perceptually registered voluntary signal for fixation opposes an unregistered involuntary signal for optokinetic nystagmus, then the illusion should reflect known dynamic properties of the optokinetic system. That is, the magnitude of induced visual motion will be proportional to the nystagmus signal being opposed. Induced visual motion should therefore vary across stimulation in the same way that nystagmus varies, but have the opposite directional sign. Our efforts to disconfirm this prediction have so far failed. Induced visual motion is correlated with OKAN of opposite directional sign across variations in stimulus illuminance and velocity (Post, 1986). The magnitude of induced visual motion increases along with the slowphase velocity of OKAN with increasing stimulus duration. The illusion also decays and reverses direction along with OKAN after stimulus termination. Further, both responses show an increased tendency to reverse direction following stimulation in the presence of a fixation target rather than after stimulation without fixation (Heckmann and Post, 1988).

Induced visual motion is not the only motion illusion involving visual fixation of moving or stationary targets which can potentially be explained by interaction of voluntary and involuntary eye-movement signals. These illusions include autokinesis, the Aubert-Fleischel effect, the Filehne Illusion, and several others (Post and Leibowitz, 1985). Induced visual motion, however, provides a particularly good model for testing the eye-movement hypothesis, since a good deal is known about the dynamics of visually induced involuntary eye movements. We have not been so much interested in "championing" a particular explanation of induced visual motion, therefore, as we have been to test the existence and applicability of a particular mechanism. Of course, since we are using a well-known illusion as our model, we must also explore the applicability of alternative explanations of induced visual motion to our results.

With further reference to the origin of induced visual motion in vection, therefore, we recently reported a dissociation between the two illusions (Post and Heckmann, 1987). Briefly, fixation of a target located 10° left of the midline during exposure to rightward-moving background contours reliably increased the magnitude of induced visual motion. This finding is consistent with the idea that extra voluntary efference is needed to maintain a leftward as compared to a straight-ahead gaze during rightward motion of background contours. Vection, however, was reduced when a fixation target was made available, and further reduced when the target was placed 10° left of the midline. We emphasize that this dissociation does not reject the idea that some form of induced visual motion originates with vection, only the idea that all of induced visual motion originates with vection.

References

Heckmann, T. and Post, R.B. (1988). Induced motion and optokinetic afternystagmus: parallel response dynamics with prolonged stimulation. *Vision Research*, **28** (6), 681–694.
Koenig, E., Dichgans, J. and Schmucker, D. (1982). The influence of circularvection (CV) on optokinetic nystagmus (OKN) and optokinetic afternystagmus (OKAN). In Lennerstrand, G., Zee, G. and Keller, E. (Eds), *Functional Basis of Oculomotor Disorders*. Oxford: Pergamon.
Post, R.B. (1986). Induced motion considered as a visually-induced oculogyral illusion. *Perception*, **15**, 131–138.
Post, R.B. and Chaderjian, M. (1988). The sum of induced and real motion is not a straight path. *Perception and Psychophysics*, **43**, 121–124.
Post, R.B. and Heckmann, T. (1987). Experimental dissociation of vection from induced motion and displacement of the apparent straight-ahead. *Investigative Ophthalomology*, **28**, Supp., 311.
Post, R.B. and Leibowitz, H.W. (1985). A revised analysis of the role of efference in motion perception. *Perception*, **15**, 131–138.

24

Sensory conflict in motion sickness: an Observer Theory approach

Charles M. Oman

Man–Vehicle Laboratory
Massachusetts Institute of Technology
Cambridge, Massachusetts

Summary

"Motion sickness" is the general term describing a group of common nausea syndromes originally attributed to motion-induced cerebral ischemia, stimulation of abdominal organ afferent, or overstimulation of the vestibular organs of the inner ear. Sea-, car-, and airsickness are the most commonly experienced examples. However, the discovery of other variants such as Cinerama-, flight simulator-, spectacle-, and space sickness in which the physical motion of the head and body is normal or absent has led to a succession of "sensory conflict" theories which offer a more comprehensive etiologic perspective. Implicit in the conflict theory is the hypothesis that neural and/or humoral signals originate in regions of the brain subserving spatial orientation, and that these signals somehow traverse to other centers mediating sickness symptoms. Unfortunately, our present understanding of the neurophysiological basis of motion sickness is far from complete. No sensory conflict neuron or process has yet been physiologically identified. To what extent can the existing theory be reconciled with current knowledge of the physiology and pharmacology of nausea and vomiting? This paper reviews the stimuli which cause sickness, synthesizes a contemporary Observer Theory view of the sensory conflict hypothesis, and presents a revised model for the dynamic coupling between the putative conflict signals and nausea magnitude estimates. The use of quantitative models for sensory conflict offers a possible new approach to improving the design of visual and motion systems for flight simulators and other "virtual environment" display systems.

Stimuli causing motion sickness: exogenous motion and "sensory rearrangement"

Motion sickness is a syndrome characterized in humans by signs such as vomiting and retching, pallor, cold sweating, yawning, belching, flatulence, decreased gastric tonus; and by symptoms such as stomach discomfort, nausea, headache, feeling of warmth, and drowsiness. It has a significant incidence in civil and military transportation, and is a common consequence of vestibular disease. Virtually everyone is susceptible to some degree, provided the stimulus is appropriate and lasts long enough. Many other animal species also exhibit susceptibility.

A century ago, physicians commonly attributed motion sickness to acceleration-induced cerebral ischemia, or to mechanical stimulation of abdominal afferents (Reason and Brand, 1975). These theories were largely discounted when the role of the inner ear vestibular organs in body movement control was appreciated, and when James (1882) noted that individuals who lack vestibular function were apparently immune. As a result, it was commonly thought that motion sickness results simply from vestibular overstimulation.

Certainly the most common physical stimulus for motion sickness is exogenous (i.e., nonvolitional) motion, particularly at low frequencies. However, when individuals are able (motorically) to anticipate incoming sensory cues, motion stimuli are relatively benign. For example, drivers of cars and pilots of aircraft are usually not susceptible to motion sickness, even though they experience the same motion as their passengers. In daily life, we all run, jump, and dance. Such endogenous (volitional) motions never make us sick. Thus, it is now recognized that motion sickness cannot result simply from vestibular overstimulation.

Many forms of motion sickness consistently occur when people are exposed to conditions of "sensory rearrangement"—when the rules which define the normal relationship between body movemenus and the resulting neural inflow to the central nervous system have been systematically changed (Reason, 1978). Whenever the central nervous system receives sensory information concerning the orientation and movement of the body which is unexpected or unfamiliar in the context of motor intentions and previous sensory-motor experience—and this condition occurs for long enough—motion sickness typically results. Thus, sickness occurs when a person moves about while wearing a new pair of glasses (spectacle sickness) or when a subject in laboratory experiments walks around wearing goggles which cause left-right or up-down reverse vision. Similarly, sickness is also encountered in flight simulators equipped with compelling visual displays (simulator sickness) and in wide-screen movie theaters (Cinerama sickness), since visual cues to motion are not matched by the usual pattern of vestibular and proprioceptive cues to body acceleration. Space sickness among astronauts is believed to result in part because the sensory cues provided by the inner ear otolith organs in weightlessness do not correspond to those experienced on Earth. Astronauts also commonly experience visual spatial reorientation episodes which are provocative. When one floats in an inverted position in the spacecraft, a true ceiling can seem somehow like a floor. Visual cues to static orientation can be ambiguous, often because of symmetries inherent in the visual scene. Cognitive reinterpretation of ambiguous visual orientation cues results in a sudden change in perceived orientation, which astronauts have found can be nauseogenic (Oman, 1988). These various forms of sickness illustrate that the actual stimulus for sickness cannot always be adequately quantified simply by quantifying the physical stimulus. The trigger for sickness is a signal inside the central nervous system (CNS) which also depends on the subject's previous sensory motor experience.

Physiological basis of motion sickness

Despite the ubiquity of motion sickness in modern society and significant research (well reviewed, collectively, by Tyler and Bard, 1949; Chinn and Smith, 1955; Money, 1970; Reason and Brand, 1975; Graybiel, 1975; and Miller, 1988), the physiological mechanisms underlying motion sickness remain poorly defined. Classic studies of canine susceptibility to swing sickness (Wang and Chinn, 1956; Bard et al., 1947) indicated that the cerebellar nodulus and uvula—portions of the central vestibular system—are required for susceptibility. Many neurons in the central vestibular system which subserve postural and oculomotor control are now known to respond to a variety of spatial orientation cues. A brain stem vomiting center was identified by Wang and Borison (1950) and Wang and Chinn (1954), which initiates emesis in dogs in response to various stimuli, including motion. Nausea sensation in humans is commonly assumed to be associated with activity in the vomiting center (Money, 1970). The integrity of an adjacent chemoreceptive trigger zone (CTZ), localized in area postrema on the floor of the fourth ventricle, was also believed to be required for motion sickness (Wang and Chinn, 1954; Brizzee and Neal, 1954). It was generally assumed that signals originating somewhere in the central vestibular system somehow traverse to the chemoreceptive trigger zone, which in turn activates the vomiting center. Wang and Chinn (1953) and Crampton and Daunton (1983) have found evidence suggestive of a possible humoral agent in cerebrospinal fluid (CSF) transported between the third and fourth ventricle. However, an emetic linkage via CSF transport does not easily account for the very short latency vomiting which is occasionally observed experimentally. The vomiting center receives convergent inputs from a variety of other central and peripheral sources, including the diencephalon and gastrointestinal tract. The possibility of multiple emetic pathways and significant interspecies differences in mechanism must be considered. Also, more recent experiments have led workers to question the notion that medullary emetic centers are discretely localizable. Attempts to verify the earlier findings by demonstrating motion sickness immunity in area postrema ablated and cerebellar nodulectomized and uvulectomized animals have not been successful (Miller and Wilson, 1983a,b; Borison and Borison, 1986; Wilpizeski et al., 1986).

The act of emesis itself involves the somatic musculature. However, many other signs of motion sickness as listed earlier and associated with vasomotor, gastric, and respiratory function suggest that areas in the reticular core of the brain stem and limbic system, which are associated with autonomic regulation are also coactivated. The limbic system and associated hypothalamus-pituitary-adrenal cortex (H-P-A) neuroendocrine outflow pathway is involved. Increases in circulating levels of such stress-related hormones as epinepherine and norepinepherine, ADH, ACTH, cortisol, growth hormone, and prolactin have been found during sickness (e.g., Eversmann et al., 1978; La Rochelle et al., 1982). Whether the limbic system and H-P-A axis simply mediate a generalized stress response, or are also involved in motion-sickness adaptation by somehow triggering stimulus-specific sensory/motor learning is unknown. The question of the site of action of antimotion-sickness drugs is also far from resolved. There is no substantial evidence that effective drugs act on the vestibular end organs. Their primary effect is probably simply to raise the threshold for sickness. Antimotion-sickness drugs could be acting on brain-stem emetic centers. Alternatively, they may shift the fundamental andrenergic-cholinergic balance in the limbic system (e.g., Janowsky et al., 1984).

Development of the sensory conflict theory

Although our physiological understanding of motion sickness is thus incomplete, analyses of the wide variety of physical stimuli which produce the same syndrome of symptoms and signs and the dynamic pattern of these responses have nonetheless given us some insight concerning possible etiologic mechanisms. Recognition that motion sickness could occur not only under exogenous motion stimulation, but also as a result of sensory rearrangement, as defined above, has led to the development of a succession of sensory conflict theories for the disorder.

The sensory conflict hypothesis for motion sickness was originally proposed by Claremont (1931), and has since been revised and extended by several authors. Implicit is the idea that a neural or humoral sensory conflict-related signal originates somewhere in the brain and somehow couples to brain centers mediating sickness symptoms. In early statements of the theory, conflict signals were assumed somehow to result from a direct comparison of signals provided by different sensory modalities (e.g., "the signals from the eye and ear do not agree"; canal-otolith, and visual-inertial conflicts). However, Reason (1978) emphasized that a direct intermodality comparison of afferent signals is simply not appropriate, because signals from the various sense organs have different "normal" behavior (in terms of dynamic response and coding type), and whether they can be said to conflict or not actually depends upon context and previous sensory-motor experience. Hence the conflict is more likely between actual and anticipated sensory signals. Extrapolating from earlier interrelated work by von Holst and Held, Reason argued that the brain probably evaluates incoming sensory signals for consistency using an "efference copy" based scheme. As motor actions are commanded, the brain is postulated continuously to predict the corresponding sensory inputs, based on a neural store (memory bank or dictionary) of paired sensory and motor memory traces learned from previous experience interacting with the physical environment. Sensory conflict signals result from a continuing comparison between actual sensory input and this retrieved sensory memory trace. Any situation which changed the rules relating motor outflow to sensory return (sensory rearrangement, a term coined by Held) would therefore be expected to produce prolonged sensory conflict and result in motion sickness. Adaptation to sensory rearrangement was hypothesized to involve updating of the neural store with new sensory and motor memory-trace pairs. Reason proposed a formal neural mismatch model which incorporated these concepts. However, the model was only qualitative, making simulation and quantitative prediction beyond its reach. Key structural elements such as the neural store and memory traces were only intuitively defined. The model did not really address the question of why the CNS should have to compute a sensory conflict signal, other than to make one sick. Reason's model dealt with sensory conflict only and did not incorporate emetic brain output pathway elements which must be present to account for the latency and order of appearance of specific symptoms.

A mathematical definition of sensory conflict

In order to address these difficulties, the author proposed a model for motion sickness (Oman, 1978; 1982) in a mathematical form, shown in block diagram format in Figures 1–3. This new model contained a statement of the conflict theory which was congruent with Reason's view, and also the emetic linkage output pathway dynamics missing from Reason's model. The conflict theory portion of the model was formally developed by application of Observer Theory concepts from control engineering to the neural information processing task faced by the CNS in actively controlling body movement using a limited set of noisy sensory signals. The conflict model formulation can be considered an extension of the optimal control model in the field of manual control (Baron and Kleinman, 1968) and in the field of spatial orientation research, an extension of Kalman filter models (Young, 1970; Borah et al., 1978). These have been used to predict orientation perception in passive observers with some success. In these previous models, however, sensory conflict was not defined in the same sense as that used by Reason and me.

In guidance, control, and navigation systems, engineers are often faced with the problem of controlling a vehicle's state vector (e.g., angular and linear position, velocity, and acceleration) when information from sensors which measure these states is noisy or is even not directly measured at all. To deal with this problem, engineers now routinely incorporate into the control system design a computational element known as an "observer", whose function is to provide an optimal estimate of the actual states of the vehicle (or other system) being controlled. Control loops are closed using the state estimate provided by the observer in lieu of direct feedback sensor measurements in the traditional way. Analytical techniques have been developed (Kalman, 1960; Wonham, 1968) for mathematically linear systems which allow designers to choose observer and control-loop parameters so that the observer state estimate is always converging with reality, and which optimizes the closed-loop performance of the entire system. In control engineering parlance, such systems are formally called "output feedback" optimal-control systems.

Of particular importance in the present context is the way in which the observer state estimate is calculated in these engineering systems. The observer contains an internal dynamic model of the controlled system and of the sensors being used. The observer element uses these models to calculate what the available feedback sensor measurements should be, assuming the vehicle state estimate of the observer is correct. The difference between the expected and the actual feedback measurements is then computed, because it is an indirect measure of the error in the observer state estimate. The difference signals play an important role in the observer. They are used continuously to steer the observer vehicle state estimate toward reality, using a method described in more detail below.

There is a direct analogy between the "expected" feedback sensor measurement and "internal dynamic model" concepts in control engineering Observer Theory, and the "efference copy" and "neural store" concepts which have emerged in physiology and psychology. From the perspective of control engineering, the "orientation" brain must "know" the natural behavior of the body, i.e., have an internal model of the dynamics of the body, and maintain a continuous estimate of the spatial orientation of all of its parts. Incoming sensory inputs would be evaluated by subtraction of an efference copy signal, and the resulting sensory conflict signal used to maintain a correct spatial orientation estimate.

The mathematical model for sensory conflict and movement control in the orientation brain is shown schematically in Figure 2, and mathematically in Figure 3. (Arrows in the diagrams represent vector quantities. For example, the actual state of the body might consist of the angular and/or linear

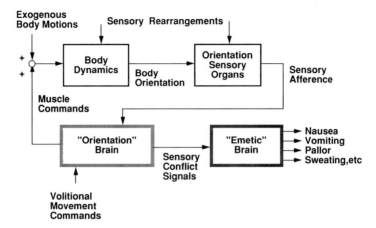

Figure 1. Schematic diagram of model for movement control, sensory conflict, and motion-sickness symptom dynamics (Oman, 1982). Under conditions of sensory rearrangement, the rules which relate muscle commands to sensory afference are systematically changed. Sensory conflict signals used spatial orientation perception and movement control in the orientation brain couple to the emetic brain.

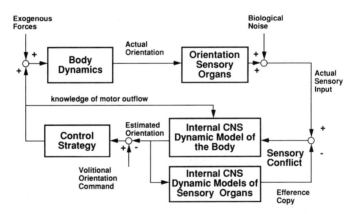

Figure 2. Observer theory model for movement control (Oman, 1982).

displacement of all the parts of the body, and higher derivatives.) The model function can be summarized as follows: the internal CNS models are represented by differential equations describing body and sense organ dynamics. Based on knowledge of current muscle commands, the internal model equations derive an estimated orientation state vector, which is used to determine new muscle commands based on control strategy rules. Simultaneously, the estimated orientation state is used by the CNS sense organ model to compute an efference copy "vector". If the internal models are correct, and there are

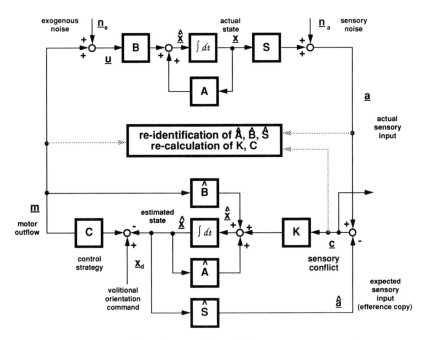

Figure 3. Mathematical formulation of model shown in Figure 2 (Oman, 1982).

no exogenous motion disturbances, the efference copy vector nearly cancels polysensory afference. If not, the difference—the sensory-conflict vector—is used to steer the model predictions toward reality, to trigger corrective muscle commands, and to indicate a need for reidentification of the internal model differential equations and steering factors.

How a sensory conflict vector might be used to correct internal model predictions is shown explicitly in Figure 3. Here, the physical body and sense organ dynamic characteristics are expressed in linearized state variable notation as a set of matrix equations of the form:

$$\dot{x} = Ax + Bu \tag{1}$$

$$a = Sx + n_a \tag{2}$$

$$u = m + n_e \tag{3}$$

The coefficients of the state differential equations for body and sense organ characteristics are thus embodied in the matrices **A**, **B**, and **S**. These equations are shown graphically in the upper half of figure 3. The internal CNS dynamic model is represented by an analagous state differential equation using hatted variables in the bottom half of the figure. This state estimator (the observer) with its matrices **Â**, **B̂**, and **Ŝ** corresponds to the neural store of

Reason's (1978) model. The sensory conflict vector c is obtained by subtracting actual sensory input a from expected sensory input $\hat{S}\hat{x}$. Sensory conflict normally originates only from exogenous motion cue inputs n_e, and noise n_a. The conflict vector is multiplied by a matrix **K** calculated using an optimization technique defined by Kalman and Bucy (1961) which lightly weights noisy modalities. When the result is added to the derivative of the estimated state, the estimated state vector is driven toward the actual state, and the component of the conflict vector magnitude due to noise is reduced. However, when exogenous motion cues inputs n_e are present, or under conditions of sensory rearrangement, such that matrices **A**, **B**, and/or **S** are changed, and no longer correspond to the matrices of the internal model, actual sensory input a will be large, and will not be cancelled by the efference copy vector. Sensory-motor learning takes place via reidentification by analysis of the new relationship between muscle commands and polysensory afference (reidentification of $\hat{A}$, $\hat{B}$, and $\hat{S}$), and internal model updating. Additional details are available in Oman (1982).

This model for sensory conflict overcomes many of the limitations of Reason's mismatch approach outlined earlier. The neural store is replaced by an internal mathematical dynamic model, so that efference copy and sensory conflict signals are quantitatively defined. Increased sensory conflict is noted to result not only from sensory rearrangement, but also from exogenous disturbance forces acting on the body. The role of active movement in creating motion sickness in some circumstances, and in alleviating them in others is clarified.

A revised model for symptom dynamics

The author's 1982 motion-sickness model included dynamic elements in the path between sensory conflict and overall discomfort and nausea in motion sickness. This model has since been altered in some important details; the current version is shown in Figures 4 and 5.

The input to the model is a sensory conflict vector. Because of the bandwidth requirements imposed on signals involved in orientation perception and posture control, it seems likely that the components of the conflict vector are neurally coded. In the nausea model, the various conflict vector components (describing the visual, vestibular, proprioceptive modalities) are rectified, and then weighted and added together. Rectification is required because sensory conflict components, as Reason and I have defined them, are signed quantities. The information carried in the sign is presumably useful in correcting orientation perception and posture control errors. However, stimuli which presumably produce sensory conflicts of opposite signs produce the same type and intensity of nausea, as far as we can tell. Hence rectification is appropriate here. In weighting the various conflict components, vestibular conflicts (i.e., semicircular canal and otolith modalities) must be

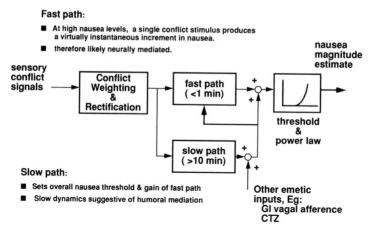

Figure 4. Schematic diagram of revised model for nausea-path symptom dynamics.

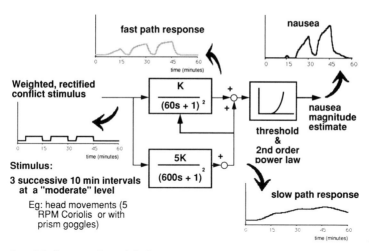

Figure 5. Mathematical model for nausea-path symptom dynamics. Insets show results of computer simulation.

weighted relatively heavily in the model, since people without vestibular function seem to be functionally immune. Visual motion inputs (as in Cinerama and simulator sickness) may thus exert their major sick-making effects indirectly: visual inputs would create illusory movement and thus expected vestibular signals, so sensory conflicts would be produced in the heavily weighted vestibular modality. However, to be consistent with our experimental evidence that visual and proprioceptive conflicts under prism goggle sensory rearrangement (Oman, 1987; Eagon, 1988) eventually become provocative while writing or when building can structures on a desktop, *absent* concomitant head motion or vestibular conflict, visual and proprioceptive modality model weighting factors are not zero.

As shown in Figures 4 and 5, rectified, weighted conflict signals then pass along two parallel, interacting dynamic pathways (fast and slow paths) before reaching a threshold/power law element and resulting in a nausea-magnitude estimate model output. Magnitude estimates are assumed to be governed by a power law relationship (Stevens, 1957) with an exponent of about 2. Susceptibility to motion sickness is determined in the model not only by the amount of sensory conflict produced, but also by the fast and slow pathway gains, time constants, and the nausea threshold. The transfer of a generalized adaptation from one different nauseogenic stimulus situation to another might result from adaptation in these output pathways.

The parallel arrangement of the fast and slow pathways and their relationship to the threshold element requires some explanation. In the past, many authors have therefore assumed that sensory conflict coupling to symptom pathways is a temporary (facultative) phenomenon. However, I have argued (Oman, 1982) that some level of subliminal sensory conflict coupling must be present in normal daily life because conflict signals seem to be continuously functionally averaged at subliminal levels, probably by the same mechanisms or processes which determine the intrinsic dynamics (latency, avalanching tendency, recovery time, etc.) of symptoms and signs when conflict exceeds normal levels. The output pathways probably consist functionally of dynamic elements followed by a threshold, and not the reverse, as would be the case if the linkage were temporary.

In the model, information flows along two paths prior to reaching the threshold. Both paths incorporate dynamic blocks which act continuously to accumulate (i.e., low pass filter or "leaky" integrate) the weighted, rectified conflict signal. One block (the fast path) has a relatively short characteristic response time, and the other (the slow path) has a relatively long one. (In the model simulations shown in the insets of Figure 5, the fast path is a second low-pass filter with 1 min time constants; the slow path is a similar filter with 10 min time constants. Second-order or higher block dynamics are required so that model predictions show characteristic overshoot when the conflict stimulus is turned off.) The slow path block normally has a higher gain (by a factor of about 5) than the fast path, and at the beginning of stimulation is functionally the more important element. Slow path output acts together with other classes of fast-acting nauseogenic inputs (e.g., vagal afference from the gut, or emetic drug stimulation) to bias the threshold of nausea response. In the present model, the slow path block output also acts as a multiplicative factor on fast path response gain. When prolonged stimulation has raised the slow path output, the response of the fast path becomes much larger, as shown in the Figure 5 simulation. Thus, the revised model mimics the much magnified response to incremental stimulation which we observe experimentally in long-duration sickness. (In the 1982 version of this model, increased response sensitivity at high symptom levels was a consequence only of the time-invariant, power-law, magnitude-estimation characteristic at the output of the model. This earlier model failed adequately to simulate the rapid rise and fall of sensation at high sickness levels.)

Physically, the fast and slow dynamic elements in the model could correspond to physiological mechanisms responsible for conveying conflict-related information from the orientation brain to the emetic brain. Since conflict signals must be rectified, and the dynamics of the fast and slow pathways are qualitatively those of a leaky integration process, it is tempting to think that at least the slow dynamics might involve a humoral mediator and/or a second messenger agent. Alternatively, the dynamics might reflect the action of some diffusion or active transport process, or instead be the intrinsic dynamics exhibited by a network of vomiting center neurons to direct neural or humoral conflict signal stimulation.

Conclusions

Over the past decade, the sensory conflict theory for motion sickness has become the generally accepted explanation for motion sickness, because it provides a comprehensive etiologic perspective of the disorder across the variety of its known forms. Motion sickness is now defined as a syndrome of symptoms and signs occurring under conditions of real or apparent motion creating sensory conflict. Symptoms and signs (e.g., nausea, vomiting) are not pathognomonic of the motion sickness syndrome unless conditions of sensory conflict are also judged to be present, since the same symptoms and signs also occur in many other nausea related conditions. Thus, the definition of sensory conflict is implicit in any formal definition of the syndrome. It is essential to define as precisely as possible what is meant by the term sensory conflict. Mathematical models for sensory conflict have sharpened our definitions considerably.

The models presented here capture many of the known characteristics of motion sickness in semi-quantitative fashion. However, they have certain limitations, e.g., the sensory conflict model posits a mathematically linear observer. Although recent experimental data are consistent with the notion that the CNS functions as an observer, there is some evidence that sensory conflict is evaluated in nonlinear ways. Also, the model can only mimic, but not predict, the adaptation process. The model for symptom dynamics does not (yet) incorporate elements which account for observed autogenous waves of nausea at high symptom levels, nor the "dumping" of the fast and slow process pathways when emesis occurs. Models for response pathways mediating other physiologic responses such as pallor, skin temperature, and EGG changes have not yet been attempted.

Do the sensory conflict pathways postulated in the models really exist? Unfortunately, to date no such sensory conflict neuron has been found which satisfies the functional criteria imposed by the current theory. The strongest evidence for the existence of a neural or humoral entity which codes sensory conflict is the ability of the conflict theory to account for and predict the many different known forms of motion sickness. One possibility is that conflict pathways or processes do not exist, but in view of the strong circumstantial evidence, this seems unlikely. There are several alternative explanations:

(1) Until recently, there has been surprisingly little discussion of exactly what one meant by the term sensory conflict, so that a physiologist would be able to recognize a "conflict" neuron experimentally. The availability of mathematical models has now changed this situation, and provided a formal definition. However, such models must be presented in ways which physiologists can understand.
(2) So far, relatively few animal experiments have been conducted with the specific objective of identifying a conflict neuron. The search has been largely limited to the vestibulo-ocular pathways in the brain stem and cerebellum. Recent evidence suggests that cortex and limbic systems are major sites for spatial orientation information processing. Real progress may be limited until orientation research focuses on these areas.
(3) Although sensory conflict signals are arguably neurally coded, the conflict linkage mechanisms may have a significant humoral component. If so, a search for the emetic link using classical anatomical or microelectrode techniques will be unsuccessful.

Mathematical characterization of the dynamic characteristics of symptom pathways is a difficult black-box, system-identification problem. The model described above was based only on the character of responses to exogenous motion and sensory rearrangements. Much can potentially be learned from the study of dynamic responses to other classes of emetic inputs, and from studying the influence of behavioral (e.g., biofeedback) and pharmacological therapies.

In other areas of systems physiology and psychology, mathematical models have proven their value by providing a conceptual framework for understanding, for interpreting and interrelating the results of previous experiments, and for planning new ones. Mathematical models can become a useful new tool in motion-sickness research. In the fields of flight simulation and virtual environment displays, simulator sickness is an important practical problem. Models for sensory conflict and motion sickness may become useful tools in the design of these systems.

Acknowledgements

Supported by NASA-JSC Grant NAG9-244. The author is indebted to Dr. J.T. Reason, who in 1977 asked: "Isn't there some way to reconcile the neural mismatch model with the Kalman filter approach?" and to Dr. R. Sivan, whose suggestion led to an answer. Dr. O.L. Bock made many contributions to our early experiments and models for human symptom dynamics. A major role in subsequent experiments was played by W.J.C. Cook, B.W. Rague, and J.C. Eagon.

References

Bard, P., Woolsey, C.W., Snider, R.S., Mountcastle, V.B. and Bromley, R.B. (1947). Delimitation of central nervous system mechanisms involved in motion sickness. *Federal Proceedings*, **6**, 72.

Baron, S. and Kleinman, D.L. (1968). *The Human as an Optimal Controller and Information Processor*. NASA CR-1151.

Borah, J., Young, L.R. and Curry R.E. (1978). *Sensory Mechanism Modelling*. USAF ASD Report AFHRL TR 78-83.

Borison H.L. and Borison, R. (1986). Motion sickness reflex arc bypasses the area postrema in cats. *Experimental Neurology*, **92**, 723.

Brizzee, K.R. and Neal, L.M. (1954). A reevaluation of the cellular morphology of the area postrema in view of recent evidence of a chemoreceptive function. *Journal of Comparitive Neurology*, **100**, 41.

Chinn, H.I. and Smith, P.K. (1955). Motion sickness. *Pharmacological Review*, **7**, 33-82.

Claremont, C.A. (1931). The psychology of seasickness. *Psyche*, **11**, 86-90.

Crampton, G.H. and Daunton, N.G. (1983). Evidence for a motion sickness agent in cerebrospinal fluid. *Brain Behavior Evolution*, **23**, 36-41.

Eagon, J.C. (1988). "Quantitative frequency analysis of the electrogastrogram during prolonged motion sickness." M.D. Thesis, Harvard-MIT Div. of Health Science and Technology. Massachusetts Institute of Technology, Cambridge.

Eversmann, T., Gottsmann, M., Uhlich, E., Ulbrecht, G., von Werder, K. and Scriba, P.C. (1978). Increased secretion of growth hormone, prolactin, antidiuretic hormone, and cortisol induced by the stress of motion sickness. *Aviation Space and Environmental Medicine*, **49**, 53-57.

Graybiel, A. (1975). "Angular velocities, angular accelerations, and coriolis accelerations." *Foundations of Space Biology and Medicine*, Ch. 7, Vol. II, Book 1. Washington, DC: NASA/USSR Acad. Sci.

Janowsky, D.S., Risch, S.C., Ziegler, M., Kennedy, B. and Huey, L. (1984). A cholinomimetic model of motion sickness and space adaptation syndrome. *Aviation Space and Environmental Medicine*, **55**, 692-696.

James, W. (1882). The sense of dizziness in deaf mutes. *American Journal Otolaryngology*, **4**, 239-25.

Kalman, R.E. (1960). Contributions to the theory of optimal control. *Biol. Soc. Mat. Mexicana*, **5**, 102-119.

Kalman, R.E. and Bucy, R.S. (1961). New results in linear filtering and prediction theory. *Journal of Basic Engineering Transactions*, ASME. Ser. D, **83**, 95-108.

La Rochelle, F.T. Jr., Leach, C.S., Homick, J.L. and Kohl, R.L. (1982). Endocrine changes during motion sickness: effects of drug therapy. *Proceedings of the 1982 Aerospace Medical Association Meeting*.

Miller, A.D. (1988). "Motion-induced nausea and vomiting." In Harding, R.K., Kucharczyk, J. and Stewart, D.J. (Eds), *Nausea and Vomiting: Recent Research and Clinical Advances*, Ch. 5. Boca Raton: CRC Press.

Miller, A.D. and Wilson, V.J. (1983a). Vestibular-induced vomiting after vestibulo-cerebellar lesions. In: Mechanisms of motion induced vomiting. *Brain Behavior Evolution*, **23**, 26–31.

Miller, A.D. and Wilson, V.J. (1983b). "Vomiting center" reanalyzed: An electrical stimulation study. *Brain Research*, **270**, 154–158.

Money, K.E. (1970). Motion sickness. *Physiological Review*, **50**, 1–39.

Oman, C.M. (1978). A sensory motor conflict model for motion sickness. Workshop III. *Space Motion Sickness Symposium*, Nov. 16, 1978, NASA Johnson Space Center, Houston, TX.

Oman, C.M. (1982). A heuristic mathematical model for the dynamics of sensory conflict and motion sickness. *Acta Otolaryngologica Supplementum*, 392.

Oman, C.M. (1987). Spacelab experiments on space motion sickness. *Acta Astronautica*, **15**, 55–66.

Oman, C.M. (1988). "The role of static visual cues in the etiology of space motion sickness." In Igarashi, M. and Nute, K.G. (Eds), *Proceedings Symposium on Vestibular Organs and Altered Force Environment*, Oct. 1987, 25–38. NASA/Space Biomedical Research Institute, USRA/Division of Space Biomedicine, Houston, TX.

Reason, J.T. (1978). Motion sickness adaptation: A neural mismatch model. *Journal of the Royal Society of Medicine*, **71**, 819–829.

Reason, J.T. and Brand, J.J. (1975). *Motion Sickness*. London: Academic Press.

Stevens, S.S. (1957). On the psychophysical law. *Psychological Review*, **64**, 153–184.

Tyler, D.B. and Bard, P. (1949). Motion sickness. *Physiological Review*, **29**, 311–369.

Wang, S.C. and Borison, H.L. (1950). The vomiting center: A critical experimental analysis. *Archives of Neurological Psychiatry*, **63**, 928–941.

Wang, S.C. and Chinn, H.I. (1953). Vestibular reflex pathway in experimental motion sickness in dogs. *Abstr. XIX Int. Physiological Congress*, 868–869. Montreal.

Wang, S.C. and Chinn, H.I. (1954). Experimental motion sickness in dogs: Functional importance of chemoreceptive emetic trigger zone. *American Journal of Physiology*, **178**, 111–116.

Wang, S.C. and Chinn, H.I. (1956). Experimental motion sickness in dogs: Importance of labyrinth and vestibular cerebellum. *American Journal of Physiology*, **185**, 617–623.

Wilpizeski, C.R., Lowry, L.D. and Goldman, W.S. (1986). Motion-induced sickness following bilateral ablation of area postrema in squirrel monkeys. *Laryngoscope*, **96**, 122.

Wonham, W.H. (1968). On the separation theorem of stochastic control. *SIAM Journal of Control*, **6**, 312–326.

Young, L.R. (1970). On visual vestibular interaction. *Proceedings of the 5th Symposium on Role of the Vestibular Organs in Space Exploration*, NASA SP 314.

25

Interactions of form and orientation

Horst Mittelstaedt

Max-Planck-Institut fur Verhaltensphysiologie,
D-8130 Seewiesen, Germany

Effect of orientation of perception of form

It is well known that the orientation of an optical pattern relative to egocentric or extraneous references affects its figural quality, that is, alters its perceived form and concomitantly delays or quickens its identification (Rock, 1973). A square presented in the frontal plane to an upright person (S), for instance, changes from a "box" to a "diamond" when it is rotated with respect to the S's median plane by 45°. This angle, that is, the angle between the orientations of the pattern in which the two apparent figures ("Gestalten") attain a summit of purity and distinctness, will be called the "figural disparity" of the pattern. If, as in this case, the S is upright, the retinal meridian and the subjective vertical (SV) are both in the viewer's median plane. The question arises with respect to which of these orientation references the two figures are identified. The answer may be found when the pattern and the S are oriented in such a way that the projections of the retinal meridian and the SV into the plane of the pattern diverge by the pattern's figural disparity or its periodic multiples; that is, in the case of a square by 45° or 135°, respectively. Similarly, which reference determines whether an equilateral triangle is seen as a "pyramid" or a "traffic warning sign" may be revealed at a divergence of SV and retinal meridian of 60° or 180°, respectively. It is generally found that for head roll tilts (ρ) and figural disparities of up to 90°, the figure whose axis coincides with the SV is seen. At head tilts of $\rho = 180°$, however, the retinal reference dominates, as a rule independently of the figural disparity (for reviews see Rock, 1973 and Howard, 1982).

Effect of form on perception of orientation

Clearly, then, orientation may determine apparent form. But conversely, form may also influence apparent orientation. This is explicitly true in

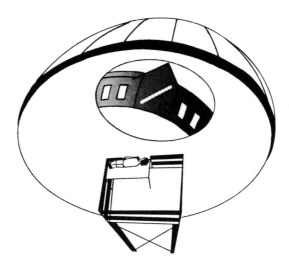

Figure 1. Experimental setup for testing the effect of tilted images on the subjective vertical. The image is projected in a sequence of static roll tilts onto a hemispherical (ϕ = 9.1 m) screen in front of the subject. The S, lying on her side, is asked to set a projected luminous line to subjective vertical.

the case of the SV (for review see Bischof, 1974; for the recent state see Wenderoth, 1976 and Mittelstaedt, 1986).

As shown in Figure 1, our method is to project the pattern within a circular frame (of 16°, 35°, or 80° visual angle) into a tilted planetarium cupola (ϕ = 9.1 m) in 24 stationary orientations presented to the S in a pseudo-random sequence. The S, lying on her side, indicates her SV by means of a rotatable luminous line, which is projected onto the cupola such that its center of rotation coincides with the center of the pattern's circular frame and the S's visual axis.

The effect of the pattern on the SV turns out to be a rather involved function of the orientation of the pattern. This relation becomes clear, however, if we assume that the luminous line is eventually oriented such that the effect of the pattern is opposite and equal to the nonvisual effect on the SV, exerted mainly by the vestibular system. Both effects are then expected to be functions of the difference between the angle β at which the luminous line is set with the pattern present and the angle β_g at which it is found in the absence of visual cues. For the nonvisual effect, fortunately, this function may be computed according to an extant theory (Mittelstaedt 1983a,b): the SV is influenced not only by information about head tilt, but also by intrinsic parameters which are independent of head tilt, notably the "idiotropic vector" (**M**). Presumably by addition of constant endogeneous discharges to the saccular output, it leads to a perpetual shift of the SV into the direction of the S's long axis and hence causes the phenomenon which is well known as the

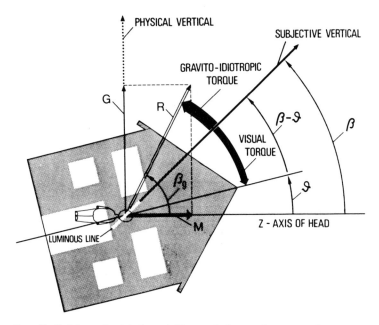

Figure 2. Definition of critical variables and their relations to hypothetical determinants of the SV: (1) It is supposed that the visual scene (here a house) exerts an attraction effect on the SV. This "visual torque" is supposed to be a function of $\beta - \vartheta$, the angle between the main axis of the tilted image and the luminous line when set subjectively vertical; (2) This visual torque is supposedly counterbalanced by a "gravito-idiotropic torque." The latter is a function of $\beta - \beta_g$, the angle between the present SV and the β_g the SV would have in the absence of visual cues. The latter function may be determined as

$$g = \sqrt{G^2 + M^2} \sin(\beta - \beta_g) = \sqrt{G^2 + M^2} \sin\left(\beta - \text{arccotan}\frac{M}{G}\right)$$
$$= M \sin \beta - G \cos \beta$$

with $G = 1$. Hence the unknown visual torque may be quantitatively described. All angles defined with respect to (long) Z-axis of head.

Aubert phenomenon. At first approximation, this relation may be represented by a vector diagram (Figures 2): in the absence of visual cues, the SV is perceived in the direction of the resultant **R** of the otolithic vector **G** and the idiotropic vector **M**.

In our case, since $\rho = 90°$, the nonvisual effect **g** becomes a particularly simple function of $\beta - \beta_g$, namely,

$$\mathbf{g} = \sqrt{G^2 + M^2} \sin(\beta - \beta_g) = \sqrt{G^2 + M^2} \sin\left(\beta - \text{arccotan}\frac{\mathbf{M}}{\mathbf{G}}\right)$$
$$= \mathbf{M} \sin \beta - \mathbf{G} \cos \beta \tag{1a}$$

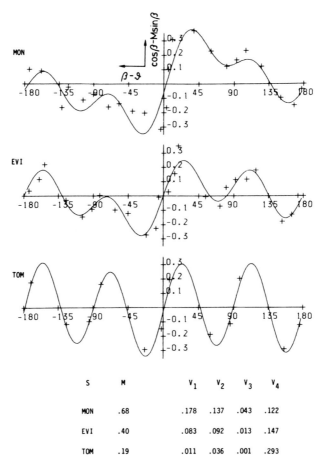

Figure 3. Effect of the same tilted scene (a house) on the SV of three Ss (MON, EVI, TOM). The gravito-idiotropic torque $-g$ is plotted as a function of $\beta - \vartheta$ (see Figure 2). Crosses: means of pairs of settings. Curves: least-square fits of summed sine functions $-g = \Sigma V_n \sin n(\beta - \vartheta)$ with amplitudes V_{1-4} to the data. Note the large variation of the amplitude V_1 of the first harmonic in contrast to the moderate variation of V_2 and V_4.

Because of the normalization of the vestibular information (which is inferred from effects of centrifugation), **g** may be computed with **G** = 1 and **M** = cotan β_g. Hence the unknown visual effect on the SV may be determined if the known quantity g is plotted as a function of the angle on which effect of the pattern depends. There seem to be only two possible candidates: the angle ϑ between the pattern's main axis and the S's long axis, or the angle $\beta - \vartheta$ between the former and the present direction β of the SV.

Figure 3 shows plots of this latter function (named SV-function) engendered in three Ss by a color slide of the house of Figure 1. It turns out that the visual effect is zero, that is, does not change the SV($\beta = \beta_g$) if and only if $\beta - \vartheta$ is zero, rather than when ϑ is zero. Hence its magnitude must be a function of the former angle. We may envisage the SV as being at equilibrium between two tendencies ("torques"), (1) the gravito-idiotropic torque g, trying to pull it toward $\beta - \beta_g = 0$, (2) the other, the visual torque v, trying to pull it toward $\beta - \vartheta = 0$ (see Figure 2). Generally, the visual torque exerted on the SV by a pattern turns out to be an antisymmetrical periodic function composed of the sine of ($\beta - \vartheta$) and the sine of the angle's multiples. Hence it may be simply and fully characterized by the amplitudes V_n of these sine components, to be called "(circular) harmonics" of the respective SV function.

$$v = -g = V_1 \sin(\beta - \vartheta) + V_2 \sin 2(\beta - \vartheta) + \ldots$$
$$= \sum_{n=1}^{max} V_n \sin[n(\beta - \vartheta)] \tag{1b}$$

With the picture of the house of Figure 1 as well as with other photographed scenes, the *first* circular harmonic is generally found to vary greatly inter- as well as intrapersonally. By contrast, the second and fourth harmonics vary but moderately (within an order of magnitude) between Ss, and are rather constant intrapersonally for a given pattern.[1] The formal difference is supposed to be due to a difference in the underlying information processing. The first harmonic expresses the effect of the picture's bottom-to-top polarity, that is, of those cues for the vertical which may be inferred from its normal orientation to gravity. The recognition of what is the top must probably be learned through personal experience, and its effect is hence expected to vary with individual visual proficiency. The even-number harmonics, by contrast, are presumably based on invariant structures of the visual system, possibly by a weighting process, from the "simple cells" of the visual cortex (Mittelstaedt, 1986).

This is highlighted by the following experimental series. If orthogonal lines are presented as a pattern, the resulting SV-function contains only circular harmonics which are multiples of four. The fourth usually is then the largest and is positive; that is, at its null-crossings with positive slope the SV coincides (is in phase) with the direction of the lines (Figure 4).

If a pictograph of a human figure is presented which consists of uniformly

---

[1] All circular harmonics higher than the fourth, except for the eight, which is sometimes found to be just above noise level, are insignificant or zero. With the sampling used, the amplitudes of the first four harmonics were about the same irrespective of whether the Fourier analysis was made with the equidistant sampling of plots over ϑ or with the, necessarily, scattered sampling of plots over $\beta - \vartheta$ as in Figure 3.

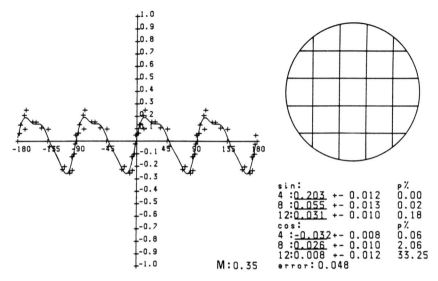

Figure 4. Effect of pattern of squared luminous lines on SV. Method and evaluation as in Figures 1–3. Inset gives numerical values of amplitudes (sines and cosines) of fourth, eighth, and twelfth harmonics of SV-function, their SD, and p (in per cent; two-tailed). The error is the mean square deviation of data from approximation.

oriented lines (Figure 5; "star man") or random dashes, the first harmonic is in phase with star man's long axis and hence is positive.

What will happen if the pictograph of a human figure is presented which consists, as in Figure 6 ("diamond man"), exclusively of lines that are oriented at 45° with respect to the figure's long axis? As a matter of fact, the two figural components are superimposed: the first harmonic is in phase and hence positive; the fourth is in counterphase and hence negative, neither "taking notice" of the other (Figure 7).

Evidently, the result falsifies the hypothesis (Bischof and Scheerer, 1970) that the CNS first computes a "resultant visual vertical" of the picture and subsequently forms an antisymmetrical periodic function *in phase* with this resultant. For then, the resultant would either coincide with the long axis of diamond man and hence the fourth harmonic would be positive, or (rather unlikely though) the resultant would coincide with one of the line directions and hence the first harmonic would be in phase with that line (or would be missing). Instead, the first harmonic results from a processing which is determined by the bottom-to-top polarity of the picture independently of its unpolarized axial features. At the same time, the even-number harmonics are determined by the pattern's unpolarized axial features independently, at least with respect to phase, of its bottom-to-top polarity.

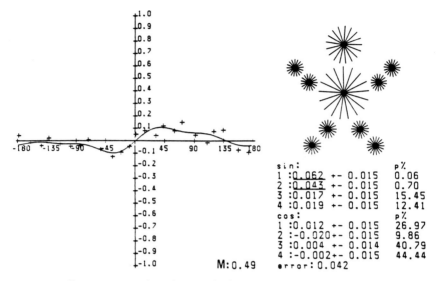

Figure 5. Effect on SV of a figure which is composed of uniformly oriented luminous lines (star man). Procedures as in Figures 1–3; symbols as in Figure 4. Note that only the first and the second harmonic are significantly different from zero (two-tailed).

Interrelations between the determinants of apparent vertical and of form perception

It shall now be examined whether, by means of the comprehensive mathematical theory of the SV, understanding the effect of perceived form on the SV may help in understanding the effect of the SV on form perception mentioned earlier.

First, the theory does indeed offer a good reason why the influence of the SV on the perception of form should decrease with an increasing tilt angle of the S. The effect of the otolithic output then decreases (besides due to comparatively small deviations from a linear response to shear) as a consequence of the addition of the idiotropic vector. Its amount is an idiosyncratic constant averaging around 50 per cent of that of **G**. The magnitude of the resultant **R** of the idiotropic vector **M** and the gravity vector **G** may be approximated as

$$|\mathbf{R}| = \sqrt{\mathbf{G}^2 + \mathbf{M}^2 + 2\,\mathbf{GM}\cos\rho} \qquad (2)$$

Evidently, **R** must decrease with increasing angle of tilt ρ, and so will its relative influence when competing with visual cues!

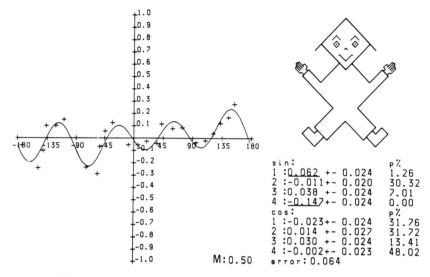

Figure 6. Effect on SV of a figure which is composed of oblique luminous lines (diamond man). Note that the first and fourth harmonics are significantly (two-tailed) different from zero, but of different sign; that is, exactly (no cosines!) in counterphase at $\beta - \vartheta = 0$.

Second, the theory may open a way to assess the relative strength of the factors that influence form perception. The influence of visual patterns on the SV is not independent of the angle of tilt (Bischof and Scheerer, 1970). This effect may be quantitatively described by weighting the visual torque **v** with the sum of the squared saccular and utricular (roll) components (for details see Mittelstaedt, 1986). Hence the effect of the visual torque is maximal at a roll tilt between 60° and 90° and declines toward the upright as well as toward the inverted posture. As a result, at small roll tilts of the S, the nonvisual torque **g** may, under certain conditions, be larger than the visual torque, about equal to the latter around $\rho = 90°$, but much smaller than the visual torque when the S is inverted ($\rho = 180°$). Which component will determine which form is perceived under which angle of divergence may be predictable, if the relative weights of the nonvisual and visual components in the determination of the SV would be correlated with the relative weights of the two reference systems in the perception of form.

Suppression or additive superposition

However, the underlying information-processing systems may be fundamentally different in the two cases. Evidently, additive superposition

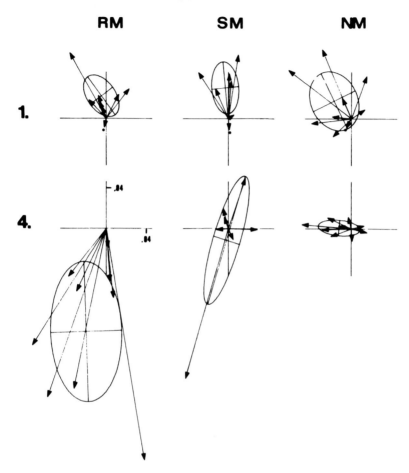

Figure 7. First and fourth harmonics of experiments of Figures 5 and 6 and in nine Ss. Location of arrowhead results from plotting sine amplitudes on ordinate and cosine amplitudes on abscissa (for scale see 4 per cent marks on fourth of RM). RM: diamond man of Figure 6; SM: star man of Figure 5; NM figure in the shape of SM, but composed of randomly oriented dashes ("needle man"). Ellipses: two-dimensional SD. Note similarity of first harmonics for all figures and in all Ss except one (dot under arrowhead), who evinces a negative first harmonic, that is, sees the polarity inverted. Furthermore, only RM engenders a significant fourth harmonic.

suffices to explain the interaction of the components in the case of the SV. But in their influence on form perception, a decision in case of conflict appears to be called for, and hence to necessitate a nonlinear interaction in that one of the competitors is suppressed.

This we have tested by using the well-known ambiguous figure of Figure 8. It is seen, by an upright S, as a "princess" P or a "witch" W when the long axis of P is aligned or reversed with respect to the S's long axis.

Figure 8. The well-known ambiguous figure appearing as witch or princess upon inversion of long axis.

If the S is tilted by 180° relative to gravity ($\rho = 180°$) the *retinal* reference determines the perception, as is generally found in comparable cases. The crucial situation arises when the S views the figure while lying on the side (at $\rho = 90°$). In this position the figure was presented at various angles ϑ with respect to the S's long axis, and the S was instructed to report whether the witch or the princess appeared more distinctly. In order to determine the point of transition between the two phenomena, their distinctness was scaled by the Ss in seven steps, which are condensed in Figure 9 into five (exclusively P; preponderantly P; ambiguous; preponderantly W; exclusively W).

Two Ss, who were well versed in psychophysical tests were chosen. In addition their SV in the absence of visual cues and their ocular counterroll at $\rho = 90°$ were determined and were found as shown in Figure 9. Clearly, in both Ss, the midline between the transition zones neither coincides with the SV nor with the retinal meridian, but assumes an intermediate direction between these two. Hence even in their influence on form perception the gravito-idiotropic and the visual effects may combine vectorially rather than suppress one another.

It is advisable, then, to reexamine those instances where an exclusive decision between the two references is found. As mentioned earlier, this happens regularly, when S and pattern are placed such that *the SV and the retinal meridian diverge by the figural disparity angle*. Now let the "salience" s (die "PRAEGNANZ") of a figure (X) vary as a symmetrical periodic function of its deviation from the respective reference such that

$$s_x = \sum_{n=0}^{max} E_{xn} \cos n\, \vartheta'_x + \sum_{n=0}^{max} V_{xn} \cos n(\beta' - \vartheta'_x) \qquad (3)$$

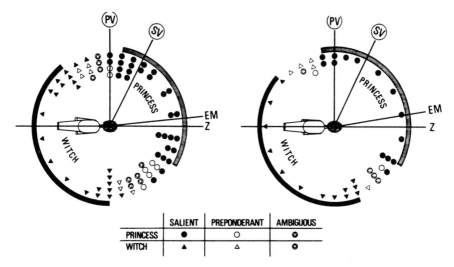

Figure 9. To S lying on the side, the princess is presented in various static orientations. Direction of long (upright) axis of princess with respect to S's long axis (Z) is shown as direction of dot or triangle (like angle ϑ in Figure 2). Type of symbol represents judgement of S on how the figure appears to her when presented in that direction. One symbol stands for one presentation. More presentations were made in directions of critical transitions than in those of complete salience (exclusive distinctness). The latter are connected by black (witch exclusive) or grey (princess exclusive) circular segments. Note that the direction of the midline of saliency coincides neither with direction of the SV nor with that of the vertical retinal meridian (EM), nor with that of the physical vertical (PV).

where ϑ' is the angle between the figure's main axis and the retinal meridian, β' is the angle between the SV and the retinal meridian, and E_{xn}, V_{xn} are the amplitudes of the figure's circular harmonics weighted by the retinal (E_{xn}) and the SV reference systems (V_{xn}), respectively. The central nervous correlate of the relative salience of figures X,Y may then be determined by the difference $s_x - s_y$. In the case of princess versus witch, because $\vartheta'_w = \vartheta'_p - 180°$ and if, for the sake of simplicity, n_{max} is assumed to be unity, the difference becomes

$$s_p - s_w = (E_{p0} + V_{p0}) - (E_{w0} + V_{w0}) + E_{p1} + E_{w1}) \cos \vartheta'_p \quad (4)$$
$$+ (V_{p1} + V_{w1}) \cos(\beta' - \vartheta'_p)$$

In the upright S ($\rho = 0$, $\beta' = 0$), with $E_{p0} + V_{p0} = E_{w0} + V_{w0}$, this becomes

$$s_p - s_w = (E_{p1} + E_{w1} + V_{p1} + V_{w1}) \cos \vartheta'_p \quad (5)$$

That is, independently of the relative weights, the princess dominates at acute angles ϑ'_p and the witch dominates at obtuse angles ϑ'_p. However, with the S inverted ($\rho = 180°$, $\beta' = 180°$):

$$s_p - s_w = [(E_{p1} + E_{w1}) - (V_{p1} + V_{w1})]\cos \vartheta'_p \qquad (6)$$

Consequently, the pattern is identified exclusively according to one of the two reference systems, if their respective weighting factors differ and $\vartheta \neq 90°$, *even though the assumed processing is purely additive.* The same holds for the other examples given above. In the case of the square, for instance, with n = 4, and the S tilted until $\beta' = 45°$ ($\rho \approx 45°$),

$$s_{b(box)} - s_{d(diamond)} = [(E_{b4} + E_{d4}) - (V_{b4} + V_{d4})] \cos 4\vartheta'_b \qquad (7)$$

This leads to a "decision" in favor of the SV-reference if—quite plausibly at that acute angle—the V factors are then larger than the E factors, whereas at $\beta' = 135°$ ($135° < \rho < 180°$) they appear to be almost equal: in that position some of our Ss refuse to decide about what they see! In the case of princess versus witch with the S at $\rho = 90°$ and $\beta = 60°$

$$s_p - s_w = (E_{p1} + E_{w1}) \cos \vartheta'_p + (V_{p1} + V_{w1}) \cos(60° - \vartheta'_p) \qquad (8)$$

Hence a compromise is to be expected depending on the relative magnitudes of the weighting factors. The relative salience ($s_p - s_w$) is then zero at $\vartheta'_{\Delta p\,zero}$, and

$$\cotan \vartheta'_{\Delta p\,zero} = \frac{\sin 60°}{-\left(\dfrac{E_{p1} + E_{w1}}{V_{p1} + V_{w1}} + \cos 60°\right)} \qquad (9)$$

as is borne out by the results mentioned in the preceeding chapter and shown in Figure 9.

In conclusion, the present state favors the notion that angular relations are represented and processed in the CNS by variables which are trigonometric functions of the respective angles. That the characteristics and the spatial arrangement of the otolithic receptors and of the simple cells in the visual cortex are well suited to implement this kind of coding (Mittelstaedt, 1983a,b; 1986; 1988) lends a neurophysiological backbone to the demonstrated descriptive and predictive powers of such a theory.

Acknowledgements

I am grateful to all members of my technical staff for their dedicated assistance and to Evi Fricke, Werner Haustein, Werner Mohren, Stephen R. Ellis, and Heiner Deubel for their valuable contributions to experimentation, evaluation and formulation.

References

Bischof, N. (1974). Optic-vestibular orientation to the vertical. In Autrum, H. et al. (Eds): *Handbook of Sensory Physiology*, pp. 155–190. Berlin, Heidelberg, New York: Springer.
Bischof, N. and Scheerer, E. (1970). Systemanalyse der optisch-vestibularen Interaktion bei der Wahrnehmung der Vertikalen. *Psychologische Forschung*, **34**, 99–181.
Howard, I.P. (1982). *Human Visual Orientation*. New York, Brisbane, Toronto: Wiley & Sons.
Mittelstaedt, H. (1983a). A new solution to the problem of the subjective vertical. *Naturwissenschaften*, **70**, 272–281.
Mittelstaedt, H. (1983b). Towards understanding the flow of information between objective and subjective space. In Huber, F. and Markl, H. (Eds): *Neuroethology and Behavioral Physiology*, pp. 382–402. Berlin, Heidelberg, New York: Springer.
Mittelstaedt, H. (1986). The subjective vertical as a function of visual and extraretinal cues. *Acta Psychologica*, **63**, 63–85.
Mittelstaedt, H. (1988). The information processing structure of the subjective vertical. A cybernetic bridge between its psychophysics and its neurobiology. In Marko, H., Hauske, G., Struppler, A. (Eds): *Processing Structures for Perception and Action*, pp. 217–263, Weinheim: VCH-Verlagsgesellschaft .
Rock, I. (1973). *Orientation and Form*. New York: Academic Press.
Wenderoth, P.M. (1976). An analysis of the rod-and-frame illusion and its variants. In Day, R.H. and Stanley, G.S. (Eds): *Studies in Perception*. Perth: University of Western Australia, Perth.

26

Optical, gravitational and kinesthetic determinants of judged eye level

Arnold E. Stoper and Malcolm M. Cohen

NASA Ames Research Center
Moffett Field, California

Summary

Subjects judged eye level, defined in three distinct ways relative to three distinct reference planes: (1) a gravitational horizontal, giving the "gravitationally referenced eye level" (GREL); (2) a visible surface, giving the "surface-referenced eye level" (SREL); and (3) a plane fixed with respect to the head, giving the "head-referenced eye level" (HREL). The information available for these judgments was varied by having the subjects view an illuminated target that could be placed in a box which: (1) was pitched at various angles, (2) was illuminated or kept in darkness, (3) was moved to different positions along the subject's head-to-foot body axis, and (4) was viewed with the subjects upright or reclining. Our results showed: (1) judgments of GREL made in the dark were 2.5° lower than in the light, with a significantly greater variability; (2) judged GREL was shifted approximately half of the way toward SREL when these two eye levels did not coincide; (3) judged SREL was shifted about 12 per cent of the way toward HREL when these two eye levels did not coincide; (4) judged HREL was shifted about half way toward SREL when these two eye levels did not coincide and when the subject was upright (when the subject was reclining, HREL was shifted approximately 90 per cent toward SREL); (5) the variability of the judged HREL in the dark was nearly twice as great with the subject reclining than with the subject upright. These results indicate that gravity is an important source of information for judgment of eye level. In the absence of information concerning the direction of gravity, the ability to judge HREL is extremely poor. A visible environment does not seem to afford precise information as to judgments of direction, but it probably does afford significant information as to the stability of these judgments.

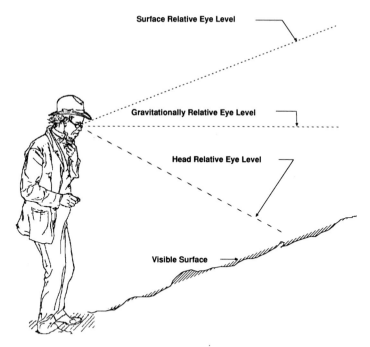

Figure 1. Three types of eye level in normal terrestrial environment. See Table 1 for description.

Introduction

A normal video display conveys fairly accurate information about exocentric directions among displayed visual objects (see Ellis, this volume), but not about egocentric directions, particularly those relative to eye level. This information is important to the observer in the natural environment, and can be used to advantage, especially in the case of a head-mounted display. The concern of the present paper is the mechanism underlying judgments of eye level, and the interactions of vision, gravitation, and bodily senses in these judgments.

There are at least three distinct meanings for visual eye level, all of which are important for the present analysis. Each meaning has associated with it a distinct reference plane with respect to which eye level can be specified. If a given reference plane passes through both the eye and a visual target, the target is said to be at that particular eye level. The three types of eye level are shown in Figure 1, and described in Table 1.

The Target/Head (T/H) system is responsible for the determination of the direction of a target relative to the head, or head-referenced eye level (HREL). This system presumably uses extra-retinal (e.g., kinesthetic or

Table 1. Types of eye level

Symbol	Type	Physiological system	Reference plane
HREL	Head-referenced eye level	Target/head (T/H)	Arbitrary plane tied to head
GREL	Gravity-referenced eye level	Target/gravity (T/G) (T/G = T/H + H/G)	Gravitational horizontal
SREL	Surface-referenced eye level	Target/Surface (T/S)	Ground surface or other visible plane surface

proprioceptive) eye position information (Matin, 1976). The Target/Gravity (T/G) system is responsible for the determination of the direction of a target relative to gravity, the gravitationally referenced eye level (GREL). It is composed of T/H and a Head/Gravity (H/G) system. The latter system presumably operates on the basis of vestibular (primarily otolithic) and postural information (Graybiel, 1973). The Target/Surface (T/S) system is responsible for determining the direction of a target relative to a visible surface, the surface-referenced eye level (SREL). In order to judge the direction of a target relative to the SREL, an observer must use optical information about the orientation of the surface; no extra-retinal, vestibular, or other proprioceptive information is necessary. The optical information involved might be in the form of depth cues which allow the observer to compare eye-to-surface distance with target-to-surface distance, or it might be in a form which allows a "direct" determination of SREL from optical information without recourse to judgments of distance (Gibson, 1950; Purdy, 1958; Sedgwick, 1980). Thus, in principle, T/S can be completely independent of T/H and T/G.

If an observer is standing on a level ground plane in a normal, illuminated, terrestrial environment, with head erect, all three eye levels (HREL, GREL, and SREL) coincide, and determination of any one automatically leads to determination of the other two. It is thus impossible, in that environment, to determine the relative contributions of the three physiological systems described. To do that, some means of separating them is necessary. Various methods to accomplish this separation were used in the following experiments.

Experiment I: the effect of illumination on judgment of GREL

Introduction and method

Our experimental paradigm consisted simply in having the subject adjust a point of light to eye level, defined in one of the three ways above. First, we

Table 2. Means and standard deviations (deg) for error in eye-level judgments in light and dark, average of 10 subjects (Stoper and Cohen, 1986)

	Light	Dark
Constant error (mean)	0.29	2.79
Variable error (standard deviation)	1.03	1.72

ask, "What contribution does optical information make to judgments of GREL?" To answer this question we simply turned off the lights. This eliminated optical information regarding orientation to the ground plane and all other environmental surfaces, and presumably eliminated information to the T/S system. The subject was seated in a dental chair which he or she could raise and lower hydraulically. (This technique minimized the possibility of the subject simply setting the target to the same visible point in each trial.) The task was to adjust the height of the chair so that the subject's eyes were "level" with a small target. (All three types of eye level are coincident in this situation.) A total of 80 trials occurred for each of 10 subjects.

Results

Constant errors (which indicate accuracy) and standard deviations (which indicate precision) were calculated individually for each subject. The averages over all subjects are shown in Table 2. The differences between light and dark are significant ($p < 0.01$ by ANOVA).

Discussion

The finding of higher constant error in the dark means that a small target appears to be about 2.5° higher in the dark than in the light. Others (MacDougall, 1903; Sandstrom, 1951) have found similar results. We have no satisfactory explanation for this effect.

The finding that eye level judgments are more variable in the dark is not surprising, nor is it easily explained. Three distinct hypotheses seem possible; the first two assume that T/S provides more accurate and precise directional information than T/G; the third makes no such assumption. The three hypotheses are:

(1) The "suppression" hypothesis assumes T/G is simply suppressed when T/S is available. If T/S is more precise than T/G, this suppression will result in improved precision.
(2) The "weighted average" hypothesis assumes that the variability of the final judgment is a weighted average of the variabilities of T/G and T/S.

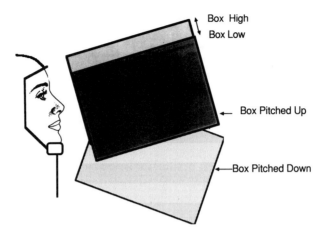

Figure 2. Orientations and positions of the pitchbox.

(3) The "stability" hypothesis assumes that the function of optical information is to minimize the drift of directional judgments made by means of nonoptical information. Thus, no directional information *per se* is necessary from T/S, and no assumptions are made about its precision.

The following experiments are intended to help decide among these three hypotheses.

Experiment 2: the effect of pitched surroundings on GREL

Introduction

Another way to study the interaction of the eye-level systems is to put them into "conflict". This effect has been extensively investigated in the roll dimension with the now classical "rod-and-frame" paradigm (Witkin and Asch, 1948).

Method

A modification of the "pitchbox" method (Kleinhans, 1970) was used. Each of twelve subjects looked into a Styrofoam box, 30 cm wide × 45 cm high × 60 cm deep. The box was open at one end, and could be pitched 10° up or down (Figure 2).

Illumination was very dim (0.5 cd/m^2) to minimize visibility of surface features, but the inside edges of the box could be seen clearly. The apparatus allowed the pitchbox to be displaced linearly up or down as well as to be changed in pitch orientation. The subject could indicate eye level by adjusting the vertical position of a small target (produced by a laser beam).

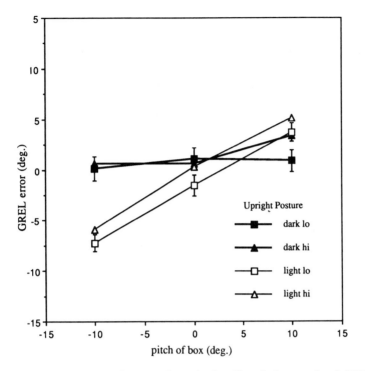

Figure 3. Mean error in judgment of gravitationally relative eye level (GREL) of 12 subjects as a function of orientation, position, and illumination of the pitchbox. Pitch of +10° means the pitchbox was pitched up. Error bars represent the standard error of the mean (between subjects).

In this experiment, the subject was instructed to set the target to the point in the pitchbox that was at his or her GREL. A 2 × 2 × 3 × 2 design with replication was used. The experiment consisted of four within-subject factors: (1) viewing condition (dark *vs.* light), (2) pitchbox position (high *vs.* low: 6 cm apart), (3) pitchbox angle (10° up, level, or 10° down), and (4) laser starting position (up *vs.* down). Each factor combination was presented twice, yielding a total of 48 trials per subject.

Results and discussion

Box pitch

Mean error of judged GREL is plotted in Figure 3 as a function of orientation, position, and illumination of the pitchbox. It is clear that a strong effect of orientation on GREL exists in the light condition, but not in the dark. This can be described as a shift of judged GREL in the direction of true

SREL. The magnitude of this shift is indicated by the slope of the judgment function. A total change in pitch (i.e., of SREL) of 20° produced a shift in GREL of 11.1° in the light, but only 1.5° in the dark. We will consider the slope of 0.55° (in the light) to be a measure of the strength of the effect of the visual environment. This effect is comparable in magnitude to that found by Matin and Fox (1986), and by Matin et al. (1987). The simple fact of compromise between SREL and GREL means that T/G is not totally suppressed, even while T/S is operating, and is strong evidence against the suppression hypothesis.

Box height

The effect of box height is clearly evident in the figure. The linear shift of the pitchbox of 6 cm (5.5° of visual angle) produced a 1.47 cm (1.35°) shift in GREL. This is comparable in magnitude to a similar linear displacement effect found by Kleinhans (1970). It may be due to the Dietzel-Roelofs effect (Howard, 1982, p. 302), where the apparent straight ahead is displaced toward the center of an asymmetrical visual display. Another possible explanation is a tendency for subjects to set eye level toward the same optically determined point on each successive trial. Whatever the cause of this effect, it may account for as much as 40 per cent of the orientation effect, since with our apparatus, a change in orientation also produced a displacement of the visual scene.

Variability

It might be expected that conflict between two systems would greatly increase variability. For example, each system could contribute a component equal to its own variability, and there would be an additional component caused by variability in combining the systems. Figure 4 shows within-subject standard deviations calculated separately for each of the three orientations, in the light and the dark.

Here it can be seen that variability of judgment in the dark is higher than in the light; however, it is not affected by orientation. There is no more variability when the systems are in conflict (at ±10°) than when they are not (when the pitchbox is level, at 0°). This finding indicates that the weighting of the systems is very stable over a series of trials for each subject.

Experiment 3: the effect of gravity on SREL judgments

Introduction and method

To observe the operation of T/S, we instructed the subject to align his or her line of sight with the floor of the movable pitchbox, thus judging the SREL.

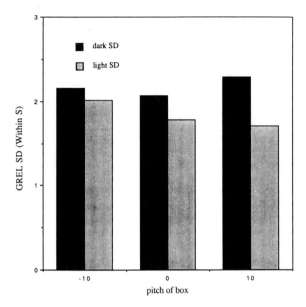

Figure 4. Standard deviations (within subjects) of GREL judgments of 12 subjects for each of three orientations, in the light and in the dark.

Just as we "turned off" T/S by extinguishing the light, we can turn off T/G by orienting the subject so that gravity does not abet the task. Each of 12 subjects judged SREL, both with upright posture, when they could presumably use gravitational information and T/G, and reclining on the left side, where gravity and T/G were of no use. (The T/H system presumably continued to operate in both conditions.) In the upright condition the method was identical to that of Experiment 2, except that the instructions were to find SREL rather than GREL. In the reclining condition the entire apparatus (shown in Figure 2) was rotated 90°.

As in Experiment 2, the pitchbox was set in two different positions displaced 6 cm along the subject's longitudinal body axis (Z axis).

Results and discussion

Results are plotted in Figures 5 and 6. ANOVA showed significant effects of box pitch and box height.

Box pitch

There is a clear shift of SREL judgments in the direction of HREL in both the upright and reclining conditions. The slope is 0.15, much less than the 0.55 found in Experiment 2. (Note that, while Experiment 2 showed an

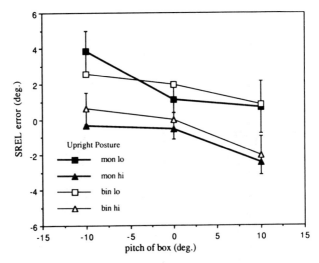

Figure 5. Mean error in judgment of surface-relative eye level (SREL) of 12 subjects as a function of orientation, position, and illumination of the pitchbox; judgments made with upright posture. Error bars represent the standard error of the mean (between subjects).

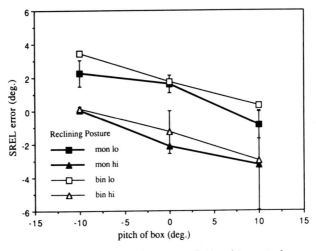

Figure 6. Mean error in judgment of SREL of 12 subjects; judgments made with reclining posture.

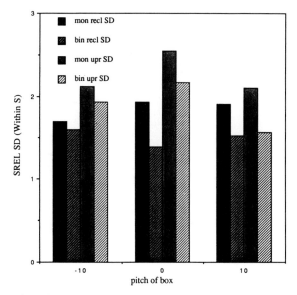

Figure 7. Standard deviations (within subjects) of SREL judgments of 12 subjects for each of three orientations, in the light and in the dark.

effect of optical variables on a nonoptical judgment, the present experiment found an effect of nonoptical variables on an optical judgment.) The fact that the slope is essentially the same for both upright and reclining body orientations implies that T/H rather than T/G is producing the bias we obtained. This result is similar to that of Mittelstaedt (1983).

Box height

The effect of the 6 cm box displacement was a shift of 2.47 cm (2.26°) in the upright and 3.5 cm (3.21°) in the reclining condition. The size of this effect implies that the subjects did not effectively use the optical orientation information available to them. Instead, they seem to have had a strong tendency to set the target near the same location on the back of the box with each trial.

Variability

Standard deviations for SREL judgments are shown in Figure 7.

SREL judgments made with the subject upright showed greater within-subject variability than those made with the subject reclining. This observation may be taken to imply that gravity does not enhance the precision of SREL judgments under upright conditions.

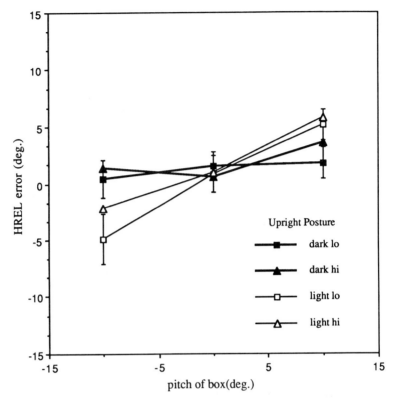

Figure 8. Mean error in judgment of head-relative eye level (HREL) of 12 subjects as a function of orientation, position, and illumination of the pitchbox; judgments made with upright posture. Error bars represent the standard error of the mean (between subjects).

Experiment 4: the effect of gravity and pitched surroundings on HREL judgments

Introduction and method

To observe the influence of T/S on T/H, we instructed the subject to set his or her eyes "straight ahead" and place the target at the fixation point, thus judging HREL. In the upright condition the method was identical to that of Experiment 2, except that the instructions were to find HREL rather than GREL. The reclining condition arrangement was identical to that of Experiment 3.

Results and discussion

Results are plotted in Figures 8 and 9. ANOVA showed significant effects of orientation and box height.

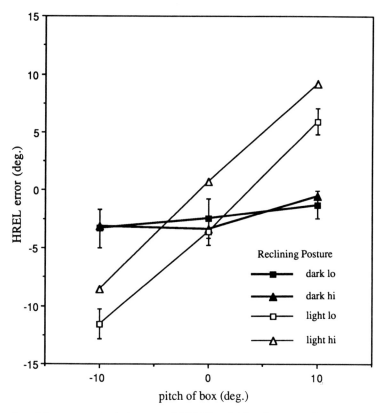

Figure 9. Mean error in judgment of HREL of 12 subjects; judgments made with reclining posture.

Box pitch

There is a clear shift of HREL judgments in the direction of SREL in both the upright and reclining conditions. The slope for the judgments of HREL with upright posture in the light is 0.45, about the same magnitude as was observed in Experiment 2. We thought that this effect could be due to a confusion of instructions when HREL and GREL were coincident, and we expected a much weaker effect in the reclining conditions, when GREL was absent. In fact, however, a much stronger effect was found (slope = 0.89). This can be explained in terms of Mittelstaedt's (1983) vector combination model. In the upright condition, both T/G and T/H indicate a more or less horizontal eye level, and T/S would be combined with both of these. In the reclining condition T/S combines with only T/H. The result in the reclining condition is thus closer to T/S.

Variability

It can be seen in Figure 10 that, for upright posture, the variabilities of HREL and GREL judgments are very similar, both in the dark and in the light. For

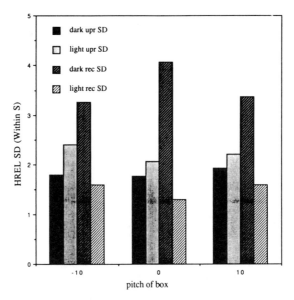

Figure 10. Standard deviations (within subjects) of HREL judgments of 12 subjects for each of three orientations, in the light and in the dark.

reclining posture, however, HREL variability is twice as great in the dark as in the light. This result indicates that the presence of gravitational information has a stabilizing effect on HREL judgments.

Conclusions

(1) Increased precision in the light. We present evidence against both the suppression and the weighted average hypotheses. Only the stability hypothesis is not contradicted by these data. This hypothesis could be tested directly by using a random dot field as a visual environment. Such a field would have no direction information, so any improvement in precision of GREL would be by means of stability information.

(2) Box displacement effect. This may be a significant factor in the orientation effect. It could be controlled in a future experiment by rotating the pitchbox around the center of its back, rather than around the subject's eye.

The large size of this effect when judging SREL indicates that ability to judge orientation of the line of sight in the pitch dimension relative to a surface on the basis of purely optical information is poor under the conditions of this experiment.

(3) Head relative information. Perhaps our most surprising result was the almost complete "visual capture" of HREL judgments in the light while

the subject was reclining on his or her side in Experiment 4, and the corresponding high variability of these judgments in the dark. Both of these results indicate very low ability to use T/H to judge eye level in the absence of gravity information. In more practical terms, this result indicates that judgment of the pitch of the observer's head (and by implication, the rest of his or her body) relative to a surface is much less precise, and subject to a much higher degree of visual capture, when gravity is not present to aid this judgment.

We wish to thank Leonard Matin who in 1985 shared with us unpublished results and his "pitch room" technique which is similar to the "pitch box" used in this study.

References

Ellis, S. (1987). Pictorial communication: Pictures and the synthetic universe. Presentation to *Spatial Displays and Spatial Instruments Conference*: Asilomar, California.
Gibson, J.J. (1950). *The Perception of the Visual World*. Boston: Houghton Mifflin.
Graybiel, A. (1973). The vestibular system. In: *Bioastronautics Data Book*. NASA SP 3006, 533–609.
Howard, I.P. (1982). *Human Visual Orientation*. New York: Wiley.
Kleinhans, J.L. (1970). "Perception of spatial orientation in sloped, slanted and tilted visual fields." Ph.D. Dissertation, Rutgers Univ., New Jersey.
MacDougall, R. (1903). The subjective horizon. *Psychological Review of Monograph Supp.*, **4**, 145–166.
Matin, L. (1976). Saccades and extraretinal signal for visual direction. In Monty, R.A. and Senders, J.W. (Eds), *Eye Movements and Psychological Processes*. New Jersey: Erlbaum, 205–220.
Matin, L. and Fox, C.R. (1986). Perceived eye level: Elevation jointly determined by visual field, pitch, EEPI, and gravity. *Investigative Ophthalmology and Visual Science* (Supp.) **27**, 333.
Matin, L., Fox, C.R. and Doktorsky, Y. (1987). How high is up? II. *Investigative Ophthalmology and Visual Science* (Supp.), **28**, 300.
Mittelstaedt, H. (1986). A new solution to the problem of the subjective vertical. *Naturwissenschaften*, **70**, 272–281.
Purdy, W.C. (1958). "The hypothesis of psychophysical correspondence in space perception." Doctoral Dissertation, Cornell Univ. (Ann Arbor: University Microfilms, No. 585–594).
Sandstrom, C.I. (1951). *Orientation in the Present Space*. Upsala: Almqvist and Wicksell.
Sedgwick, H.A. (1980). The Geometry of Spatial Layout in Pictorial Representation. M. Hagen (Ed.), *The Perception of Pictures*, pp. 33–88. New York: Academic Press.
Stoper, A.E. and Cohen, M.M. (1986). Judgments of eye level in light and darkness. *Perception and Psychophysics*, **40**, 311–316.
Witkin, H.A. and Asch, S.E. (1948). Studies in space orientation. IV. Further experiments on perception of the upright with displaced visual fields. *Journal of Experimental Psychology*, **38**, 762–782.

27

Voluntary influences on the stabilization of gaze during fast head movements

Wolfgang H. Zangemeister

*Hamburg University Neurological Clinic,
Hamburg, Germany*

Summary

Normal subjects are able to change voluntarily and continuously their head-eye latency together with their compensatory eye movement gain. A continuous spectrum of intent-latency modes of the subject's coordinated gaze through verbal feedback could be demonstrated. It was also demonstrated that the intent to counteract any perturbation of the head-eye movement, i.e., the mental set, permitted the subjects to manipulate consciously their vestibular ocular reflex (VOR) gain. From our data we infer that the VOR is always "on." It may be, however, variably suppressed by higher cortical control. With appropriate training, head-mounted displays should permit an easy VOR presetting that leads to image stabilization, perhaps together with a decrease of possible misjudgments.

Introduction

For some time it has been known that visual and mental effort influence the vestibular ocular reflex (VOR). Besides visual long-and short-term adaptation to reversing prisms (Melvill-Jones and Gonshor, 1982) and fixation suppression of the VOR (Takemori and Cohen, 1974; Dichgans *et al.*, 1978; Zangemeister and Hansen, 1986), the mental set of a subject can influence the VOR, e.g., through an imagined target (Barr *et al.*, 1976; Melvill-Jones *et al.*, 1984) or anticipatory intent only (Zangemeister and Stark, 1981).

Barr *et al.*, (1976) showed that given a constant vestibular input the oculomotor output can be modulated according to the conscious assessment of the prevailing functional need, even without visual feedback (see Figure 1b, left).

The VOR velocity gain as a function of five different conditions could be altered highly significantly just by imagining respectively earth-fixed or head-fixed targets.

If real retinal afferent signals derived from a slipping retinal image can be substituted through imagined targets, it appears likely that a comparison can be made between the vestibular input and an efferent copy of the concurrent oculomotor output as Miles and Eighmy have proposed (1980). Therefore, it appeared only reasonable to test this hypothesis by using the ability of mental VOR suppression to produce prolonged alteration of sensory-motor relations in the reflex without the aid of vision (Melvill-Jones et al., 1984).

In their experiment (Figure 1a), the subject had to look for three hours to an imagined head fixed target while actively oscillating its head in a self paced mode (1.0–0.2 Hz). After 3 hours the stationary subject had to fixate a stationary target. Then the lights were extinguished (shaded area) while subject still tried to "look" at the target. After a rapid passive horizontal head rotation (passive rotation for exclusion of active preprogramming or neck dependent influences on the result) the lights were switched on and the subject had to refixate the now visible target. The eye movement in darkness represented the VOR gain, (that would normally be close to unity), during a mental, non-visual suppression of the VOR. In Figure 1a on the right the adaptive change is clearly to be seen by the generation of consistently large undershoots, followed by correspondingly large saccadic corrections of gaze back onto target, whereas before adaptation (left) the eyes were usually close to being on target when the lights went on. These findings most elegantly showed that external visual stimulation is not a necessary condition for the generation of a modified reflex. Apparently, internal neurally coded reference signals can serve an effective role instead of error activated feedback control of central parameters that are referenced to the real physical world.

A striking example of a longterm adaptation to the deficit of one visual hemifield is given as a lifelong "experiment" of the nature in patients with congenital hemianopia (Zangemeister et al., 1982; Figure 1b right). The VOR gain is very high (1.5) when the patient's eyes move to the side of the blind hemifield; however the VOR gain becomes zero only two seconds later, when the gaze shifts back towards the side of the seeing hemifield. This is a perfect example of long term adaptive behaviour enforced by a lifelong deficit, that has induced an asymmetrical change of the VOR gain due to a symmetric non-visual frame of reference. In an asymmetrical visual world, this keeps the seeing fovea off the blind hemifield and close to the target using respectively a suppressed or augmented VOR.

In contrast to animals, human head and eye movements are governed by the conscious will of the human performer that includes verbal communication. Thus in a given experimental setup, the synkinesis of active human gaze maybe changed according to instruction. Generally small gaze amplitudes yield predominantly an eye-dependent strategy of gaze shift in conjunction

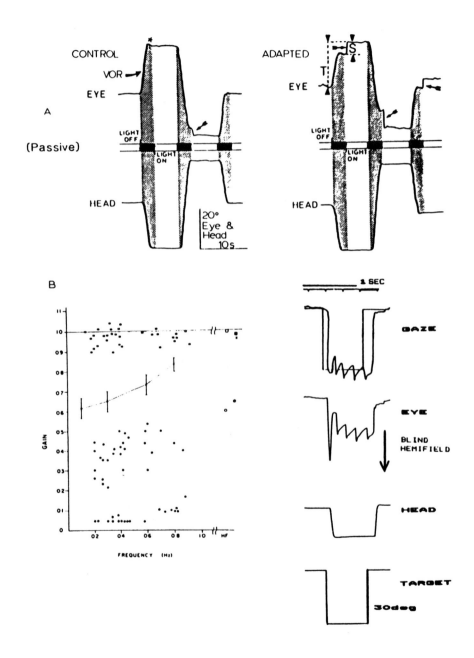

with low VOR suppression; large gaze amplitudes yield a head-dominant strategy with high VOR suppression (Zangemeister and Stark, 1981 and 1989). The verbal feedback to the subject might permit a whole range of gaze types, even with amplitude and prediction of a visual target being constant. The gaze types (Zangemeister and Stark, 1982a) are defined by head minus eye latency differences (Table 1). This has been demonstrated particularly by looking at the timing of the neck electromyogram as the head movement control signal (Zangemeister et al., 1982b; Zangemeister and Stark, 1983; Stark et al., 1986). In this study, we compared the voluntarily changeable human gaze types performed during the same experiment with and without the addition of a randomly applied perturbation to the head-eye movement system. We tried to answer three questions in particular:

(1) Are we able to modulate continuously the types of coordinated gaze through conscious intent during predictive active head movements?
(2) What is the gaze (saccade and VOR/CEM (compensatory eye movement)) response to passive random head rotation from zero head velocity with respect to the preset intent of a given subject?

◀ *Figure 1 (a) Eye and head movements using an adaptive test for mental control of the VOR before (left) and 3 hours after (right, adapted) mental VOR suppression in the dark. Three successive passive step changes of head angle, each conducted in darkness (shaded bands), with the subject trying to maintain ocular fixation on the earth fixed target. Inadequacy of VOR leads to a corrective saccade (S) shortly after switching the lights on. After refixation the total eye excursion (T) is equated to total head excursion. VOR gain (i.e. amplitude gain in this case) = $(T - S)/T$ and the estimate is self calibrated. Note the greatly enlarged corrective saccades in the "adapted" state, reflecting the systematic adaptive reduction of VOR gain, even though the subject's conscious goal is to achieve unit gain throughout these tests (modified after Melvill-Jones et al., 1984).*

(b) Left: Dependence of VOR gain (i.e. velocity gain, eye vel./head vel.) upon intended goal. Filled points obtained during rotation in the dark (VOR) when trying to 'look' at an earth fixed (squares) or head fixed (circles) target. Corresponding open points obtained with real visual targets. The drawn curve shows the intermediate effect of mental arithmetic, which has no goal directed relevance for the oculomotor system (modified after Barr et al., 1976).

Right: Instantaneous change of VOR gain (i.e. amplitude gain in this case) in a patient with congenital hemianopia, i.e. not a vestibular but a cortical visual hemideficit. Note high gain VOR (1.5) with gaze to right (down = blind hemifield). But a suppression of the VOR, i.e. gain of zero with gaze to left (up = seeing hemifield) only two seconds later. The asymmetric optokinetic input is not sufficient to explain this VOR change that takes place in a frequency range beyond ($\geqslant 1.0$ Hz) the performance of the optokinetic system.

Table 1. Gaze types defined by latency: eye minus head latency. Type II: early prediction of eye, late head movement; eye movement dominates gaze. Type I: head follows eye shortly before eye has reached target; classical gaze type. Type III: head and eye movements start about simultaneously. Predictive gaze type. Type IV: early prediction of head, late eye saccade; head movement dominates gaze. Suppression of VOR/CEM in Type III and Type IV. See also Figure 1b.

Type	Eyelatency-headlatency, msec	Average rate of success in generating intentionally different gaze types through verbal feedback, per cent
I	+50	76
II	<50	56
IIIa	>50–200	69
IIIb	>200–550	
IV	>550	16

(3) Does random perturbation of the head during the early phase of gaze acceleration generate responses that are the sum of responses to experiments (1) and (2)?

Methods

Eye movements were recorded by monocular DC Electrooculography, head movements by using a horizontal angular accelerometer (Schaevitz) and a high-resolution ceramic potentiometer linked to the head through universal joints (Zangemeister and Stark, 1982b). Twelve normal subjects (age 22–25) attended a semicircular screen sitting in a darkened room. While they actively performed fast horizontal (saccadic) head rotations between two continuously lit targets at 30° amplitude with a frequency around 0.3 Hz, they were instructed to focus on the following tasks: (1) "shift your eyes ahead of your head", (2) "shift your head ahead of your eyes". During (1) they were instructed to shift eyes "long before" (i, type II), or "shortly before" (ii, type I) the head. During (2) they were instructed to shift head "earlier" (i, type IIIA), or "much earlier" (ii, type IIIB) than the eye, eventually "with the intent to suppress any eye movement" (type IIIB or IV). Each task included 50 to 100 head movements.

Perturbations were done pseudorandomly, (1) from a zero P,V,A (position, velocity, acceleration) initial condition of the head-eye movement system, and (2) during the early phase of head acceleration. They consisted of (1) fast passive head accelerations, of (2) short decelerating or accelerating impulses during the early phase of active head acceleration and were recorded by the head-mounted accelerometer.

Perturbation impulses were generated through an apparatus that permitted manual acceleration or deceleration of the head through cords that were tangentially linked directly to the tightly set head helmet.

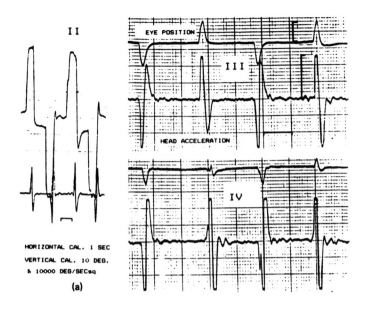
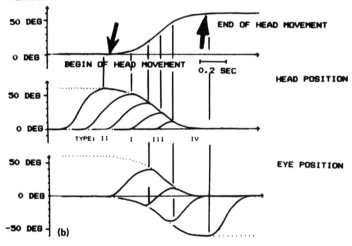

Figure 2. (a) (upper). Gaze types II, III, IV generated intentionally through verbal feedback with change of the subjects intent to suppress the reflex response. On left, gaze type II with early prediction of eye and late head movement; eye latency $(E_1 - H_1)$ minus head latency < 50 msec; note very late head acceleration. On right upper, predictive gaze type III, head and eye movement start about simultaneously; $E_1 - H_1 > 50$ to 200 msec. On right lower; early prediction of head and late prediction of eye saccade; $E_1 - H_1 > 200$ to 550 msec (III) or >550 msec (IV). (b) (lower). Explanatory schema for the continual change of gaze types. The schema demonstrates that with increasing lead of head movement, because it is predicted, (upper trace) the synkinetic eye movement, i.e., the saccade and the CEM/VOR, gets either suppressed (middle trace) or "reversed", i.e., the saccade occurs after the CEM/VOR.

Results

The subjects demonstrated their ability (Figure 2) to switch between gaze types in the experimentally set predictive situation of constant and large-amplitude targets. The respective gains (eye/head velocity) were: ty.II 0.9–1.1, ty.III 0.13, ty.IV 0.06–0.09. This result was expected from our earlier studies (Zangemeister and Huefner, 1984; Zangemeister and Stark, 1982a,c). The subjects showed differing amounts of success in performing the intended gaze type, with type IV being the most difficult to perform, supposedly because of the high concentration necessary (Table 1).

Random perturbation of the head while in primary position, with head velocity and acceleration being zero (Figure 3), resulted in large saccades/quick phases of long duration, and a large and delayed VOR/CEM, if the subject had low preset intent to withstand the perturbation; in this case head acceleration showed a long-lasting damped oscillation. Respective gains were: Figure 3 (upper): 0.35 (upper) 0.45 (lower); Figure 2 (lower left): 0.5 (upper), 0.17 (lower). With increasing intent of the subject (Figure 3 left, middle, and lower), head acceleration finally became highly overdamped, but still with comparable initial acceleration values, and eye movements showed increasingly smaller and shorter quick phases as well as an early short VOR response. In addition, with the highest intent a late anticompensatory eye movement was obtained.

Random perturbations of the accelerating head, i.e., sudden acceleration or deceleration of gaze in flight (Figure 4), were characterized by small VOR responses after the perturbation in case of high intent of the subject as in gaze type IIIB, or much higher VOR/CEM gain in case of low intent comparable to gaze type I. Respective gains were: Figure 4 (left) ty.I 0.55, ty.3 0.06, ty.IV 0.08 (left), 0.09 (right); Figure 4 (right): 0.13 (upper), 0.90 (lower).

Random perturbations were also applied during coordinated head-eye movements in pursuit of a sinusoidally moving target (maximum velocity 50 deg/sec) with the VOR being suppressed through constant fixation of the pursuit target. Figure 5 (left) demonstrates the different amount of VOR fixation suppression as a function of changing intent during fixation of a sinusoidal target of the same frequency. With perturbation (Figure 5, right) a response was obtained that was comparable to the result of experiment (2). That is, depending on the subject's intent and concentration, the VOR response was low for high intent and vice versa (gain Figure 5, right: 0.044).

Therefore, the three initial questions could be answered as follows:

(1) In nonrandom situations subjects can intentionally and continuously change their gaze types.
(2) Gaze responses to passive random head accelerations depend on the subject's preset intent.
(3) Perturbation of predictive gaze saccades in midflight results in the sum of tasks one and two.

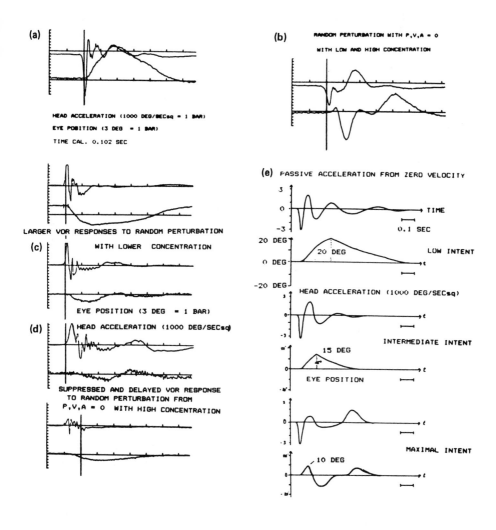

Figure 3. (a) Random perturbation from primary position with very low intent, (b) very high intent, (c) low to intermediate intent, (d) intermediate to high intent. For the technical part of this experiment see methods section. As shown in the explanatory schema (e), an increasing inhibition of the initial saccade can be obtained, seen by the decreasing amplitude from 20 to 10 deg with increasing intent to suppress the VOR response. The eye movement response with maximal intent (b) looks as if it were a delayed sample of the acceleration curve.

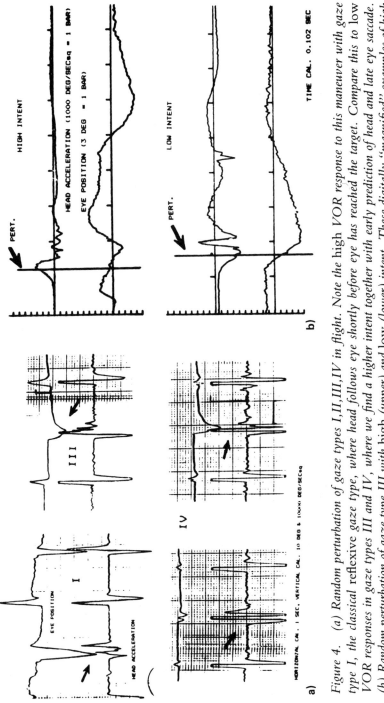

Figure 4. (a) Random perturbation of gaze types I,II,III,IV in flight. Note the high VOR response to this maneuver with gaze type I, the classical reflexive gaze type, where head follows eye shortly before eye has reached the target. Compare this to low VOR responses in gaze types III and IV, where we find a higher intent together with early prediction of head and late eye saccade. (b) Random perturbation of gaze type III with high (upper) and low (lower) intent. These digitally "magnified" examples of high and low intent gaze types III clearly demonstrate the suppressed and delayed VOR response with high intent that compares to the response shown in Figure 3b.

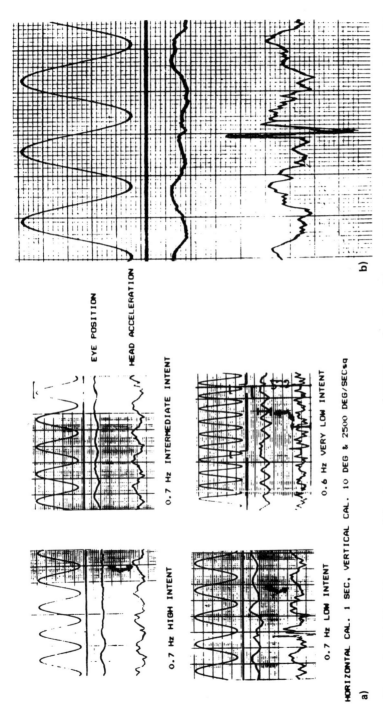

Figure 5 (a) Variable amount of fixation suppression of VOR as a function of intent. Although the subject's gaze was always following the target, with decreasing intent the VOR would be less suppressed. This is also of importance in clinical testing. (b) Random perturbation of coordinated gaze pursuit: the VOR response is suppressed during concentrated pursuit as seen in the middle part.

Discussion

The input-output characteristics of the VOR are subject to major moment-to-moment fluctuations depending on nonvisual factors, such as state of "arousal" (Melvill-Jones and Sugie, 1972) and mental set (Zangemeister and Stark 1981, 1989; Collins, 1962). More recently, it has been found that the influence of "mental set" depends explicitly upon the subject's conscious choice of intended visual goal (Barr et al., 1976; Sharpe et al., 1981; Baloh et al., 1984; Fuller et al., 1983), i.e., following earth-fixed or head-fixed targets during head rotation. Consistent alteration of the mentally chosen goal can alone produce adaptive alteration of internal parameters controlling VOR gain (Berthoz and Melvill Jones, 1985). Obviously, comparison of afferent retinal slip detectors with concurrent vestibular afferents can be substituted by a "working" comparison made between the vestibular input and an efferent feedback copy of either the concurrent, or the imagined or anticipated oculomotor output, as proposed by Miles and Eighmy (1980).

Our results here demonstrate the ability of the subjects to perform short-term adaptation during verbal feedback instructing for eye-head latency changes that changed the types of active gaze. These results are comparable to the data from Barr et al. (1976), in that an almost immediate change between different VOR gains with constant visual input could be generated. In addition, our perturbation experiment expanded these data, demonstrating the task- (or gaze-type) dependent attenuation of the VOR. This is in contrast to results in animals, where perturbation of visually triggered eye-head saccades resulted in an acceleration of the eye (Guitton et al., 1984; Fuller et al., 1983), because a conscious task-influence of the VOR is impossible. Therefore not only can a representation of the target's percept (Barr et al., 1976) be created, but also an internal image of the anticipated VOR response in conjunction with the appropriate saccade.

We hypothesize that through the cortico-cerebellar loop a given subject is able continuously to eliminate the VOR response during predictive gaze movements, which is not any more possible in patients with cerebellar lesions (Zangemeister et al., 1990) and their "asynergie cerebelleuse" of Babinski's. This is done internally by generating an image of the anticipated VOR response in conjunction with the appropriate saccade, and then subtracting it from the actual reflex response. This internal image can be manipulated intentionally and continuously *without* a VOR on/off switch. In this way a flexible adaptation of the conscious subject to anticipated tasks is performed.

References

Baloh, R.W., Lyerly, R., Yee, R.D. and Houmstra, V. (1984). Voluntary control of the human vestibulo-ocular reflex. *Acta Otolaryngology*, **97**, 1–6.

Barr, C.C., Schultheis, L. and Robinson, D.A. (1976). Voluntary, non-visual control of the human vestibulo-ocular reflex. *Acta Otolaryngology*, **81**, 365–375.

Berthoz, A. and Melvill-Jones, G. (1985). *Adaptive Mechanisms in Gaze Control*. Amsterdam-New York: Elsevier.

Collins, W.E. (1962) Effects of mental set upon vestibular nystagmus. *Journal of Experimental Psychology*, **63**, 191–197.

Dichgans, J., Reutern, V., Reutern, G. and Rommelt, U. (1978). Impaired suppression of vestibular nystagmus by fixation. *Archiv fuer Psychiatrie und Nervenkrankheiten*, **226**, 183–199.

Fuller, J.H., Maldonado, H. and Schlag, J. (1983). Vestibular-oculomotor interaction in cat eye-head movements. *Experimental Brain Research*, **271**, 241–250.

Guitton, D., Douglas, R. and Volle, M. (1984). Eye-head coordination in cats. *Journal of Neurophysiology*, **52**, 1030–1050.

Melvill Jones, G. and Gonshor, A. (1982). Oculomotor response to rapid head oscillation after prolonged adaptation to vision reversal. *Experimental Brain Research*, **45**, 45–58.

Melvill Jones, G. and Sugie, N. (1972). Vestibulo-ocular responses during sleep in man. *EEG Clinical Neurophysiology*, **32**, 43–53.

Melvill Jones, G., Berthoz, A. and Segal, B. (1984). Adaptive modification of the vestibulo-ocular reflex by mental effort in darkness. *Experimental Brain Research*, **56**, 149–153.

Miles, F.A. and Eighmy, B.B. (1980). Long-term adaptive changes in primate vestibulo-ocular reflex. *Journal of Neurophysiology*, **43**, 1406–1425.

Sharpe, J.A., Goldberg, H.J., Lo, A.W. and Hensham, Y.O. (1981). Visual-vestibular interaction in multiple sclerosis. *Neurology*, **31**, 427–433.

Stark, L., Zangemeister, W.H., Hannaford, B. and Kunze, K. (1986). Use of models of brainstem reflexes for clinical research. In Kunze, K., Zangemeister, W.H., Arlt, A. (Eds), *Clinical Problems of Brainstem Disorders*, 172–184. Stuttgart-New York: Thieme.

Takemori, S. and Cohen, B. (1974). Loss of visual suppression of vestibular nystagmus after flocculus lesions. *Brain Research*, **72**, 213–224.

Zangemeister, W.H. and Hansen, H.C. (1986). Fixation suppression of the vestibular ocular reflex and head movement correlated EEG potentials. In O'Reagan, J.K. and Levy-Schoen, A. (Eds), *Eye Movements*, 247–256. Amsterdam-New York: Elsevier.

Zangemeister, W.H. and Huefner, G. (1984). Saccades during active head movements: interactive gaze types. In Gale, A. and Johnson, F. (Eds), *Theoretical and Applied Aspects of Eye Movement Research*, 113–122. Amsterdam-New York: Elsevier.

Zangemeister, W.H. and Stark, L. (1981). Active head rotation and eye-head coordination. In Cohen, B. (Ed.), *Vestibular and Oculomotor Physiology*, **374**, 540–559, Annals of New York Academy of Science.

Zangemeister, W.H. and Stark, L. (1982a). Gaze types: interactions of eye and head movements in gaze. *Experimental Neurology*, **77**, 563–577.

Zangemeister, W.H. and Stark, L. (1982b). Gaze latency: variable interactions of eye and head in gaze. *Experimental Neurology*, **75**, 389–406.

Zangemeister, W.H. and Stark, L. (1983). Pathological types of eye and head gaze coordination. *Neuro-ophthalmology*, **3**, 259–276.

Zangemeister, W.H. and Stark, L. (1989). Gaze Movements: patterns linking latency and VOR gain. *Neuro-ophthalmology*, **9**, 112–123.

Zangemeister, W.H., Dee, J. and Arlt, A. (1990). Abnormal timing of antagonist splenius burst in cerebellar patients' head movements. In Berthoz, A., Graf, W., Vidal, P.P. (Eds), *The Head-Neck Sensory-Motor System*. IBRO Series, New York: Wiley Publications, in print.

Zangemeister, W.H., Meienberg, O., Stark, L. and Hoyt, W.F. (1982a). Eye head coordination in Homonymous Hemianopia. *Journal of Neurology*, **225**, 243–254.

Zangemeister, W.H., Stark, L., Meienberg, O. and Waite, T. (1982b). Motor control of head movements. *Journal of Neurological Science*, **55**, 1–14.

PART IV
Seeing

Seeing

Stephen R. Ellis

Introduction

Human interaction with environments is primarily visuo-motor and vision is the dominant source of feedback concerning the effects of actions. The papers in the following section are concerned with mechanisms and consequences of the three dimensional interpretation of visual information presented on artificial displays. Three different forms of this interpretation of 3D space may be distinguished. They may be described as varieties of virtualization.

The first form, construction of a virtual space, refers to the process by which a viewer perceives a three-dimensional layout of objects in space when viewing a flat surface presenting the pictorial cues to space, i.e., perspective, shading, occlusion, and texture gradients. This process, which is akin to map interpretation, is the most abstract of the three. It is abstract because the stimuli for many of the physiological reflexes associated with the experience of a real three dimensional environment are either missing or inappropriate for the information on a flat picture which depicts a space differing from the picture surface. Some of the consequences of this inappropriateness are discussed by Sedgwick (Chapter 30), Goldstein (Chapter 31), Cutting (Chapter 32), Ellis *et al.* (Chapter 34) and by Enright (Chapter 38).

The second form of virtualization is the perception of a virtual image. In conformance with the use of this term in geometric optics, it is the perception of a three dimensional space in which the accommodative[1], vergence[2], and stereoscopic cues are appropriate for the objects that are depicted in the display used to make the image. These more direct cues to the spatial layout are discussed in Chapters 36 and 39.

The final form is the virtualization of an environment. In this case the key sources of information are observer-slaved motion parallax, depth-of-focus variation keeping only a single depth plane in focus, and wide field of view without a prominent frame. If properly implemented, these additional features can provide visual stimulation consistent with major physiological

---

[1] Focusing required of the eye to make a sharp image on the retina
[2] Convergence or divergence of the eyes to produce an apparently single image

reflexes such as accommodative vergence[3], vergence-accommodation[4] (See Chapters 38, 39), the optokinetic reflex[5], and the vestibulo-ocular reflex[6] (Chapters 22–24, 27). These features can substantially contribute to the illusion of actually being present in the synthetic environment.

Mechanisms of virtualization

As emphasized by Gregory (Chapter 21), we see with significant preconceptions that he described as "object hypotheses". These hypotheses may be thought of as mathematical functions which require arguments for their evaluation. Vision provides these arguments. But they are not provided as free, unfettered variables.

The first two papers in this section discuss mechanisms by which three dimensional interpretations may be given to the pattern of contour and connectivity in the visual image that has been described as the "2½D" sketch by Marr (1982). Lappin and Wason (Chapter 28) in the spirit of J.J. Gibson describe how the "ordinal topography of the visible surface is ... recoverable from its optical image". In their terms the optical image is the pattern of contours and lines in the perspective projection which a viewer sees. To the extent that surface shape and position is recoverable totally from this projection, visual preconceptions or object hypotheses would be unnecessary for spatial interpretation. But even in their analysis some preconceptions are required. For example, in their discussion of the "isometries of moving objects" which provide the basis for developing a spatial metric, they identify implicit assumptions concerning the nature of the objects, e.g., two non-composite objects cannot be as the same place at the same time. Furthermore, the elements of the Jacobian matrix that they associate with the perspective transformation cannot be computed without knowledge of the viewing direction. This knowledge is also a form of visual preconception that governs removal of ambiguities in the visual image and establishes, for example, the scale of the image.

Stevens (Chapter 29) in contrast to Lappin and Wason, is concerned with the apparent three dimensionality of *static* images of curved contours. The three dimensional interpretation of such images is particularly striking since such stimuli are well known to be difficult to interpret (Metzger, 1953; Ellis and Grunwald, 1987). Thus, his observation that such patterns give rise to perception of a particular surface indicates that human spatial perception is subject to specific, strongly constraining assumptions which may be identified in a complete theory of spatial perception.

---

[3] Reflexive changes in the convergence of the eyes triggered by changes in the required focus
[4] Reflexive changes in the focusing of the eye triggered by changes in convergence
[5] Reflexive tracking eye movements triggered by movement of objects subtending large visual angles
[6] Tracking eye movements triggered by vestibular stimulation normally associated with head movement

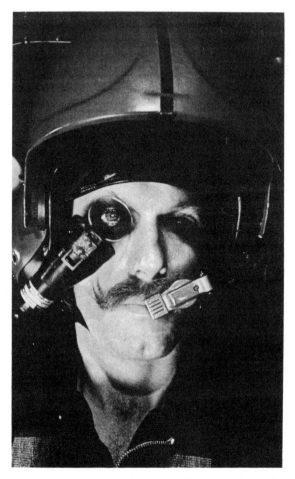

Figure 1. Helmet-mounted displays such as this night vision system for use in helicopters can provide head-slaved, virtual image, perspective displays to extend the pilot's ability to night fly or to fly during low visibility conditions. Though these systems already have been used in the field, they significantly increase the difficulty of the pilot's flying task and can benefit from improvements to match their visual requirements better to normal human capacities to virtualize an artificially displayed space (Hart and Brickner, 1989).

Consequences of geometrical errors

The five papers in the next group, Chapters 30–34, are concerned with the consequences of errors in estimating the geometrical characteristics of a virtual image or space. They all tacitly share the assumption that errors in subjective metrics of a virtualized space may be traced to implicit errors in determining the geometric parameters that define the visible projection of the space. For example, since the stimuli used to present a virtual space or virtual image are generally produced through perspective projection, recovery of the depicted space requires operations equivalent to determining the viewing parameters for the projection. Faulty recovery of these parameters can lead to errors which may be predicted by using these incorrect values in otherwise correct geometric calculations. As noted in several papers below, this approach may not completely work, but it is a useful starting point for computationally explicit theorizing.

Sedgwick (Chapter 30), Goldstein (Chapter 31) and Cutting (Chapter 32) are specifically concerned with the consequences of viewing a picture from some position other than the correct geometric eye point. Such a viewpoint presents interpretation problems uniquely associated with picture viewing. Perrone and Wenderoth (Chapter 33) and Ellis, Smith, McGreevy and Grunwald (Chapter 34) explore some of the consequences and mechanisms of erroneous interpretation of the geometric information in the picture, in particular the estimation of the orientation of the "camera" used to make the picture. Perrone and Wenderoth discuss a possible mechanism for the well known fact that viewers typically overestimate the slants of surfaces depicted in pictures. Ellis et al., use a generalized model of slant overestimation to model patterns in estimating exocentric direction. In conclusion Nagata (Chapter 35) discusses practical measures that may be taken to increase the appearance of three dimensions in two-dimensional pictures.

Primary depth cues and reflexes

The accuracy and precision of common experience in real environments naturally leads to the supposition that human perception must have access to unambiguous cues to the spatial positions of objects. Stereopsis is frequently thought to be such a cue since the eyes appear to have access to all the information needed to triangulate the position of objects. The papers in this section, however, point out that even stereopsis has significant ambiguities in the depth signal it provides (Schor in Chapter 36) and that even if these ambiguities are resolved, the stereoscopic space that is constructed may be biased as if the position of the point of convergence of the two eyes was misjudged (Foley in Chapter 37). Thus, our accurate perception of our natural environment must be a construction based on multiple sources of information that provide mutually supporting redundant information which also benefits from corrective feedback. Foley's approach to a theory to explain patterns of perceptual errors in terms of errors in assumptions concerning the viewing geometry is directly analogous the approach of Ellis et al. in Chapter 34.

Many of the primary physiological responses to objects varying in depth are linked. Accommodation causes reflexive vergence and vice versa. The pupil is also linked so that when an object approaches, the eyes focus by adding refractive power to the lens, the eyes converge to maintain binocular vision, and the pupil constricts to increase depth of field, all mutually supporting adaptive effects called the near response. Enright (Chapter 38) examines conditions in which the pictorial cues to depth are able alone to produce vergence. He notes that since not all pictures seem to be equally effective at producing this "perspective vergence", he may be able to use it as a measure of the effectiveness with which a picture stimulates a sensation of depth.

Practical difficulties

The final paper in this section concerns difficulties viewers have when required to interpret the space depicted in a virtual image. Virtual image displays have been used in aircraft for many years as the platform for optically collimated Head-Up Displays (HUDs) in which the virtual image is presented near optical infinity. This format was thought to aid the pilot by relieving him of the need to shift his attention constantly between the cockpit instruments inside and the outside environment. Research has shown, however, that even though the HUD may bring the pilots line of sight out the window, his attention may be directed to the HUD symbology and though he is simultaneously *looking* at targets "out the window", he may not *see* them (Haines et al., 1980). Furthermore, Roscoe and others have reported that pilot's eyes do not automatically focus at optical infinity, but rather inward at a resting accommodative level which is about arms length away. Associated with this misaccommodation is a reported shrinking of the apparent visual angle of objects in view which Roscoe attributes to the accommodative error. This shrinking is claimed to be the cause of overestimates of the subjective distance to objects, an error with potentially disastrous consequences for a pilot using a HUD. The presumed perceptual mechanism is simple: objects of known physical size appear further away as their apparent angular size shrinks.

In fact, as Roscoe reports, (Chapter 39), there is some evidence from aircraft operations that HUDs used in military tactical aircraft may contribute to disorientation and other problems. Whether the increased incidence of problems with HUD is due to the HUD symbology, focusing requirements or the more difficult flight regimes flown with HUDs, may still need to be determined. Furthermore, the misaccommodation to which the apparent shrinking of the visual image is attributed may not be the correct explanation for the misjudgment of apparent distance. Though it is true that focusing in front of a target would increase the power of the lens and hence decrease the size of the retinal image, there are alternative causes. For example an alternative cause of the error in size could be a side effect of the efference-copie or perceptual expectations associated with the entire near response, i.e., the coordinated change in accommodation, vergence, and pupil diameter that occurs when an object is viewed up close (Mittelstaedt, personal communication, 1987; Enright, 1989; Lockhard and Wolbarsh, 1989). In any case, the phenomena and issues raised by Roscoe, regardless of their ultimate explanation, are certain to have increasing practical importance as the aircraft HUD technology is introduced to the automobile industry.

References

Ellis, S.R. and Grunwald, A.J. (1987). A new illusion of projected three dimensional space. *NASA TM 100006*, NASA Ames Research Center.

Enright, J.T. (1989). The eye, the brain, and the size of the moon: toward a unified oculomotor hypothesis for the moon illusion. In *The Moon Illusion*, Hershenson, M. (Ed.), Hillsdale, N.J.: Erlbaum.

Haines, R.F., Fischer, E. and Price, T. (1980). Head-up transition behavior of pilots with and without head-up display in simulated low visibility approaches, *NASA TP 1720* (December, 1980).

Hart, S. and Brickner, M. (1989). Helmet mounted pilot night vision systems: Human Factors issues. *Spatial Displays and Spatial Instruments, NASA CP 10032*, 14-1–14-14

Lockhard, G.D. and Wolbarsh, M.L (1989). The moon and other toys. In *The Moon Illusion* Hershenson, M. (Ed.), Hillsdale, N.J.: Erlbaum.

Marr, D. (1982). *Vision*, New York: Freeman.

Metzger, W. (1953). *Gesetze des Sehens*, Frankfort am Main, Germany: Kramer.

28

The perception of geometrical structure from congruence[1]

Joseph S. Lappin
Vanderbilt University
Nashville, Tennessee

and

Thomas D. Wason
Allotech Inc.
Raleigh, North Carolina

Introduction

The principal function of vision is to measure the environment. As demonstrated by the coordination of motor actions with the positions and trajectories of moving objects in cluttered environments and by the rapid recognition of solid objects in varying contexts from changing perspectives, vision provides real-time information about the geometrical structure and location of environmental objects and events.

Information about the geometrical structure of scenes, objects, and motions may be visually acquired not only by the exploration of natural environments, but also from artificial, human-designed displays. Photographs, drawings, movies, computer graphics, and other such artificial 2D displays are widely and effectively used tools for communicating information about spatial structures. Understanding the basis for the effectiveness of such tools

---

[1] Work on this paper and on related experimental and theoretical research was supported in part by a Small Business Innovative Research Grant from NASA to T.D. Wason, by NIH Grant EY-05926 to J.S. Lappin, and by the University Research in Residence Program of the Air Force Office of Scientific Research which enabled several extended visits by Lappin to Wright-Patterson Air Force Base. The mathematical ideas outlined in this paper have benefitted significantly from discussions with Jan Koenderink and Andrea van Doorn, State University of Utrecht, The Netherlands, R. Alan Peters, Dept. of Electrical Engineering, Vanderbilt, Steven Tschantz, Dept. of Mathematics, Vanderbilt, and especially John G. Ratcliffe, Dept. of Mathematics, Vanderbilt.

poses a special theoretical challenge, because the trigonometric mapping from the 3D structures and motions portrayed in these displays to the optical patterns on the observer's retinae differs from the perspective projections that normally hold for vision in natural environments. Cutting (1987) has recently discussed the theoretical difficulties posed by this discrepancy between the projective geometry of movies versus that of natural vision, and he has also provided experimental demonstrations of the abilities of humans to perceive 3D structure in movies viewed "from the front row side aisle".

The purpose of this paper is to examine the geometric information provided by 2D spatial displays. We propose that the geometry of this information is best understood not within the traditional framework of perspective trigonometry, but in terms of the structure of qualitative relations defined by *congruences* among *intrinsic* geometric relations in images of *surfaces*. The mathematical details of this theory of the geometry of vision are presented elsewhere (Lappin, 1990); the present paper outlines the basic concepts of this geometrical theory.

Traditionally, the structure of space—both the 3D space of the environment and the 2D space of the image—has been regarded as defined *a priori*, independently of the objects and motions contained within it. Indeed, the geometric structure of objects and motions is typically described by reference to extrinsic standards that define parallel and perpendicular directions and quantify relative magnitudes of distance extrinsic to the objects themselves.

When described in terms of this extrinsic framework, however, the geometry of vision is quite complicated: metric[2] relations in the 2D image plane cannot be isomorphic with metric relations in the 3D environment; the perspective projection from 3D spatial structures in the environment onto the 2D image plane does not have a well-defined inverse. Therefore, the recovery of information about the geometric structures and locations of the environmental objects has often been thought to require supplementary information about the perspective position of the observer or about the structure and location of the objects. The 2D optical images alone have seemed insufficient.

But the assumption that vision begins with an abstract structure of space as a prior standard for describing environmental objects begs the question. The basic problem of vision is to find a measurement structure for representing the spatial characteristics of observed scenes, objects, and events. Such a measurement structure is generally not given beforehand, but must be discovered in the organization of the empirical observations themselves.

---

[2] The term *metric* is used in a conventional mathematical sense, referring in this context to measures of distance over a potentially curved surface. A relation m(a,b) between two elements a and b is said to be a metric relation if it satisfies the following axioms for all elements a, b, and c:
positivity: $m(a,b) \geq 0$
symmetry: $m(a,b) = m(b,a)$
reflexivity: $m(a,a) = 0$
triangle inequality: $m(a,c) \leq m(a,b) + m(b,c)$.

Intrinsic geometry of surfaces and images

When described in terms of the intrinsic geometry of *surfaces*, the geometry of vision becomes much simpler. In the first place, the mapping of a visible region of an environmental surface onto its optical image is a mapping from one 2D manifold onto another. The derivatives and singularities of the surface—its slopes, peaks and valleys, inflections, saddlepoints, and occluding edges—are isomorphic with the derivatives and singularities of the image. This is true for images described by gradients of texture, motion parallax, or stereoscopic disparity (Koenderink and van Doorn, 1975, 1976a,b,c, 1977). Although the isomorphism does not hold for images described by luminance gradients, partly because of the additional influence of the direction of illumination, it is still true that the intrinsic surface structure (in particular, the parabolic lines, which are inflections of curvature that separate regions of convexity and concavity) is systematically related to the differential structure of the image (Koenderink and van Doorn, 1980). Because the differential structures of the two manifolds are essentially isomorphic with one another, the ordinal topography of the visible region of an environmental surface is fully described and recoverable by its optical image.

Furthermore, the specific mapping between cu ves and forms on the environmental surface and their corresponding images on an observer's retina may be locally described simply by a *linear coordinate transformation* between the derivatives on the two manifolds. This linear approximation holds for "infinitely small" surface patches that may be locally approximated by a tangent plane at that location. This linear mapping of the surface onto its image also has a well-defined inverse. Accordingly, the local structure of the surface may be obtained from the local structure of its image by a linear coordinate transformation.

These simple relationships between the surface and its image involve the *derivatives* on the two manifolds. The linear transformation that best describes the relationship between these two manifolds at any given point is given by the partial derivatives of the two coordinate systems. Thus, if O^2 represents the 2D manifold of the object surface, and if R^2 represents the 2D manifold corresponding to the observer's retina, then the linear differential map **v**: $O^2 \rightarrow R^2$ is specified by the following Jacobian matrix of partial derivatives:

$$\mathbf{v} = \begin{bmatrix} \dfrac{\partial r^1}{\partial o^1} & \dfrac{\partial r^1}{\partial o^2} \\ \dfrac{\partial r^2}{\partial o^1} & \dfrac{\partial r^2}{\partial o^2} \end{bmatrix}$$

Suppose that $[dO] = [do^1, do^2]^T$ is a 2×1 column vector that specifies an infinitesimal displacement on the surface in terms of two intrinsic coordinates on the object surface, and suppose that $[dR] = [dr^1, dr^2]^T$ is a corresponding description of the image of this vector in terms of the intrinsic coordinates of the retina. Then the transformation between these two coordinate systems produced by the optical projection from the object to its image on the retina is given by the linear equation

$$[dR] = \mathbf{V}[dO]$$

and the inverse map is given by

$$[dO] = \mathbf{V}^{-1}[dR]$$

where $\mathbf{V}$ is the Jacobian matrix given above. (The form of this equation is independent of the specific coordinate systems used to specify positions on the two manifolds. The coordinates need not intersect at right angles nor even be straight lines; they need only be differentiable and provide a unique specification of each position on the manifold. The generality of this representation seems especially relevant to vision, where no specific coordinate system can be assumed beforehand for any particular environmental surface.[3]) The important point is that *the local structure of the retinal image of a given surface is described by this Jacobian matrix of partial derivatives*, **v**. The entries in this matrix vary as a function of position on the surface, with variations in the values of these entries reflecting variations in the orientation and curvature of the surface.

The same approach can also be used to describe the relationships with a third 2D manifold associated, for example, with an intervening display image such as a movie or photograph. Suppose that I^2 represents the manifold of such an intervening image, that $\mathbf{a}: O^2 \to I^2$ represents the differential map between these two manifolds, and that $\mathbf{b}: I^2 \to R^2$ is the visual map from the display image onto the retinal manifold. Then, using the chain rule, the two successive maps can be combined by a composition of the two functions, $\mathbf{v} = (\mathbf{b} \circ \mathbf{a}): O^2 \to R^2$. Similarly, the coordinate transformation corresponding to this chain would be given by a linear equation of the following form:

$$[dR] = \mathbf{BA}[dO],$$

where the matrix product $\mathbf{BA} = \mathbf{V}$ again provides a linear coordinate transformation functionally equivalent to the previous construction.

---

[3] For concreteness, we may assume that the coordinates reflect the spatial arrangement of the gradients and singularities of the surface—e.g., tending to run parallel and perpendicular to the gradients of curvature of the surface and to the boundary contours, corners, and parabolic lines (which separate structurally distinct regions). We need not assume that these coordinates have specific numerical values, only that they are differentiable and uniquely label every location on the surface.

Representation of the *metric* structure of the surface requires an embedding of the 2D manifold of the surface or its image into the 3D manifold of Euclidean 3-space, E^3. Suppose that $[dX] = [dx^1, dx^2, dx^3]^T$ is a 3×1 column vector giving the three orthogonal cartesian coordinates of an infinitesimal displacement on the object surface. Then the perspective coordinate embedding of the image of the surface into E^3, **p**: $R^2 \to E^3$, is given by a linear coordinate transformation of the following form:

$$[dX] = \mathbf{PV}[dO]$$

where **P** is a 3×2 matrix of partial derivatives, $\mathbf{P} = [\partial x^k/\partial r^i]$, with $k = 1,2,3$ and $i = 1,2$. Measures of metric relations require a quadratic expression similar to the Pythagorean formula for distance in E^3. The metric tensor that provides the measure of distance on the surface is obtained by substituting from the above equation for the vector $[dX]$ in the Pythagorean formula:

$$\begin{aligned} ds^2 &= [dX]^T [dX] \\ &= [\mathbf{PV}[dO]]^T \mathbf{PV}[dO] \\ &= [dO]^T \mathbf{V}^T \mathbf{P}^T \mathbf{PV}[dO] \\ &= [dO]^T \mathbf{V}^T \mathbf{P}^* \mathbf{V}[dO], \end{aligned}$$

where $\mathbf{P}^* = \mathbf{P}^T \mathbf{P}$ is a symmetric 2×2 matrix with quadratic entries of the form

$$\mathbf{P}^* = \left[\sum_k (\partial x^k/\partial r^i)(\partial x^k/\partial r^j) \right].$$

Thus, the entries in this matrix provide a measure of squared distance on the object surface at a particular position on the retina corresponding to the image of the surface. The length of any arbitrary curve on the surface is obtained by integrating the quantities as defined in the preceding equation at each position along the curve.

The three independent parameters of the matrix $\mathbf{P}^*$ are not given directly by a single stationary image of an isolated local surface patch. In certain special cases these perspective parameters and therefore the metric structure of the local surface patch are determined, up to a scalar, simply by the motion of the local patch. More generally, however, these perspective parameters must be derived from more global constraints on the image structure associated with the observer's position and motion within the 3D environment. In general, the perspective embedding of the image into E^3 is revealed by actual or implied motions of objects within the space.

Metric structure from congruence

Although geometric relations are often described in terms of extrinsic coordinate systems in which directions and distances are defined *a priori*, it is important in many applications to derive the structure of space from more fundamental qualitative relationships among the objects and events contained within it. This was the case, for example, in the development of relativistic physics, where the symmetries of observations associated with the velocity of light and with gravitation were used to construct spaces in which the lawful relations among observed variables could be expressed in simpler and more general form (Einstein et al., 1923, 1952; Misner et al., 1973). The same strategy has also been employed in formulating the theoretical foundations of measurement (Krantz et al., 1971; Luce, 1978; Luce and Narens, 1983). That is, symmetries of qualitative relations under various physical operations and under varying conditions of observation may often be used as a foundation for quantitative equations that describe empirical laws of nature.

Analogously, the geometry of vision may also rest upon the symmetries of intrinsic qualitative relations in the spatio-temporal optical images rather than on the prior metric structure of an extrinsic coordinate system. Because metric relations in the 3D environment are not isomorphic with those in the 2D image, and because the optical projections of environmental objects onto the retinae change with the perspective positions of the displays and observers, the extrinsic framework of space is neither constant nor readily accessible to vision. Instead, we hypothesize, the metric structure of environmental objects and spaces may be induced from the isometries of moving objects.

This conception of the geometry of vision is a continuation of ideas developed by Gibson (1950, 1957) about the importance of the concepts of invariance and transformations for perception (Lombardo, 1987). Gibson's (1950) conception of the visual information provided by such "higher order variables" as a texture density gradient was based on the idea that gradients of repeated structural relations specified the projective transformation of a surface onto an image and also specified an intrinsic scaling of the 3D space in which the surface texture was homogeneous. The same conception was subsequently expanded (e.g., Gibson, 1957) to emphasize the information provided by the continuous transformations of optical flow produced by moving objects and moving observers. These deformations of the optical images were believed to enable the perception of both the structural invariants and the projective transformations associated with the motions of objects and observers in 3D space.

The essential ideas underlying this conception of geometry were described by the mathematician Killing (1892)[4]:

---

[4] We are grateful to Jan Koenderink for bringing this paper to our attention and to Bernd Rossa for translating the paper from the original German.

'Every object covers a space at every time. The space covered by one object cannot simultaneously be covered by another object.'

'Every object can be moved. If an object covers the space of a second object at any time, then the first object can cover any space covered by the second object at any (other) time.'

'Every space (object) can be partitioned. Each part of a space (object) is again a space. If A is a part of B and B is a part of C, then A is a part of C, where A, B, and C may be either spaces or objects.' [p. 128]

These three principles, which are the first of eight principles from which Killing derives a general theory of geometry, provide qualitative criteria for defining the equality or congruence of spaces and objects: two spaces are congruent if and only if they can be covered by the same object. Two objects are congruent if and only if they can cover the same space. Thus, objects and spaces constitute mutually interdependent relational structures. The metric structure of both may be derived from elementary qualitative properties of *differentiability* and *congruence under motion*. (By definition, "motions" are isometric transformation groups.) (Also see Weyl, 1952, and Guggenheimer, 1963, Section 11-2.)

This conception of form and space provides a basis for understanding how visual information about the metric structure and dimensionality of objects and spaces may be gained from "motions" or transformations which bring objects at one position in space into congruence with those at other positions. The metric equality of neighboring spaces successively occupied by the same object and the equality of separate parts of an object which successively occupy the same space may be determined from the motions of objects. Accordingly, the dimensionality of visible spaces and objects need not be restricted to the two coordinate dimensions of the image. Rather, the dimensionality may be associated with the number of parameters needed to bring an object at one location in space into congruence with an object at another location.

In certain special cases the metric structure of a given surface patch may be *locally* determined (up to a scalar) by its moving images, independent of global properties of the retinal image as a whole: if the *trajectory* of the moving patch is also a *surface in space-time* with constant curvature equal to that of the object patch, then of course the metric tensor for this spatio-temporal surface remains constant over the surface. The shape of this surface in space-time will be a *surface of revolution* (Guggenheimer, 1963, pp. 272–273). Rotation of the object patch over this spatio-temporal surface from one region of the surface to another does not change the mapping of the surface onto the retina, and the contravariant tensor coefficients for this projective mapping of the object patch and its trajectory onto the retina vary only as one-parameter functions of time, where this transformation parameter corresponds to the angle of rotation. Accordingly, the perspective parameters for embedding the retinal images of this surface into E^3 also vary as one-parameter functions of time or of retinal position (which are correlated in this case). The simplicity of these relationships between the differential structure of the object surface, its trajectory in space-time, and the retinal images of these surfaces involves sufficiently few unknown perspective parameters that these are determined by the invariance of the metric tensor of the surface patch under motion. That is, suppose that $\mathbf{V}_o$ and $\mathbf{P}_o$ are the Jacobian matrices for the visual and perspective coordinate transformations, respectively, for an initial retinal image of the surface patch, and suppose that $\mathbf{V}_t$ and $\mathbf{P}_t$ are the corresponding matrices for a second retinal image of the same surface patch following a rotation onto another position along its constant-curvature trajectory. The equivalence of the geometric structure of the two retinal images can be expressed by the equation

$$\mathbf{V}_o^T \mathbf{P}_o^* \mathbf{V}_o = \mathbf{V}_t^T \mathbf{P}_t^* \mathbf{V}_t$$

where $\mathbf{P}_t = r(\mathbf{P}_o)$ is a one-parameter transformation of $\mathbf{P}_o$ corresponding to a rotation in E^3 around an axis that does not pass through the focal point of the observer's eye. This matrix equation involves four independent quadratic equations in four unknowns—the three independent perspective parameters of $\mathbf{P}_o$ and the transformation of these by the parameter t.

Specific examples of this special case include a sphere that rotates around an axis (different from the direction of gaze) through its center (e.g., Lappin et al., 1980; Doner et al., 1984) and planar patterns that rotate within the same plane (Lappin and Fuqua, 1983) tilted with respect to the retinal image. In both of these cases the time-varying set of positions of the surface patch form a surface of revolution in space-time generated by a one-parameter transformation group (the magnitude of the rotation). In general, the metric tensor for the images of the moving surface patch remains invariant under the motion (i.e., its Lie derivative is zero) if and only if the vector field of this group of isometries (the "Killing vector") is a one-parameter group that generates a surface of revolution (Guggenheimer, 1963, pp. 272–273). Thus, because the moving object forms a surface whose images are generated by a one-parameter transformation, the perspective parameters for embedding this spatio-temporal surface into E^3 are determined up to a scalar by the invariant metric structure of the given surface patch. Indeed, the experimental results of Lappin et al. (1980) and Doner et al. (1984)—for the perceived shape of a random-dot sphere rotating about a vertical axis through its center—and of Lappin and Fuqua (1983) for the perceived inter-point distances among three collinear points rotating in a plane—demonstrated just this invariance of visually perceived metric structure under motion even though the optical displays contained unnaturally exaggerated amounts of polar projection.

In general, however, the trajectory of a moving object does not constitute a surface but a *volume* in space-time. (Consider, for example, the region of space occupied by a face of a rotating cube.) As a result, the perspective embedding of the images of the surface into E^3 necessarily varies over time from one image to the next. Moreover, because this volume constitutes a three-dimensional rather than a two-dimensional manifold, the projective visual mapping of this manifold onto its retinal images is no longer diffeomorphic and no longer has a well-defined inverse. Despite this apparent mathematical difficulty, the trajectory of an *infinitesimal* surface *patch* usually approximates a section of a surface of revolution for at least a brief interval of time. The accuracy of this approximation improves as the area of the patch and the interval of time are reduced. More importantly, the connectivity and smoothness of the moving surface patch impose considerable constraints on the values of the angular rotation velocities and radii of rotation for neighboring surface patches: for a rigidly rotating surface, the angular velocities are equivalent for all neighboring patches, and the radial distances of the patches from the axis of rotation vary smoothly over the surface. Presumably, these constraints among the rotational and perspective parameters of neighboring patches serve to disambiguate the parameter values of the individual surface patches. Thus, we hypothesize that this rotational transformation is employed generally by vision to describe the retinal image deformations produced by the motions of objects and observers in environmental 3-space.

A second apparent limitation of this geometric approximation is that the group of motions in E^3 includes translations as well as rotations. Translational motions of objects and of the observer are common visual events, and these appear to provide an important source of information about spatial structure of objects and the environment. Two different directions of such translations are relevant: (a) translations approximately perpendicular to the direction of gaze, producing the classical motion parallax transformations, and (b) translations approximately parallel to the direction of gaze, producing divergence or "looming" of the retinal images. Presently available evidence indicates that neither of these classes of translational transformations constitutes a basic source of information about the surface structure of environmental objects.

The motion parallax transformations produced by translations approximately perpendicular to the direction of gaze have long been known to constitute a source of information about depth. The geometrical nature of this information has often been misunderstood, however. Especially in the traditional psychological literature, the relevant optical information was often believed to consist of the relative velocities of a neighboring pair of moving points: when the observer is fixated on the horizon and translating in a perpendicular direction, as might occur in gazing out the side window of a car or train, then the relative image velocities of stationary environmental objects are inversely proportional to their distances in E^3 from the observer. If the observer maintains fixation on an intermediate object, then the image velocities are inversely proportional to their distances in E^3 from the point of fixation, with points nearer than the fixation point moving across the image in the direction opposite from that of the observer's translation, and with the images of points lying beyond the fixation point moving in the same direction as the observer's translation. Thus, the relative image velocities have been presumed to constitute the visibly detected optical information about depth from motion. Psychophysical evidence obtained by Kottas (1977), however, indicates that the relative image velocities of discrete visually disconnected points do not yield perceptual information about relative distance. In displays consisting of three discrete points moving at different velocities, any one of the three points was as likely as another to be seen as farther from the observer if the points were subjectively perceived as disconnected. The subjective impression of connectedness was necessary for the observer to discriminate reliably which one of the points was between the other two, and even then the faster moving point was as likely as the slower point to be seen as farther from the observer. Rather than the relative velocities of pairs of disconnected points, the optical information about structural relations in environmental three-space is probably provided by the image deformations of the spatial relations among three or more neighboring points on the same surface.

When the retinal images of environmental structures are deformed by such translations, however, these deformations may be locally approximated by rotations rather than translations. The global patterns of image deformations produced by these two different groups of transformations differ from each other, but their local effects are very similar. If, as we hypothesize, the optical information for perceiving the structure of moving objects is provided by the local deformations of images of smooth surfaces, then these local approximations by rotations are sufficient; the global pattern of the transformation might not be visually detected anyway. Indeed, currently available evidence suggests that this is the case—that image transformations produced by translations are typically seen as rotations—although the question has not yet been rigorously tested by a direct experiment. We as well as other investigators have noticed that the motion parallax produced by translations perpendicular to the direction of gaze often appears subjectively to be rotating around a vertical axis parallel to the frontal-parallel plane (G.J. Andersen and M.L. Braunstein, personal communications, 1989). This same effect also seems to be involved in the "stereokinetic effect"—where local image deformations similar to translations are perceived as if produced by rotation in depth. (The rationale and empirical justification for this interpretation lie outside the domain of this paper, but will be presented in a forthcoming paper.) In any case, rotation in depth seems to be generally used by vision to describe the local image deformations produced by motions of the observer and of environmental objects, even though this local approximation is often not geometrically valid, especially for the global pattern of the transformation.

When the direction of translation (of either the observer or the object) coincides with the observer's viewing direction, producing a divergence or "looming" of the optical pattern, vision is probably insensitive to the optical information about surface shape that may be provided by these divergence patterns. Theoretically, the velocity field associated with the optical flow patterns produced by such motions might provide visible information about the egocentric distance of any given point—about its "time to contact" as Lee (1974) pointed out: the image trajectory of any point is in a radial direction away from the so-called "focus of expansion" with a velocity that increases with its nearness to the observer and with its angular deviation from the observer's direction of gaze. Although this velocity field (but not the radial direction of the image trajectories) might in principle provide information about the orientation of a surface relative to the observer, as Prazdny (1983) and Perrone (1989) have pointed out, human observers are probably not sufficiently sensitive to the differential structure of these velocity fields to discriminate environmental structure. Indeed, the effects of variations in surface slant on the image velocities are quite small until the surface is very close to the observer. Experiments on this question were conducted recently at the NASA Ames Research Center and found that human observers were strikingly insensitive to the slants of surfaces portrayed by such optical patterns (Kaiser et al., 1990). In contrast, the same variations in surface slant were discriminated quite accurately when the direction of translation was perpendicular to the direction of gaze and parallel to the surfaces in question. (As noted above, however, the direction of motion appeared to be rotational rather than translational.)

Even though the optical divergence patterns produced by translation in the direction of gaze seem to be a poor source of visual information about the shapes of environmental objects, such translations probably do provide effective visual information about the locations of environmental objects relative to the observer and about the observer's location and motion within the environment. We consider this second aspect of the geometry of vision in the next section.

Perspective embedding of images in 3-space

The preceding geometrical development accounts only for perception of the surface structure of a single object. The same geometrical relationships are not sufficient to describe the 3D environmental space that contains the object. Neither the distance of the object from other environmental objects or from the observer, nor the orientation of the object relative to the observer, nor the location of the observer within the environmental space are determined by the local optical information that specifies the metric structure of the given object surface.

As shown in the preceding section, the metric structure of a given surface patch is locally determined by just three metric tensor coefficients (given by the matrix $\mathbf{P}^*$); and these three parameters can be determined from the image deformation produced by a rotation in E^3. In representing this rotation by a single parameter (the angle of rotation), the orientation of the axis of rotation relative to an independent 3D coordinate system is not determined. The perspective projection from E^3 onto the retinal image, however, is determined by six parameters—three parameters to specify the location of the observer's station point within the environmental 3-space, and three additional parameters to specify the location of the fixation point or direction of gaze. The optical information that determines these six perspective parameters is defined by global rather than local properties of the retinal images, and is associated with image transformations different from the rotations that produce the image deformations which provide information about local surface structure.

The perspective projective mapping from E^3 onto the retinal image yields a hyperbolic geometry in the image: for any given line segment in the image and any given point not in the line segment, an infinite number of line segments could in principle pass through the given point which are parallel to the given line segment; and an infinite number of parallel line segments might intersect at any given point in the image.[5] Lines that are regarded as parallel in the hyperbolic geometry of the image are the images of lines that are parallel in E^3. The retinal images of lines that are parallel in E^3 converge in the image at a point which is the retinal image of a point infinitely distant from the observer in the direction of the given parallel lines. Thus, for example, if the observer were standing in a rectangular room, the retinal images of the four lines defined by the intersections of the ceiling and floor with two opposite walls would converge in the image at a single point; if the ceiling and floor are parallel to the gravitational ground plane, then this "vanishing point" where the lines converge would lie on the "horizon line" and would specify the observer's eye height relative to the objects in the image.[6]

The nature of this projective model of hyperbolic geometry may be better understood by considering the retinal images as regions of a sphere centered at the observer's station point (at the focal point of the observer's eye). This sphere is what has been called the "optic array" (Gibson, 1966; Cutting, 1986). By definition, this spherical optic array is uniquely specified by the optical projections through a given station point within a given environment and is invariant under rotations of the observer's eye around the focal point of the eye. If the horizon line were extended a full 360 degrees around the observer it would define a great circle in the spherical optic array—the

---

[5] This multiplicity of potentially parallel lines through any given point not on the given line segment is possible only if the vanishing point of the given line segment has not already been specified. Each of the lines that is potentially parallel to the given line segment would converge with the given line at a different vanishing point in the image.

[6] We are grateful to Steven Tschantz of the Department of Mathematics at Vanderbilt for providing this example.

intersection of the sphere with a plane passing through the center of the sphere parallel to the ground plane. This great circle horizon line is the set of vanishing points of all lines that are parallel (in both E^3 and in the hyperbolic geometry of the retinal images) to the ground plane. The position and orientation of this horizon line on the retina provide direct optical information about the height and orientation of the observer's eye within the environment. But this horizon line is not a uniquely specified optical structure; it is no more explicitly defined than the ground plane.

Any set of mutually parallel lines that are not parallel to the ground plane will converge in the image at a vanishing point that does not lie on the horizon line. Every great circle in the optic array is the image of points that are infinitely distant from the observer in the set of directions parallel to a plane containing the observer's station point. Any set of lines that are parallel in E^3 is parallel to two orthogonal planes which pass through the observer's station point; and these parallel lines would converge in the image at a vanishing point given by the intersection of the corresponding two orthogonal great circles.

Consider the set of images of stationary objects projected onto the optic array of an observer translating through the environment in a constant direction. The image trajectories of all points on these objects would define a set of globally parallel lines that would converge in the image at a vanishing point corresponding to the image of an infinitely distant point in the direction of the observer's translational motion. Thus, the image transformations produced by an observer's translational motion through the environment are especially informative about the relative location and motion of the observer within the environment.

Next consider a subset of these images of objects moving in relation to the observer—those whose linear trajectories are normal to the spherical optic array, producing images on the optic array which differ by a divergence (or contraction) with no deformation. Such a set of images of a given object are *congruent* and therefore *isometric* with one another in the hyperbolic geometry of the optic array, even though the Euclidean sizes of these images differ from one another. Congruence is a property possessed by hyperbolic as well as by Euclidean geometry. The hyperbolic isometry of the images under this set of motions in E^3 serves to scale the projective correspondence between the optical images and the environmental 3-space in which these rigid object motions occur. Thus, these optical divergence patterns are a particularly rich source of information about the global perspective embedding of the retinal images into the Euclidean 3-space of the environment.

In general, however, the image transformations produced by rotations around axes that do not contain the observer's station point and by translations in directions that do not pass through the observer's station point produce deformations as well as divergences of the images. Although these deforming images do not remain strictly congruent in the hyperbolic geometry of the image, their obvious symmetry does serve to specify a

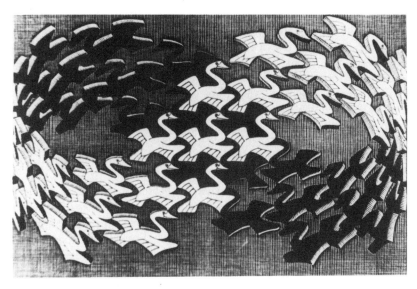

Figure 1. "Swans," etching by M.C. Escher, 1956. Copyright 1988, M.C. Escher heirs/Cordon Art-Baarn-Holland.

Euclidean 3-space in which the objects remain congruent under rotation. We consider illustrations of this generalized congruence property in the next section.

Congruences in images

The potential for constructing spaces from such implied congruences among imaged forms has been wonderfully illustrated by M.C. Escher. For example, he has often used translational symmetry of a replicated form to define a 2D plane. Both the metric structure of this space and also its orientation parallel to the image plane are specified by its translational symmetry. The elementary component form is also defined by its recursion in the image rather than by the familiarity of the form itself.

Congruences in Euclidean 3D space are exhibited in Figure 1, where the congruence of swan-like component forms is obtained by translations and rotations in a 3D space. The 3D metric structure of the space is defined by the depicted congruence under 3D rotation and translation of the forms recurring in separate regions of the space. The perspective mapping of this space onto the 2D image plane is also induced by this congruence of the component forms. Thus, the perspective trigonometry is derived from the congruence; the fundamental property is the congruence rather than the trigonometry.

In the preceding example, congruence is defined among stationary and concurrent forms. The "motion" that brings a form in one location into congruence with a form in another location is abstract, rather than an actual

Perception of geometrical structure 439

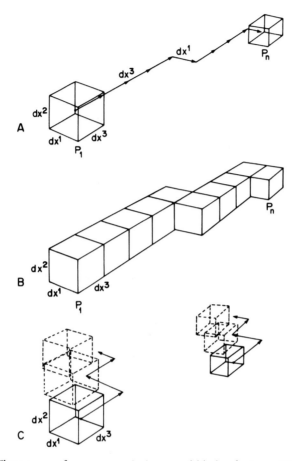

Figure 2. Three types of congruences in images. (A) A cube in position P_1 is moved in a temporal sequence of displacements through 3D space to position P_n. A single object appears in a trajectory through space-time. (B) The same cubic form as in A appears simultaneously in positions P_1 and P_n, connected in this case by a spatial series of cubes. A 3D space is defined by the congruences of the spatial series of repeated component forms. (C) Two objects are moved concurrently by a sequence of displacements as if rigidly connected. The 3D structure of the space is indicated in this case by the congruence of the motions in the separate spatial regions rather than by the congruences of the spatial forms as in the other two panels.

trajectory in space-time. If one generalizes the concept of an image from a stationary 2D spatial array to a space-time volume in which the spatial structures are extended in time, then the same principle of congruence illustrated in Escher's art can be applied to the specification of spaces by the motions of single forms.

The schematic diagram in Figure 2 illustrates three conceptually different types of congruence in images. Figure 2B is like that in the Escher print, where the image is a stationary 2D pattern in which a single cube-like structure is recursively positioned at a sequence of neighboring spatial positions. The 3-dimensionality of the space is induced by the continuous linear change in the 2D lengths of the contours of the cube as a function of its position in the image plane. This linear relation between 2D length and position corresponds to a particular perspective mapping of 3D space onto the image plane. Thus, the continuous linear relation among neighboring regions of the image of a single connected surface specifies the perspective mapping of a 3D space onto the 2D image.

In Figure 2A the same perspective mapping is defined by a temporal sequence of spatial images as the cube is translated through space from position P_1 to position P_n. The linear transformation that corresponds to the perspective projection of a plane slanted in depth is now specified by a function in space-time, though the geometric relation between the image and the depicted space obviously is essentially the same as in Figure 2B. In both cases, relationships among neighboring image regions correspond to relationships among neighboring regions of a smooth surface. The perspective relation between the image and the 3D space in which the surfaces, objects, and motions reside is specified by the linear relationship between the lengths of the contours and their positions in the image.

Figure 2C illustrates a slightly different case in which the structure of a space is specified by congruences among simultaneous *motions* of separate forms at separate locations in the image, as if the forms were connected and moved in 3D space. This situation might be produced, for example, by motions of the observer or image plane (e.g., a movie or video camera) within a 3D environment. In this example two cubes, at positions P_1 and P_n in 3D space, are simultaneously displaced in a sequence of four successive translations. The perspective mapping from the 3D space in which these events occur onto the 2D image of the events may be specified by the functional relation between the magnitudes of the velocities and their locations in the image plane. Although the forms at positions P_1 and P_n in this particular illustration are both cubes that are potentially congruent under the same transformations that would bring the motions of the two cubes into congruence, this spatial congruence is not necessary and provides in this case an additional redundant specification of the perspective transformation of the 3D space onto the 2D image.

The geometric relation between the concurrent motions of just two forms as in Figure 2C is not generally sufficient to specify the perspective transformation that has yielded the observed spatio-temporal image. By the fundamental theorem of plane perspectivity (Delone, 1963), the perspective mapping of four points in general position (where no three points are collinear) in one image plane onto a corresponding set of four points in another image plane is necessary and sufficient to ensure that all of the remaining points are in isometric correspondence in the two planes. Thus, for a set of four or more points in a single plane, the concurrent motions of the images of these points in another plane seem to be sufficient in principle to specify the perspective transformation between these two planes and to specify the metric structure of the spatial relations within these planar images.

This geometric relationship endows spatial as well as moving images with considerable capacity for carrying information about the geometric structure of the environmental surfaces depicted in the images: the geometric structure of an infinitesimally small patch on any arbitrarily curved but smooth surface may be locally approximated by a tangent plane at that location, and the perspective mapping of this tangent plane onto an image plane may be described by a linear coordinate transformation. The parameters of this linear transformation vary with the relative 3D orientation (the direction of tilt and the magnitude of slant) and distance of the environmental surface in relation to the image plane. The perspective parameters which embed the image of the surface into E^3 and thereby determine the metric structure of the surface are those parameters that will yield the isometry of the same object under motion to different locations within the depicted scene.

Experimental evidence

In addition to the evidence provided by the preceding illustrations, by everyday visual experience in viewing both natural environments and artificial spatial displays, and by the capabilities of moving observers to coordinate their actions with the identities, positions, and trajectories of environmental objects, the hypothesis that perceived geometric structure derives from the congruences of moving and movable objects is also supported by experimental evidence. A vast amount of experimental evidence appears consistent with this hypothesis, but we mention here only a few experiments that seem to provide more direct support for this hypothesis.

One of the relevant investigations is that of Cutting (1987). Judgments of the apparent rigidity of rotating rectangular solids were evaluated in a variety of experimental display conditions, including both rigidly and nonrigidly rotated figures and displays that simulated varying degrees of polar versus parallel projection, and varying degrees of slant of the projection screen relative to the direction of the perspective convergence point. He found good discrimination of rigid versus nonrigid figures in displays with approximately parallel projection, essentially independent of the degree of simulated screen slant (90°, 67°, 45°, or varying between 80° and 55°), even when the simulated slant was varied sinusoidally during a given trial. Although the figures appeared to move nonrigidly in conditions with polar projection onto screens slanted at 45°, the results generally demonstrated the robustness of perceived structural rigidity under at least moderate screen slants and moderate viewing distances. These results challenge many conventional assumptions about the geometrical information for perceiving the spatial structure of form. Cutting concludes that these results probably reflect the insensitivity of vision to the distortions produced by optical projections, but this interpretation rests upon assumptions about the definition of visual space by the metric structure of 2D display screens and retinae. An alternative interpretation is that vision is very sensitive to spatial relations defined in another way—by the congruence of form under perspective transformations.

Evidence that vision is indeed very sensitive to the spatial structure of moving forms and that this structure is associated with invariant spatial relations in depth rather than the projected 2D positions is provided by experiments reported previously by Lappin and Fuqua (1983). They evaluated observers' acuities in detecting a displacement (a stationary offset in 3D space) of a point from the 3D center of an imaginary line segment defined by moving patterns of three collinear points. The points were rotated in computer-controlled CRT displays as if around an axis slanted in depth by amounts varying between trials from 0° (no slant) to 60°. Very small displacements were accurately detected—displacements greater than 1 per cent of the 3D distance between the two outer points could be detected above chance, and displacements of 4 per cent were detected at approximately 90 per cent accuracy. The essential 3-dimensionality of the perceived spatial relations was demonstrated by the following findings: (1) Detection accuracy was independent of either the magnitude or variability of the slant of the axis of rotation in depth. (2) Distance-like measures of the detection accuracy (similar to the signal detectability measure d') were linearly related to the physical distance of the displacement in 3D space, with discriminability being proportional to physical displacement distances above about 1 per cent. (3) The accuracy for detecting any given displacement was the same in displays with parallel and with polar perspective, although in the latter displays points centered in 3D depth were not centered in the projected 2D images. The differences in spatial positions between the parallel and polar displays were visually easily resolvable, however. (4) When the task required detection of displacements from the projected 2D centers of the line segments in displays with polar projections, accuracies were not significantly above chance. The subjective appearance of the latter displays was that the three points were still seen as rotating in depth, but the middle point appeared neither centered nor rigidly attached to the two outer points.

Thus, these findings suggest that vision may often be unaffected by the 2D optical "distortions" in cinema not merely because these spatial differences cannot be resolved by vision, but because they do not constitute the geometrical information for perceiving the spatial structure of moving patterns. Apparently, perceived spatial structure derives from congruences of form under perspective transformations.

Evidence about the role of such congruences in *stereoscopic* form perception has been provided by recent experiments described by Lappin and Love (in preparation). The purpose of these experiments was to determine whether the stereoscopic perception of 3D structure might be shaped by the congru-

Perception of geometrical structure

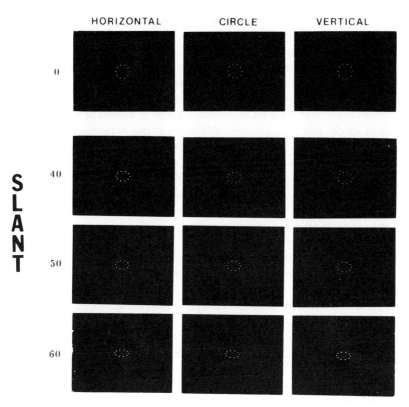

Figure 3. Photographs of three different shapes slanted by four different magnitudes from the frontal-parallel plane. In the ellipse labelled "horizontal" the vertical axis was 94 per cent of the horizontal axis; in the shape labelled "circle" the horizontal and vertical axes were equal; and in the "vertical" ellipse the vertical axis was 6 per cent greater than the horizontal axis.

ences of form associated with motion in depth, rather than by the binocular disparities as such. The experiments were motivated by the theoretically challenging fact that for any given magnitude of binocular disparity between the horizontal separations of a pair of points in each eye, the associated separation in depth increases rapidly and nonlinearly with the viewing distance from the observer to the points in question: how then is the stereoscopic perception of form and depth calibrated for variations in viewing distance? Does this require "interpretations" of retinal disparities based on extra-retinal information about the viewing distance? Alternatively, might the perceived geometric structure of surfaces in depth be based on the invariance of the intrinsic geometric structure of the surface under the perspective transformations associated with stereoscopic disparities and with motions in depth? The theoretical problem is related to those in understanding the apparent "paradoxes" of cinema.

In one of these experiments, observers were presented with two very slightly different ellipses, in which the vertical axis was either 3 per cent greater or less than the length of the horizontal axis. These ellipses were displayed as if in a plane slanted in depth by either 50° or 60° varying randomly from one trial to the next. Figure 3 shows a set of three such shapes (a circle and two ellipses, differing in the magnitude of the vertical axis) at each of four different magnitudes of slant from the frontal-parallel plane, as photographed on the computer-controlled CRT screen. Thus, the projected forms were always elliptical, depending on the magnitude of the slant as well as on the shape of the ellipse as measured in its own plane in depth. Stereoscopic information about the shapes and slants of these patterns was also manipulated by random variations in the magnitude of the disparities with which the forms were displayed, using disparities that were appropriate for either one-half or one-quarter of the actual viewing distance at which the patterns were seen. Thus, there were eight alternative stimulus patterns which randomly varied between trials.

There were four main experimental conditions—in which the forms were either rotated in depth or were stationary, and in which the experimental task was either *shape-discrimination* between the two alternative ellipses or *disparity-discrimination* between the two alternative magnitudes of disparity magnification. If stereoscopic information about 3D structure is scaled by the congruences of moving forms, then shape discrimination should be accurate when the forms are rotated in depth, independently of the distortions and variability produced by the exaggerated binocular disparities. Indeed, this is just what happened: shape discriminations were very accurate when the forms were moving, and were uncorrelated with the variations in either slant or disparity. Not surprisingly, shape discriminations were near chance accuracy when the forms were stationary, because of the perceptually inseparable conjoint effects of variations in slant and disparity. For the disparity-discrimination task, however, motion had the opposite effects: discriminations between the two alternative disparity values were more accurate for the stationary than for the moving forms, evidently because the congruence of the moving forms tended to obscure differences between the stationary disparity spaces. The combined results for four observers are given in Figures 4 and 5.

Thus, these results indicate that the visual scaling of 3D structure from stereoscopic disparity derives from the congruences of the perspectively changing forms. Analogous to the case for stationary pictures and optic flow patterns, binocular disparity *per se* may have only an indirect relation to the perceived depths.

Perception of geometrical structure 445

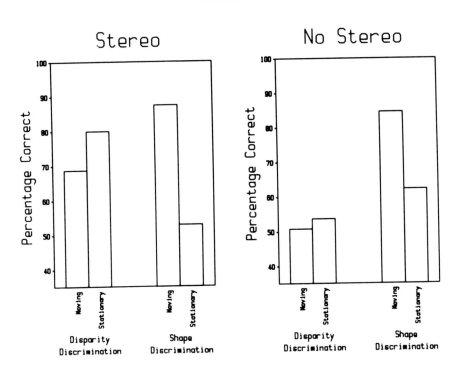

Figure 4. The averaged accuracies of four observers in eight different experimental conditions—involving discriminations of either the shapes or of the magnitudes of the disparity magnification (two alternative values), for shapes that were either moving or stationary and were displayed either with or without stereoscopic disparity. (Discrimination of the magnitude of disparity, which was manipulated by varying the simulated viewing distance, should of course have been impossible when the shapes were displayed with no stereoscopic disparity; this condition was included merely as a control to evaluate whether these discriminations might be based on the very small differences in perspective or on some other unintended cue.)

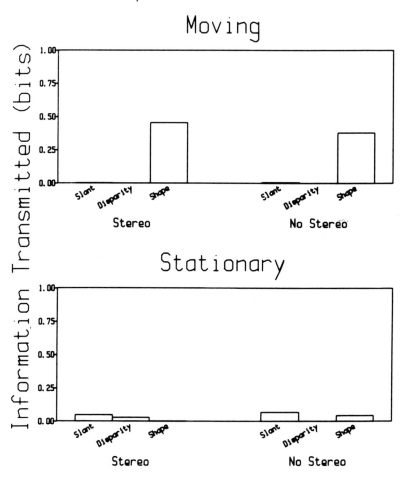

Figure 5. The correlations of the shape-discrimination judgments with each of the three independent stimulus variables—slant, disparity, and shape—when the forms were either moving or stationary. All eight possible combinations of slant, disparity, and shape occurred equally often in random order within each block of trials. The moving and stationary patterns were presented in separate blocks of trials. The amount of information transmitted by the judgments is used here as a measure of the correlation between the binary response categories and each of the three binary stimulus variables.

References

Cutting, J.E. (1986). *Perception with an Eye for Motion*. Cambridge, MA: MIT Press, Bradford Books.
Cutting, J.E. (1987). Rigidity in cinema seen from the front row, side aisle. *Journal of Experimental Psychology: Human Perception and Performance*, **13**, 323–334.
Delone, B.N. (1963). Analytic geometry. In Aleksandrov, A.D. Kolmogorov, A.N. and Lavrent'ev, M.A. (Eds) *Mathematics: Its Content, Methods, and Meaning*. Cambridge: MIT Press.
Doner, J., Lappin, J.S. and Perfetto, G. (1984). Detection of three-dimensional structure in moving optical patterns. *Journal of Experimental Psychology: Human Perception and Performance*, **10**, 1–11.
Einstein, A., Lorentz, H.A., Weyl, H. and Minkowski, H. (1923/1952). *The Principle of Relativity*. New York: Dover.
Gibson, J.J. (1950). *The Perception of the Visual World*. Boston: Houghton-Mifflin.
Gibson, J.J. (1957). Optical motions and transformations as stimuli for visual perception. *Psychological Review*, **64**, 288–295.
Gibson, J.J. (1966). *The Senses Considered as Perceptual Systems*. Boston: Houghton-Mifflin.
Guggenheimer, H.W. (1963). *Differential Geometry*. New York: McGraw-Hill.
Killing, W. (1892). Ueber die Grundlagen der Geometrie. *Journal für die reine und angewandt Mathematik*, **109**, 121–186.
Kaiser, M.K., Perrone, J., Andersen, G.J., Lappin, J.S. and Proffitt, D.R. (1990). The effect of motion on surface slant perception. *Investigative Opthalmology and Visual Science*, **31**, 520 (ARVO abstract).
Koenderink, J.J. and van Doorn, A.J. (1975). Invariant properties of the motion parallax field due to the movement of rigid bodies relative to the observer. *Optica Acta*, **22**, 773–791.
Koenderink, J.J. and van Doorn, A.J. (1976a). Local structure of movement parallax of the plane. *Journal of the Optical Society of America*, **66**, 717–723.
Koenderink, J.J. and van Doorn, A.J. (1976b). Geometry of binocular vision and a model for stereopsis. *Biological Cybernetics*, **21**, 29–35.
Koenderink, J.J. and van Doorn, A.J. (1976c). The singularities of the visual mapping. *Biological Cybernetics*, **24**, 51–59.
Koenderink, J.J. and van Doorn, A.J. (1977). How an ambulant observer can construct a model of the environment from the geometrical structure of the visual inflow. In Hauske, E. and Butenandt, G. (Eds) *Kybernetik*. Munich: Oldenberg.
Koenderink, J.J. and van Doorn, A.J. (1980). Photometric invariants related to solid shape. *Optica Acta*, **27**, 981–996.
Kottas, B.L. (1977). "Visual 3D information from motion parallax". Unpublished doctoral dissertation. Vanderbilt University.
Krantz, D.H., Luce, R.D., Suppes, P. and Tversky, A. (1971) *Foundations of Measurement*. New York: Academic Press.
Lappin, J.S. (1990). Perceiving the metric structure of environmental objects from motion and stereopsis. In Warren, R. and Wertheim, A.H. (Eds) *The Perception and Control of Self-Motion*. Hillsdale, N.J.: Lawrence Erlbaum.
Lappin, J.S. and Fuqua, M.A. (1983). Accurate visual measurement of three-dimensional moving patterns. *Science*, **221**, 480–482.
Lappin, J.S. and Love, S.R. (in preparation). Metric structure of stereoscopic form from congruence under motion.
Lappin, J.S., Doner, J.F. and Kottas, B.L. (1980). Minimal conditions for the visual detection of structure and motion in three dimensions. *Science*, **209**, 717–719.

Lee, D.N. (1974). Visual information during locomotion. In MacLeod, R.B. and Pick, H.L. (Eds) *Perception: Essays in Honor of J.J. Gibson*. Ithaca, NY: Cornell University Press.

Lombardo, T.J. (1987). *The Reciprocity of Perceiver and Environment: The Evolution of James J.Gibson's Ecological Psychology*. Hillsdale, N.J.: Lawrence Erlbaum.

Luce, R.D. (1978). Dimensionally invariant numerical laws correspond to meaningful numerical relations. *Philosophical Science*, **45**, 1–16.

Luce, R.D. and Narens, L. (1983). Symmetry, scale types, and generalizations of classical physical measurements. *Journal of Mathematical Psychology*, **27**, 44–85.

Misner, C.W., Thorne, K.S. and Wheeler, J.A. (1973). *Gravitation*. San Francisco: Freeman.

Perrone, J.A. (1989). In search of the elusive flow field. *Proceedings of the Workshop on Visual Motion*. Washington, D.C.: Computer Society of the IEEE.

Prazdny, K. (1983). On the information in optical flows. *Computer Vision, Graphics, and Image Processing*, **22**, 239–259.

Weyl, H. (1952). *Space-Time-Matter*. New York: Dover.

29

The perception of three-dimensionality across continuous surfaces[1]

Kent A. Stevens

Department of Computer Science
University of Oregon
Eugene, Oregon

Abstract

The apparent three-dimensionality of a viewed surface presumably corresponds to several internal perceptual quantities, such as surface curvature, local surface orientation, and depth. These quantities are mathematically related for points within the silhouette bounds of a smooth, continuous surface. For instance, surface curvature is related to the rate of change of local surface orientation, and surface orientation is related to the local gradient of distance. It is not clear to what extent these 3D quantities are determined directly from image information rather than indirectly from mathematically related forms, by differentiation or by integration within boundary constraints. An open empirical question, for example, is to what extent surface curvature is perceived directly, and to what extent it is quantitative rather than qualitative. In addition to surface orientation and curvature, one derives an impression of depth, i.e., variations in apparent egocentric distance. A static orthographic image is essentially devoid of depth information, and any quantitative depth impression must be inferred from surface orientation and other sources. Such conversion of orientation to depth does appear to occur, and even to prevail over stereoscopic depth information under some circumstances.

Introduction

One can derive a compelling impression of three-dimensionality from even static, monocular surface displays. Figure 1, for example, suggests an un-

---
[1] Supported by Office of Naval Research Contract N00014-87-K-0321.

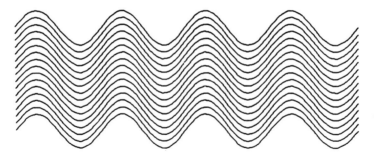

Figure 1. Undulating lines.

dulating surface. The three-dimensionality of this figure can be dramatically enhanced when one removes the visual evidence about the surface on which the figure is printed. If, say, the pattern is viewed on a graphics display, in a darkened room, monocularly and without head movements, the apparent three-dimensionality is particularly vivid, sufficiently so that one could replicate the apparent surface by curving a ruled sheet of paper and holding it in a particular attitude.

On reflection, it is actually quite curious that a pattern of lines such as those in Figure 1 provides so fixed and stable a percept. There is, after all, an infinity of possible 3D surfaces containing lines that would project to that 2D pattern. To posit that the pattern corresponds to a *particular* surface requires certain, specific, strongly constraining assumptions. A theory has been developed of the geometric constraints that support such inferential 3D percepts, one that explains how a range of 3D qualities, such as local surface orientation and curvature might be derived in principle (Stevens 1981a, 1983b, 1986). But it is difficult to extend such theories to explain more precisely *what* 3D information is extracted and internally represented in the process of deriving apparent three-dimensionality from such a 2D stimulus. It is one thing to discuss perception in terms of "affordances", "cues", or other characterizations of incident information, and quite another thing to determine the specific course of processing that takes incident information into explicitly represented perceptual quantities.

The remarkable ability to derive surface information from simple monocular configurations has been quite difficult to explain adequately within any of the traditional psychological paradigms. The difficulty stems, I believe, from the lack of basic understanding about what constitutes "apparent three-dimensionality". Depth perception is an often-used term that refers to the perception of surfaces and points in 3D. What differentiates the perception of mere 2D patterns of stimulation from 3D arrangements, seemingly, is perception of the third dimension, namely depth or distance from the viewer to points in space. Gibson insightfully proposed that "visual space perception is reducible to the perception of visual surfaces, and that distance, depth, and orientation ... may be derived from the properties of surfaces" (Gibson 1950). To Gibson, the term "apparent three-dimensionality" refers to the perception of more than merely the "third dimension". Visual perception clearly developed to operate in the richly redundant visual world. But the very little 3D information in Figure 1 hardly compares to the redundant and seemingly unambiguous wealth of incident information afforded by a natural scene. It might justifiably be relegated to the domain of so-called "picture perception".

Approaches toward understanding surface perception that attempt to isolate the contribution provided by a particular cue, such as texture or contours, or motion or stereopsis, have often been criticized as failing to address enough of the problem. By not embracing the complexity of natural scenes, it is argued, one fails to examine the system in the environment for which it was designed. But while one might well fail to observe important phenomena when only examining components in isolation or in simple combination, by *not* doing so one might equally fail to observe effects central to the strategies that allow the system to deal effectively with complexity and redundancy.

If vision is regarded computationally as the construction of internal descriptions of the visual world, there is no particularly compelling reason to expect qualitatively different modes of visual processing depending on whether the retinal image derives from a picture or a real scene. If one does not expect a different mode for "picture perception", one must then explain how an ambiguous and obviously underspecified 2D stimulus can result in a definite and stable 3D percept.

The challenge, then, is to understand our seeming ability to perceive more specifically than is objectively specified by the stimulus. To Helmholtz, Gregory, and others, this ability stems from the basic perceptual strategy of "unconscious inference". To mix terminology from traditionally antagonistic schools of thought on this matter: higher-order variables in the incident optical array are cues that afford particular 3D inferences. After a while such word play is seen for what it is, and we should go on to more constructive explorations. Substantial progress will likely come only with understanding of the nature of the 3D *percept*, something that has been given remarkably little attention over the entire history of perceptual studies.

As will be discussed, this task is difficult in theory, because of various mathematical equivalences among different representational forms, and difficult in practice, because of the robustness of the visual observer in performing psychophysical judgments. Despite the intrinsic difficulty, however, there is some evidence that surface perception is sufficiently modular and restricted in its ability to extract and combine 3D information as to be amenable to study using traditional psychophysical methods.

Quantifying apparent three-dimensionality

Following the usage by Foley (1980), *absolute distance* will refer to the egocentric range from an observer to a specific 3D point, which might be a point on a visible surface. *Relative distance* refers to a ratio of absolute distances (without knowing the absolute distances, one might know that one distance is twice another). In this usage *depth* refers specifically to the *difference* of absolute distances to a given point and a reference point. (Hence the depth of a given point relative to a reference point might be known in absolute units without knowing the overall absolute distances involved. Also, if the depth at a point were known and the absolute distance to the reference point were known, their algebraic sum would specify the absolute distance to the given point.)

In addition to scalar distance information at a point, derivatives of distance information specify the orientation of the tangent plane and curvature of the surface in the vicinity of a point. Surface orientation has two degrees of freedom, and is readily described as a vector quantity related to the normal to the tangent plane (Stevens 1983c). The psychological literature has long used the magnitude quantity *slant* to refer to the angle between the line of sight and the local surface normal (slant varies from 0° to 90°). The other degree of freedom, the *tilt* of the surface, specifies the direction of slant, which is the direction to which the normal projects onto the image plane, and also the direction of the gradient of distance (Stevens 1983a). Since the slant-tilt form aligns with the direction and magnitude of the local depth gradient, it provides many advantages for encoding surface orientation, such as allowing for simultaneous representation of precise tilt and imprecise slant, being closely related to various monocular cues such as shading, texture foreshortening, motion parallax, and perspectivity, and providing for (Necker-type) ambiguity in local surface orientation as reversals in tilt direction (see Stevens, 1983c).

Derivatives of surface orientation, or higher derivatives of distance, are related to surface curvature (across a continuous, twice-differentiable region). Surface curvature also has two degrees of freedom in the neighborhood of a surface point, which might be encoded as principle curvatures, or their image projections.

The central problem, which I will illustrate momentarily, is that across a continuous surface it is possible to convert among these different forms by differentiation (in one direction) and integration (in the other). One source of information about local slant might be used to infer both surface curvature and depth, and another might indicate curvature information directly. With sufficient boundary constraints the information provided by any source might be converted to a form comparable with another across a continuous surface. In general, then, it is difficult to determine whether a given 3D quantity M is derived directly from the image or indirectly from derivatives or integrals of M.

The mathematical equivalences among these various forms of 3D information leave quite open the empirical question of to what extent surface curvature is registered directly versus converted internally (Stevens 1981b; Cutting and Millard 1984; Stevens 1984), and furthermore, the question of the extent to which this information is represented quantitatively rather than qualitatively (Stevens 1981a, 1983b, 1986).

The 3D information content of a simple stimulus

Returning to Figure 1, what sorts of 3D information can be extracted feasibly? Observe that the figure consists merely of a family of parallel curves; we impose the interpretation that they correspond to parallel curves across a continuous surface in three dimensions. Given the nature of orthographic projection, this pattern is devoid of information about the third dimension (distance). And yet, one sees measurable depth as well as slant in monocular stimuli consisting of line-drawing renditions of continuous ruled (developable) surfaces (Stevens and Brookes, 1987a). Both orthographic (as in Figure 1) and perspective projection were used. Using a randomized-staircase forced-choice paradigm, apparent slant was measured by varying the aspect ratio of an ellipse that was briefly superimposed on the monocular surface stimulus. Observers readily interpreted the ellipse as a foreshortened circle slanted in depth, and by adjusting the aspect ratio it could be made to appear flush on the surface. The resulting slant judgments were in close correspondence to the predicted geometric slant of the stimuli.

The apparent depth in these stimuli was then tested by superimposing a stereo depth probe over the monocular surface. Apparent depth was probed stereoscopically using a device similar to Gregory's (1968, 1970) "Pandora's Box". A Wheatstone-style stereoscope provided near-field (38 cm) convergence and accommodation, well within the range of acute stereopsis. After first fixating a binocular point on an empty field, the monocular stimulus was presented briefly (for as little as 100 msec) to the dominant eye only, after which a binocular probe was superimposed at a given stereo disparity over the monocular stimulus for an additional brief interval. Subjects performed a randomized-staircase forced-choice experiment in which the depth of the stereo probe was compared with that of the monocular surface at various locations. Just as Gregory (1970) found measurable apparent depth in a variety of illusion figures, minimal renditions of monocular surfaces, such as Figure 1, are also perceived quite measurably in the third dimension.

The experiments suggest that in orthographic projection the visual system can compute from local surface orientation a depth quantity that is commensurate with the relative depth derived from stereo disparity. Apparent slant is a measure of the local gradient of depth, i.e., the rate of change of depth (and being the derivative of distance, slant is independent of the absolute distance to the surface). Depth might be integrated from slant across the surface, but only up to a constant of integration. How, then, are monocular and stereo depth coupled so that they can be compared? The perceptual assumption used to link these two spaces, apparently, is that the absolute distance of the monocular surface at the given fixation point equals that of the stereoscopic horopter at that point. This hypothesis seems sound in that whatever surface location is fixated in sharp focus is likely to lie at zero disparity, since in the near field at least, there is close coupling between vergence and accommodation that brings into sharp focus the (zero disparity) fixation point. The fixated point (seen monocularly in our stimuli but binocularly in normal vision) is thus assumed to be at the absolute distance of the horopter. With the two depth measures sharing a common zero intercept, monocular depth from slant, appropriately scaled by the reference distance, could then be compared to depth from stereo disparity. This conjecture remains to be confirmed empirically.

Depth from gradient, curvature and discontinuity information

In addition to demonstrating the perception of three-dimensionality from highly underspecified stimuli, these observations suggest to us that the visual system has a robust ability internally to convert one form of 3D information into another mathematically equivalent form. The perception of depth from the various so-called monocular depth cues (such as shading, contours, and texture gradients) may well provide "direct" information about surface curvature and shape, and only indirect information about depth.

More generally, we propose that shape properties associated with derivatives of distance, specifically surface orientation, curvature, and loci of discontinuity, both in depth (edge boundaries) and tangent plane (creases), are the primary percepts, and that smoothly varying depth across continuous regions is recovered subsequently and indirectly (Stevens and Brookes, 1987b, 1988).

This proposal explains various phenomena involving the perception of apparent depth from stereopsis. The apparent depth of an isolated bar or point is predicted quite well by the geometry of the binocular system, with depth a straightforward function of stereo disparity and a reference binocular convergence signal (Foley, 1980). But various depth phenomena have been reported recently in the perception of more complicated surface-like stimuli that are not predicted by such a direct functional relationship (Gillam et al., 1984; Mitchison and Westheimer, 1984). Gillam et al., (1984) argue that depth derives most readily from disparity discontinuities, and Mitchison and Westheimer (1984) show that coplanar arrangements of lines result in elevated thresholds for depth detection. In a series of experiments in which binocular stimuli presented contradictory monocular and stereo information, we found instances where the stereo information was dramatically ineffective in influencing the 3D percept (Stevens and Brookes, 1987c). The patterns were wire-frame stereo depictions of planar surfaces, rendered orthographically and in perspective, and devoid of disparity discontinuities and disparity contrast (e.g., with a surrounding frame or background). Constant gradients

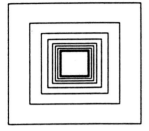
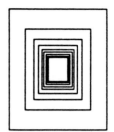

Figure 2. *Stereogram of coplanar lines.*

of stereo disparity, consistent with slanted planes, were introduced that were orthogonal to or opposite to the monocularly suggested depth gradients. The monocular interpretation dominated in judgments of apparent surface slant and tilt and in 2-point relative depth ordering. Figure 2, for example, is a stereogram of coplanar lines, with disparities varying linearly in accordance with a slanted plane. The dominant depth impression is the monocular interpretation of a perspective view of a corridor extended in depth. With scrutiny, however, many observers can also discern the true depth of the component lines in the pattern, as if the monocular interpretation can be selectively disregarded. The strategy which integrates these two independent sources of information into a coherent whole does not appear fixed, a point to which I shall return.

We hypothesize that stereo disparity influences the monocular 3D interpretation primarily where the distribution of disparities indicates curvature and depth discontinuities (i.e., where disparity varies discontinuously or has nonzero second spatial derivatives). Stereo depth across surfaces, by this view, is substantially a reconstruction from disparity contrast and curvature (see also Rogers 1986; Rogers and Cagenello 1989). A Craik-O'Brien-Cornsweet analog (Anstis *et al.*, 1978), and various depth induction effects (e.g., Werner, 1938) suggest that the derivation of depth and of brightness are analogous; both being reconstructions from higher spatial derivatives (Stevens and Brookes, 1987b; Brookes and Stevens 1989, 1990). While simultaneous contrast effects are seen in binocular depth, which are suggestive of depth being derived from second spatial derivative information, the analogy does not extend to the depth versions of the artifacts expected to result from derivative-like operators: Mach bands, the Hermann grid phenomenon, and other lateral inhibition effects are not found in the corresponding stereo displays (see Brookes and Stevens, 1990, for further discussion).

To illustrate, Figure 3 shows a ring at zero disparity embedded in background dots that comprise a linear disparity gradient. This stereogram is analogous to the familiar brightness contrast demonstration in which a constant luminance gradient background causes a constant luminance ring to appear to have unequal brightness around its perimeter. In the stereo version,

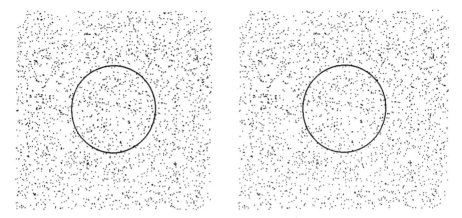

Figure 3. A ring of constant disparity set against a constant-gradient background of dots.

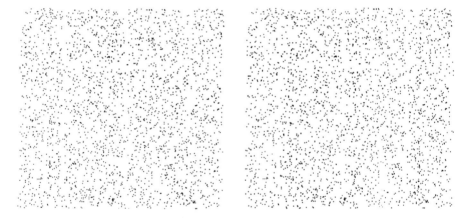

Figure 4. The stereo analogue to the Hermann grid.

the constant disparity gradient induces slant in the constant-disparity ring. The analogy fails, however, for contrast effects attributed to lateral inhibition. Figure 4, for example, shows the stereo counterpart to the Hermann grid. The original Hermann grid consists of a pattern of solid rectangles set against a contrasting background; illusory spots are seen in the background at the corners between rectangles, an effect usually attributed to lateral inhibition associated with circularly symmetric receptive fields. In the stereo version, the analogue of illusory brightness spots would be bump-like distortions in depth in the background. But as Julesz (1965) earlier noted, no analogous effects are seen in binocular depth. Likewise, we tried without success to generate the stereo analogue of the traditional Mach band, which, like the Hermann grid, is attributed to lateral inhibition of operators per-

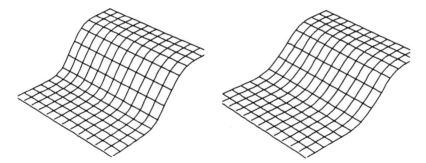

Figure 5. A stereogram in which a stereo edge and a monocular edge conjoin to produce an impression of a trough. Use uncrossed (free-fusion) viewing or a stereoviewer.

forming derivative-like operations. We conclude that stereo depth is analogous to brightness to the extent that both are reconstructed from contrast (of luminance or disparity), but that the analogy does not extend to the underlying contrast-detection mechanisms. On the other hand, it is not surprising that stereo disparity processing does not exhibit evidence of second derivative operators. Positional instability between the two retinae likely preclude the visual system having access to reliable absolute disparities in the first place (see Brookes and Stevens, 1990 for further discussion).

Returning to the broader issue of the perception of continuous surfaces, there is the central question of how information from different cues is combined. Algebraic summation of the strength of evidence given by independent cues has been suggested by Dosher et al. (1986) for proximity-luminance covariance and stereopsis, and by Bruno and Cutting (1988) for several monocular cues to the depth: relative size, height in the projection plane, occlusion and motion parallax. Does a summation model also apply to the integration of cues across a continuous surface? As discussed, there is little evidence for the summation of depth from stereo and depth from monocular cues, at least in a direct pointwise manner across the surface. In (Stevens et al., in press) we generated a series of stimuli in which different planar and curved patterns were independently defined by surface contours and by binocular disparity. Extending the approach (Stevens and Brookes, 1988), the stereo and monocular information was spatially superposed: e.g., where the monocular surface information might suggest a planar surface, the stereo might suggest smooth curvature, and so forth. Figure 5 shows a simple example in which stereo and monocular edges of opposite polarity are superimposed in proximity. This produces an "emergent surface feature": a trough defined by the adjacency of the two curvature features. While this example is roughly consistent with both algebraic summation and winner-take-all models, the results are more complicated when geometrically inconsistent features are directly superposed. The perceptual effects which we found resulting from the combination of these conflicting cues may be summarized as follows.

The monocular interpretation of a set of surface contours, whether planar or curved, tends to dominate the combined percept at locations in the display where the disparity pattern indicates planarity. Next, the binocular interpretation tends to dominate where the disparity pattern indicates curvature and where the monocular pattern indicates planarity. On the other hand, where both stereo and monocular interpretations indicate inconsistent surface curvature features, more complex resolution strategies are suggested, sometimes involving conscious attention to either the stereo or the mono interpretation, sometimes involving a compromise between both, but varying among observers and among presentations for the same observer. Finally, where both stereo and monocular interpretations indicate surface curvature features which are qualitatively consistent, but differ in amplitude, different observers show markedly different response patterns in a quantitative comparison task. More complex resolution strategies are suggested by these cases than have been proposed earlier. These strategies may vary among different observers, and involve conscious attentive processing.

References

Anstis, S., Howard, I.P. and Rogers, B. (1978). A Craik-O'Brien-Cornsweet illusion for visual depth. *Vision Research*, **18**, 213–217.
Brookes, A. and Stevens, K.A. (1989). Binocular depth from surfaces versus volumes. *Journal of Experimental Psychology: Human Perception and Performance*, **15**(3), 479–484.
Brookes, A. and Stevens, K.A. (1990). The analogy between stereo depth and brightness. *Perception*, in press.
Bruno, N. and Cutting, J.E. (1988). Minimodularity and the perception of layout. *Journal of Experimental Psychology: General*, **117** (2), 161–170.
Cutting, J.E. and Millard, R.T. (1984). Three gradients and the perception of flat and curved surfaces. *Journal of Experimental Psychology: General*, **113**, 217–220.
Dosher, B.A., Sperling, G. and Wurst, S.A (1986). Tradeoffs between stereopsis and proximity luminance covariance as determinants of perceived 3D structure. *Vision Research*, **26**, 973–990.
Foley, J.M. (1980). Binocular Distance Perception. *Psychological Review*, **87**, 411–434.
Gibson, J.J. (1950). *The Perception of the Visual World*. Boston: Houghton Mifflin.
Gillam, B.J., Flagg, T. and Finlay, D. (1984). Evidence for disparity change as the primary stimulus for stereoscopic processing. *Perception and Psychophysics*, **36**, 559–564.
Gregory, R. (1968). Visual illusions. *Scientific American*, **218** (11), 66–76.
Gregory, R. (1970). *The Intelligent Eye*. London: Weidenfeld and Nicolson.
Julesz, B. (1965). Some neurophysiological problems of stereopsis. In Nye, P.W. (Ed.) *Proceedings of a Symposium on Information Processing in Sight Sensory Systems*, 135–142. Pasadena, CA.
Mitchison, G.J. and Westheimer, G. (1984). The perception of depth in simple figures. *Vision Research*, **24**, 1063–1073.
Rogers, B. (1986). The perception of surface curvature from motion and parallax

cues. *Investigative Ophthalmology and Visual Science* ARVO Abstracts Supplement, **27**, 181.

Rogers, B. and Cagenello, R. (1989). Disparity curvature and the perception of three-dimensional surfaces. *Nature*, **339**, 135–137.

Stevens, K.A. (1981a). The visual interpretation of surface contours. *Artificial Intelligence*, **217**, Special Issue on Computer Vision, 47–74.

Stevens, K.A. (1981b). The information content of texture gradients. *Biological Cybernetics*, **42**, 95–105.

Stevens, K.A. (1983a). Surface tilt (the direction of surface slant): A neglected psychophysical variable. *Perception and Psychophysics*, **33**, 241–250.

Stevens, K.A. (1983b). The line of curvature constraint and the interpretation of 3-D shape from parallel surface contours. *Eighth International Joint Conference on Artificial Intelligence*, August.

Stevens, K.A. (1983c). Slant-tilt: The visual encoding of surface orientation. *Biological Cybernetics*, **46**, 183–195.

Stevens, K.A. (1984). On gradients and texture "gradients". Commentary on: Cutting and Millard, 1984. Three gradients and the perception of flat and curved surfaces. *Journal of Experimental Psychology: General*, **113**, 217–220.

Stevens, K.A. (1986). Inferring shape from contours across surfaces. In Pentland, A.P. (Ed.) *From Pixels to Predicates: Recent Advances in Computational Vision*, pp. 93–110. Norwood, NJ: Ablex.

Stevens, K.A. and Brookes, A. (1987a). Probing depth in monocular images. *Biological Cybernetics*, **56**, 355–366.

Stevens, K.A. and Brookes, A. (1987b). Depth reconstruction in stereopsis. *Proceedings of the First IEEE International Conference on Computer Vision*, London, June.

Stevens, K.A. and Brookes, A. (1988). Integrating stereopsis with monocular interpretations of planar surfaces. *Vision Research*, **28** (3), 371–386.

Stevens, K.A., Lees, M. and Brookes, A. Combining stereo and monocular curvature features. *Perception*, in press.

Werner, H. (1938). Binocular depth contrast and the conditions of the binocular field. *American Journal of Psychology*, **51**, 489–407.

30

The effects of viewpoint on the virtual space of pictures

H.A. Sedgwick

Schnurmacher Institute for Vision Research
S.U.N.Y. College of Optometry, New York

Introduction

Pictures are made for many different purposes (Hagen, 1986; Hochberg, 1979). This discussion is about pictorial displays whose primary purpose is to convey accurate information about the three-dimensional spatial layout of an environment. We should like to understand how, and how well, pictures can convey such information. I am going to approach this broad question through another question that seems much narrower. We shall find, however, that if we could answer the narrow question, we should have made a good start on answering the broader question as well.

Every pictorial display that presents a precise perspective view of some three-dimensional scene has a single geometrically correct viewpoint.[1] In most viewing situations, however, the observer is not constrained to place his or her eye precisely at this correct viewpoint; indeed the observer generally has no explicit knowledge of the location of this viewpoint.[2] My "narrow" question is: "What effect does viewing a picture from the wrong location have on the virtual space represented by that picture?"

This question is in itself of theoretical as well as practical importance. It has received considerable attention, but its answer is still far from being clear. The research literature is fragmentary and conflicting. I believe that a more vigorously applied theoretical analysis can clarify the issues and can help in evaluating the existing literature.

---

[1] For a camera image, this point is determined by the optics of the imaging system; for a display created by a draftsman or a computer, this point is determined by the relation between the center of projection and the projection plane (Carlbom and Paciorek, 1978; Sedgwick, 1980).

[2] A complex pictorial display generally does contain sufficient information, under certain constraints, to specify its own correct viewpoint. This issue is discussed by Green (1983), Jones and Hagen (1978), and Sedgwick (1980).

My theoretical analysis follows the approach developed by J.J. Gibson (1947, 1950, 1954, 1960, 1961, 1971, 1979). I shall be referring frequently to the *optic array*, which is Gibson's term for the structured array of light reflected to a point of observation by the surfaces of the environment. I shall also be relying on Gibson's concept of *available visual information*. Information is said to be available in the optic array when some projective structure in the optic array mathematically specifies, with appropriate constraints, some structure in the environment. The optic array typically contains multiple, redundant sources of information for the spatial layout of the environment.

The theoretically determined availability of visual information of course does not guarantee that such information will be used by a human observer. The extent to which any such information actually influences perception is a separate question that must be addressed empirically. The contention of Gibson's approach is simply that we are not in a proper position to formulate or interpret empirical investigations of human visual perception until we understand the underlying available information on which any successful perception must be based.

This discussion will concentrate on theoretical analysis. At several points, however, I shall briefly indicate how well this analysis accords with the empirical work that has been done on human pictorial perception. More detailed reviews of this subject are offered elsewhere (Cutting, 1986a; Farber and Rosinski, 1978; Hagen, 1974; Kubovy, 1986; Rogers, 1985; Rosinski and Farber, 1980).

To simplify the discussion I am going to consider separately the effects of deviating from the correct viewpoint in each of three orthogonal directions: deviations perpendicular to the picture plane (that is, being too close or too far from the picture), lateral deviations parallel to the picture plane, and vertical deviations parallel to the picture plane. Any possible viewing position can then be interpreted as some combination of these three deviations.

Theoretical analysis

Viewing from too close or too far

What is the theoretical effect of viewing a pictorial display from too close or too far[3]? As we approach or withdraw from the picture, its projection in the

---

[3] A number of analyses of this problem have been offered. The first systematic analysis appears to come from La Gournerie (1859), whose work has been discussed more recently by Pirenne (1970, 1975), Kubovy (1986), and Cutting (1987). Other analyses, apparently independent of La Gournerie, have been given by Purdy (1960), Farber and Rosinski (1978), Lumsden (1980), and Rosinski and Farber (1980).

Obtaining an unambiguous three-dimensional interpretation of a pictorial display requires that some constraints be placed on the possible interpretations. In the above analyses, those referring to La Gournerie and that of Farber and Rosinski (1978) do not make these constraints explicit.

optic array expands or contracts around the center of the picture, which is the point at which a perpendicular from the viewpoint pierces the picture plane. If we let z be the correct distance from the picture and z' be our actual distance, and let A and A' be the angular separations from the center at these two distances, respectively, of some other point on the picture, then

$$\tan A/\tan A' = z'/z = m$$

where m is a constant. Thus the optic array projection of the picture is magnified or minified by 1/m, where m measures how close or how far we are, relative to the correct distance.[4]

What, in theory, is the effect on the virtual space of the picture of magnifying or minifying its projection in the optic array? We can begin to answer this question by looking at the available visual information that is present in the *perspective structure of the optic array*, by which I mean the vanishing points of straight edges in the environment and the vanishing lines of planar surfaces.[5]

Let us imagine a picture of a flat, endless ground plane covered with a regular texture represented by a grid of lines. The horizon, or vanishing line, of the ground, will be located at eye level on the picture plane. If our point of observation is located at a height h above the ground, then the distance d along the ground to any particular grid line parallel to the picture plane is given by the simple expression

---

The other analyses use explicit constraints derived from analyses of normally viewed pictures. Purdy (1960) bases his analysis on gradients of texture, Lumsden bases his on familiar size, and Rosinski and Farber base theirs on linear perspective. I offer two analyses here, one based on the ground plane and the other based on perspective structure, as suggested in Sedgwick (1980). All of these analyses converge on the same results.

A different analysis, reaching different results, has been offered recently by McGreevy and his colleagues (Ellis *et al.*, 1985; McGreevy and Ellis, 1984, 1986; McGreevy *et al.*, 1987). McGreevy's analysis proceeds by arbitrarily constraining all virtual distances from the picture plane to be unchanged by viewing position. This analysis has the weakness that it assumes a knowledge of these distances without indicating how they could be determined by an observer of the display, either when viewing from the wrong viewpoint or when viewing from the correct viewpoint. The question of how virtual layout could be determined here is made difficult because the constraint that is imposed leads to violations of all of the other constraints mentioned in the preceding paragraph.

Another kind of analysis, based on optimizing the match between a noisy registration of the projection and a noisy *a priori* internal model of the spatial layout has been offered recently by Grunwald and Ellis (1986). There is not room here to consider the interesting question of how such a model-based approach to spatial layout might be reconciled with the constraint-based approach taken in this paper.

[4] Approaching a picture is optically equivalent to viewing the pictured scene through a telephoto lens, and withdrawing from the picture is optically equivalent to viewing the scene through a wide-angle lens (Lumsden, 1980; Rosinski and Farber, 1980).

[5] Perspective structure is usually only implicit in the optic array. The available visual information that specifies this perspective structure is not discussed in this paper, but I have analyzed it in detail elsewhere. Not all pictorial displays contain sufficient information to completely specify their perspective structure (Sedgwick, 1983, 1986, 1987a).

$$d = h(1/\tan G)$$

where G is the optic array angle subtended between the horizon of the ground plane and the grid line.

We can now combine these two expressions to derive the theoretical effect of magnification or minification. If we let d' be the geometrically specified distance of the grid line when the picture is seen from the incorrect viewpoint and let G' be the new optic array angle corresponding to G, then

$$d' = h(1/\tan G')$$

substituting for G',

$$d' = h(m/\tan G)$$

and substituting again,

$$d' = md$$

Next, if we let s be the specified separation in depth between any two successive grid lines, at distances d_1 and d_2, when the picture is seen from the correct viewpoint and let s', d_1', and d_2' be the specified separation and distances when seen from the incorrect viewpoint, then

$$\begin{aligned} s' &= d_2' - d_1' \\ &= md_2 - md_1 \\ &= m(d_2 - d_1) \\ &= ms \end{aligned}$$

Thus as we approach the picture, the geometrically specified depths in the picture are compressed proportionally to the closeness of our approach and as we move away from the picture, depths are expanded proportionally (Figure 1).[6]

---

[6] There is an invariant associated with the optic array gradient projected from equally spaced grid lines parallel to the picture plane. If s is the separation in depth between any two successive grid lines, then

$$s = d_2 - d_1 = h(1/\tan G_2 - 1/\tan G_1)$$

Thus, for any two successive optic array angles G_1 and G_2 in this gradient

$$1/\tan G_2 - 1/\tan G_1 = k$$

where k is a constant. The presence of this invariant in the optic array specifies that the grid lines are equally spaced. It can be shown that this invariant is preserved when the picture is viewed from too close or too far.

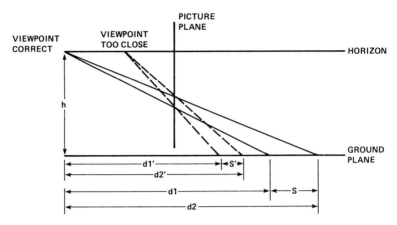

Figure 1. Close viewing compresses geometrically specified virtual depth.

Consider now what happens to frontal plane dimensions. The tangent of the angle F subtended by a width w that is parallel to the picture plane is inversely proportional to its distance from the point of observation (assuming for simplicity that the width is measured from the center of the picture)

$$w = d \tan F$$

As we approach the picture, the specified distance of w decreases, but its optic array angle F increases in the same proportion, so that w remains constant (Figure 2)

$$w' = d' \tan F' = (md)(\tan F/m) = d \tan F = w.$$

The depth of the pictured scene is thus compressed relative to its frontal dimensions. Shapes that are not in the frontal plane are distorted. The square grid covering the ground plane, for example, becomes a grid of rectangles whose depth to width ratio is m (Figure 3).

We may note here that all distances specified in the virtual space of the picture depend on h, the height of the viewpoint above the ground plane, which thus provides a scale factor for all distances, as well as sizes, in the picture. Because h itself is not geometrically specified in the picture, its value may be indeterminate.[7] This indeterminancy of h puts in doubt the appropriateness of comparing absolute distances or sizes across different pictures or across different views of the same picture. The ratio, however, of depth to width, s/w or s'/w', does not depend on h; thus, geometrically specified compression of shape by the factor m is an invariant effect of too close viewing.

---

[7] The value of h can be determined by assuming that the ground plane of the picture is coextensive with the ground plane of the real environment, but such an assumption may for some pictures be neither appropriate nor perceptually compelling.

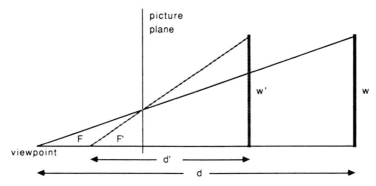

Figure 2. Close viewing leaves geometrically specified virtual frontal dimensions unchanged.

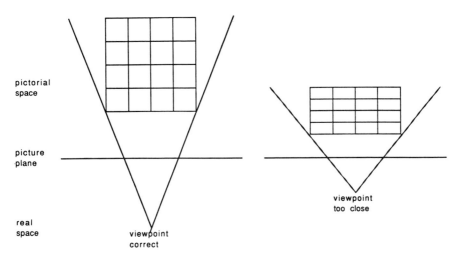

Figure 3. Close viewing distorts geometrically specified virtual shape.

Geometrically specified angles and orientations in the pictured scene are also changed by approaching the picture. This result follows directly from the compression that occurs, but it is instructive to derive the result in a different way.

Every set of parallel lines in the pictured scene has a vanishing point on the picture plane (lines parallel to the picture plane have their vanishing points at infinity on the picture plane). The three-dimensional orientation of a set of parallel lines is equal to the orientation of a line from the point of observation to their vanishing point. This very simple optic array relation specifies the pictured orientation of any edge once its vanishing point is known (Hay, 1974; Sedgwick, 1980).

Edges perpendicular to the picture plane have their vanishing point at the

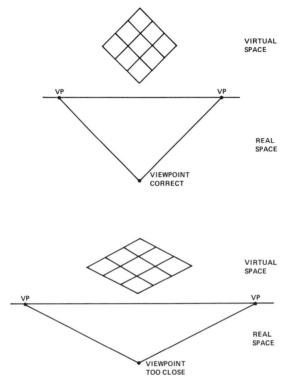

Figure 4. Vanishing points geometrically specify distortions in virtual orientation with close viewing.

center of the picture. As we approach the picture, every vanishing point except for the one at the center of the picture increases its optic array separation from the central vanishing point. Thus the specified orientations of all nonperpendicular edges move closer to being parallel to the picture plane. For example, a square ground plane grid oriented at 45° to the picture plane becomes a grid of squashed diamonds (Figure 4).

If we let E be the angle, measured relative to the straight-ahead, that a vanishing point subtends at the correct viewpoint, and let E' be the angle that it subtends when the viewpoint is too close or too far, then the distortion D in the specified orientation of edges having that vanishing point is given by E minus E'.[8] The relation between E and E' is the same as for any other optic array angles measured from the center of the picture, namely

$$\tan E / \tan E' = m$$

---

[8] Throughout this paper, orientations are specified in environment-centered terms (i.e., relative to the fixed framework of the environment), rather than in viewer-centered terms (i.e., relative to the observer's line of regard). I have discussed this distinction and its significance at length elsewhere (Sedgwick, 1983; Sedgwick and Levy, 1985).

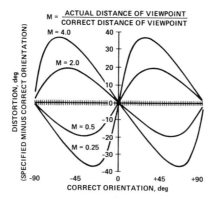

Figure 5. Geometrically specified distortion in virtual orientation as a function of viewing distance.

Calculating D as a function of E for several values of m, we obtain a family of curves showing no distortion for orientations perpendicular (0°) or parallel (−90° or 90°) to the picture plane, with maximum distortion at intermediate values (Figure 5). For example, for m equal to either 2 or 0.5, the maximum distortion approaches 20°.

A similar analysis can be made for the orientations of planar surfaces. The angle subtended between the vanishing line of a slanted surface and the vanishing line of the ground plane is equal to the three-dimensional angle between the depicted surface and the ground (Sedgwick, 1980). As we approach the picture plane, geometrically specified surface orientations are distorted in just the same way as are edge orientations.

Perceptually, effects qualitatively similar to those predicted theoretically here can be seen by a careful observer moving closer to or farther from a picture containing strong linear perspective. If the perspective information in the picture is weaker, the distortions may be much harder to see. Most empirical investigations, but not all, have found such distortions in human picture perception, although not always at the magnitude predicted.[9] I shall say a bit more about the reasons for the discrepancies between investigations later.

---

[9] Empirical evidence that is at least qualitatively consistent with the analysis presented here has been reported by Bartley (1951), Bartley and Adair (1959), Bengston et al. (1980), Farber (1972), Lumsden (1983), Purdy (1960), Smith (1958a, 1958b), Smith and Gruber (1958), and Smith, et al. (1958). Anecdotal supporting observations are also reported by MacKavey (1980) and Pirenne (1970). On the other hand, Rosinski and Farber (1980) briefly report failing to find distortions when the frame of the display is visible, and Hagen and Elliott (1976) and Hagen and Jones (1978) report that adults' choice of the most "realistic looking" display was essentially independent of their actual viewing distance.

It is important to distinguish between the presence of measurable distortions in the perception of spatial layout and the detection of these distortions by the observer. Observers' perceptions may contain distortions of which the observers themselves are unaware. A number of researchers have suggested that observers are often not very sensitive to the presence of such distortions (Gombrich, 1972; Pirenne, 1970; Cutting, 1986a, 1986b).

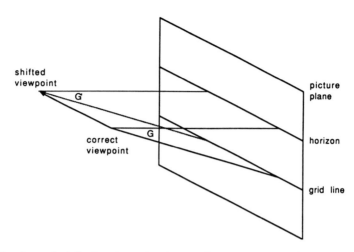

Figure 6. Lateral shifts in viewpoint do not change geometrically specified virtual depth.

Viewing from the side

Let us now consider what happens when we view a pictorial display from the side.[10] It is easy to see that when the viewpoint is displaced laterally, maintaining the same distance from the picture plane, the horizon of the ground and all of the grid lines parallel to the picture plane simply slide along themselves in the optic array. Thus the angular separation of each of these grid lines from the ground horizon remains unchanged. Consequently, the geometrically specified distance of each of these grid lines, relative to the height of the viewpoint, also is unchanged (Figure 6).

As the viewpoint slides to the right, for example, each point in the geometrically specified virtual space of the picture slides to the left, with its projected point on the surface of the picture acting as a stationary fulcrum. This lateral shift in virtual space is thus directly proportional to, but opposite in sign from, the amount of the viewpoint's displacement; it is also directly proportional to the distance of the point from the picture plane, and is inversely proportional to the viewpoint's distance from the picture plane (Figure 7). The overall effect of this viewpoint displacement is to produce a lateral shear in the geometrically specified virtual space of the picture (Figure

---

[10] Systematic analysis of this problem is again offered by La Gournerie (1859), whose work has been put to use by Cutting (1987). More recent analyses are offered by Farber and Rosinski (1978) and Rosinski and Farber (1980), who explicitly base their second analysis (1980) on linear perspective constraints. I again offer two analyses, one based on the ground plane and the other, following Sedgwick (1980), based on perspective structure. All of these analyses agree in the distortions that they predict.

Viewpoint and the virtual space of pictures 469

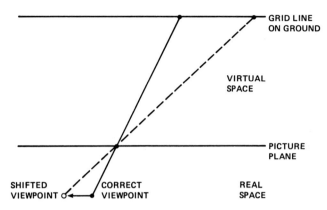

Figure 7. Lateral shifts in viewpoint geometrically specify lateral shifts in virtual space.

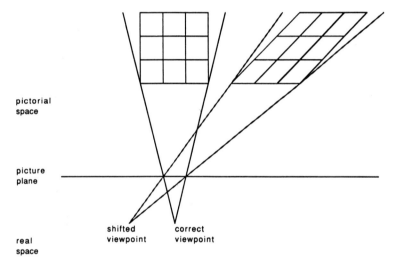

Figure 8. Lateral shifts in viewpoint geometrically specify a shearing of virtual space.

8). Frontal plane dimensions and orientations are unchanged, but shapes and orientations extending in depth are all distorted.

We can readily determine the specified shifts in the orientations of pictured edges and surfaces by again making use of the perspective structure of the picture. Let us consider, as an example, the orientations of horizontal edges, whose vanishing points lie on the horizon of the ground plane. As the viewpoint shifts laterally, its angular relation to each of these vanishing points changes. We shall let E again be the angle, measured relative to the straight-ahead, that the vanishing point makes with the correct viewpoint, and let E' be the angle that it makes after the vanishing point has shifted laterally. We can express this lateral shift as the ratio, k, between the amount, r, of the shift, and the distance, z, of the viewpoint from the picture plane. It is easy to see that (Figure 9)

$$\tan E' = \tan E + k$$

If we express the position of the shifted viewpoint in terms of its angular deviation, V, from the correct viewpoint, then

$$\tan V = k$$

so that

$$\tan E' = \tan E + \tan V$$

We can use this relation to determine the specified distortion of orientation, E' minus E, as a function of the correct orientation E, for a variety of angular shifts V of the viewpoint (Figure 10). The resulting family of curves shows that the specified distortions in orientation can be very large, approaching 180° as V approaches ±90°, which is parallel to the picture plane, and that the orientation E at which the distortion is maximal increases as V increases.

We may note that the same distortions in orientation would also be specified for vertical planes in the virtual space of the picture when the viewpoint is displaced laterally.

Perceptually, again, a careful observer comparing the appearance of a picture seen from one side or the other can notice differences in apparent orientation if the picture contains sufficient perspective information. Some empirical investigations have also found results that are qualitatively similar to those derived here, although others have not.[11] Again, I shall refer back to these discrepancies a little later.

Viewing from too high or too low

Let us now briefly consider what happens when the viewpoint is too high or too low. This is again a displacement parallel to the picture plane, so the

---

[11] Anecdotal reports of these distortions are common (Pirenne, 1970, 1975; Wallach, 1976, 1985). Experimental evidence that such distortions occur perceptually under some circumstances is offered by Goldstein (1979, 1987), Rosinski et al. (1980), Rosinski and Farber (1980), and Wallach and Marshall (1986), although all of these authors also report conditions under which the analytically predicted distortions do not occur. Cutting (1987) has analyzed some of the data of Goldstein (1987) in detail and has shown it to be in generally good accord with the theoretical predictions. Perkins (1973) finds some distortion from lateral viewing, but much less than this analysis would predict.

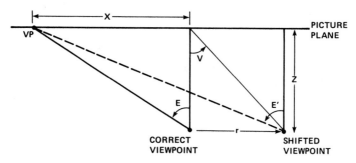

Figure 9. Vanishing points geometrically specify distortions in virtual orientation with lateral shifts in viewpoint.

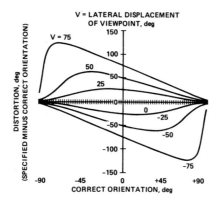

Figure 10. Geometrically specified distortion in virtual orientation as a function of lateral shift in viewpoint.

geometrically specified distortions in the virtual space of the picture are identical in form to those produced by lateral shifts, except that here the virtual space is sheared vertically instead of laterally.

Thus, for example, if we consider a plane in virtual space that is rotated around a horizontal axis so that it makes an angle E with the ground, its specified slant E', when seen from an incorrect viewpoint having a vertical angular deviation V, is given by the same relation

$$\tan E' = \tan E + \tan V$$

Notice that if we are considering the ground plane itself, then $E = 0$, so that $E' = V$. That is, if we must look down by a certain angle to see the pictured horizon, then the ground plane is specified as slanting down by that same angle.

Theoretical complications

So far we have seen how we can use the perspective structure of the optic array to determine the geometrically specified sizes, distances, and orientations of surfaces and edges in the virtual space of a picture. We have also seen how this visual information, when it is present, specifies distortions in the pictured layout when we observe the picture from the wrong viewpoint. Unfortunately for our ease of understanding, there are theoretical complications that are not taken into account by this straightforward analysis. We need to consider some of these complications now.

Resolving multiple sources of visual information

In a normally complex pictorial display, there are other sources of visual information for spatial layout available besides those arising from the perspective structure of the picture. How these multiple sources of information, which are normally partially redundant and partially complementary, may be combined into a single perceptual interpretation is a difficult and as yet unsettled question.[12] The difficulty is increased when the picture is observed from the wrong viewpoint because these different sources of information do not all predict the same distortions; nor is it always easy to tell what they do predict.

As an example, consider some of the information arising from surface texture (Gibson, 1950; Sedgwick, 1983, 1986). If several edges are resting on a surface that is uniformly textured, then the relative lengths of the edges are specified by the relative amounts of texture that they cover; likewise, the relative distances between the edges are specified by the relative amounts of texture between them. This texture scale information is as valid for edges that extend into depth as for those in the frontal plane; it thus serves to specify the shapes and the relative sizes and distances of objects resting on a common textured surface such as the ground plane.

It is easy to see that all such texture scale information is completely invariant over changes in viewpoint because such changes do nothing to alter the depicted amounts of texture between or under the objects in the picture. If, for example, we approach the picture of a square object resting on the textured ground, the specified object remains square because each of its edges continues to cover an equal amount of texture. On the other hand, according to the analysis based on perspective structure, the specified object is compressed into a rectangle whose width is greater than its depth.

This apparent contradiction between the distortions predicted by these two sources of visual information can be resolved, but only in a way that further

---

[12] An expert system that I have developed to study the interaction of multiple sources of visual information is described elsewhere (Sedgwick, 1987a, 1987b).

complicates our analysis. I mentioned earlier that any visual information entails constraints on the environment; if these constraints are violated, then the information is no longer valid. In the case of texture scale information, an essential constraint is that the texture's distribution across a surface be at least statistically uniform. Yet, in the example that we are considering now, when we come too close to the picture, perspective analysis specifies that the texture of the ground is itself compressed in the depth dimension. Thus the uniform distribution constraint is violated and texture scale information is no longer valid.

A visual system might do any of a number of things when faced with this situation. It might simply reject texture scale information as being invalid. It might go ahead and use texture scale information anyway. It might recognize that the viewpoint is incorrect. It might abandon the attempt to find a consistent virtual space for the picture. It might adopt a modified version of texture scale information using compressed texture. It might do something intermediate between some of these options. Analysis only indicates the possibilities without specifying which one will be adopted by any particular visual system.

A number of other sources of visual information, such as right-angle constraints (Perkins, 1972, 1976) and orientation-distribution constraints (Witkin, 1980), present similar difficulties when the viewpoint is incorrect, but there is not space to consider these additional difficulties here. Careful analysis of the interactions between these different sources of information should give us a basis for manipulating the information content of pictures so as better to determine the perceptual effects they produce.

Constancy and the dual nature of pictures

A second set of theoretical complications arises from what has often been referred to as the "dual nature" of pictures (Gibson, 1954; Haber, 1979, 1980a, 1980b; Hagen, 1974, 1986; Hochberg, 1962, 1979; Pirenne, 1970). In addition to being a representation of a spatial layout existing in a three-dimensional virtual space that lies beyond the plane of the picture, a pictorial display is also a real object consisting of markings of some sort, usually on a flat surface. Normally, visual information for the flat surface of the picture is made available by binocular stereopsis, by motion parallax, by the oculo-motor adjustments of convergence and accommodation, by the frame of the picture, and by the surface texture of the picture.

To perceive pictures, a perceptual system must be able, to some extent, to differentiate its response to the picture's virtual layout from its response to the real layout of the picture's surface. The human visual system seems able to make this differentiation, but not without some interaction, or "cross talk", between its responses to these two classes of information.

We can get some understanding of one effect of the picture surface by examining the relation between the picture plane and the optic array. If x

measures a separation in the picture plane from the center of the picture, which we have already defined as the point where a perpendicular from the viewpoint pierces the picture plane, and A measures the optic array angle subtended by this separation, then x is related to A by the relation

$$x = z \tan A$$

where z is the distance from the viewpoint to the picture plane. Near the center of a picture there is a close congruence between the optic array projection and the flat picture plane projection. This is because the tangent function is nearly linear for small angles. For larger angles, however, the tangent function becomes highly nonlinear, and consequently the optic array projection and the picture place projection become strongly noncongruent.

Perceptually, the cross talk between the picture surface and the virtual space of the picture, as specified in the optic array, becomes most noticeable when the picture plane projection and the optic array projection are noncongruent. Toward the edges of wide-angle pictorial displays, for example, the projections on the picture plane and in the optic array are still geometrically correct, but objects in the virtual space of the picture often appear to be distorted (Pirenne, 1970, 1975; Kubovy, 1986).[13] It seems that the noncongruent shape on the surface of the picture takes on a perceptual salience that interacts with the virtual space of the picture.

A similar noncongruence between the picture plane and the optic array is produced when the viewpoint is displaced laterally or vertically from the correct viewpoint. Again, the noncongruent shape on the surface of the picture may interact perceptually with the virtual space of the picture, but here its effect would be to diminish the distortion that is specified in the optic array. This would result in some degree of "constancy" in the virtual space of the picture in the sense that the virtual layout would not be as distorted as the optic array information would predict.

These effects of the picture's surface on the perceived virtual space of the picture could be eliminated, in principle, by removing the visual information for the picture's surface. Using a monocular display, restricting head movements relative to it, hiding the frame of the display, and so on, would all contribute to this result (Ames, 1925; Enright, 1987; Schlosberg, 1941; Smith and Smith, 1961).

The hypothesis of pictorial compensation

Finally, many theorists have suggested that when information for the picture surface is available, the human visual system may be able to compensate for

---

[13] This assumes that the perpendicular from the correct viewpoint pierces the picture plane somewhere near the center of the pictorial display, as it usually does.

being at the wrong viewpoint and so avoid distortions in the virtual space of the picture (Cutting, 1987; Farber and Rosinski, 1978; Hagen, 1974, 1976a, 1976b; Kubovy, 1986; Perkins, 1973, 1980; Pirenne, 1970; Rosinski, 1976; Rosinski and Farber, 1980; Rosinski et al., 1980; Wallach and Marshall, 1986). This compensation process would operate by either detecting or assuming a "correct" position of the viewpoint. The optic array information would then be adjusted to determine the virtual layout as it would be seen from this correct viewpoint.

Although a number of experiments have been offered in support of this view, it seems to me that, on balance, the compensation hypothesis is neither necessary nor sufficient to account for the bulk of the empirical results. It is not necessary because, as we have just seen, however sketchily, there are other explanations available for some of the disparities that exist between the distortions predicted by perspective structure and those actually found. Moreover, these other explanations are more parsimonious, in that they are derived from the analysis of general perceptual processes without having to postulate special processes that exist solely for perceiving pictures from the wrong viewpoint. The compensation hypothesis is not sufficient because it does not account for the considerable number of experimental results that find distortions in virtual space even when there is information available for the surface of the picture (Bengston et al., 1980; Goldstein, 1979, 1987; Wallach, 1976, 1985). Finally, it seems to me that a careful reading of several of the key experiments offered in favor of the compensation hypothesis casts some doubt on the firmness of their conclusions.[14]

Conclusion

As a conclusion to this brief discussion, I would suggest that picture perception is not best approached as a unitary, indivisible process. Rather, it is a complex process depending on multiple, partially redundant, interacting sources of visual information for both the real surface of the picture and the

---

[14] Kubovy (1986) is critical of many of the stimuli used by Hagen and Elliott (1976) and Hagen and Jones (1978) in their demonstration that adults at various distances from a picture do not choose the correct perspective as being most realistic. Perkins' (1973) demonstration of compensation for lateral viewing uses such minimal stimuli that the applicability of his results to more complex displays may reasonably be questioned. Hagen's (1976b) study, which claims to find evidence of compensation for lateral viewing in adults, has been criticized at length on logical grounds by Rogers (1985), who also failed to replicate Hagen's results. In the carefully controlled study of Rosinski et al. (1980) on the effects of frame visibility on perceived surface slant with lateral viewing, the interpretation of results is clouded by a confusion in the description of the experiment, and possibly in the experiment itself, about the frame of reference for their observers' judgments. Finally, Wallach and Marshall (1986, exp. 2) find evidence of compensation in pictorial shape perception from a lateral viewpoint, but their results, as they note, could be due to ordinary shape constancy because their stimulus shape was nearly parallel to the picture plane.

virtual space beyond. Each picture must be assessed for the particular information that it makes available. This, I would suggest, will determine how accurately the virtual space represented by the picture is seen, as well as how it is distorted when seen from the wrong viewpoint.

References

Ames, A. (1925). The illusion of depth from single pictures. *Journal of the Opthalmic Society of America*, **10**, 137–148.
Bartley, S.H. (1951). A study of the flattening effect produced by optical magnification. *American Journal of Optometry*, **28**, 290–299.
Bartley, S.H. and Adair, H.J. (1959). Comparisons of phenomenal distance in photographs of various sizes. *Journal of Psychology*, **47**, 289–295.
Bengston, J.K., Stergios, J.C., Ward, J.L. and Jester, R.E. (1980). Optic array determinants of apparent distance and size in pictures. *Journal of Experimental Psychology: Human Perception and Performance*, **6**, 751–759.
Carlbom, I. and Paciorek, J. (1978). Planar geometric projections and viewing transformations. *Computing Survey*, **10**, 465–502.
Cutting, J.E. (1986a). *Perception with an Eye for Motion*. Cambridge, MA: Bradford Books/The MIT Press.
Cutting, J.E. (1986b). The shape and psychophysics of cinematic space. *Behavorial Research Methods, Instruments, Computers*, **18**, 551–558.
Cutting, J.E. (1987). Affine distortions of pictorial space: Some predictions for Goldstein (1987) that La Gournerie (1859) might have made. *Journal of Experimental Psychology: Human Perception and Performance*, **14**, 305–311.
Ellis, S.R., Tyler, M., Kim, W.S., McGreevy, M.W. and Stark, L. (1985). Visual enhancements for perspective displays: Perspective parameters. *Proceedings from 1985 IEEE International Conference Systems, Man, and Cybernetics*, Tucson, AZ, November 12–15.
Enright, J.T. (1987). Perspective vergence: oculomotor responses to line drawings. *Vision Research*, **27**, 1513–1526.
Farber, J.M. (1972). "The effects of angular magnification on the perception of rigid motion", Doctoral dissertation, Cornell University.
Farber, J. and Rosinski, R.R. (1978). Geometric transformations of pictured space. *Perception*, **7**, 269–282.
Gibson, J.J. (1947). Motion picture testing and research. *Army Air Forces Aviation Psychology Program Research Reports*, (7). Washington, DC: GPO.
Gibson, J.J. (1950). *The Perception of the Visual World*. Boston: Houghton Mifflin,.
Gibson, J.J. (1954). A theory of pictorial perception. *Audio-Visual Communication Review*, **2**, 3–23.
Gibson, J.J. (1960). Pictures, perspective, and perception. *Daedalus*, **89**, 216–227.
Gibson, J.J. (1961). Ecological optics. *Vision Research*, **1**, 253–262.
Gibson, J.J. (1971). The information available in pictures. *Leonardo*, **4**, 27–35.
Gibson, J.J. (1979). *The Ecological Approach to Visual Perception*. Boston: Houghton Mifflin.
Goldstein, E.B. (1979). Rotation of objects in pictures viewed at an angle: Evidence for different properties of two types of pictorial space. *Journal of Experimental Psychology: Human Perception and Performance*, **5**, 78–87.
Goldstein, E.B. (1987). Spatial layout, orientation relative to the observer, and per-

ceived projection in pictures viewed at an angle. *Journal of Experimental Psychology: Human Perception and Performance*, **13**, 256–266.

Gombrich, E.H. (1972). The "what" and the "how": Perspective representation and the phenomenal world. In Rudner, R. and Scheffler, I. (Eds), *Logic and Art: Essays in Honor of Nelson Goodman*. Indianapolis: Bobbs-Merrill.

Green, R. (1983). Determining the preferred viewpoint in linear perspective. *Leonardo*, **16**, 97–102.

Grunwald, A.J. and Ellis, S.R. (1986). Spatial orientation by familiarity cues. In Patrick, J. and Duncan, R.D. (Eds), *Training, human decision-making and control*, 257–279, Amsterdam: North-Holland.

Haber, N.H. (1979). Perceiving the layout of space in pictures: A perspective theory based upon Leonardo da Vinci. In Nodine, C.F. and Fisher, D.F. (Eds), *Perception and Pictorial Representation*, New York: 84–89, Praeger.

Haber, N.H. (1980a). How we perceive depth from flat pictures. *American Scientist*, **68**, 370–380.

Haber, N.H. (1980b). Perceiving space from pictures: A theoretical analysis. In Hagen, M.A. (Ed.), *The Perception of Pictures, vol. 1*, 3–31. New York: Academic Press.

Hagen, M.A. (1974). Picture perception: Toward a theoretical model. *Psychology Bulletin*, **81**, 471–497.

Hagen, M.A. (1976a). Problems with picture perception: A reply to Rosinski. *Psychology Bulletin*, **83**, 1176–1178.

Hagen, M.A. (1976b). Influence of picture surface and station point on the ability to compensate for oblique view in pictorial perception. *Developmental Psychology*, **12**, 57–63.

Hagen, M.A. (1986). *Varieties of Realism: Geometries of Representational Art*. Cambridge: Cambridge University Press.

Hagen, M.A. and Elliott, H.B. (1976). An investigation of the relationship between viewing condition and preference for true and modified linear perspective with adults. *Journal of Experimental Psychology: Human Perception and Performance*, **2**, 479–490.

Hagen, M.A. and Jones, R.K. (1978). Differential patterns of preference for modified linear perspective in children and adults. *Journal Experimental Child Psychology*, **26**, 205–215.

Hay, J.C. (1974). The ghost image: A tool for the analysis of the visual stimulus. In MacLeod, R.B. and Pick, Jr. H.L. (Eds), *Perception: Essays in Honor of James J. Gibson*, 268–275. Ithaca, NY: Cornell University Press.

Hochberg, J.E. (1962). Psychophysics of pictorial perception. *Audio-Visual Communication Review*, **10**, 22–54.

Hochberg, J.E. (1979). Some of the things that paintings are. In Nodine, C.F. and Fisher, D.F. (Eds), *Perception and Pictorial Representation*, 17–41, Praeger.

Jones, R.K. and Hagen, M.A. (1978). The perceptual constraints on choosing a pictorial station point. *Leonardo*, **10**, 1–6.

Kubovy, M. (1986). *The Psychology of Linear Perspective in Renaissance Art*. New York: Cambridge University Press.

La Gournerie, J. (1859). *Traite de perspective lineaire contenant les traces pour les tableaux plans et courbes, les bas-reliefs, et les decorations theatrales, avec une theorie des effets de perspective*. 1 vol. and 1 atlas of plates. Paris: Dalmont et Dunod; Mallet-Bachelier.

Lumsden, E.A. (1980). Problems of magnification and minification: An explanation of the distortion of distance, slant, shape, and velocity. In Hagen, M.A. (Ed.), *The Perception of Pictures*, **1**, 91–135. New York: Academic Press.

Lumsden, E.A. (1983). Perception of radial distance as a function of magnification

and truncation of depicted spatial layout. *Perception and Psychophysics*, **33**, 177–182.
Mackavey, W.R. (1980). Exceptional cases of pictorial perspective. In Hagen, M.A. (Ed.), *The Perception of Pictures*, **1**, 213–223. New York: Academic Press.
McGreevy, M.W. and Ellis, S.R. (1984). Direction judgement errors in perspective displays. *Proceedings on the 20th Annual Conference on Manual Control*, NASA CP 2341, **1**, 531–549.
McGreevy, M.W. and Ellis, S.R. (1986). The effect of perspective geometry on judged direction in spatial information instruments. *Human Factors*, **28**, 439–456.
McGreevy, M.W., Ratzlaff, C.R. and Ellis, S.R. (1987). Virtual space and two-dimensional effects in perspective displays (manuscript).
Perkins, D.N. (1972). Visual discrimination between rectangular and nonrectangular parallelopipeds. *Perception and Psychophysics*, **12**, 396–400.
Perkins, D.N. (1973). Compensating for distortion in viewing pictures obliquely. *Perception and Psychophysics*, **14**, 13–18.
Perkins, D.N. (1976). How good a bet is a good form? *Perception*, **5**, 393–406.
Perkins, D.N. (1980). Pictures and the real thing. In Kolers, P.A., Wrolstad, M.E. and Bouma, H. (Eds) *Processing of Visible Language 2*, 259–278. New York: Plenum.
Pirenne, M.H. (1970). *Optics, Painting, and Photography*. Cambridge: Cambridge University Press.
Pirenne, M.H. (1975). Vision and art. In Carterette, E.C. and Friedman, M.P. (Eds) *Handbook of Perception, vol. 5, Seeing*, 433–490. New York: Academic Press.
Purdy, W.C. (1960). The hypothesis of psychophysical correspondence in space perception. Ithaca, NY: General Electric Advanced Electronics Center, September (NTIS R60ELC56).
Rogers, S. (1985). "Representation and reality: Gibson's concept of information and the problem of pictures". Doctoral dissertation, Royal College of Art.
Rosinski, R.R. (1976). Picture perception and monocular vision: A reply to Hagen. *Psychological Bulletin*, **83**, 1172–1175.
Rosinski, R.R. and Farber, J. (1980). Compensation for viewing point in the perception of pictured space. In Hagen, M.A. (Ed.), *The Perception of Pictures*, **1**, 137–176. New York: Academic Press.
Rosinski, R.R., Mulholland, T., Degelman, D. and Farber, J. (1980). Picture perception: An analysis of visual compensation. *Perception and Psychophysics*, **28**, 521–526.
Schlosberg, H. (1941). Stereoscopic depth from single pictures. *American Journal of Psychology*, **54**, 601–605.
Sedgwick, H.A. (1980). The geometry of spatial layout in pictorial representation. In Hagen, M.A. (Ed.) *The Perception of Pictures*, **1**, 33–90. New York: Academic Press.
Sedgwick, H.A. (1983). Environment-centered representation of spatial layout: Available visual information from texture and perspective. In Beck, J., Hope, B. and Rosenfeld, A. (Eds), *Human and Machine Vision*. New York: Academic Press.
Sedgwick, H.A. (1986). Space perception. In Boff, K., Kaufman, L. and Thomas, J. (Eds), *Handbook of Perception and Human Performance*, **1**. New York: Wiley.
Sedgwick, H.A. (1987a). A production system modeling high-level visual perspective information for spatial layout. Technical Report No. 298. New York Univ. Depart. of Computer Science.
Sedgwick, H.A. (1987b). Layout 2: A production system modeling visual perspective information. *Proceedings IEEE First International Conference on Computer Vision*, London.
Sedgwick, H.A. and Levy, S. (1985). Environment-centered and viewer-centered

perception of surface orientation. *Computer Vision, Graphics, and Image Processing,* **31**, 248–260.

Smith, O.W. (1958a). Comparison of apparent depth in a photograph viewed from two distances. *Perception of Motor Skills,* **8**, 79–81.

Smith, O.W. (1958b). Judgments of size and distance in photographs. *American Journal of Psychology,* **71**, 529–538.

Smith, O.W. and Gruber, H. (1958). Perception of depth in photographs. *Perception of Motor Skills,* **8**, 307–313.

Smith, O.W., Smith, P.C. and Hubbard, H. (1958). Ball throwing responses to photographically portrayed targets. *Journal of Experimental Psychology,* **62**, 223–233.

Smith, P.C. and Smith, O.W. (1961). Perceived distance as a function of the method of representing perspective. *American Journal of Psychology,* **71**, 662–674.

Wallach, H. (1976). The apparent rotation of pictorial scenes. In Henle, M. (Ed.) *Vision and Artifact,* 65–69. New York: Springer.

Wallach, H. (1985). Perceiving a stable environment. *Scientific American,* **252**, 118–124.

Wallach, H. and Marshall, F.J. (1986). Shape constancy in pictorial representation. *Perception and Psychophysics,* **39**, 233–235.

Witkin, A.P. (1980). Shape from contour. MIT technical report AI-TR-589, November.

31

Perceived orientation, spatial layout and the geometry of pictures

E. Bruce Goldstein

Department of Psychology
University of Pittsburgh
Pittsburgh, Pennsylvania

The purpose of this paper is to discuss the role of geometry in determining the perception of spatial layout and perceived orientation in pictures viewed at an angle. This discussion derives from Cutting's (1988) suggestion, based on his analysis of some of my data (Goldstein, 1987), that the changes in perceived orientation that occur when pictures are viewed at an angle can be explained in terms of geometrically produced changes in the picture's virtual space. Before dealing with Cutting's idea, let's first consider the paper that stimulated it.

Goldstein (1987) distinguishes between three different perceptual attributes of pictures:

(1) Perceived orientation. The direction a pictured object appears to point when extended out of a picture, into the observer's space.
(2) Perceived spatial layout. The perception of the layout in three-dimensional space of objects represented in the picture.
(3) Perceived projection. The perception of the projection of the picture's image on the observer's retina.

One basis for making these distinctions is that the perception of these attributes is affected differently by changes in the observer's viewing angle. Perceived orientation and perceived spatial layout, the two attributes we will focus on in this paper, differ in the following way:

(1) Perceived spatial layout remains relatively constant with changes in viewing angle. This "layout constancy" is demonstrated by presenting photographs of triangular arrays of dowels like the ones in Figure 1, and asking subjects to reproduce the layout this array would have if viewed from directly above. The results of these experiments, indicated by the general correspondence between the shapes of the solid triangles in

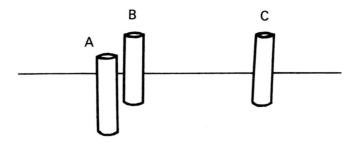

Figure 1. Stimulus used to determine the perceived spatial layouts of Figure 2 and the perceived orientations in Figure 3. In the actual photographic stimuli the dowels had horizontal black and white strips to distinguish them clearly from the background.

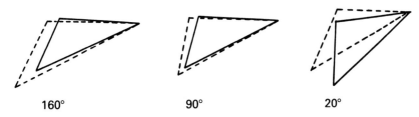

Figure 2. Solid triangles—average spatial layouts produced by four observers viewing the array of rods in Figure 1 from a distance of 8 in at viewing angles of 20°, 90°, and 180°. Dashed triangles—average spatial layouts produced by the same observers from a viewing distance of 64 in. Viewing angle is the angle between the observer's line of sight and the picture plane, with a viewing angle of 0° occurring when the observer is looking at the right edge of the picture and a viewing angle of 180°, occurring when the observer is looking at the left edge. (See Goldstein (1987) for further details of stimulus specification and procedures.)

Figure 2, indicate that changing viewing angle causes only small changes in a subject's ability to reproduce spatial layout. This relative constancy has also been observed for other arrays and for pictures of environmental scenes (Goldstein, 1979, 1987).

(2) Perceived orientation, on the other hand, undergoes large changes with changes in viewing angle. Figure 3 shows the average perceived orientations for four observers judging the orientations defined by pairs of dowels BA and BC of Figure 1. When the picture is viewed at an angle of 20° (far to the right side of the picture plane), the relationship between the two orientations is different than when it is viewed at 160° (far to the left side of the picture plane). These differences are manifestations of the *differential rotation effect*—the fact that pictured objects oriented more parallel to the picture plane rotate less in response to an observer's

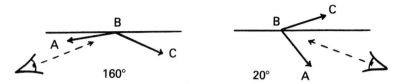

Figure 3. Averaged perceived orientations defined by dowels BA and BC of Figure 1, when viewed at viewing angles of 20° and 160°. The picture plane is indicated by the horizontal line and the observer's position is shown by the schematic eye. Perceived orientations are indicated by the direction of the arrows. Note that for a viewing angle of 20°, the orientation of BC points behind the picture plane. This is a typical result, which has been previously reported (Goldstein, 1979, 1987).

change in viewing angle than do pictured objects that are oriented more perpendicular to the picture plane. (See Goldstein, 1979, 1987, for a more detailed graphical presentation of similar data for a number of viewing angles.)

In my paper I presented evidence that the subject's awareness of the picture plane is one of the causes of these changes in the perceived orientation of different objects relative to one another. Cutting (1988) has offered an alternate explanation—that perceived orientation is controlled by the geometrical changes associated with the affine shear that accompanies changes in viewing angle. His analysis is based on an analysis of the virtual space defined by a picture—that is, the three-dimensional space that corresponds to the picture's geometrical array. Cutting's original analysis was based on a formula developed by Rosinski et al. (1980), but it is also possible to use the graphical method illustrated in the top part of Figure 4 (see Cutting, 1986, p. 36, for an illustration of the geometrical method used to construct this figure) to determine how the picture's virtual space is affected by changes in viewing angle. This figure shows the virtual space defined by the array in the center top of the figure, for viewing angles of 20°, 90°, and 160°.

After determining the virtual space defined by my triangular array at different viewing angles, Cutting used the orientations defined by this space to predict perceived orientations at each viewing angle. The resulting predictions for perceived orientations fit the data well at some viewing angles and not as well at others. Consider, for example, his prediction for a viewing angle of 160°. We can compare the predicted orientations shown at the top right of Figure 4 to those determined empirically by constructing a triangle based on the empirically determined perceived orientations. Such a triangle, calculated from the data in Figure 3 of Goldstein (1987)[1] and shown on the

---

[1] The data on which these triangles are based were collected using a stimulus with the same layout as the stimulus shown in Figure 1, but the photograph of the dowels was taken from a slightly lower angle (see Goldstein, 1987, for a picture of this stimulus).

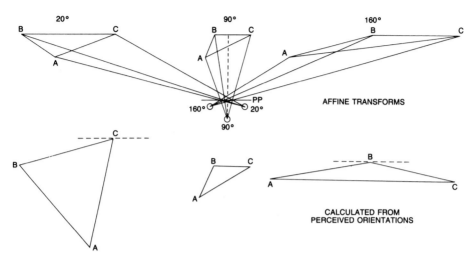

Figure 4. Top: center—layout of the triangular array used in Goldstein (1987); left—affine transformed array for a viewing angle of 20°; right—affine transformed array for a viewing angle of 160°. PP = picture plane; open circles = positions of observers at viewing angles of 20°, 90°, and 160° relative to the PP. The dashed vertical line is the observer's line of sight for the 90° viewing angle.

Bottom: layouts calculated from the empirically determined perceived orientations in Figure 3 of Goldstein (1987). Since perceived orientations do not provide information regarding size, the sizes of these triangles were determined by setting the length of side BC equal to the length of side BC of the corresponding triangle above. The triangles were constructed by drawing each line so its orientation matches its empirically determined perceived orientation relative to the picture plane. The orientation of the picture planes for the lower triangles are indicated by dashed lines for the 20° and 160° viewing angles. The picture plane is omitted for the 90° viewing angle for clarity, since the angle between BC and the picture plane is 2°.

lower right of Figure 4, is oriented slightly differently than Cutting's predicted triangle, but has the same general shape. The fit is not, however, as good for a viewing angle of 20°; at that angle Cutting's predicted orientations for the directions defined by B → C and C → A differ from those determined empirically.

Although these differences between geometrically predicted and empirical results suggest that geometry cannot supply the entire explanation for the changes in perceived orientation that occur with changes in viewing angle, Cutting's model does succeed in predicting the differential rotation effect. Geometry may, therefore, play at least some role in determining perceived orientation, and it is this role I wish to focus on now.

Let's assume for the moment that perceived orientations *are* linked to the changes that occur in virtual space with changes in viewing angle. This possible linkage between changes in virtual space and perceived orientation becomes particularly significant when we consider that *these same changes in virtual space* cause little change in the observer's perception of spatial layout. This constancy of spatial layout occurs not only for changes in viewing angle, as illustrated by the solid triangles in Figure 2, but also for changes in viewing distance, as indicated by comparing the solid and dashed triangles in Figure 2. The solid triangles were produced by subjects viewing the array in Figure 1 from a distance of 8 in, whereas the dashed triangles were produced from a viewing distance of 64 in. Despite this eight-fold difference in distance, which causes a large expansion of virtual space,[2] there are only small differences between the triangles.

What we have here, therefore, is a situation in which large changes in virtual space cause little or no change in the perception of spatial layout, but which, to the extent that the geometrical hypothesis is correct, cause large changes in perceived orientation. This situation raises the possibility that perceived orientation may result directly from stimulus geometry, whereas the perception of spatial layout may involve a processing step to compensate for the geometrical changes caused by viewing at an angle.

This idea of a compensation mechanism is not new. Pirenne (1970), Rosinski et al. (1980) and Kubovy (1986) have linked such mechanisms to the subject's awareness of the picture plane; however, the exact operation of this compensation mechanism has never been specified. The first question that should be asked to help elucidate the nature of this hypothetical mechanism is: what stimulus manipulation will cause a subject's perception of layout to correspond to the picture's virtual space—or, put another way, what stimulus manipulation will eliminate layout constancy?

It is also possible that layout constancy is the outcome, not of a compensation mechanism, but of the subject's attention to information in the picture that remains invariant with changes in virtual space. While it is easy to talk glibly about invariant information, we need to identify this information if, in fact, it exists.

Finally, returning to perceived orientation, the suggestion that this percept may result directly from stimulus geometry cannot be the whole story. It seems clear that the observer's awareness of the angle of view is also important (Goldstein, 1987), although exactly how this factor interacts with stimulus geometry remains to be determined.

Obviously, many questions remain to be answered before we fully understand the mechanisms underlying perceived orientation and perceived spatial layout. These questions are important, not only because they suggest possibilities for future research that could yield answers that will greatly enhance our understanding of picture perception, but also because they acknowledge

---

[2] The use of the graphical method to determine how virtual space is changed by this increase in distance indicates that the expansion of the space caused by changing the viewing distance from 8 to 64 in produces an elongated triangle in which side BA is stretched to four times the length of side BC.

an important fact about picture perception: perceived orientation and perceived spatial layout are affected differently by changes in viewing angle, are probably controlled by different mechanisms, and should, therefore, be clearly distinguished from one another in future research on picture perception.

References

Cutting, J.E. (1986). *Perception with an Eye for Motion.* Cambridge: MIT Press.
Cutting, J.E. (1988). Affine distortions of pictorial space: Some predictions for Goldstein (1987) that LaGournerie (1859) may have made. *Journal of Experimental Psychology: Human Perception Performance*, in press.
Goldstein, E.B. (1979). Rotation of objects in pictures viewed at an angle: Evidence for different properties of two types of pictorial space. *Journal of Experimental Psychology: Human Perception Performance*, **5**, 78–87.
Goldstein, E.B. (1987). Spatial layout, orientation relative to the observer, and perceived projection in pictures viewed at an angle. *Journal of Experimental Psychology: Human Perception Performance*, **13**, 256–266.
Kubovy, M. (1986). *The Psychology of Perspective and Renaissance Art.* Cambridge, England: Cambridge University Press.
Pirenne, M.H. (1970). *Optics, Painting, and Photography.* Cambridge, England: Cambridge University Press.
Rosinski, R.R., Mulholland, T., Degelman, D. and Farber, J. (1980). Picture perception: An analysis of visual compensation. *Perception and Psychophysics*, **28**, 521–526.

32

On the efficacy of cinema, or what the visual system did not evolve to do

James E. Cutting

*Department of Psychology
Cornell University, Ithaca, New York*

My topic concerns spatial displays, and a constraint that they do not place on the use of spatial instruments. Much of the work done in visual perception by psychologists and by computer scientists has concerned displays that show the motion of rigid objects. Typically, if one assumes that objects are rigid, one can then proceed to understand how the constant shape of the object can be perceived (or computed) as it moves through space. Many have assumed that a rigidity principle reigns in perception; that is, the visual system prefers to see things as rigid. There are now ample reasons to believe, however, that a rigidity principle is not always followed. Hochberg (1986), for example, has outlined some of the conditions under which a rigid object ought to be seen, but is not. Some of these concern elaborations of some of the demonstrations that Adelbert Ames provided us more than 35 years ago.

There is another condition of interest with respect to rigidity and motion perception. That is, not only must we know about those situations in which rigidity ought to be perceived, but is not, we also must know about those conditions in which rigidity ought not to be perceived, but is. Here I address one of these conditions, with respect to cinema. But before discussing cinema, I must first consider photography.

When we look at photographs or representational paintings, our eye position is not usually fixed. A puzzle arises from this fact: linear perspective is mathematically correct for only one station point, or point of regard, yet almost any position generally in front of a picture will do for object identity and layout within the picture to appear relatively undisturbed. Preservation of phenomenal identity and shape of objects in slanted pictures is fortunate. Without them the utility of pictures would be vanishingly small. Yet the efficacy of slanted pictures is unpredicted by linear perspective theory.

This puzzle was first treated systematically by La Gournerie in 1859 (see Pirenne, 1970). I call it La Gournerie's paradox; Kubovy (1986) has called it the robustness of perspective. The paradox occurs in two forms: the first

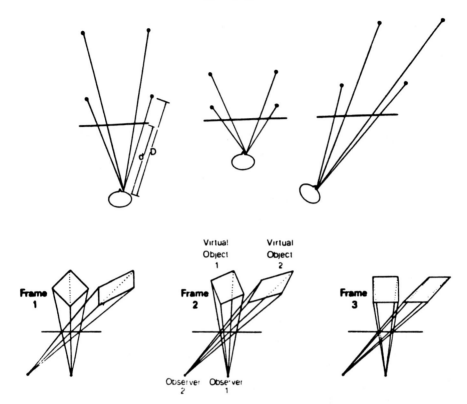

Figure 1. Reconstructive geometry and images. The upper panels show the reconstruction of four pillars in depth. Consider the left-most panel a representation of the real depth relations projected onto the image plane. If that plane is now a photograph, the pillars are fixed in position on the image plane. Thus, when an observer moves toward the plane, depth must be compressed, as in the upper middle panel. When the viewer moves to the side, all pillars slide over by differing amounts. The bottom panels show reconstructions of a moving square across three frames, from two viewpoints. Notice that the reconstruction for Observer 1 is rigid, but that that for Observer 2 is not (from Cutting, 1986a).

concerns viewing pictures either nearer or farther than the proper station point; the second and more dramatic concerns viewing pictures from the side. Both are shown in the top panels of Figure 1.

To consider either distortion one must reconstruct, as La Gournerie did, the geometry of pictured (or virtual space) behind the picture plane. The premise for doing so is that the image plane is unmoving, but invisible, and that observers look through it into pictured space to make sense out of what is depicted. Invisibility is, in many cases, obviously a very strong, if not false, assumption, but it yields interesting results. Possible changes in viewing position are along the z axis, orthogonal to the picture plane, and along the x or y axes, parallel to it. Both generate affine transformations in depth in all xz planes of virtual space. Observer movement along x or y axes also generates perspective transformations of the image, but these will not be considered here (Cutting 1986a, 1986b).

In the upper left panel, four points are projected onto the image plane as might be seen in a large photograph taken with a short lens. When the observer moves closer to the image, as in the upper middle panel, the projected points must stay in the same physical locations in the photo. Thus, the geometry of what lies behind must change. Notice that the distance between front and back pairs of points of this four-point object is compressed, a collapse of depth like that when looking through a telephoto lens. All changes in z axis location of the observer create compression or expansion of the object in virtual space. When an observer moves to the side, as seen in the upper right panel, points in virtual space must shift over, and do so by different amounts. Such shifts are due to affine shear. All viewpoints of a picture yield additive combinations of these two affine effects—compression (or expansion) and shear.

Such effects are compounded when viewing a motion sequence, as shown in the lower panels of Figure 1. In particular, an otherwise rigid object should appear to hinge and become nonrigid over the course of several frames for a viewer seated to the side. Theoretically, the problem this poses for the cinematic viewer is enormous—every viewer in a cineauditorium has an eye position different than the projector and camera position, and thus, by the rules of perspective, no moving object should ever appear rigid. This is, I claim, the fundamental problem of the perception of film and television.

Most explanations for the perception of pictures at a slant are in sympathy with Helmholtz. Pirenne (1970, p. 99), for example, suggested that "an unconscious intuitive process of psychological compensation takes place, which restores the correct view when the picture is looked at from the wrong position". Pirenne's unconscious inference appears to unpack the deformations through some process akin to mental rotation (Shepard and Cooper, 1982). According to this view, the mind detransforms the distortions in pictured space so that things may be seen properly, and although Pirenne didn't discuss film, it might hold equally for film seen from the front row, side aisle. The force of my presentation is to show that this view is not necessary in the perception of slanted cinema. But first consider how this account might proceed.

Pirenne and others have suggested at least three sources of image surface information that might be used to "correct" slanted images—(1) the edges of the screen, which yield a trapezoidal frame of reference; (2) binocular disparities, which grade across the slanted surface; and (3) projection surface information such as texture and specularities. Since I am interested in none of these, I removed them from my displays through a double projection scheme, as shown in Figure 2. If one considers the situation of viewing slanted cinema, one has the real, slanted surface and one can measure a cross

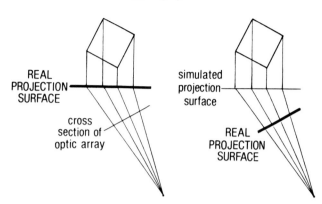

Figure 2. Arrangements of real and simulated projection surfaces that can remove image information from objects projected onto slanted screens (from Cutting, 1987).

section of that optic array from it. This would be an imaginary projection surface. Once considered this way, one can reverse the two, placing the real surface in front of the imaginary, and this is what I did.

In this manner, although the display frame was always rectangular for the observer, the shapes of rotating stimuli were like those seen from the side, with the right-edge elements in each frame longer than the left-edge ones and with the z axis compressed. This simulation yields a perspective transformation of the image screen, and a nonperspective transformation of the stimulus behind it in virtual space. I presented viewers with computer-generated, rotating, rectangular solids. Two factors are relevant to this discussion. (For a more complete analysis see Cutting, 1987.)

First, half the solids presented were rigid, half nonrigid. Nonrigid solids underwent two kinds of transformation during rotation—one affine, compressing and expanding the solid like an accordion along one of its axes orthogonal to the axis of rotation during rotation, and one nonaffine, with a corner of the solid moving through the same excursion. Deformations were sinusoidal and were accomplished within one rotation of the stimulus. It was relatively easy to see the large excursions as making the solid nonrigid; it was more difficult in smaller excursions. This nonaffine deformation was much easier to see than the affine deformation, but there were no interactions involving types of nonrigidity, so here I will collapse across them (see Cutting, 1987, for their separate discussion).

Second, stimuli were presented with cinematic viewpoint varied; in Experiment 1, half were projected as if viewed from the correct station point, half as if seen from the side, with the angle between imaginary and real projection surfaces set at 23°. The latter condition allows investigation of La Gournerie's paradox, and compounds the nonrigid deformations of the stimulus in pictorial space with an additional perspective transformation of the image.

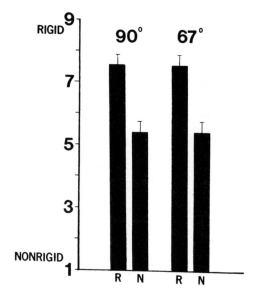

Figure 3. Selected results from Experiment 1. 90° and 67° are the two viewing conditions of interest, where 67° is the simulated screen slant as indicated in Figure 2. R = rigid stimuli, N = nonrigid stimuli.

Viewers looked at many different tokens of all stimuli, and used a bipolar graded scale of rigidity and confidence, from 1 to 9—with 1 indicating high confidence in nonrigidity, 9 high confidence in rigidity, and 5 indicating no confidence either way.

Figure 3 shows the results of the first experiment for rigid and nonrigid stimuli, at both 90° and simulated 67° viewing angles. Two effects are clear. First, rigid stimuli were seen as equally rigid regardless of simulated viewpoint in front of the screen, and second, nonrigid stimuli were seen as equally nonrigid regardless of simulated viewpoint.

The lack of difference in the slanted and unslanted simulated viewing conditions is striking, but it could be due to the fact that the screen slant was relatively slight. Experiment 2, then, introduced a third viewing condition, a steeper angle—45°. A fourth condition was also introduced. Its impetus came from structure-from-motion algorithms in machine vision research. Several people suggested to me that screen slant could be another parameter in rigidity-finding algorithms and that only a few more frames or points might be needed to specify slant. To test for this idea, I introduced a variable screen-slant condition, where the simulated slant of the screen oscillated between 80° and 55°, with a mean of 67°. It seemed highly unlikely that an algorithm could easily solve for both rigidity and a dynamically changing projection surface.

This time stimuli were generated in near-parallel and polar perspective. Again, stimuli could be rigid or nonrigid. Selected results for the nonrigid

The efficacy of cinema

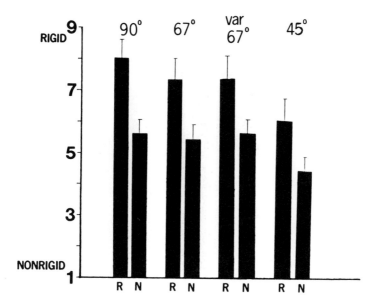

Figure 4. Selected results from Experiment 2. The added conditions are simulated screen slants of 45°, and one of variable slant (between 80° and 55°), averaging 67°. R = rigid stimuli, N = nonrigid stimuli.

stimuli are shown in Figure 4, and show two striking effects. First, the variable 67° screen slant condition was not different from the nonvarying condition, and the lack of difference would seem to be embarrassing for any structure-through-motion approach to the perception of these stimuli that includes screen slant as a variable to be solved for. Second, if simulated screen slant is great enough, all stimuli begin to look nonrigid.

A more interesting result is an interaction concerning near-parallel and polar-projected stimuli, as shown in Figure 5, with the two 67° conditions collapsed, and all rigid and nonrigid trials collapsed. The near-parallel projected stimuli show no difference in perceived rigidity from any angle that they are viewed; the more polar-projected stimuli, on the other hand, show a sharp decrease in perceived nonrigidity as the angle of regard increases.

This latter effect adds substance to other results in the literature. For example, Hagen and Elliott (1976) found what they called a "zoom effect"— the general preference for static stimuli seen is more for parallel than polar projection. Here, in cinematic displays, stimuli that are near-parallel-projected are seen as more rigid from more places in a cineauditorium.

In conclusion, let us be reminded that photographs and cinema are visual displays that are also powerful forms of art. Their efficacy, in part, stems from the fact that, although viewpoint is constrained when composing them, it is not nearly so constrained when viewing them. The reason that viewpoint is relatively unconstrained, I claim, is not that viewers "take into account" the slant of the screen, but that the visual system does not seem to compute

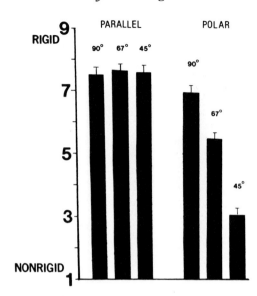

Figure 5. Another description of the results of Experiment 2, parsed according to projection.

the relatively small distortions in the projections, at least for certain stimuli that are projected in a near-parallel fashion.

It is obvious that our visual system did not evolve to watch movies or look at photographs. Thus, what photographs and movies present to us must be allowed in the rule-governed system under which vision evolved. Slanted photographs and cinema present an interesting case where the rules are systematically broken, but broken in a way that is largely inconsequential to vision. Machine-vision algorithms, to be applicable to human vision, should show the same types of tolerances.

But with regard to the use of the camera lens in movies, it becomes quite clear why long lenses—those that are telephoto and nearly telephoto—are so popular and useful. First, and known for nearly a century, standard lenses tend to make people look like they have bulbous noses. Second, and corroborated by my results, long lenses provide a more nearly parallel projection of objects, and the distortions seen in these objects when a viewer looks at a slanted screen are significantly diminished. This enhances their efficacy considerably, despite the fact that it introduces the nonnatural situation of collapsing the apparent depth of a scene.

References

Cutting, J.E. (1986a). *Perception With an Eye for Motion.* Boston: MIT Press/Bradford Books.

Cutting, J.E. (1986b). The shape and psychophysics of cinematic space. *Behavorial Research Methods for Institutions and Computers*, **18**, 551–558.

Cutting, J.E. (1987). Rigidity in cinema seen from the front row, side aisle. *Journal of Experimental Psychology: Human Perception and Performance*, **13**, 323–334.

Hagen, M.A. and Elliott, H.B. (1976). An investigation of the relationship between viewing conditions and preference for true and modified perspective with adults. *Journal of Experimental Psychology: Human Perception and Perfermance*, **4**, 479–490.

Hochberg, J. (1986). Representation of motion and space in video and cinematic displays. In Boff, K.R., Kaufman, L. and Thomas, J.P. (Eds), *Handbook of Perception and Human Performance*. **1**, 22:1–63. New York: John Wiley.

Kubovy, M. (1986). *The Psychology of Linear Perspective and Renaissance Art*. Cambridge, England: Cambridge University Press

Pirenne, M.H. (1970). *Optics, Painting, and Photography*. Cambridge, England: Cambridge University Press.

Shepard, R.N. and Cooper, L.A. (1982). *Mental Images and Their Transformation*. Boston: MIT Press/Bradford Books.

Appendix

Mathematical proof of a formula for predicting perceived slant angle in a picture seen from the side

When one views a picture from the side, the virtual space behind the picture plane undergoes an affine transformation measured in any given xz plane (horizontal plane of the pictorial space), and a perspective transformation of any given yz plane (Cutting, 1986a). Since these two effects are additive, and since we are not concerned with the latter, only the affine effects will be considered here. These concern the predicted direction of perceived slant of an object in pictorial space as measured horizontally against the picture plane.

I. Consider how a picture is composed, as in Figure 6.

Let: O = center of projection of the picture (station point or composition point).
  α = right angle, by definition (except in anamorphic art).
  d = distance from the picture plane to the station point, O. Thus, d is the principal ray.
  x = the point of intersection of the principal ray with the picture plane.
  p = distance from the picture plane to the object that is slanted in the picture (or to a line of slant extended from that object). Thus, p is the continuation of the principal ray to the slanted object (or its line of continuation; and $d + p$ is the reconstructed distance from the station point to the slanted object (or its continuation).
  σ = true slant angle of the pictured object, as measured against the picture plane.

II. Consider a new viewpoint, as in Figure 6.

Let: O' = new viewpoint of the picture.
  σ' = nonright angle between the picture plane and the new viewpoint at x.
  d' = angled distance of new viewpoint to x.

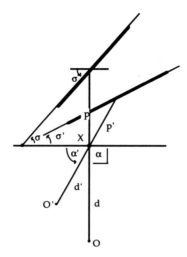

Figure 6. Geometry of a picture viewed along an oblique.

III. To derive:

σ' = predicted angle of perceived slant.

Step of Proof

(1) Since the pictorial space is affine, changes along the length of a line are proportional. Thus:

$$d/d' = p/p'$$

For further calculations we need p', thus

$$p' = d'p/d$$

(2) Since the pictorial space is affine, and since nothing in the picture plane moves, the line of the slanted object extended to the picture plane (or an extension of it) remains in place. Thus, we need to know, n, the distance from x to the point where the extension of the slanted object meets the picture plane.

$$n = p/\tan \sigma$$

(3) Since we now know n, p', and α', we can solve for σ'.

$$\sigma' = a \tan [(p' \sin \alpha')/(n + p' \cos \alpha')]$$

or, going back to the terms defined originally:

$$\sigma' = a \tan [((d'p/d)\sin \alpha')/((p/\tan \sigma) + (d'p/d)\cos \alpha')]$$

rearranging and simplifying the equation yields:

$$\sigma' = a \tan [(\sin \alpha' \cdot (d'/d))/((1/\tan \sigma) + (\cos \alpha'(d'/d))]$$

Thus, the predicted angle of perceived slant from affine geometry is a function of four variables (d', d, α', and σ) known about the original spatial relations in the composition of the picture.

33

Visual slant underestimation

John A. Perrone

NASA Ames Research Center
Moffett Field, California

and

Peter Wenderoth

University of Sydney
Sydney, Australia

Summary

Observers frequently underestimate the in-depth slant of rectangles under reduction conditions. This also occurs for slanted rectangles depicted on a flat display medium. Perrone (1982) provides a model for judged slant based upon properties of the two-dimensional trapezoidal projection of the rectangle. Two important parameters of this model are the angle of convergence of the sides of the trapezoid and the projected length of the trapezoid. We tested this model using a range of stimulus rectangles and found that the model failed to predict some of the major trends in the data. However, when the projected width of the base of the trapezoidal projection was used in the model, instead of the projected length, excellent agreement between the theoretical and obtained slant judgements resulted. The good fit between the experimental data and the new model predictions indicates that perceived slant estimates are highly correlated with specifiable features in the stimulus display.

Introduction

Attempts at depicting surfaces slanted in depth on a flat display medium are often hampered by a common perceptual illusion which results in underestimation of the true depth. Surfaces appear to lie closer to the fronto-parallel plane than the perspective projection dictates. This has been a common finding in a wide range of experiments involving slant perception, starting with Gibson's study (1950) on texture gradients (e.g., Clark et al., 1955; Gruber and Clark, 1956; Smith, 1956; Flock, 1965; Freeman, 1965; Braunstein, 1968; Wenderoth, 1970).

The mode of viewing slanted surfaces under the conditions used in slant perception experiments differs from the way we normally encounter visual slant in our environment (Perrone, 1980). Cutting and Millard (1984) also questioned the use of slant as a variable in the understanding of surface perception. However, slant underestimation remains an interesting phenomenon because the information is present in the stimulus display for the veridical perception of slant (Perrone, 1982), yet apparently the human visual system does not use that information correctly.

Theories attempting to explain the underestimation are rare. Gogel (1965) applied his "equidistance tendency" theory to slant underestimation effects and Lumsden (1980) speculated that truncation of the visual field by the use of an aperture may be a factor causing underestimation.

Perrone (1980, 1982) has proposed several models of slant perception which attempt to account for the slant underestimation. This paper tests and modifies one of these models. Our aim is to pinpoint the stimulus features used by observers when making visual slant estimates. This would provide useful insights into areas such as spatial orientation, picture perception, and pilot night-landing errors (Perrone, 1984).

Model of slant underestimation

The slant angle θ, is obtainable from the two-dimensional projection of the surface onto the retina. (For a technique using perspective lines, see Freeman, 1966; Perrone, 1982.)

The slant angle is found from the two-dimensional variables given in Figure 1 using:

$$\theta = \tan^{-1} (\tan \pi/X)f \qquad (1)$$

This equation states that the slant angle, θ, can be derived from the angle of convergence (π) of the perspective line in the projection, and the distance, X, from the center of the projection out to the perspective line. In Equation 1, f is a known constant and it is the arbitrary distance from the eye to the theoretical projection plane used to analyse the array of light reaching the eye.

The convergence angle of perspective lines, π, can give the slant angle θ as long as the correct distance X is used. Using a value of X greater than the true value will result in a calculated slant angle less than the actual slant angle, i.e., slant underestimation. Perrone (1980, 1982) proposed a model which suggested that deviation of the perceived straight-ahead direction results in a judgment of slant based on an incorrect value of X.

Two versions of the model have been proposed:

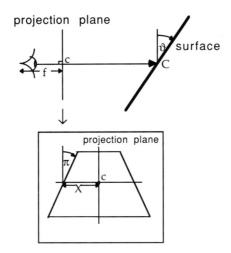

Figure 1. The two-dimensional information reaching the eye is analysed on a theoretical projection plane an arbitrary distance f from the eye. All measurements on the projection plane are made within the plane of the page.

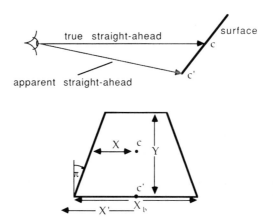

Figure 2. Deviation of the perceived straight-ahead results in the analysis being carried out about c' instead of c. Model A states that the length X' (equal to Y) is used instead of X. Model B proposes that X_b is used instead of X.

Model A. Perrone (1982) suggested that because of the reduced viewing conditions and because of the unusual form of the presenting slant, the observer's perceived straight-ahead direction deviates from the true straight-ahead (Figure 2) and that the visual system uses the length X' (equal to the projected length Y) instead of X.

It is proposed that the visual system is attempting to measure the change in width over a square area of the projection plane, determined by Y, but because there are no perspective lines a distance X' out from c', the outside edge of the rectangle is used instead. When X' is substituted into Equation 1 instead of X, the equation for perceived slant becomes $\beta = \tan^{-1}(\tan \pi/X')f$. However, in order to use this equation for predicting perceived slant, we need to replace the two-dimensional variables (π and X') with the three-dimensional parameters of the stimulus situation. This gives the following equation for perceived slant:

$$\beta = \tan^{-1}\left[\frac{W \sin \theta (D^2 - L^2 \sin^2 \theta)}{4 L D^2 \cos^2 \theta}\right] \qquad (2)$$

θ = actual slant
W = actual width of rectangle
L = half the total length of rectangle
D = distance from eye to center of rotation

To date, Perrone (1982) has shown how this sort of analysis provides acceptable fits to data collected by others (e.g., Clark et al., 1955; Smith, 1956), but these studies were designed to investigate other aspects of slant perception and so did not involve direct manipulation of the variables integral to the model.

One problem with this version of the model is that it predicts that slant *overestimation* will occur when the projected height of the test rectangle (Y) becomes less than the projected half-width at the axis of rotation (X). However, there have been no published accounts of slant overestimation occurring, but this may simply be because nobody has used test rectangles with the appropriate length-to-width ratio.

Model B. (Modified version of Model A). This version proposes that the total base width of the rectangle (X_b) is used in the evaluation of the slant angle instead of X. This new form of the model can be interpreted as saying that the observers are basing their slant estimates on the convergence angle, π, of perspective lines which they believe to be twice the true distance out from the center. It may be that it is a difficult and unnatural task for the observer to judge the slant of a surface which is centered on the median plane of the eye. It is easier if we have a side view or at least a more oblique view of the slanted surface. The observers may resort to making their judgments on the basis that they have a more extreme or displaced viewpoint than is in fact the case. Their interpretation of the slant of the rectangle may be based on an assumed view of the rectangle which is displaced or rotated relative to its true position.

When this error is combined with the proposed deviation of the perceived straight-ahead (Perrone, 1982), the result may be the erroneous use of the total base width of the projected trapezoid rather than the correct half-width at the axis of rotation. When the total projected base width of a slanted rectangle is used to estimate θ from Equation 1, the predicted perceived slant angle is found using:

$$\beta = \tan^{-1}\left[\frac{\tan\theta\,(D - L\sin\theta)}{2D}\right] \quad (3)$$

θ = actual slant
L = half the total length of rectangle
D = distance from eye to center of rotation

Testing the model

An experiment was designed to verify which of the two cases (Equation 2 or Equation 3) best models the data from human observers in the slant perception task. If it can be established that specific features of the stimulus display are being used in the slant estimation process, then the more difficult task of discovering why these particular variables are being used can be attempted. The model provides a means of narrowing down the choice of possible variables and the combinations in which they are used.

Experiment

The stimuli were computer-generated two-dimensional perspective representations of rectangular outline figures, presented on a CRT and viewed monocularly through an aperture. These figures represented rectangles measuring 25 cm wide with the following lengths: 50 cm (condition 1), 25 cm (condition 2), and 15 cm (condition 3). These were depicted to be at a distance of 57 cm from the subject's eye and slanted backwards away from the observer by varying angles of slant. The actual slant angles used were 20°, 40°, 60° and 80° measured from the vertical.

The subject reproduced the judged slant of the rectangle on a response device which was located 90° to the right and positioned at eye level. The response device consisted of a thin black line inscribed on a clear plexiglass strip which was mounted on a circular white metal disk 23 cm in diameter. Vertical and horizontal black lines were drawn on the disk to provide anchor points (Wenderoth, 1970). Subjects were ten paid volunteers, naive as to the aims of the experiment.

Predictions

If Model A is correct, then the slant estimates for the three different conditions should lie along three distinct curves given in Figure 3a. For some of the stimulus conditions, the subjects should judge the rectangle to be slanted farther back from the fronto-parallel plane than the true position (slant *overestimation*). This corresponds to any region of the curves which lies above the dotted line in Figure 3a. If a Model B is correct, the slant estimates for all three conditions should all lie on approximately the same curve of the shape shown in Figure 3b. No slant overestimation should occur.

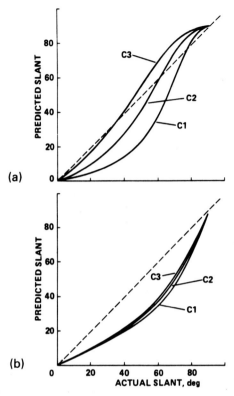

Figure 3. Plots showing (a) predictions from Model A for each of the three experimental conditions and (b) predicted slant versus actual slant for Model B. No slant overestimation is predicted to occur.

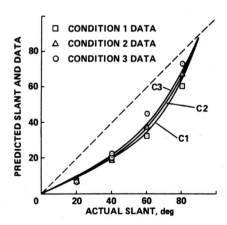

Figure 4. Data are plotted from conditions 1, 2, and 3 along with predictions from Model B. Error bars have been omitted for clarity, but the largest standard error was 4.5° for the 80° slant angle.

Results

The data from the ten subjects have been plotted in Figure 4 along with the predictions from Model B. For the case in which a tall narrow rectangle was used (condition 1), the results are similar to those obtained in past slant perception experiments which used rectangles with a length-to-width ratio greater than one (e.g., Smith, 1956). For this condition, both Model A and B give reasonable predictions for the smaller test angles (see C1 predictions in Figure 3a). However, for the remaining conditions, the data depart greatly from the Model A predictions and none of the predicted overestimation of slant occurred.

The mean absolute error between the Model A predictions and the data over the three conditions was 13.9° (sd = 8.1). For Model B, on the other hand, the mean absolute error was only 2.6° (sd = 1.9). The mean absolute errors from Model A are significantly greater than those from Model B, ($t = 4.5$, $p < 0.05$, $df = 22$) and represent a worse fit between the model predictions and data.

Conclusions

Slant underestimation Model A (Perrone, 1982) incorrectly predicts overestimation to occur for rectangles which have a projected length less than half of the base width. In fact, the influence of the projected length of the rectangle on slant judgments is minimal. However, Model B provides an excellent fit between the experimental data and the predictions. These predictions are based on measurable features of the experimental configuration. There are no free parameters. Model B states that the total projected base width of the rectangle is used instead of half the projected width at the axis of rotation. Two parameters of the two-dimensional projection are important in the slant estimation process: (1) the angle of convergence of perspective lines and (2) the distance of the perspective lines from the center of the projection. The success of Model B suggests the human observers make errors in slant estimates because they misperceive this second parameter.

The question remains as to why human observers use "incorrect" features of the stimulus in their assessment of the slant angle. It has been shown that the correct slant angle is obtainable from the appropriate use of the variables given in Equation 1. These variables are known to be present in the two-dimensional stimulus reaching the observer's eye. The experimental data are consistent with the proposal that the total base width of the trapezoidal projection is used instead of half the projected width at the axis of rotation. However, it does not shed any light as to why this should be the case.

Further research is required before we can conclude the actual mechanisms used by the human visual system in making slant estimates. In the meantime, sufficient evidence exists to conclude that slant judgments by an observer are highly correlated with specific measurable features in the two-dimensional array of light reaching the observer's eye. The slant estimates exhibit a large amount of error and often greatly underestimate the true slant angle. This paper shows that such errors cannot be attributed to the fact that insufficient information exists in the stimulus for veridical slant judgments. The information is available, but is incorrectly used.

References

Braunstein, M.L. (1968). Motion and texture as sources of slant information. *Journal of Experimental Psychology*, **8**, 584–590.
Clark, W.C., Smith, A.H. and Rabe, A. (1955). Retinal gradients of outline as a stimulus for slant. *Canadian Journal of Psychology*, **9**, 247–253.
Cutting, J.C. and Millard, R.T. (1984). Three gradients and the perception of flat and curved surfaces. *Journal of Experimental Psychology: General*, **113**, 198–216.
Flock, H.R. (1965). Optical texture and linear perspective as stimuli for slant perception. *Psychology Review*, **72**, 505–514.
Freeman, R.B., Jr. (1965). Ecological optics and visual slant. *Psychology Review*, **72**, 501–504.
Freeman, R.B., Jr. (1966). Function of cues in the perceptual learning of visual slant. *Psychology Monographs*: Gen. Appl., **80** (2) whole no. 610.
Gibson, J.J. (1950). The perception of surfaces. *American Journal of Psychology*, **63**, 367–384.
Gogel, W.C. (1965). Equidistance tendency and its consequences. *Psychology Bulletin*, **64**, 153–163.
Gruber, H.E. and Clark, W.C. (1956). Perception of slanted surfaces. *Perception and Motor Skill*, **16**, 97–106.
Lumsden, E.A. (1980). Problems of magnification and minification: An explanation of the distortions of distance, slant, shape and velocity. In Hagen, M.A. (Ed.) *The Perception of Pictures*, **1**, 91–134. New York: Academic Press.
Perrone, J.A. (1980). Slant underestimation: A model based on the size of the viewing aperture. *Perception*, **9**, 285–302.
Perrone, J.A. (1982). Visual slant underestimation: A general model. *Perception*, **11**, 641–654.
Perrone, J.A. (1984). Visual slant misperception and the "black-hole" landing situation. Aviation space and environment. *Medicine*, **55**, 1020–5.
Smith, A.H. (1956). Gradients of outline convergence and distortion as stimuli for slant. *Canadian Journal of Psychology*, **10**, 211–218.
Wenderoth, P.M. (1970). A visual spatial after-effect of surface slant. *American Journal of Pyschology*, **83**, 576–590.

34

Direction judgement error in computer generated displays and actual scenes[1]

Stephen R. Ellis, Stephen Smith, Arthur Grunwald and Michael W. McGreevy

NASA Ames Research Center
Moffett Field, CA 94035

Introduction

Shape constancy

One of the most remarkable perceptual properties of common experience is that the perceived shapes of known objects are constant despite movements about them which transform their projections on our retina. This perceptual ability is one aspect of shape constancy (Thouless, 1931; Metzger, 1953; Borresen and Lichte, 1962). It requires that the viewer be able to correct for his relative position and orientation with respect to a viewed object. This discounting of relative position may be derived directly from the ranging information provided by stereopsis, motion parallax, vestibularly sensed rotation and translation, or corollary information associated with voluntary movement. Some correction may even be possible directly based on purely gibsonian higher order psychophysical variables.

Significantly, shape constancy, which usually involves requesting that the viewer make some estimate of the geometric properties of an object, does not disappear during static, monocular viewing. Its basis under these conditions must be different since sensed motion is not involved. In a static image shape constancy amounts to the recognition that each of a variety of views of the objects in the scene are all views of the same objects. This perceived constancy may be based on consciously or unconsciously accessed information concerning alternative views of the objects. These "memories", however, need not be of complete objects since perceived constancy may be based on

---

[1] Preliminary versions of the results included in this paper have been reported at the *NASA Ames—U.C. Berkeley Conference on Spatial Displays and Spatial Instruments*, Asilomar, California, September, 1987.

recall of only some salient features, such as parallelism of significant planes of the object.

In situations in which information directly providing range and orientation is absent, as during viewing realistic pictures, the viewer's relative position with respect to an object can only be indirectly inferred from the projection of the object itself and its surround. But the information in the projected lines-of-sight in the optic array can be used to infer the relative position of the viewer with respect to the pictured objects only if the viewer has at least a partial internal 3D model of the viewed objects and their surround (Grunwald and Ellis, 1986; Grunwald et al., 1988; Wallach, 1985). Thus, "shape constancy" in static, monocular scenes is somewhat circular since the necessary shape information required to infer relative viewing position is itself the shape of the object in question. Nevertheless, shape constancy can be obtained through an interactive process if the viewer has a variety of static views of the same scene or object from different viewing positions and is able to construct correct hypotheses regarding the shapes. Due to inherent regularities in the world, viewers are usually quite good at forming appropriate shape hypotheses in natural environments (Gregory, 1966). But they can be tricked (Ittelson, 1952; Hochberg, 1987).

Position constancy

Shape constancy may be generalized to constancy of interrelations among objects in a spatial layout. Just as the shape of an object ordinarily appears constant when a viewer moves with respect to it, so too do the spatial interrelations among objects generally appear constant during corresponding movement of a viewer (Pirenne, 1970; Wallach, 1985; but also see Ellis et al., 1987; Goldstein 1987). Piaget's decentering task which requires that one imagine how a scene would appear from an external view point is an experimental scenario that particularly exercises this type of constancy (Piaget, 1932; Flavell, 1963).

The Piaget decentering judgement is formally similar to that required of someone using a map to establish his orientation with respect to some exocentric landmark. When based on a map in which there is a marker representing the viewer's position, i.e., the "you-are-here" marker (Levine, 1984) this judgement constitutes an exocentric direction judgement (Howard, 1982). In recent experiments we have examined a specific instance of this judgement by presenting subjects with computer-generated, perspective views of three dimensional maps that have two small, marker cubes on them (see Figure 1). One marker represented the subject's assumed position on the map, i.e., his reference position. The other represented a target position. The subject's task was to make an exocentric direction judgement and estimate the relative azimuth of the target direction with respect to a reference direction parallel to one axis of the ground reference. In the previous experiments this reference was typically a full grid of two sets of orthogonal parallel lines.

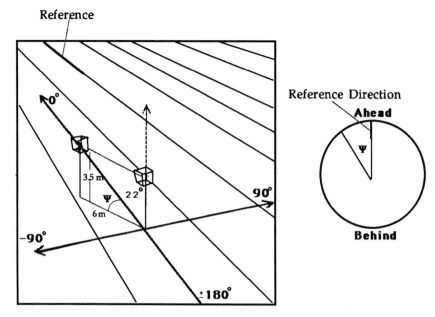

Figure 1. Schematic illustration of the direction judgement task. The subject adjusted the angle Ψ shown on the dial at the right until it appeared equal to the azimuth angle Ψ of the target cube. Dotted lines, labels and arrows did not appear on the display.

Position constancy during judgement of exocentric direction

Interpretations of recent systematic measurements of these exocentric judgements have suggested that the observed patterns of error could be analytically described in terms of an external world coordinate system rather than a viewing coordinate system centered and aligned with the view direction (McGreevy and Ellis, 1986; McGreevy et al., 1985). In these experiments in which scenes were viewed from the center of projection direction, errors were observed in which the subjects exhibited a kind of equidistance tendency in that they judged the target cubes to be closer to the axis crossing the reference axis than they actually were. The same bias appeared independent of viewing direction and thus the patterns of direction judgement error exhibited a kind of position constancy; that is, the errors were functions of the physical positions of the targets and not the subject's view of them.

Geometric mechanisms

Since the subjects were not allowed freedom to move the display's eye point during the individual judgements, position constancy would have to be based on assumed properties of the objects and features of the scene. The most likely feature that could provide the basis for this constancy is the ground reference grid. Since the subjects may correctly assume that the grid axes are orthogonal and in the same plane, the grid can provide information about the compressive and expansive perspective effects of the viewing parameters and allow the viewer to determine them.

Alternative sources of the same information could be sought in the convergence angle between parallel lines at the horizon. Since the horizon was generally not visible due to placement of the clipping planes, this angle would be hard to determine. Thus, the information sufficient to determine the view direction pitch, Θ, and yaw, Ψ, is provided most directly in the projected angle between the reference axis and the crossing axis. For objects viewed along the principal direction of the projection this angle, A'O'B' in Figure 8, may be expressed as:

$$\angle A'O'B' = \sin(2\Psi)\tfrac{1}{2}(\sin^2 \Theta - 1)((\cos^2 \Psi + \sin^2 \Theta \sin^2 \Psi) \\ (\sin^2 \Psi + \sin^2 \Theta \cos^2 \Psi))^{-1/2} \qquad (1)$$

This projected angle is invariant with distance and magnification distortion and also approximately describes the projected angles for right angles near the principal direction of the projection and in the ground reference plane (see Appendix for development of this equation; for related work see Attnaeve and Frost, 1969; Ellis et al., 1987). When the pitch down is greater than about -70 deg, the right most term is weakly modulated by changes in yaw and approximately is equal to 1 so that the resulting equation is simplified to: $\tfrac{1}{2}(\sin \Theta - 1)\sin (2\Psi)$. In fact, this simplified expression is the dominant factor for most of the range of Θ and Ψ and allows the geometry to provide a relatively fixed association between the grid crossing angle, assumed to be 90 deg, and its projection. The interpretation of the projected angle only breaks down for very shallow depression angles when the projection can become indeterminate due to small amounts of measurement noise. Thus, measurement of the projected angle can provide a means to infer and possibly discount view direction.

Experimental manipulation

Deletion of the crossing axis should remove this information that directly allows the viewer to correct for the geometric consequences of his particular viewing direction. Thus, the direction errors from a display used for the same kind of exocentric direction judgements but lacking the crossing axis should exhibit weakened position constancy. With such a display direction judgement errors should depend upon the viewing direction since the principal source of information that allowed the subject directly to determine the direction of the viewing vector has been removed. Experiment 1 examines this conjecture.

Experiment 1

Methods

Subjects

Eight paid subjects participated in the experiment, six of whom were aircraft pilots. All were selected from the Ames Research Center subject pool of aircraft pilots and non-pilots and had normal or corrected to normal visual acuity.

Apparatus and stimuli

The images presented to the subjects showed a spatial layout made from a ground plane reference and two slowly and irregularly tumbling wire-frame cubes (< 1 rpm) used to mark positions on the reference plane. One marked the reference position at the center and the other was the target. The viewing and display parameters of the geometric projection were made identical to those used in previous analytical and experimental studies (McGreevy and Ellis, 1986; Grunwald and Ellis, 1986; Grunwald et al., 1988). Most notably the reference cube was centered in the frustum of vision and subtended an average of about 5 deg of visual angle. The entire image was 28 × 28 cm and viewed from a 48 cm viewing distance. It was generated with a Silicon Graphics IRIS 2400 color raster graphics workstation controlled by mouse and keyboard input.

Figure 1 provides a schematic image of the stimuli illustrating the ground reference made only of randomized parallel line segments which was used to remove cues provided by compressive and expansive projection effects evident in the angles between orthogonal grid axes. The constrained randomization of the placement of the line segments and the slow tumbling of the cube markers were features intended to defeat specific object-based judgement strategies which might favour particular target positions. The subjects were thus encouraged to make a subjective estimate of the spatial layout.

Viewing stimulus geometry and procedure

The ground reference of parallel lines aligned with the reference direction was constructed with randomized modeled spacing at an average of 5 m and a modeled viewing distance of 28 m to the reference cube. To assure presentation of the correct lines of sight, the subject's eye was located at the center of projection of the image 48 cm from the screen. Two symmetrically placed viewpoint locations rotated clockwise and counterclockwise 22 deg with respect to a reference direction were used. Hereafter these viewing directions are referred to as "left" station and "right" station, respectively. Both had a viewing pitch of −22 deg, i.e., pitch down. The target cubes were randomly placed at each of 72 target azimuths with respect to the reference direction ranging between −177 deg (ccw) and +178 deg (cw) in 5 deg increments.

Each experimental series contained one set of 72 azimuth angles for the left and one set for the right station. The viewing station (left or right) and target azimuth were picked randomly without replacement from the series. Each subject performed two series of 144 trials each, or two repetitions per azimuth angle, per station in a 2 × 72 factorial design with repeated measures.

The subject was instructed to show his estimates of the target cube azimuth angle with respect to the reference direction by adjusting a dial drawn on the CRT to the right of the display. The 10 cm diameter dial, which was drawn electronically adjacent to the perspective viewport, was provided with a vertical red line corresponding to a red line in the perspective display that indicated the reference direction. Mouse buttons were used to rotate a pointer on the dial clockwise and counter clockwise for a method of adjustment. The dial adjustment required that the subject judge the target azimuth through a subjective compensation involving an inverse perspective transformation. The subjects were instructed to produce an angle that would be needed if they had to correlate the display information with a 2D map of the layout. To insure that all subjects understood the task, the frame of reference for the judgement and the meaning of the dial adjustment subjects were shown a demonstration program in which dial position and the azimuth of a target cube were both slewed to the y axis of the mouse.

Although no time limit was set for the response to each trial, the subjects were told not to take more than about 30 seconds per judgement. Azimuth direction error was calculated as estimated azimuth target minus true azimuth so that clockwise errors were positive.

Results

The subjects' estimates of depicted target azimuth were subjected to an analysis of variances, with repeated measures on subjects. The analysis showed a statistically significant interaction between viewing station and true azimuth, (F = 2.413, df = 71,497, p < 0.001), hence the azimuth error curves of left and right station appear to depend upon viewpoint.

Figure 2 shows the error in the azimuth angle, averaged over all eight subjects, for the left and for the right stations plotted on circular graphs in which each arc length corresponds to both the magnitude and direction of an error. The across subject means are good summaries of the data since the associated standard errors were only 1–4 deg. For both stations a systematic relationship between the azimuth error and the true azimuth angle, is clearly recognized. The errors are virtually zero on the reference axis. Secondary zeros are not exactly where an actual grid crossing axis would be but are shifted. Those zero crossings on the side in the direction of the view vector

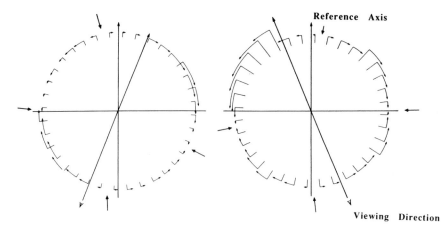

Figure 2. Circular plot of mean azimuth error in Experiment 1. The eye symbols show the subject's view directions with respect to the reference and crossing axes. The length of each directed arc corresponds to the mean error ($N = 8$) in target azimuth at the position marked by the tail of the arc. Reversals of arc directions show target azimuths where azimuth errors were at local minimum.

rotation tend to be rotated towards a position perpendicular to the view direction. As in previous experiments, the largest direction errors are near ±45 and ±135 deg azimuth.

In order to investigate the existence of symmetry about the reference axis, the right station data in the set were reflected and replotted in ordinary cartesian form. As may be seen from Figure 3, the reflection largely superimposes the error data from both view stations confirming the expected symmetry in the error pattern. This observation of symmetrical response patterns provides a control for dial-specific response biases.[2] Had these been a dominant effect, the distinctive features of the error in the data from the left viewpoint would not have symmetrical counterparts in the data from the right viewpoint.

Discussion

The generally symmetrical pattern of mean error clearly shows a dependency on view direction and demonstrates a breakdown of position constancy in the error pattern. This result confirms the initial hypothesis that removal of the crossing axis should break down this constancy. The breakdown is particularly evident near ±90 target azimuths since these are generally not minimums near zero as they were for left and right viewing directions in previous

---

[2] Control calibration experiments with the adjustment dial have shown that the error in across subject means range +/−2 deg with less than a 2 deg clockwise bias and with a pattern uncorrelated with observations in experimental results described in this paper.

experiments with fully grided ground references (McGreevy and Ellis, 1986; Grunwald and Ellis, 1986).

Alternative hypotheses

The breakdown of position constancy would be consistent with an alternative hypothesis which arises from previous analyses of errors in estimation of depicted directions in pictures (Ellis et al., 1987; Gogel and Da Silva, 1987; Grunwald and Ellis, 1986) and which raises the classical question of the extent to which perception of an object's true geometric properties can be made to depend upon its projected retinal image (Thouless, 1931; Beck and Gibson, 1955; Gilensky, 1955; Gogel and Da Silva, 1987). According to this hypothesis, errors in judged direction in pictures are modeled as functions of the interrelations of actual lines of sight to contours and vertices of viewed objects. For viewing situations in which pictures are viewed from the geometric center of projection, this analysis may be restricted to hypothesizing that the error, e, in estimated target azimuth is proportional to the difference between the depicted and projected azimuth angles Ψ and Ψ' respectively, i.e., $e = k(\Psi' - \Psi)$. This formulation makes clear that not only should viewing direction affect the pattern of direction estimation but also that symmetrically placed viewpoints should produce the observed symmetrical patterns of direction errors.

In fact, though the actual error data does exhibit symmetry, it departs in significant ways from that expected based on this hypothesis. For example, the hypothesis implies that all direction errors for a view from the left station should be clockwise (see Figure 4). The actual error data corresponding to this condition are both clockwise and counter-clockwise as shown by the circular plots of the error data. This projected angle model could be improved, as previously suggested (McGreevy and Ellis, 1986), by introducing a 22 deg shift in the assumed view direction to align it with the reference axis. This kind of assumption causes a vertical shift in the theoretical function in Figure 4 which can bring it into better correspondence with the data (McGreevy and Ellis, 1986; McGreevy et al., 1985). This shift is equivalent to asserting that the subject is responding to a potential projection rather than the one he actually sees and amounts to modeling position constancy. Since the data show evidence of symmetrical viewpoint dependence, the use of a theoretical function that models a viewpoint-independent, position constancy seems inappropriate. Accordingly, alternative theoretical explanations may be sought.

Binocular conflict

One possible influence on the direction judgements that the subjects were requested to make is the binocular stimulus which they viewed. This stimulus was essentially the picture surface which provided fixed accommodative

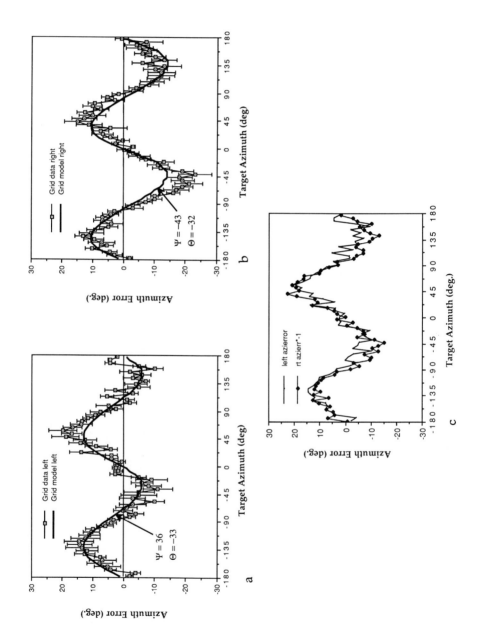

and vergence demands as well as disparity and motion parallax cues to its physical distance, since no head restraint was used. These cues tell the viewer that all objects are at an approximately equal egocentric distance, i.e., on the picture surface. Thus, if exocentric direction were to be based solely on egocentric ranges estimated from the binocular information, all targets would be at the same distance as would be the case if the pitch of view vector were overestimated to be −90 deg. In the reference system used, all targets would appear at azimuth positions perpendicular to the view direction e.g., for a left view station they would appear either at 68 or −112 deg. Some evidence for this is found in Figure 2 which shows for both view stations that the apparent azimuth of targets located on the side of the direction of view rotation is rotated towards a plane orthogonal to the view direction.

The binocular information possibly causing this apparent rotation is at odds with the monocular information that is drawn on the display, e.g., the decreasing projected size of the cube as its depicted distance increases. The viewer is in a sense being presented with two simultaneous but conflicting stimuli: one binocular and the other monocular. One may suppose that the resulting perception is a combination of the two. Conflicts of this type have been studied in classical experiments (Beck and Gibson, 1955; Gogel, 1977) in which monocular and binocular stimuli are superimposed and viewed. Significantly, the finding has been that for some simple stimuli, the binocular depth sensation spreads to determine the apparent position of a visually proximate, monocularly viewed component of the visual field.

Accordingly, it is reasonable to suspect a similar process could influence the judgements in this experiment. In this case the binocular information in the picture surface causes the apparent positions of all targets to be attracted to a plane normal to the view direction and induce an overestimate of the view vector pitch. This process provides a hypothetical mechanism for the equidistance tendency observed in the first experiment. Its effects could be expected to be dominating were it not for the opposing influence of the numerous monocular depth cues provided by familiar shapes in the image. Since the monocular cues are well developed from familiar shapes in these images, the binocular cues would not be expected to determine totally the

---

◀ *Figure 3. Cartesian plots of mean azimuth errors in Experiment 1 are plotted for both left (a) and right (b) viewing stations. Error bars are ±1 standard error, N = 8. The heavy traces are theoretical functions derived from the assumption that the subjects misjudge the view vector. The estimated viewing parameters for the theoretical trace are at the tail of the arrow and show the expected overestimation of the true values ($\Psi = \pm 22$, $\theta = -22$). These functions have been fitted to the data as described in the discussion of Experiment 2 and correlate fairly well with the data (left station: $r = 0.896$, $p < 0.001$, $df = 72$; right station: $r = -0.925$, $p < 0.001$, $df = 70$). Part (c) illustrates the symmetry between the pattern of error from the left and right view stations by reflecting the data from the right station and replotting it with that from the left station.*

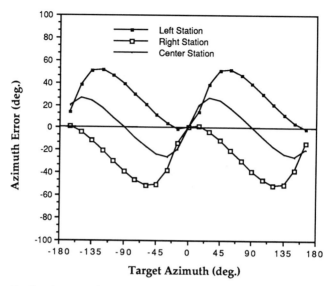

Figure 4. Predicted azimuth errors. If the subject's direction errors were entirely determined by the difference between the true depicted value of a target's azimuth angle and its projection, errors like those shown in this figure would be expected. The three traces show the expected error pattern can be vertically shifted if the depicted targets are assumed to be observed from a left (22.5 deg), right (−22.5 deg), or center (0 deg) viewing station.

apparent distances to the objects in the images. Furthermore, the rotated viewing direction would introduce asymmetries into these monocular features of the image such as the texture gradient that could be expected to introduce corresponding asymmetries into their interaction with binocular cues to distance. Examples of this kind of feature in the direction judgement data are the mismatches between the theoretical functions and those data that occur at symmetrical target positions for data from the left and right view stations (see Figure 3).

The overestimation of pitch discussed above is a form of classical error called "slant overestimation" (Sedgwick, 1986)[3] and may provide a mechanism for the incorrect estimate or use of the viewing parameters. Figure 5 shows a family of theoretical azimuth error curves for different overestimates of the viewing vector pitch together with the data from Experiment 1. These curves are constructed on the assumption that the viewer correctly measures the line of sight angles to all contours and vertices but makes an error in the

---

[3] Interestingly, the hypothesis that azimuth error could be influenced by the difference between depicted target angle and its projection, which was described in the discussion of Experiment 1, really is a special case of this kind of slant overestimation. It is equivalent to asserting that the overestimation is equal to the complement of the actual pitch angle.

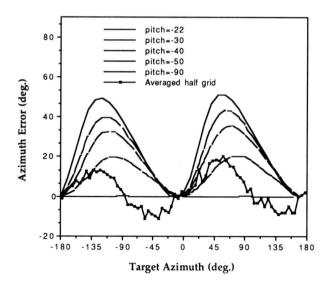

Figure 5. Plot of expected azimuth error if a subject misjudged the depression angle of the viewing direction. Errors are calculated for a left viewing station (azimuth = 22.5 deg) with a true depression angle of −22.5 deg assuming that the subject misjudged the depression by the parameter of each of the curves. Average error data from Experiment 1 are also plotted for comparison. These data are the average of the left station with those from the right station which have first been reflected to correct for symmetry differences.

interpretation of the projected target angle by in a sense looking up its 3D characteristics in the wrong table. For example, the trace labeled "pitch = −40" shows the expected azimuth errors from a subject who when looking at a scene from a left viewing station ($\Psi = 22.5°$, $\theta = -22.5°$) incorrectly assumes that the actual pitch is −40 deg. He then looks up the 3D interpretation of the projected angles that he does see in the wrong table, i.e., the one for a −40 degree pitch, and finds incorrect corresponding depicted angle values. These curves show that errors in pitch estimation alone can not account for the data from Experiment 1.

In addition to the pitch, the yaw and roll of the view vector may be incorrectly estimated and cause a similar kind of error in a more generalized look-up table that associates depicted angles with their projections. Though the roll of the view vector also can influence the geometric relationship between depicted and projected angles; the kinematics of the head constrain the amount of roll associated with head rotations in pitch and yaw. For constrained ranges of rotation, the cross-coupling of yaw onto roll has been estimated, for example, at only about 2 per cent (Chouet and Young, 1974; Larsen and Stark, 1988), but in any case, any roll around the view direction after a pitch and yaw will not influence the size of projected angles with vertices along the view vector, only the orientation of the projected angle. Accordingly, a look-up table associating view direction with projected angles as a first approximation need only have two indices associated with the pitch and yaw of the view vector.[4]

To determine the actual depicted angle corresponding to a measured projected angle, the visual system need only search a database of depicted-angle-projected-angle associations for the given viewing parameters. If the viewing assumptions were correct, that recorded projected angle which most closely matched the projected angle currently measured would point to the correct depicted angle. In an alternative use if a corresponding depicted angle can be assumed, the measured projected angle can be used to infer the view direction by recovering the viewing direction indices. Interestingly, neither of these searches will necessarily yield unique results because the viewing constraints may be under specified, but since additional assumptions concerning size and habitual viewing positions can remove the ambiguities, the lack of geometric uniqueness is not necessarily a problem.

General theories of errors in determining depicted target directions may then be expressed in terms of errors the viewer makes in determining the view direction. A model of this sort (Tharp, 1989; Tharp and Ellis, 1990) was fitted to the data from Experiment 1 by conducting a grid search of a range of possible errors in pitch and yaw estimates to find a pair that minimized the RMS error between the inferred target azimuth errors that such erroneous viewing assumptions would produce and the actual observed azimuth errors. These fitted models plotted in Figure 3 show that this two parameter theory provides a pretty good fit to the data of Experiment 1.

As shown in Figure 3, the view vector parameters estimated from the data in Experiment 1 indicate that not only is pitch overestimated by 10–11 deg, as conjectured in the discussion of the experiment, but so is yaw, in fact yaw has a greater error of 14–21 deg. This difference makes sense in retrospect since proprioception would have been a good cue to view direction and the subject's heads were correctly pitched to view the display surface. They were not rotated in yaw to the left or right. Though the view vector model fits fairly well for a two parameter model, i.e., Pearson r is around 0.9, systematic deviations from it near the obliques suggest that further refinement may be necessary. Interestingly, an alternative model which was based on a "telephoto" error in measuring the divergence of lines of sight into a picture also had the same systematic difficulty matching the azimuth error data along the oblique axes (Grunwald et al., 1988).

---

[4] If the head kinematics mimicked eye kinematics and obeyed Listing's and Donder's Laws, the look-up table would not in principle require more than two indices. We have made some rough measurements of the amount of roll associated with head clockwise and counterclockwise yaws of 30 deg and downward pitches of −15 deg and found the amount of associated roll to be of the order of 1 deg. It can, however, be up to about 30 per cent of the pitch or yaw for more extreme positions such as a 60 deg yaw associated with a 60 deg pitch up. But the amount of roll produced under these circumstances is very variable, depending upon the constraints placed on the torso, and is under the voluntary control of the subject.

We have reanalysed previously reported data in which the ground reference was a full grid, in order to determine if errors in assumed direction of the view vector could also model these. In fact the model works on the older data about as well as that in the current experiment with a correlation with the observed data of 0.89. In this case the best fit corresponds to an overestimate of pitch of only 4 deg and yaw of 9 deg and, as is the case with the present data, there is a tendency for the model to underestimate the errors near the oblique axes. The error in yaw compares roughly with independent direct measurements of perceived yaw of fully grided ground references. These data predict about a 6 deg overestimation of yaw for approximately comparable viewing conditions and image content (Ellis and Grunwald, 1988).

Cause of error in estimated viewing direction

Assumptions regarding the physical properties of objects or elements of pictures are necessary for picture perception because of the inherent ambiguity of the monocular pictorial information. These assumptions provide the means for quantitative interpretation of that information. Examples would be: that the reference lines dropped from the cube markers are parallel, equal and themselves perpendicular to the ground reference; that the marker cubes remain equal in depicted size and that the lines in the ground reference are all parallel and coplanar. Other examples would be assumptions regarding the regularities of background textures that would allow geometric interpretations of the texture gradients present in perspective projections.

It is probably correct to argue that for monocular perspective displays these shape assumptions are the principal basis for the construction of a perceived space from the provided line of sight information. The properties of this inferred virtual space are opposed, however, by the properties of the physical space of the picture surface which provide a mechanism to produce the pattern of direction errors that have been recorded. A simple test of this hypothetical distorting mechanism would be to repeat the previous experiment in a real scene, a situation where there is no binocular conflict and in which there are an abundance of cues to the correct view direction. Experiment 2 investigates this possibility.

Experiment 2

Methods

Subjects

Four nonpilots and four pilots with normal or corrected to normal visual acuity participated in the study. Five of the subjects were graduate psychology students one of whom was a pilot. Three of the nonpilots were experienced psychophysical observers.

Apparatus and stimuli

The geometric conditions in Experiment 1 were generally duplicated although electronically produced apertures and dials were replaced by actual equivalent sized objects with similar functions. The stimuli and viewing geometry used in Experiment 1 were physically reproduced in a parking lot adjacent to the Life Science Building at the Ames Research Center which was viewed from observation stations on top of this building. Two large cubes (91 × 91 × 91 cm) were constructed from 1.27 cm white polyvinyl chloride pipe. Each cube was suspended from its center 3.6 m above the ground by a single aluminum pole and allowed to tumble irregularly in the breeze. One of the two cubes marked the position of the reference cube, and the other marked the location of the target cube. An assistant, in radio contact with the experimenter, moved the target cube. During the experiment the subjects viewed the stimuli through two different sized windows in the two observation stations. Each station was 104 cm square × 61 cm deep. The angular sizes of the windows measured from a distance equal to the 61 cm depth of the station were 30 and 60 deg.

Immediately adjacent to the windows in each observation station was a circular, clear plastic angle indicator dial for collecting angular data geometrically equivalent to that used in Experiment 1. The face of the dial was normal to the subjects line-of-sight to the reference cube and had two lines on it, one fixed and one moveable. The fixed one was parallel to the vertical axis of the window. Subjects in the experiment rotated the dial to adjust the moveable line to match the angle on the face of the dial with a specified azimuth angle of the target cube. Response recording and stimuli selection were controlled by a microcomputer.

The subjects viewed the stimulus scenes binocularly from about 61 cm behind and centered in the viewing windows. At the 28 m viewing distance the reference cube subtended on average about 5 deg; its suspension allowed it to rotate irregularly. The cubes markers provided a significant stereoscopic stimulus since the binocular disparity of the target varied between 6.6' to 9.8' as it was positioned around the reference cue. This maximum disparity difference of 3.2' is at least about thirty times a typical stereo threshold but within normal values of fusion area for the range of retinal eccentricities experienced by the subjects.

Subjects made the same exocentric direction judgments used in Experiment 1. Positions of the target cube were unobtrusively marked on the ground to allow an assistant to position the target cube accurately at the planned azimuth angles. The zero-degree-azimuth reference axis always pointed away from the center of the reference cube and was parallel to the white lines painted on the black asphalt parking lot to indicate parking places.

Prior to participation, each subject was shown the actual location of this axis. In conformance to earlier experiments the reference cube was not centered between nor positioned on any of the painted lines. Except for the immediate vicinity of the cubes, the parking lot was often filled with cars.

A $24 \times 2 \times 2 \times 2 \times 2$ (Target Position $\times$ Window Size $\times$ Direction of Viewing $\times$ Replications $\times$ Flight Experience) mixed factorial design was used in the experiment, with repeated measures on the first four variables. The between subjects variable, flight experience, referred to membership in either the pilot or nonpilot group.

Subjects were required to make azimuth judgments of 24 equally spaced (15 deg) target positions for two directions of viewing (± 22 deg) and two window sizes (30 and 60 deg FOV). Each subject proceeded through the design twice for a total of 192 judgments of target azimuth (24 Target Azimuths $\times$ 2 Window Sizes $\times$ 2 Directions of Viewing $\times$ 2 Replications). As in Experiment 1, the dependent variable was the subjects' error in judging target direction, azimuth error. Azimuth error was computed as in Experiment 1. Decision time was also recorded but will not be discussed in this paper.

Procedure

The subjects were seated so that their heads were centered in the windows. Subjects were discouraged from moving their heads toward or away from the stimulus scene but no head restraint was used in order to preserve naturalistic viewing conditions.

A method of adjustment was used in which the subjects manually adjusted the display angle indicator to match accurately the depicted horizontal angle shown by the position of the target cube, reference cube, and the reference axis. They signaled the computer to take the data by pressing a button adjacent to the dial. No premium was placed on rapid judgements but the subjects were told not to take more than about 30 seconds per judgement. Each subject was given written instructions describing the task, was shown how to manipulate the apparatus, and was then allowed up to 10 practice trials to become familiar with both the equipment and the task. No feedback was given concerning the accuracy of his judgments.

The distance between the two observation stations was 21 m. Rather than have subjects walk this distance as often as a completely random schedule would dictate, each subject stayed at one direction of viewing for at least 16 trials (one block). For each direction of viewing, the factorial combination of 24 target cube directions, two window sizes, and two repetitions were randomly assigned to six blocks of 16 trials. Each subject was presented with 12 blocks of trials (six at each direction of viewing). The total of 192 trials required about three hours to complete. Short rest periods (about 2–5 min) were provided when the subject was required to change direction of viewing (about every 1 or 2 blocks). For each subject, directions of viewing and the order in which blocks were presented were random. To balance possible hardware biases in data collection the data collection equipment at one station was switched with the equipment at the other station halfway through the experiment (i.e., after four subjects).

Results

The azimuth error data were analysed by analysis of variance with repeated measures on target azimuth, window aperture, and viewing direction. Variation in the amount of background information by changing window size did not significantly affect judgments of azimuth error nor did it interact with any other factor. As in Experiment 1, the two-way interaction between azimuth of the target cube and view direction was statistically significant ($F(23,138) = 3.861$, $p < 0.001$).

The nature of the statistical interaction that was observed between viewpoint and target azimuth is again clarified by circular plots in Figure 6. This figure illustrates the underlying symmetry in the error data, which is similar to that in Experiment 1 but it also shows the expected generally smaller size of the errors and the absence of the "equidistance tendency" since there is no pronounced tendency for the errors to be towards the crossing axis.

Discussion

The absence of the equidistance tendency in Experiment 2 confirms the supposition that the full set of spatial cues in a natural viewing situation would remove the bias in Experiment 1 hypothetically introduced by the binocular conflict or other related picture surface cues. In that experiment the azimuth errors were generally away from the reference axes and toward the crossing axis. In contrast to the relatively large bias in Experiment 1, the expected smaller errors in Experiment 2 are less consistent, and frequently away from the crossing axes rather than towards them. The residual error pattern, however, does continue to exhibit a symmetrical dependence on view positions supporting the observation in Experiment 1 that the error pattern still does not exhibit position constancy.

As in Experiment 1, the observed error pattern in Experiment 2 is not similar to what would be expected if it were due to the difference between the size of the projected and depicted azimuth angles. If the difference between depicted and projected angle were the cause of the observed error, the errors would be expected to resemble the traces in Figure 4. As in Experiment 1, the results do not closely resemble these curves so new alternatives need to be considered to explain both the smaller average size of the error and the particular pattern itself.

Though fitting the view vector error model to the data from Experiment 2 yields the expected nearly correct estimate of the view vector parameters (Figure 7) the error patterns are not markedly sinusoidal and do not fit the

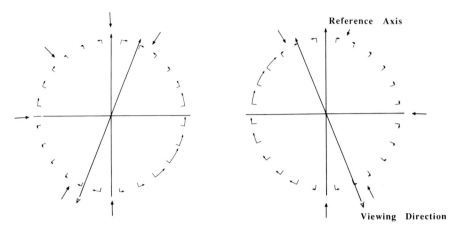

Figure 6. Circular plot of mean azimuth error for direction judgement experiment in Experiment 2.

theoretical curves well. Since the estimated view parameters from these curves are close to the correct values, one may surmise that the full panoply of depth cues available, texture gradients, motion parallax, stereopsis, shape assumption, etc., allow the viewer to estimate accurately these parameters. The residual direction errors therefore may reflect other estimation processes, perhaps smaller second order effects, associated with the use of the dial as a response measure or those producing the small but systematic differences between the theoretical curves and the data near the obliques. These kind of smaller effects may contaminate the measurements in both Experiments 1 and 2 to some degree but their presence is emphasized when the error in the 3D interpretive process is small.

The results from Experiment 2 provide an interesting contrast with Wagner's (1985) studies of the metrics of visual space in which he reported significant compression of the space in depth, a result like that reported as an "equidistance tendency" in Experiment 1. Reading of his paper shows that his subjects viewed their spatial layouts while standing on the ground and therefore at low inclination, of the order of 3 deg or less depending upon their specific eye height. At this inclination the relationship between projected and depicted angles begins to explode as illustrated for right angles by the third term in equation (1) getting close to zero. Accordingly, the corrective influence that perception of familiar projected angles could have on spatial perception would not be expected to work well. The kind of compression in depth that Wagner reported is consistent with Gilensky's (1955) earlier work, but it is also noteworthy that for scenes in which familiar shape and size cues are present, distance estimates to objects can be surprisingly accurate (IATSS, 1983; Loomis et al., 1988). In contrast, the judgmental errors that we have measured in pictorial viewing situations in which the viewing direction may be misjudged may be fairly well modeled by assuming that the error in estimating the view direction causes subjects to look up

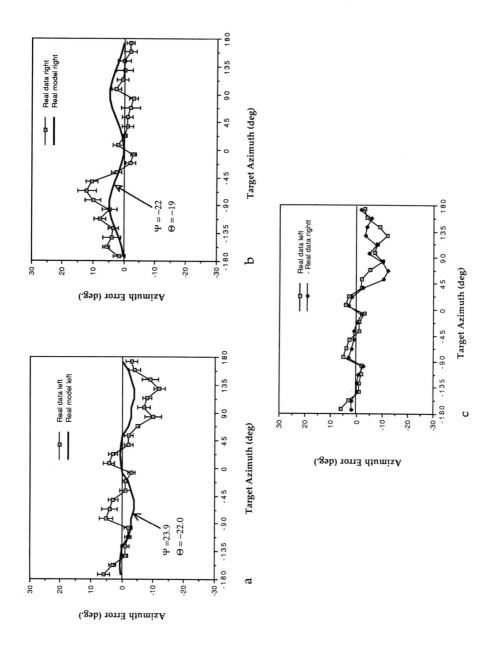

external shapes in incorrectly indexed look-up tables associating distal shapes with their proximal projections.

Summary

Two experiments have been conducted in which subjects indicate the apparent exocentric azimuth direction of a marker with respect to a reference position and direction. This judgement constitutes a precise, systematic version of the Piaget "decentering" task. The task was presented either as a perspective projection onto a binocularly viewed, flat computer display or as a geometrically equivalent physical space. Elimination of binocular conflict between picture surface cues and monocular cues to the display's virtual space, markedly reduced a judgement bias in Experiment 1 resembling a spatial compression in depth. The azimuth errors observed in Experiment 1 can be modeled by a generalization of classic slant overestimation in which the viewer is assumed to overestimate both the pitch and yaw of the viewing direction. When this model is applied to the data of Experiment 2 space, it correctly recovers the true pitch and yaw of the viewing direction thus indicating that in the physical space the subjects are able to use a much better estimate of how they view the scene to estimate exocentric direction than in perceptually degraded displays.

References

Attneave, F. and Frost, R. (1969). The determination of perceived tridimensional orientation by minimum criteria. *Perception and Psychophysics*, **6**, 391–396.

Beck, J. and Gibson, J.J. (1955). The relation of apparent shape to apparent slant in the perception of objects. *Journal of Experimental Psychology*, **50**, 125–133.

Borresen, C.R. and Lichte, W.H. (1962). Shape constancy: dependence upon stimulus familiarity. *Journal of Experimental Psychology*, **63**, 92–97.

◀Figure 7. Cartesian plots of mean azimuth errors in Experiment 2 are plotted in ordinary cartesian form for both left (a) and right (b) viewing stations. Error bars are ±1 standard error, $N = 8$. The heavy traces are theoretical functions derived from the assumption that the subjects misjudge the view vector. The estimated viewing parameters for the theoretical trace are at the tail of the arrow. These functions have been fitted to the data as described in the discussion of Experiment 2 and show the fitting procedure almost exactly estimates the actual viewing parameters ($\Psi = \pm 22$, $\theta = -22$). Nevertheless, though the fits are significant the theoretical traces do not correlate well with the data (left station: $r = 0.408$, $p < 0.05$, $df = 22$; right station: $r = -0.394$, $p < 0.05$, $df = 22$). (c) illustrates the symmetry between the pattern of error from the left and right view stations by reflecting the data from the right station and replotting it with that from the left station.

Chouet, B.A. and Young, L.R. (1974). Tracking with head position using an electro-optical monitor. *IEEE Transactions in Systems, Man and Cybernetics.* SMC-4, 192–204.

Ellis, S.R. and Grunwald, A.J. (1988). Egocentric direction estimates in pictorial displays. *Proceedings of the 1988 IEEE International Conference on Systems, Man, and Cybernetics.* Peking, China. IEEE CAT 88CH2556-9, 8–11.

Ellis, S.R., Smith, S. and McGreevy, M.W. (1987). Distortions of perceived visual directions out of pictures. *Perception and Psychophysics*, **42**, 535–544.

Flavell, J.H. (1963). *The Developmental Psychology of Jean Piaget.* Princeton: Van Norstand.

Gilensky, A.S. (1955). The effect of attitude upon perception of size. *American Journal of Psychology*, **68**, 173–192.

Gogel, W.C. (1977). The metric of visual space. In W.F. (Ed.) *Stability and Constancy in Visual Perception.* New York: Wiley.

Gogel, W.C. and Da Silva, J.A. (1987). Familiar size and the theory of off-sized perceptions. *Perception and Psychophysics*, **41**, 318–428.

Goldstein, E.B. (1987). Spatial layout, orientation relative to the observer, and perceived projection in pictures viewed at an angle. *Journal of Experimental Psychology*, **13**, 256–266.

Gregory, R.L. (1966). *Eye and Brain: The Psychology of Seeing.* New York: World University Library.

Grunwald, A.J. and Ellis, S.R. (1986). Spatial orientation by familiarity cues. In Patrick, J. and Duncan, K.D. (Eds) *Training, Human Decision Making and Control*, 257–279. Amsterdam: North-Holland.

Grunwald, A.J., Ellis, S.R. and Smith, S. (1988). Spatial orientation in pictorial displays. *IEEE Transactions on Systems, Man and Cybernetics*, **18**, 425–436.

Hochberg, J. (1987). Machines should not see as people do, but must know how people see. *Computer Vision, Graphics, and Image Processing*, **37**, 221–237.

Howard, I. (1982). *Human Visual Orientation.* New York: Wiley.

IATSS 527 project team (1983). Characteristics of visual reaction to dynamic environment. *International Association of Traffic and Safety Sciences*, **9**, 162–172.

Ittelson, W.H. (1951). Size as a cue to distance. *American Journal of Psychology*, **64**, 54–67.

Ittelson, W.H. (1952). *The Ames Demonstrations in Perception.* Princeton: Princeton University Press.

Larsen, J. and Stark, L. (1988). Difficulties in calibration of eye movements. *Proceedings of the 1988 IEEE International Conference on Systems, Man and Cybernetics*, CH2556-9, 297–298.

Levine, M. (1984). The placement and misplacement of you-are-here map. *Environment and Behavior*, **16**, 139–157.

Loomis, J., Da Silva, J.A, Marque, S.L. and Fukusima, S.S. (1989). Visual matching and visually directed walking: a comparison involving exocentric intervals. *29th Annual Meeting of the Psychonomics Society*, November 10–12, 1988 Chicago.

Metzger, W. (1953). *Gesetze des Sehens.* Frankfort am Main: Waldeemar Kramer.

McGreevy, M.W. and Ellis, S.R. (1986). The effect of perspective geometry on judged direction in spatial information instruments. *Human Factors*, **28**, 439–456.

McGreevy, M.W., Ratzlaff, C. and Ellis, S.R. (1985). Virtual space and two-dimensional effects in perspective displays. *Proceedings of the 21st Annual Conference on Manual Control.* June, 1985.

Piaget, J. (1932). *The Origins of Intelligence in Children.* London: Kegan Paul.

Pirenne, M. (1970). *Optics, Painting, and Photography.* Cambridge: Cambridge University Press.

Sedgwick, H.A. (1986). Space perception. In Boff, K.R., Kaufman, L. and Thomas, J.P. (Eds) *Handbook of Perception and Human Performance vol 1*, 21:19–21:23. New York: Wiley.

Smith, S. (1986). "Viewing in three-dimensional visual space: the two-dimensional effect and direction judgement errors." Unpublished MA thesis, San Jose State University, San Jose, CA.

Tharp, G. (1989). "Modeling human perception errors using perspective displays and the response to error display." Master's Report, Department of Mechanical Engineering, University of California, Berkeley.

Tharp, G. and Ellis, S.R. (1990). Training effects on exocentric direction judgement errors made with perspective displays. *NASA TM 102792*, NASA Ames Research Center, Moffett Field CA.

Thouless, R.H. (1931). Phenomenal regression to the real object. *British Journal of Psychology*, **21**, 338–59.

Wagner, M. (1985). The metric of visual space. *Perception and Psychophysics*, **33**, 483–495.

Wallach, H. (1985). Perceiving a stable environment. *Scientific American*, **252**, 118–124.

Appendix

Figure 8 shows the detailed geometry of a view of a right angle AOB in the XZ plane with a viewing elevation of Θ. The right angle is rotated in the plane through an azimuth of Ψ. Each of its arms of unit length are projected onto the axes of the XZ

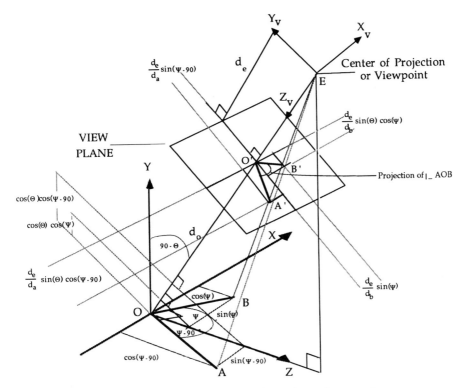

Figure 8. *Perspective projection of a right angle in a plane.*

plane and these components in turn projected onto the view plane d_o from 0. Thus, the projected vectors corresponding to vectors **A** and **B** in are $\mathbf{A}_v$ and $\mathbf{B}_v$ in the view plane $X_v Y_v$.

$$\mathbf{A}_v = \left[\frac{d_e}{d_a} \sin(\Psi - 90), \frac{d_e}{d_a} \sin(\Theta) \cos(\Psi - 90) \right]$$

$$\mathbf{B}_v = \left[\frac{d_e}{d_b} \sin(\Psi), \frac{d_e}{d_b} \sin(\Theta) \cos(\Psi) \right] \tag{2a}$$

The projected angle A'O'B' in the view plane can be directly calculated from:

$$\angle A'O'B' = \cos^{-1}\left[\frac{\mathbf{A}_v \cdot \mathbf{B}_v}{|\mathbf{A}_v||\mathbf{B}_v|} \right] \tag{2b}$$

The distances drop out and trigonometric identities allow reduction of the right side of equation (2a) to:

$$\sin(2\Psi)\tfrac{1}{2}(\sin^2\Theta - 1)((\cos^2\Psi + \sin^2\Theta \sin^2\Psi)(\sin^2\Psi + \sin^2\Theta \cos^2\Psi))^{-1/2} \tag{2c}$$

35

How to reinforce perception of depth in single two-dimensional pictures[1]

Shojiro Nagata

NHK Science & Technical Research Laboratories
Tokyo, Japan

The physical conditions of the display of single two-dimensional pictures, which produce images realistically, were studied by using the characteristics of the intake of the information for visual depth perception. "Depth sensitivity," which is defined as the ratio of viewing distance to depth discrimination threshold, has been introduced in order to evaluate the availability of various cues for depth perception: binocular parallax, motion parallax, accommodation, convergence, size, texture, brightness, and air-perspective contrast. The effects of binocular parallax in various conditions, the depth sensitivity of which is greatest at a distance of up to about 10 m, were studied with the new versatile stereoscopic display. From these results, four conditions to reinforce the perception of depth in single pictures were proposed, and these conditions are met by the old viewing devices and the new high-definition and wide-angle television displays.

The sensation of reality in a picture occurs because of visual depth perception. Therefore, in order to display pictures as if the observer were looking at real objects in three-dimensional space, the physical conditions of the pictures must be matched to the characteristics of the process involved in the intake of information relative to depth perception. The objectives of this paper are to report the results of an investigation on the availability of many cues for visual depth perception, using a common evaluating scale, and to propose ways to reinforce the perception of depth in single two-dimensional pictures.

It is well known that a pair of pictures taken from two laterally separated positions creates the effect of stereoscopic depth perception with binocular cues, such as binocular parallax and convergence cues of the eyeball shown in Figure 1 and Table 1. However, there are other monocular cues shown

---

[1] A preliminary version of this chapter appeared in *Proceedings of the Society for Information Display*, Vol. **25**, No. 3, 1984, pp. 239–246.

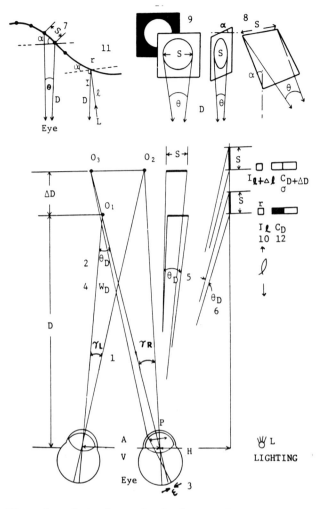

Figure 1. Illustration of visual cues for depth perception.
1: Binocular parallax $\gamma_L - \gamma_R = \theta_D - \theta_{D+\Delta D}$ at the distance A between pupils.
2: Convergence cue $\theta_D - \theta_{D+\Delta D}$.
3: Blurring cue ε of accommodation on pupil diameter P.
4: Motion parallax $\gamma_L - \gamma_R$ or $\omega_D - \omega_{D+\Delta D}$ at monocular moving vision with distance A or speed V.
5: Transverse size cue $\theta_D - \theta_{D+\Delta D}$.
6: Longitudinal size cue on depth direction axis at distance H.
7: Density cue $[(S/D) \cos \alpha]^{-1}$ of texture on surface at slant α.
8: Shape cue at slant.
9: Intersection cue.
10: Brightness cue $I_l - I_{l+\Delta l}$, $I = r \cdot L/l^2$ of the object with refractory factor r at lighting distance l under lighting L.
11: Shade cue $I \cos \alpha$ on slanted surface.
12: Air-perspective contrast cue $C_D - C_{D+\Delta D}$ of air scattering constant σ.
13: Color effect.

Table 1. Depth perception cues and bases of character for transformation[1]

Cue	Objective change in 3D	Image change in retina	Base of character for transformation
Binocular parallax	R.B. Relative distance	Position disparity	Unity
Bincoular convergence	O.B. Absolute distance	Position	Optimum
Accommodation (blurring)	O.M. Absolute	Blurring	Optimum
Motion parallax	O.M. Absolute	Position disparity	Unity
	R.M. Relative	Velocity	Uniformity
Transverse size	R.M. Relative, absolute (familiar)	Size	Identity
Longitudinal size	R.M. Relative	Size	Identity
Vertical position	R.M. Absolute	Position	Unity
Size, density	R.M. Slant in depth	Size, density	Uniformity
Shape	R.M. Slant	Shape	Simpleness Uniformity
Intersection	R.M. Front and back	Shape	Simpleness
Motion	R.M. Motion	Velocity flow	Uniformity
Luminance	R.M. Relative	Illumination	Uniformity
Shade	R.M. Slant	Illumination	Uniformity
Air-perspective	R.M. Relative	Contrast, blurring color	Identity
Color	R.B.M. Aberration	Color disparity	Unity

[1] O: oculomotor cue, R: retinal image cue, B: binocular vision, M: monocular vision.

in Figure 1, such as the accommodation cue of a crystalline lens, motion parallax on moving vision, and pictorial cues. The pictorial cues include transverse size, longitudinal size, texture density and shape, intersection, position of horizon, brightness and shade, air-perspective, and color effect. The study of the comparison of the effectiveness of each of the cues and the study of the interaction between different cues are necessary.

The availability of cues for visual depth perception has been investigated. Künapas (1968) studied the subjective absolute distance by the method of magnitude estimation as a function of viewing distance up to 4 m and five viewing conditions, where the cues (retinal size, binocular parallax, accommodation and brightness) were fully provided or partially reduced.

He found that accommodation did not permit any accurate perception of distance, and that retinal image size was one of the most important cues in the judgement of absolute distance from the observer. He also pointed out the similarity of his result and the result of Holway and Boring (1941) that the apparent size at a fixed viewing distance varies with the viewing condition. However, Künapas did not study motion parallax and relative depth perception.

When we view a picture which contains many objects, the space perception in the picture depends on the results of the relative depth perception among the objects.

Stubenrauch and Leith (1971), using holograms, found the interposition cue to dominate over most combinations of other cues (binocular parallax, motion parallax, and retinal size) for perception of normal relief or reversed relief. However, these effectiveness estimations were not measured at large viewing distances.

Furthermore, since the cues on the retina, such as parallax, size, brightness, etc., have different physical attributes, the threshold values of each cue change for depth perception cannot be directly compared with each other.

The author proposes a common scale for evaluating the availability of depth cue, which is defined as the ratio $D/\Delta D$ of the viewing distance D to the detection threshold ΔD of depth difference (depth threshold). We call this ratio scale "depth sensitivity" (Nagata, 1977, 1981) of vision.

In this way, the effectiveness of various cues can be quantitatively compared with each other as a function of viewing distance.

Methodology

Hypothesis

First, the relationship between depth sensitivity and the detection of quantitative cue change for depth perception of the object's image on the retina was considered from the viewpoint of the hypothesis that the change of cues is transformed into perception depth information while at the same time conserving the information concerning the character of the object on the base of the character as shown in Table 1.

For example, when a value $R(D)$ of the cue of binocular viewing direction is inversely proportional to the viewing distance D and is proportional to the constant A (where A is the distance between two pupils), the detection threshold ΔR of the change of the value of a cue of convergence or binocular parallax is obtained from the depth threshold ΔD as follows:

$$\Delta R = R(D) - R(D + \Delta D) = A/D - A/(D + \Delta D) \tag{1}$$

Then the depth sensitivity is deduced as

$$D/\Delta D = A/(\Delta R \cdot D) - 1 \tag{2}$$

Second, by dilating on Fechner's Law (Fechner, 1889), it was proposed that depth sensation is based on the sum of the small depth sensation unit $dS = K$ corresponding to the depth thresholds. The depth sensation $S(D)$ is obtained by

$$S(D) = \int_{D_0}^{D} \left(\frac{dS}{dD}\right) dD = \int_{D_0}^{D} (K/\Delta D) dD \tag{3}$$

where K is a transformation constant.

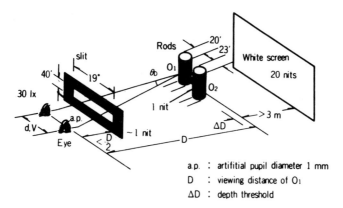

Figure 2. *Apparatus for measuring depth thresholds.*

Psychophysical experiment

The depth sensitivities of the cues of binocular parallax, motion parallax, and accommodation were obtained from the depth thresholds in psychophysical experiments. The characteristics of the cue-change threshold ΔR were induced from the depth threshold ΔD measured under the limited condition, and the depth sensitivities of these cues were calculated. Furthermore, the depth sensitivities of other cues were also calculated by estimating the detection threshold of cue change.

Experiment I

For measuring the depth threshold, an observer, by using a remote-wire system, moved one of the two black rods (20 arc/min in width, 1 cd/m²) as illustrated in Figure 2, so the difference of depth can be noticed through a slit (40 arc/min in height, 19 arc/deg in width).

Two males (SN 33 years of age, left V.A. 1.2, right V.A. 1.2 both corrected; KI 23, 1.5, 1.5) and one female (NW 23, 1.2, 0.6) having normal streoscopic vision served as subjects in these experiments.

The depth thresholds on the binocular parallax, the motion parallax, and the cue from accommodation were measured as a function of viewing conditions.

The viewing conditions for controlling the depth cues were obtained by combining binocular observation or monocular observation and static observation or lateral moving observation, and observation with natural pupils or artificial pupils (1 mm diameter).

In moving observations, the observer moved the upper body rhythmically to the right and left at different distances and velocities which were measured in real time by an electronic scale wired to the head.

The viewing distance to the fixed rod was 1, 2, 5 and 18 m, respectively. The brightness, the retinal size, and the interval distance of two stimuli were not changed as a function of observation distance.

The measurements were taken for eight trials a day for three days for each person under each condition.

Results I

Binocular parallax, motion parallax and accommodation cues

The depth thresholds with static binocular vision through natural pupils were obtained as shown in Table 2, and the symbol (o) in Figure 3 indicates the depth sensitivities as a function of distance obtained from the depth threshold of the typical subject (SN).

The cue-change threshold ΔR on binocular parallax shown in Figure 4 was calculated from the depth threshold ΔD with binocular vision from Equation (1). It was considered that the binocular threshold neither changed as a function of viewing distance, i.e., convergence angle, nor as a function of the size of the pupils.

This was in agreement with the other two observers' results and with those reported by Ogle (1958), Zoth (1923) and Nishi (1933). But Amigo (1963) and Lit and Finn (1976) reported that the threshold slightly increases as the distance decreases to less than 1 m because of the instability of the oculomotor.

The depth sensitivities of this cue shown by the solid line in Figure 3 are calculated from Equation (2), where a constant value is substituted for ΔR. The maximum distance D_{max} for which the sensitivity falls to zero is $A/\Delta R$.

The depth sensation S_D on binocular parallax in Figure 5 is deduced from Equation (4) and may be saturated at about 10 m (Nagata, 1975).

$$S_D = \int_{D_0}^{D} \left(\frac{K}{\Delta D}\right) dD = K \int_{D_0}^{D} \left(\frac{A}{\Delta\theta \cdot D^2} - \frac{1}{D}\right) dD$$
$$= K \left[\frac{A}{\Delta\theta}\left(\frac{1}{D_0} - \frac{1}{D}\right) + \log_n \frac{D_0}{D}\right] \quad (4)$$

The depth thresholds for motion parallax with moving monocular vision with a natural pupil at the speed at which the subject could detect the depth are shown by the symbol (■) in Figure 3. The depth threshold at a viewing distance of 3 m was measured for different conditions, and it was dependent on the velocity ω_D, but not on the distance d of movement as shown in Figure 6A. The optimum velocity ω_a was 6–8° of arc/sec, and at velocities

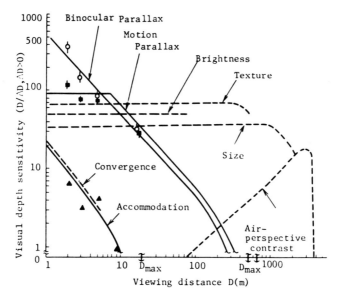

Figure 3. Depth sensitivities of various cues for visual depth perception as a function of viewing distance. Symbols (○, ■, ▲) indicate the averages of five measurements of subject SN and bars on the symbol indicate standard deviations.

Binocular parallax: $A = 0.065$ m, $\Delta\theta = 25''$
Motion parallax: $V_{max} = 0.8$ m/sec, $\Delta\omega = 4'$/sec, $\omega a = 6°$/sec
Accommodation: $P = 0.005$ m of the natural pupil, $\Delta\theta_A = [1/1.2]'$
Air-perspective: $C_0 = 1$, $\sigma = 1$ km, $\Delta C = 11$ per cent of C_D [± 1 dB], $C_{min} = 0.02$
Transverse size: $\Delta\theta_s = 2.5$ per cent of retinal size θ_s
Texture/Longitudinal size: $\Delta\theta_s = 2.5$ per cent of retinal size θ_s
Convergence: $\Delta\theta_s = 10$ min
Brightness: $I/\Delta I = 0.02$

Table 2. Depth thresholds on binocular vision as a function of viewing distance

Viewing distance, m	2	3	5	18
Sub. SN	0.5	1.9	5.7	5.1 (cm)
Sub. KI	0.2	0.4	1.2	2.9
Sub. NW	0.8	2.0	6.1	2.9

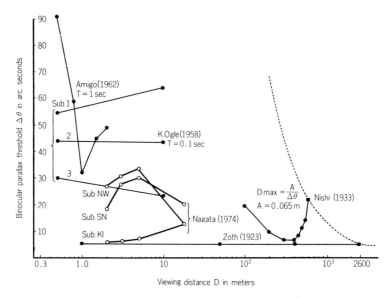

Figure 4. Thresholds of binocular parallax as a function of viewing distance.

lower than the optimum velocity, the threshold velocity of motion parallax is constant.

Graham et al. (1948) and Zegers (1948) reported the increase in the threshold as the velocity increases from about 6° to 20° of arc/sec. But in our results shown in Figure 6B, the velocity threshold of motion parallax $\Delta\omega$ is constant at velocities lower than the optimum. This constancy is deduced from the detection model where the minimum parallax is sampled at a constant interval time.

The depth sensitivity of motion parallax calculated from Equation (2) is represented by the solid line in Figure 3. The sensitivity is $\omega_a/\Delta\omega$ and is constant up to the distance at which the optimum velocity of the body movement is obtained, and when the distance is exceeded the sensitivity decreases. The descending curve is obtained by substituting the maximum velocity V_{max} of body movement for A, and $\Delta\omega$ for ΔR in Equation (2).

This motion parallax is produced not only by the absolute motion of the observer, but also by the relative motion of the objects, and in the case of moving vision on some riding machine with a speed higher than the motion of the body, the sensitivities of motion parallax at large distances are maintained at the same level as that for short distances and are higher than that for binocular parallax.

The depth thresholds for the blurring cue of accommodation with static monocular vision through a natural pupil are represented by the symbol (▲) in Figure 3, and the depth threshold with vision through the artificial pupil was nearly equal to or slightly greater than the viewing distance.

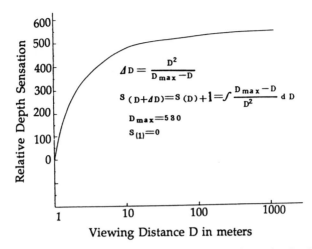

Figure 5. Relative sensation of depth distance deduced from the depth sensitivities with binocular parallax as a function of visual distance (Nagata, 1975).

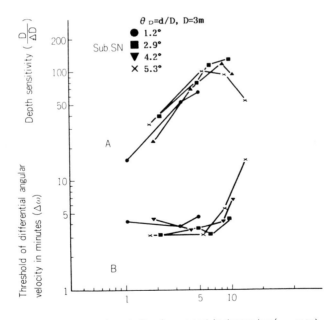

Figure 6. Depth sensitivities $D/\Delta D$ (curves of A) and the threshold of parallax velocity $\Delta \omega$ (curves of B) as functions of angular velocity of movement $\omega_D = V/D$ and movement distance A at a viewing distance D of 3 m.

The depth sensitivity of the natural accommodation cue was also calculated by substituting the pupil diameter during observation for A in Equation (2) and by substituting the blurring threshold resulting from Equation (1) (similar to the reciprocal of his visual acuity) for ΔR in Equation (2).

Other cues

The depth sensitivities relative to binocular parallax, motion parallax, and accommodation cues obtained from Equation (2) are satisfied by the data resulting from the experiment. Therefore, we applied the same method of analysis in obtaining these sensitivity data to the sensitivity data relative to the other cues: convergence, size, slanted shape, texture density, brightness, and air-perspective contrast.

In Figure 2, when two objects positioned at a large visual angle are observed in binocular vision, convergence of the line of sight of two eyes results in depth perception. However, the detection threshold of convergence change is larger than the detection threshold of binocular parallax, and the depth sensitivity of convergence was obtained from Equation (2) and is represented by a dashed line in Figure 3.

The depth sensitivity to the cue of the object transverse size shown in Figure 3 was calculated from Equation (2), where size S is substituted for A and the ratio of the size change detection threshold $\Delta\theta (\equiv \Delta R)$ to size $\theta = S/D$ in visual angle is constant as reported by Ogle (1962).

This sensitivity agrees with the depth threshold under monocular observation of two square targets (1.8 m^2) measured by Teichner et al. (1955). The maximum distance D_m is determined by the absolute detection threshold of size perception.

The depth sensitivities on the shape of a rectangular object whose upper part inclines at larger distances is represented by

$$\frac{D}{\Delta D} = \frac{S}{\Delta\theta \cdot D} - 1 = \frac{D}{L \cdot \sin \alpha_t} \qquad (5)$$

where S is the horizontal length of object, $\Delta\theta$ is the size-cue threshold, L is the height of object, and α_t is the slant threshold.

Freeman (1966) measured the slant threshold of 14 different rectangles without texture by monocular vision. The depth sensitivities calculated from these data varied with height. The optimal depth sensitivity was 78 when $D = 135$ cm and $L = 8$ cm. This sensitivity is larger than the data of Teichner, resulting in the difference between the shape cue of one object and the size cue of two separate objects.

In viewing a textured pattern, there are different sizes or density of texture: one is the transverse size or density in a plane rectangular to the depth direction as mentioned above, and the other is the longitudinal size or density along the depth-directional line (Nagata, 1978, see Figure 7).

Perception of depth in 2D pictures 537

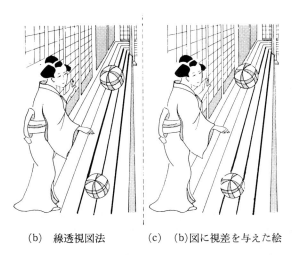

(b) 線透視図法 　　(c) (b)図に視差を与えた絵

Figure 7. Three 3D drawing methods and depth-size perception. (a) Parallel perspective from an ancient picture by Haseo-Soushi. Note how the corridor and wall, drawn the same width on the picture, appear to broaden toward their end in the upper right. (b) Linear perspective drawn by Tokuko Nagata. Notice how the upper ball appears bigger than the ball on the floor. (c) Stereo-pair for (b) with appropriate binocular parallax added. To view the stereo-pair (b)–(c), place a small sheet of paper as a perpendicular septum on the dotted lines separating them. Then diverge or converge your eyes to achieve fusion. When viewed stereoscopically, both balls should appear the same size due to size constancy effects (Nagata, 1978).

The depth sensitivity on the latter was calculated from Equation (6) and is shown in Figure 3:

$$\frac{D}{\Delta D} = 2\left(\frac{S \cdot H}{\Delta\theta \cdot D^2} - 1\right) = 2\left(\frac{\theta}{\Delta\theta}\right) \quad (6)$$

where S is the object's longitudinal size on the depth direction, H is the distance between the visual line and the object plane, and the ratio of the longitudinal size θ in visual angle to the size cue threshold $\Delta\theta$ is the same as the ratio of the transverse size cue threshold. This sensitivity is twice as large as that for the transverse size.

The depth sensitivity on the brightness cue shown in Figure 3 is deduced from Equation (7):

$$\frac{D}{\Delta D} = 2\frac{L \cdot r}{\Delta I \cdot D^2} = 2\frac{I}{\Delta I} \quad (7)$$

where L is luminous intensity, D is the lighting distance, r is the refractory factor of the object, I is the luminance of objects, and ΔI is the cue-change threshold of luminance. This sensitivity is satisfied even at a very small stimulus level at which point Ricco-Piper's law is applied.

When the observer or viewing objects move in three-dimensional space, the projected retinal image changes in position, size, shape (Hochberg and Brooks, 1960), density, and luminance, and the depth perception is affected by the changing velocity.

The depth sensitivity of the air-perspective contrast cue results from the contrast-diminishing function of Equation (8), except for the case of blurring or color effect:

$$C = C_0 \exp\left(-\frac{D}{\sigma}\right) \quad (8)$$

where C_0 is the luminance contrast at very small distances and σ is the length constant determined by the air-scattering coefficient.

The sensitivity on this cue illustrated in Figure 3 is calculated from Equation (9):

$$\frac{D}{\Delta D} = -\frac{C_0}{\Delta C} \cdot \frac{D}{\sigma} \exp\left(-\frac{D}{\sigma}\right) = -\frac{C}{\Delta C} \cdot \frac{D}{\sigma} \quad (9)$$

where ΔC represents the differences in threshold for the brightness contrast deduced from the variation of detection threshold relative to the sine-wave grating pattern given by Watanabe et al. (1968).

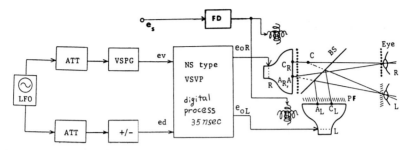

Figure 8. Diagram of experiments for binocular parallax and size changing cue. LFO: low frequency oscillator. ATT: attenuator. VSPG: variable size square pattern generator. VSVP: versatile stereoscopic video processor. FD: fixed delay. e_v: television video signal of original picture. e_O: phase-modulated video signal for left or right eye. e_d: depth signal for modulation of binocular parallax. e_s: synchronous signal. BS: beam splitter. PF: polarizing filter. $A_{R,L}$; $C_{R,L}$: position of pictures. A,C: perceived positions. A/D, D/A: converters; VSVP is designed to control the parallax or the minimum shift of 34.8 ns even on the clock frequency of 14.35 Hz.

Experiment II

Because of the above-mentioned result that the depth sensitivity of binocular parallax was very high in comparison with other cues, the effects of binocular parallax in other conditions and the interaction effect between binocular parallax and other monocular cues were studied. In Experiment I the change of binocular parallax and retinal size corresponding to moving objects in depth could not be controlled independently. To measure the effects of two coexistent cues, the new versatile stereoscopic display (Nagata, 1982) of the standard TV system in conjunction with a special video processor (fast phase modulation) were used.

In this system, as shown in Figure 8, the stereoscopic pictures have been produced with binocular parallax and convergence, controlled temporally and spatially with depth signals in a manner comparable to brightness control signals of video signals—all independent of pictorial cues; for example, size of pattern. The brightness or the size of the picture is also changed independent of the depth signal.

Results II

The subjects viewed the square pattern in streoscopic vision, of which the size and binocular parallax were changed temporally and simultaneously by the pattern-size and depth-control sine-wave synchronous signals, with variable amplitude and polarity of depth direction, so that the conditions of

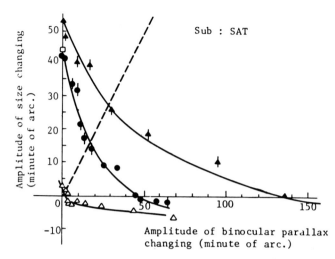

Figure 9. Interactive effects of depth sensation from two kinds of cue, binocular parallax and changing size cue with oscillating amplitude.
Δ: *conditions for the threshold of depth motion perception*
● ▲: *conditions for equal depth sensation at two levels of suprathreshold*
□: *condition of only monocular vision with size cue for equal depth sensation with that of ▲ (not threshold but suprathreshold in binocular vision)*
---: *condition in actual moving*
Sine-wave oscillation frequency, 1 Hz. Middle size, 6.4 cm (2.71°) × 6.4 cm. Back luminance, 1 cd/m^2; Pattern, 30 cd/m^2. Viewing distance, 1.35 m.

equally felt depth sensations of motion could be measured. In Figure 9, the horizontal axis represents the amplitude in arc-minutes peak-to-peak of oscillation of binocular parallax and the vertical axis represents the amplitude of oscillation of size. The smoothed curves indicate the conditions of those two cues for which equal depth sensation occurred at three levels; these are, depth threshold (Δ) and two suprathresholds (●, ▲, ■). The data show that the depth sensation from two coexistent cues, changing size and binocular parallax, is a combination of the individual effects of each cue, and when binocular parallax is zero, the changing size cue in monocular vision has more effect than the changing size cue in binocular vision. In other experiments, it was found that the effect of binocular parallax decreased when the objects moved in depth or in the lateral direction.

Discussion and conclusions

The following conclusions were derived from the comparison of the depth sensitivities of various cues and from the interactive effects of depth sensation from two different cues, size changing and binocular parallax:

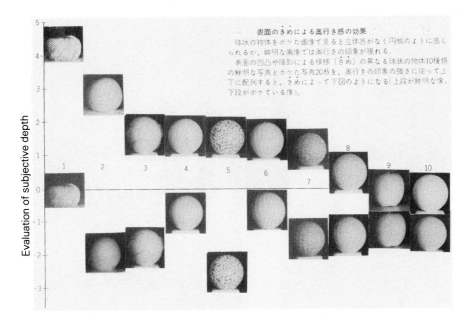

Figure 10. The evaluation of subjective depth as a function of variation of texture conditions in a sharp image (upper panel) and blurred image (lower panel) and texture patterns of form, density, shade, or polish (varying horizontally).

(1) The depth sensitivity relative to binocular parallax is maximum at a distance of up to about 10 m.
(2) The depth sensitivity to motion parallax is effective, and this sensitivity on motion at the optimum velocity exceeds that of the binocular parallax at a distance greater than 10 m.
(3) The cues from accommodation and convergence are effective for the relative depth perception only at a distance of less than 1 m, but are effective for the absolute depth perception at longer distances.
(4) The pictorial cues are effective even at long distances, and the sharp edge of pictures, and clear texture, shade, and gloss of the surface on objects strengthen the sensation of depth (Nagata, 1979, see Figure 10).
(5) The effects of these cues work together and combine spatially on the wide visual field.

From the investigation of these sensitivities, the following conditions to decrease the sensation of flatness of the display plane of single two-dimensional pictures and to reinforce the depth perception in the picture were found:

(1) The effects of binocular parallax must be decreased.
(2) The distance of convergence and accommodation must be close to the actual distance of the objects in the picture.

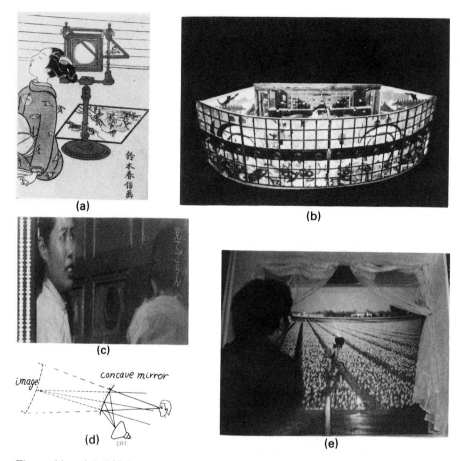

Figure 11. (a) Old Japanese viewing device called vue d'optique (nozoki-karakuri) with one lens-mirror. (b) A Japanese type with 24 lenses allowing many viewers to peer into the apparatus at the same time. (c) A similar Chinese type picked up from Chinese film. (d) A wide-angle television system using a parabolic mirror for viewing (Ninomiya, 1979). (e) Scenery simulator with perspective and large convex fresnel lens (1 m × 1 m) for accommodative relief. When viewed from a position 50 cm from the eyes, the simulator portrays a realistic scene (Nagata, 1975).

(3) The frame of the display must be separated from the images peripherally or depth-wise to be defused.
(4) There must be many monocular pictorial cues including the variety of motion cues of three-dimensional moving objects.

Conditions 1, 2, and 3 are attained by viewing with monocular vision or by positioning the picture image at a distance of about 5 m; conditions 3 and 4, by making the visual angle of the picture wide; and condition 4, by using a high-definition and moving picture.

So, we can point out that the new high-definition and wide television displays (Fujio, 1983) meet these conditions, and these displays produce more realistic picture images than the conventional television.

It is well known that one of the important conditions for space perception is the size of the viewing field of the display, which gives self-motion perception to an observer, such as when one stands in a "Wander-Room" where the wall and ceiling surrounding one rotate; nevertheless, one feels self-motion.

It was found that a visual wide-angle display over 30° induces the sensation of reality (Hatada et al., 1980) because of the integration of the depth cue effects mentioned above.

The old viewing device called reflectorscope or vue d'optique (in Japanese, nozoki-karakuri, which means "peeking device"), shown in Figure 11, in which pictures were viewed through a convex lens or a concave mirror, produces images of the picture realistically.

Concerning the reasons why this device produces realistic images, Valyus (1966) and Schwartz (1971) pointed out that because of the aberration of the lens or reflector, binocular parallax occurs and results in stereoscopic pictures, and also the difference between the illumination intensities of the binocular images, because of the difference of the diffusion of the screen, results in stereoscopic vision. If these explanations are correct, the disparity and the difference of illumination between the binocular images would increase with the distance from the median line of the picture, and then the depth sensation would depend on position.

However, according to the results of our observations, the depth sensation depends on the nature of objects in the picture, and the depth sensation in monocular vision is equal to or better than that in binocular vision.

Therefore, the actual reason why realistic images are produced on the old viewing device is that it fulfills conditions proposed in our results in the case of pictures without movement.

Summary

It is important to make the observer feel the depth sensation for the display of realistic single two-dimensional pictures. Considering the display conditions to reinforce the depth sensation, "depth sensitivities as a function of viewing distance" of various cues were obtained and the interactions between the effects of the different cues were measured. Of the different cues, binocular parallax is most effective at distances of less than 1 m. The effect of the depth sensation with binocular parallax is additive (negative in the case of this subject) with the other effects. Therefore, to improve depth sensation the images of pictures should be set at a large distance to decrease the effects of binocular vision (parallax and convergence). The picture should also be displayed in wide-angle formats to decrease the effect of the frame of the picture on the binocular parallax. The scenery simulator with convex lens/

concave mirror and HDTV picture or superwide film formats, i.e., IMAX or OMNIMAX, meets these conditions.

References

Amigo, G. (1963). Variation of stereoscopic acuity with observation distance. *Journal of the Optical Society of America*, **53**, 615–630.
Fechner, G.T. (1936). Elemente der Psychophysik. Leipzig: Breikopf and Hartel (1889). In Guilford, J.P. (Ed.) *Psychometric Methods*, 20–42. New York: McGraw-Hill.
Freeman, R.B. (1966). Absolute threshold for visual slant. *Journal of Experimental Psychology*, **71-2**, 170–170.
Fujio, T. (1983). Picture display systems for future high-definition television. *Proceedings of the 3rd International Display Research Conference*, 554–557.
Graham, C.H., Baker, K.E., Hecht, M. and Lloyd, V.V. (1948). Factors influencing thresholds for monocular movement parallax, *Experimental Psychology*, **38**, 205–233. In Graham, C.H. (Ed.) *Vision and Visual Perception*. New York: John Wiley and Sons. 513–516.
Hatada, T., Sakata, H. and Kusaka, H. (1980). Psychophysical analysis of the "sensation of reality" induced by a visual wide-field display. *SMPTE Journal*, **89**, 560–569.
Hochberg, J. and Brooks, V. (1960). In Harber, R.N. (Ed.), *The Psychology of Vision Perception*, 195–197. New York: Holt Inc.
Holway, A.H. and Boring, E.G. (1941). Determinants of apparent visual size with distance variant. *American Journal of Psychology*, **51**, 21–37.
Künapas, T.M. (1968). Distance perception as a function of available visual cues. *Journal of Experimental Psychology*, **77**, 523–529.
Lit, A. and Finn, J.P. (1976). Variability of depth-discrimination threshold as a function of observation distance. *Journal of the Optical Society of America*, **66**, 740–742.
Nagata, S. (1975). Depth information on vision and depth sensation, *Proceedings of the Image Engineering Conference*, **6** (5-2), 81–84. (In Japanese.)
Nagata, S. (1977). Depth sensitivities of various cues for visual depth perception. *Journal of the Institute of Television Engineers Japan*, **31**(8), 649–655. (In Japanese.)
Nagata, S. (1978). Visual depth information and depth perception. *NHK Giken Geppou*, **21-4**, 153–158. (In Japanese.)
Nagata, S. (1981). Visual Depth Sensitivities of Various Cues for Depth Perception. *NHK LABS. NOTE*, **266**, 1–11.
Nagata, S. (1982). New versatile stereo display (NS type) system and measurement of binocular depth perception. *Japanese Journal of Medical Electronics and Bioengineering*, **20**, (3), 16–3.
Nagata, S. (1983). Effects of luminace-tone quantitizing and texture to solidity of picture. *Proceedings of General Meeting of Institute of Television Engineers of Japan*, **1-7**, 17–18. (In Japanese.)
Nagata, S. (1989). Scenery viewing simulator with perspective and convex lens effect. *Journal of Three Dimensional Images*, **3-4**, 11–14. (In Japanese.)
Ninomiya, Y. (1979). Parabolic mirror made by the rotation method: Its fabrication and defects. *Applied Optics*, **18**, (11), 1835–1841.
Nishi, I. (1933). Limits of depth perception, proceedings IV of psychology Tokyo. *Iwanami Pub.*, 160–165.
Ogle, K.N. (1958). Note on streoscopic acuity and observation distance. *Journal of the Optical Society of America*, **48**, 794–798.

Ogle, K.N. (1962). In Davson, H. (Ed.) *The Eye IV*, 233–234. New York and London: Academic Press.
Schwartz, A.H. (1971). Stereoscopic perception with single pictures. *Optical Spectra*, September, 25–27.
Stubenrauch, C.F. and Leith, E.N. (1971). Use of hologram in depth-cue experiments. *Journal of the Optical Society of America*, **61**, 1268–1269.
Teichner, K. and Wehrkamp (1955). *American Journal of Psychology*, **68**, 193–208.
Valyus, N.A. (1966). *Stereoscopy*, 80–82. London: Focal Press.
Watanebe, A., Mori, T., Nagata, S. and Hiwatashi, K. (1968). Spatial sine-wave responses of the human visual system. *Vision Research*, **8**, 1245–1263.
Zegers, R. (1948). Monocular movement parallax thresholds as functions of field size, field position and speed of stimulus movement. *Journal of Psychology*, **26**, 477–498.
Zoth (1923). *Lehrbuch der Experimentellen Psychologie, Frobes Journal*, Freiburg: Herder.

36

Spatial constraints of stereopsis in video displays

Clifton Schor

*University of California,
School of Optometry,
Berkeley, California*

Recent developments in video technology, such as the liquid crystal displays and shutters, have made it feasible to incorporate stereoscopic depth into the three-dimensional representations on two-dimensional displays. However, depth has already been vividly portrayed in video displays without stereopsis using the classical artists' depth cues described by Helmholtz (1866) and the dynamic depth cues described in detail by Ittelson (1952). Successful static depth cues include overlap, size, linear perspective, texture gradients, and shading. Effective dynamic cues include looming (Regan and Beverly, 1979) and motion parallax (Rogers and Graham, 1982).

Stereoscopic depth is superior to the monocular distance cues under certain circumstances. It is most useful at portraying depth intervals as small as 5–10 arc seconds. For this reason it is extremely useful in user-video interactions such as in telepresence. Objects can be manipulated in 3D space, for example, while a person who controls the operations views a virtual image of the manipulated object on a remote 2D video display. Stereopsis also provides structure and form information in camouflaged surfaces such as tree foliage. Motion parallax also reveals form; however, without other monocular cues such as overlap, motion parallax can yield an ambiguous perception. For example, a turning sphere, portrayed as solid by parallax, can appear to rotate either leftward or rightward. However, only one direction of rotation is perceived when stereo-depth is included. If the scene is static, then stereopsis is the principal cue for revealing the camouflaged surface structure. Finally, dynamic stereopsis provides information about the direction of motion in depth (Regan and Beverly, 1979). When optical flow patterns seen by the two eyes move in phase, field motion is perceived in the fronto-parallel plane. When optical flow is in antiphase (180°) motion is seen in the saggital plane. Binocular phase disparity of optical flow as small as 1° can be discri-

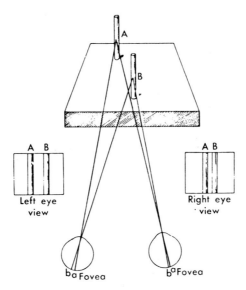

Figure 1. Retinal image disparity based on horizontal separation of the two eyes.

minated as changes in visual direction of motion in a 3D space (Beverly and Regan, 1975). This would be a useful addition to the visual stimuli in flight simulators.

Several spatial constraints need to be considered for the optimal stimulation of stereoscopic depth. The stimulus for stereopsis is illustrated in Figure 1. Each peg subtends a visual angle at the entrance pupils of the eyes, and this angle is referred to as binocular parallax. The difference in this angle and the angle of convergence forms an absolute disparity. In the absence of monocular depth cues, perceived distance of an isolated target, subtending an absolute disparity is biased toward 1.5 m from the physical target distance. Gogle and Teitz (1973) referred to this as equidistance tendency. If the target moves abruptly from one distance to another, convergence responses signal the change of depth (Foley and Richards, 1972); however, smooth continuous changes in binocular parallax, tracked by vergence eye movements do not cause changes in perceived distance (Erkelens and Collewijn, 1985; Guttmann and Spatz, 1985). Once more than one disparate feature is presented in the field, differences in depth (stereopsis), stimulated by retinal image disparity become readily apparent. Stereothresholds may be as low as 2 sec arc, which ranks stereopsis along with vernier and bisection tasks among the hyperacuities.

Stereo-sensitivity to a given angular depth interval varies with the saggital distance of the stimulus depth increment from the fixation plane. Sensitivity to depth increments is highest at the horopter or fixation plane where the disparity of one of the comparison stimuli is zero (Blakemore, 1970). This optimal condition for stereopsis was used by Tschermak (1930) as one of four

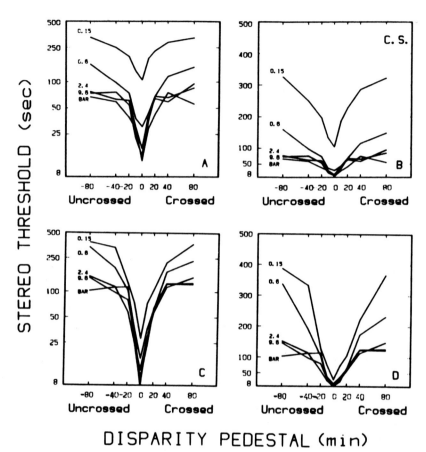

Figure 2. Threshold depth increments obtained, for observer D.B., as a function of pedestal size in both the convergent and divergent directions. Functions illustrate results obtained with a thin bar and DOGs whose center spatial frequencies ranged from 0.15 to 9.6 c/deg. Panels C and D plot the performance measured when the comparison stimulus was a thin bright bar and the test stimulus was a DOG. Panels A and B show the results obtained when a DOG was used both as a comparison and as a test stimulus. Panels A and C plot stereothreshold on a log scale. The data are replotted on a linear scale in panels B and D.

criteria for defining the empirical longitudinal horopter. The Weber fraction describing the ratio of increment stereothreshold (arc sec) over the disparity pedestal (arc min) (3 sec/min) is fairly constant with disparity pedestal amplitudes up to 1°. This fraction was derived from Figure 2, which plots stereothreshold in seconds of arc at different saggital distances in minutes arc from the fixation point for targets consisting of vertical bars composed of coarse or fine features. A two-alternative, forced choice is used to measure a just-noticeable difference between a depth increment between an upper test bar and a lower standard bar, both seen at some distance before or behind the fixation plane. The bar used was a narrow-band, spatially filtered line produced from a difference of Gaussians (DOG) whose center spatial frequency ranges from 9.5 to 0.15 cycles/deg (Badcock and Schor, 1985). When these thresholds are plotted, the slopes of these functions found with different width DOGs are the same on a logarithmic scale. However, thresholds for low spatial frequencies (below 2.5 cpd) are elevated by a constant disparity which illustrates they are a fixed multiple of thresholds found with higher spatial frequencies. These results illustrate that depth stimuli should be presented very near the plane of fixation, which is the video screen.

Stereo-sensitivity remains high within the fixation plane over several degrees about the point of fixation. Unlike the rapid reduction of stereo-sensitivity with overall depth or saggital distance from the horopter, stereo-sensitivity is fairly uniform and at its peak along the central 3° of the fixation plane (Blakemore, 1970; Schor and Badcock, 1985). Figures 2 and 3 illustrate a comparison of stereo-depth increment sensitivity for this fronto-parallel stereo and the saggital off-horopter stereothreshold. Also plotted in Figure 3 are the monocular thresholds of detecting vernier offset of the same DOG patterns at the same retinal eccentricities. Clearly, stereopsis remains at its peak at eccentricities along the horopter and there is a percipitous fall of visual acuity (Wertheim, 1894) and, as shown here, of vernier acuity over the same range of retinal eccentricities where stereo increment sensitivity is unaffected (Schor and Badcock, 1985). Thus, stereoacuity is not limited by the same factors that limit monocular vernier acuity because the two thresholds differ by a factor of 8 at the same eccentric retinal locus.

In addition to the threshold or lower disparity limit (LDL) for stereopsis, there is an upper disparity limit (UDL), beyond which stereo depth can no longer be appreciated. This upper limit is small, being approximately 10 arc min with fine (high-frequency) targets, and somewhat larger (several degrees) with coarser (low spatial frequency) fusion stimuli (Schor and Wood, 1983). This depth range can be extended either by briefly flashing targets (Westheimer and Tanzman, 1956) or by making vergence movements between them (Foley and Richards, 1972) to a UDL of approximately 24°. The UDL presents a common pitfall for many stereo-camera displays that attempt to exaggerate stereopsis by placing the stereo-cameras far apart. Paradoxically, this can produce disparities that exceed the UDL and results in the collapse of depth into the fronto-parallel plane.

Diplopia is another problem that accompanies large disparities. The diplopia threshold is slightly smaller than the UDL for static stereopsis, and depth stimulated by large flashed disparities is always seen diplopically. Normally, this diplopia can be minimized by shifting convergence from one target to another. However, this is not as easily done with a stereo-video monitor. In real space the stimulus for vergence is correlated with the stimulus for accommodation. With video displays, the stimulus for accommodation is fixed at the screen plane while vergence is an independent variable. Because

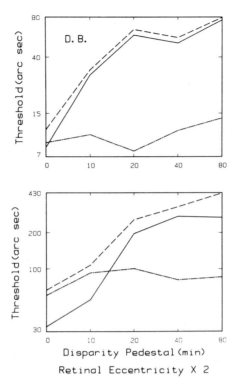

Figure 3. A comparison is made of extra-foveal vernier threshold (solid line) with extra-foveal (mixed dashed line) and extra-horopteral (long dashed line) stereothresholds for a high spatial frequency stimulus (upper plot) and a low spatial frequency stimulus (lower plot). Note that retinal eccentricity has been doubled to be comparable to disparity pedestal. Over a 40 arc min range of retinal eccentricity, stereoacuity remained unchanged and vernier acuity increased moderately. A marked increase in stereothreshold occurred over a comparable (80 arc min) disparity pedestal range.

there is cross-coupling between accommodation and vergence, we are not completely free to dissociate these motor responses (Schor and Kotulak, 1986). With some muscular effort, a limited degree of vergence can be expected while accommodation is fixed, depending on the accommodative-convergence ratio (AC/A). When this ratio is high, a person must choose between clearness and singleness.

Additional problems for stereoscopic depth occur with abstract scenes containing high spatial frequency surface texture. This presents an ambiguous stimulus for stereopsis and fusion which can have an enormous number of possible solutions as illustrated by the wallpaper illusion or by a random-dot stereogram. The visual system uses various strategies to reduce

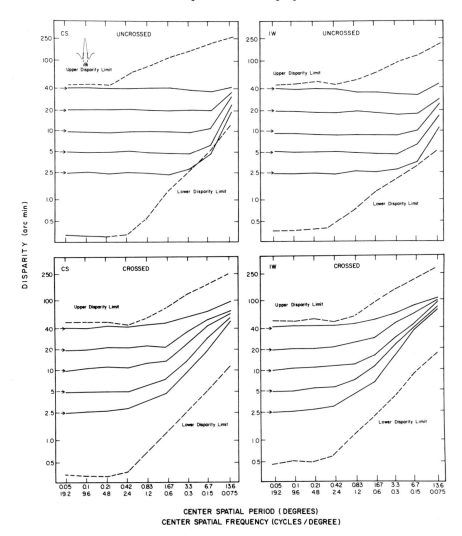

Figure 4. Upper and lower limits for stereopsis are plotted for two subjects as a function of DOG center spatial period along dashed curves at the top and bottom of data sets for uncrossed and crossed disparities respectively. Stereothreshold was lowest at small spatial periods (<0.42 arc min) and increased according to a 6° phase disparity between stereo-half images as spatial period increased. The upper limit increased proportionally to the square root of spatial period over the same range of broad spatial periods. Depth matching curves (solid lines) for several standard suprathreshold disparities (horizontal arrows) have flatter frequency responses than the upper and lower dashed threshold curves. Their breakaway point occurs at a higher spatial period for crossed than for uncrossed disparities. The luminance profile of the difference of two Gaussian functions is inset in the upper left corner.

the number of potential fusion combinations and certain spatial considerations of targets presented on the visual display can help implement these strategies. A common technique used in computer vision is the coarse-to-fine strategy. The visual display is presented with a broad range of spatial frequency content. The key idea here is that there is little confusion or ambiguity with coarse features like the frame of a pattern. These can be used to guide the alignment of the eyes into registration with finer features that present small variations in retinal image disparity. Once in registration, small disparities carried by the fine detail can be used to reveal the shape or form of the depth surface. An essential condition for this algorithm to work is that sensitivity to large disparities be greatest when they are presented with coarse detail and that sensitivity to small disparities be highest with fine (high spatial frequency) fusion stimuli. This size-disparity correlation has been verified for both the LDL and UDL by Schor and Wood (1983). Figure 4 illustrates the variation of stereothreshold (LDL) and the UDL with spatial frequency for targets presented on a zero disparity pedestal at the fixation point. Stereothresholds are lowest and remain relatively constant for spatial frequencies above 2.5 cycles/deg. Thresholds increase proportionally with lower spatial frequencies. Even though stereothreshold varies markedly with target coarseness, suprathreshold disparities needed to match the perceived depth of a standard disparity are less dependent on spatial frequency. This depth equivalence constitutes a form of stereo-depth constancy (Schor and Howarth, 1986). Similar variations in the diplopia threshold or binocular fusion limit are found by varying the coarseness of fusion stimuli (Schor et al., 1984b).

Figure 5 illustrates that the classical vertical and horizontal dimensions of Panum's fusion limit (closed and open symbols, respectively) are found with high spatial frequency targets, but the fusion limit increases proportionally with the spatial width of targets at spatial frequencies lower than 2.5 cycles/ deg. When measured with high-frequency DOGs, the horizontal radius of PFA (Panum's fusional area) is 15 min; and when measured with low-frequency stimuli, PFA equals a 90° phase disparity of the fusion stimulus.

The increase in Panum's fusion limit appears to be caused by monocular limitations to spatial resolution. For example, if the same two targets that were used to measure the diplopia threshold are both presented to one eye to measure a two-point separation threshold, such as the Rayleigh criterion, then the monocular and binocular thresholds are equal when tested with spatial frequencies lower than 2.5 cpd. At higher spatial frequencies we are better able to detect smaller separations between two points presented monocularly than dichoptically. This difference at high spatial frequencies reveals a unique binocular process for fusion that is independent of spatial resolution. With complex targets composed of multiple spatial frequencies, at moderate disparities such as 20 min arc, a diplopia threshold may be reached with high spatial frequency components while stereopsis and fusion may continue with

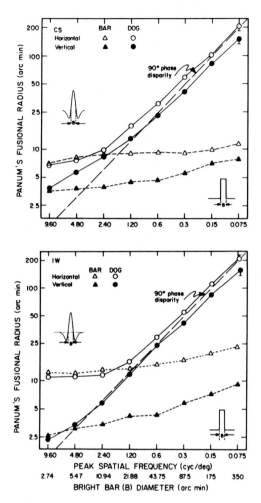

Figure 5. Diplopia thresholds for two subjects are plotted as a function of bright bar width (B) of bar and difference of two Gaussian functions (DOG). Luminance profiles of these two test stimuli are inset below and above the data respectively. A constant phase disparity of 90° is shown by the dashed diagonal line. Horizontal and vertical Panum's fusion ranges (solid lines) coincide with the 90° phase disparity for DOG widths greater than 21 arc min. At the broadest DOG width, the upper fusion limit equals the upper disparity limit for stereoscopic depth perception (bold dashed line). The standard deviation of the mean is shown for the broadest DOG stimulus. At narrow DOG widths, both horizontal and vertical fusion limits approach a constant minimum threshold. Panum's fusion ranges remain fairly constant when measured with bar patterns (dotted lines) and resemble values obtained with high spatial frequency DOGs.

the low spatial frequency components. An example of this simultaneous perception can be seen with the diplopic pixels in a random dot stereogram whose coarse camouflaged form is seen in vivid stereoscopic depth (Duwaer, 1983).

In addition to target coarseness, there are several other aspects of spatial configuration that influence stereopsis and fusion. The traditional studies of stereopsis, such as those conducted by Wheatstone (1838), mainly consider the disparity stimulus in isolation from other disparities at the same or different regions of the visual field. It is said that disparity is processed locally in this limiting case, independent of other possible stimulus interactions other than the comparison between two absolute disparities to form a relative disparity. However, recent investigations have clearly illustrated that in addition to the local processes, there are global processes in which spatial interaction between multiple relative disparities in the visual field can influence both stereopsis and fusion. Three forms of global interactions have been studied. These are disparity crowding, disparity gradients, and disparity continuity or interpolation. These global interactions appear to influence phenomena such as the variation in size of Panum's fusional area, reductions and enhancement of stereo-sensitivity, constant errors or distortions in depth perception, and resolution of a 3D form that has been camouflaged with an ambiguous surface texture.

Spatial crowding of visual targets to less than 10 arc min results in a depth averaging of proximal features. This is manifest as an elevation of stereothreshold as well as a depression of the UDL (Schor et al., 1983). The second global interaction, disparity gradient, depends upon spacing between disparate targets and the difference in their disparities (Schor and Tyler, 1981). The disparity gradient represents how abruptly disparity varies across the visual field. The effect of disparity gradients upon the sensory fusion range has been investigated with point targets by Burt and Julesz (1980), and with periodic sinusoidal spatial variations in horizontal and vertical disparity by Schor and Tyler (1981). Both groups demonstrate that the diplopia threshold increases according to a constant disparity gradient as the separation between adjacent fusion stimuli increases. Cyclofusion limits are also reduced by abrupt changes in disparity between neighboring retinal regions (Kertesz and Optican, 1974). Stereothresholds can also be described as a constant disparity gradient. As target separation decreases, so does stereothreshold, up to a limit of 15 arc min separation. Further reduction in separation results in crowding, which elevates the stereothreshold. The UDL is also limited by a constant disparity gradient (Figure 5). As spacing decreases, there is a proportional decrease in the UDL. These gradient effects set two strict limitations on the range of stereoscopic depth that can be rendered by the video display. As crowding increases, the UDL will decrease. The effect is that targets exceeding the UDL will appear diplopic and without depth. For example, a top-down picture of a forest which has trees of uneven height will

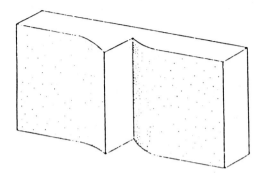

Figure 6. Perspective sketch of the illusory depth surface. Left part looks apparently nearer than the right part.

not be seen as uneven depth if the trees are imaged too closely. To remedy this problem, the depth should be reduced by moving the stereocameras closer together. In the other extreme, a shallow slope will not be seen in depth unless it exceeds the gradient for stereothresholds. Even if it does, it may still not be seen if it extends across the entire visual display. Normally there can be unequal optical errors of the two eyes which produce unequal magnification of the two retinal images. This anisotropic magnification produces an apparent tilt of the stereoscopic frame reference referred to as the fronto-parallel plane. However, this constant depth error is normally corrected or compensated for perceptually (Morrison, 1977). This perceptual compensation could reduce sensitivity to wide static displays of a shallow depth gradient.

A third form of global interaction is observed under conditions where disparity differences between neighboring regions occur too gradually to be detected, such as in the 3D version of the Craik-O'Brien Cornsweet illusion (Figure 6 by Anstis *et al.*, 1978), when stereo patterns are presented too briefly to be processed fully (Ramachandran and Nelson, 1976; Mitchison and McKee, 1985), or when several equally probable, but ambiguous, disparity solutions are presented in a region neighboring an unambiguous disparity solution (Kontsevich, 1986). Under all of these conditions, the depth percept resulting from the vague disparity is similar to or continuous with the depth stimulated by the more visible portion of the disparity stimulus. This illustrates the principle of depth continuity formulated by Julesz (1971) and restated later by Marr and Poggio (1979), which recently was shown by Ramachandran and Cavanagh (1985) to include the extension of depth to subjective contours in which no physical contour or disparity exists.

Clearly there are many spatial constraints, including spatial frequency content, retinal eccentricity, exposure duration, target spacing, and disparity gradient, which—when properly adjusted—can greatly enhance stereodepth in video displays.

References

Anstis, S.M., Howard, J.P. and Rogers, B. (1978). A Craik-Cornsweet illusion for visual depths. *Vision Research*, **18**, 213–221.

Badcock, D.R. and Schor, C.M. (1985). Depth-increment detection function for individual spatial channels. *Journal of the Optometry Society of America A.*, **2**, 1211–1216.

Beverley, K.I. and Regan, D. (1975). The relation between sensitivity and discrimination in the perception of motion-in-depth. *Journal of Physiology (London)*, **249**, 387–398.

Blakemore, C. (1970). The range and scope of binocular depth discrimination in man. *Journal of Physiology*, **211**, 599–622.

Burt, P. and Julesz, B. (1980). A disparity gradient limit for binocular fusion. *Science*, **208**, 615–617.

Duwaer, A.L. (1983). Patent stereopsis and diplopia in random-dot stereograms. *Perception and Psychophysics*, **33**, 443–454.

Erkelens, C.J. and Collewijn, H. (1985). Motion perception during dichoptic viewing of moving random-dot stereograms. *Vision Research*, **25**, 583–588.

Foley, J.M. and Richards, W. (1972). Effects of voluntary eye movement and convergence on the binocular appreciation of depth. *Perception and Psychophysics*, **11**, 423–427.

Gogle, W.G. and Teitz, J.D. (1973). Absolute motion parallax and the specific distance tendency. *Perception and Psychophysics*, **13**, 284–292.

Guttmann, J. and Spatz, H. (1985) Frequency of fusion and of loss of fusion, and binocular depth perception with alternating stimulus presentation. *Perception*, **14**, 5–13.

Helmholtz, H.C. (1866). *Handbuch der Physiologische Optik*. Hamburg: Voas.

Ittelson, W.H. (1952). *The Ames Demonstration in Perception*. Princeton: Princeton University Press.

Julesz, B. (1971). *Foundations of Cyclopean Perception*. Chicago: University of Chicago Press.

Kertesz, A.E. and Optican, L.M. (1974). Interactions between neighboring retinal regions during fusional response. *Vision Research*, **14**, 339–343.

Kontsevich, L.L. (1986). An ambiguous random-dot stereogram which permits continuous change of interpretation. *Vision Research*, **26**, 517–519.

Marr, D. and Poggio, T. (1979). A computational theory of human stereo vision. *Proceedings of the Royal Society*, **204**, 301–328.

Mitchison, G.J. and McKee, S.P. (1985). Interpolation in stereoscopic matching. *Nature*, **315**(6018), 402–404.

Morrison, L.C. (1977). Stereoscopic localization with the eyes asymmetrically converged. *American Journal Optometry and Physiological Optics*, **54**, 556–566.

Ramachandran, V.S. and Cavanagh, P. (1985). Subjective contours capture stereopsis. *Nature*, **314**, 527–530.

Ramachandran, V.S. and Nelson, J.I. (1976). Global grouping overrides point-to-point disparities. *Perception*, **5**, 125–128.

Regan, D. and Beverley, K.I. (1979). Binocular and monocular stimuli for motion-in-depth: changing disparity and changing size inputs feed the same motion in depth stage. *Vision Research*, **19**, 1331–1342.

Rogers, B. and Graham, M. (1982). Similarities between motion parallax and stereopsis in human depth perception. *Vision Research*, **22**, 261–270.

Schor, C.M. and Badcock, D.R. (1985). A comparison of stereo and vernier acuity within spatial channels as a function of distance from fixation. *Vision Research.*, **25**, 1113–1119.

Schor, C.M. and Howarth, P.A. (1986). Suprathreshold stereo-depth matches is a function of contrast and spatial frequency perception. *Perception*, **15**, 249–258.

Schor, C.M. and Kotulak, J. (1986). Dynamic interactions between accommodation and convergence are velocity sensitive. *Vision Research*, **26**, 927–942.

Schor, C.M. and Tyler, C.W. (1981) Spatio-temporal properties of Panum's fusional area. *Vision Research*, **21**, 683–692.

Schor, C.M. and Wood, I. (1983). Disparity range for local stereopsis as a function of luminance spatial frequency. *Vision Research*, **23**, 1649–1654.

Schor, C.M., Bridgeman, B. and Tyler, C.W. (1983) Spatial characteristics of static and dynamic stereoacuity in strabismus. *Investigative Ophthalmology and Vision Science*, **24**, 1572–1579.

Schor, C.M., Wood, I.C. and Ogawa, J. (1984a). Spatial tuning of static and dynamic local stereopsis. *Vision Research*, **24**, 573–578.

Schor, C.M., Wood, I. and Ogawa, J. (1984b). Binocular sensory fusion is limited by spatial resolution. *Vision Research*, **24**, 661–665.

Tschermak, A. (1930). Beitrage zur physiologishen Optik 111; Raimsinn. In Bethe, A., Bergmann, G.V., Embolen, G. and Ellinger, A. (Eds), *Handbuch der Normalen und Pathologischen Physiologic*. Berlin: Julius Springer, **12**, 833–1000.

Wertheim, T. (1894). Uber die indirekte sekscharfe. *Zeitschrift Psychologie Physiologie Sinnesory*, **7**, 172–187.

Westheimer, G. and Tanzman, I.J. (1956). Qualitative depth localized with diplopic images. *Journal of the Optical Society of the America*, **46**, 116–117.

Wheatstone, C. (1838). Some remarkable phenomena of binocular vision. *Philosophical Transactions of the Royal Society*, **128**, 371–394.

37

Stereoscopic distance perception

John M. Foley

*Department of Psychology,
University of California,
Santa Barbara, California*

Introduction

Most of this article is concerned with limited cue, open-loop tasks in which a human observer indicates distances or relations among distances. By open-loop tasks I mean tasks in which the observer gets no feedback as to the accuracy of responses. At the end of the article, I will consider what happens when cues are added and when the loop is closed, and what the implications of this research are for the effectiveness of visual displays.

Errors in visual distance tasks do not necessarily mean that the percept is in error. The error could arise in transformations that intervene between the percept and the response. I will argue, however, that the percept is in error. I will argue further that there exist post-perceptual transformations that may contribute to the error or be modified by feedback to correct for the error.

Methods

First, I will describe some experiments on binocular distance perception. The stimuli were points of light viewed in dark surroundings. These were in or near the horizontal eye-level plane. The variables that I use are illustrated and defined in Figure 1. The angle subtended by straight lines from a stimulus point to the rotation centers of the eyes is the binocular parallax of that point. (It is sometimes called the convergence angle or stimulus to convergence.) The binocular parallax and the horizontal direction, θ_i, serve as coordinates that specify the positions of points in the plane. The binocular disparity of one point relative to another is defined as the binocular parallax of the first, minus the binocular parallax of the second. Note that binocular disparity is a signed quantity; a farther point has a negative disparity relative to a nearer one. The two open dots correspond to the perceived positions of r and i. The binocular parallax of the perceived position of a point is called the effective

Stereoscopic distance perception

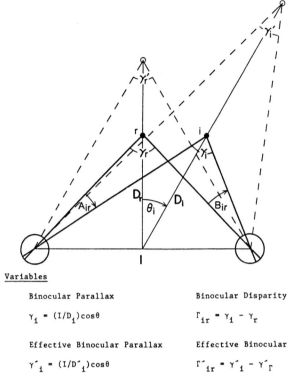

Variables

Binocular Parallax

$$\gamma_i = (I/D_i)\cos\theta$$

Binocular Disparity

$$\Gamma_{ir} = \gamma_i - \gamma_r$$

Effective Binocular Parallax

$$\gamma'_i = (I/D'_i)\cos\theta$$

Effective Binocular

$$\Gamma'_{ir} = \gamma'_i - \gamma'_r$$

Figure 1. Variables used in this article. The figure is a top view of the horizontal eye-level plane. The large circles at the bottom represent the two eyes and the solid dots labeled r and i correspond to two stimulus points. The expressions at the bottom of the figure define the four variables. I is interocular distance. D is radial distance to a point. θ_i is horizontal direction of a point relative to straight ahead. D' is perceived radial distance.

binocular parallax of the point. The difference between two effective binocular parallaxes is an effective binocular disparity. These perceptual variables are defined in the same way as the corresponding physical variables except that perceived distance, D', is substituted for physical distance, D, in each equation. I assume that perceived horizontal direction equals physical horizontal direction. There is evidence that this is correct under the conditions of my experiments.

Some of the experiments I will describe were done with stimulus points at different distances. Others were done by simulating the distance dimension stereoscopically. If the stimulus to vergence is not grossly different than the stimulus to accommodation, the results are very similar. Some of the experiments employed a fixation point; others allowed the observers to move their eyes freely. When disparities are small, the results are again very similar.

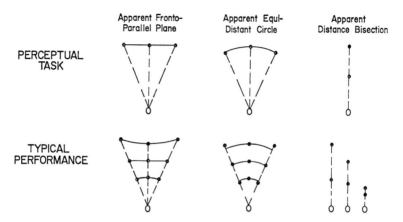

Figure 2. Illustration of three relative distance tasks (top) and typical performance for observers who show no skewing (bottom). The physical configuration corresponds to the perceptual criterion only at one distance, which is typically between 1 and 4 m. The diagram is not to scale.

Relative distance tasks

I will describe performance on two classes of distance tasks. The first are called relative distance tasks; they are tasks in which an observer adjusts the position of light points by remote control until they satisfy some relative distance criterion (Foley, 1978, 1980). Examples of such criteria are shown in Figure 2. In each case the view is from above; the oval represents the observer's head and the dots represent stimulus lights. In the apparent fronto-parallel plane (AFPP) task, one point of light is fixed and the observer moves other lights so that they appear to lie in the vertical plane through the fixed light that is parallel to the vertical plane through the eyes or, in other words, a plane that is perpendicular to straight ahead. The apparent equidistant circle (AEDC) task is very similar, except that the lights are set so that they are perceived to lie on a circle with the observer at the center. In the apparent distance bisection (ADB) task, one point is fixed and the observer adjusts a second point so that the distance between the two points is perceived to equal the distance from the observer to the near point.

Typical performances in these tasks are illustrated in the second row for three distances of the fixed point. In each task there is one distance at which the physical configuration corresponds to the perceived configuration. This distance is generally within the range of 1–4 m. At other distances, there are systematic errors in the settings. At far distances, variable points are set too far, and at near distances, they are set too near, relative to accurate performance. Although there are individual differences in the magnitude of the errors, errors of this kind are reliably found. (For many observers, one side

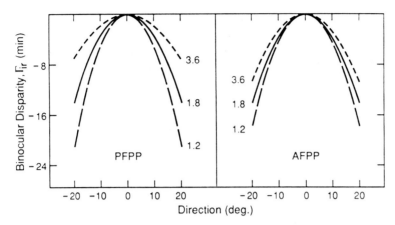

Figure 3. Binocular disparity as a function of horizontal directions for PFPP and AFPP; the smooth curves describe the results of a typical observer. Each function is shown for three distances of the fixed center point: 1.2, 1.8, and 3.6 m. For this observer the functions correspond at 1.8 m. As distance becomes greater or less than this, the disparities that correspond to the AFPP change less than those corresponding to a PFPP.

of the configuration is set closer than the other (skewing). This can be accounted for by a very small difference in magnification in the two eyes. This is incorporated in a general theory of binocular distance perception (Foley, 1980), but it is not considered in this article.)

I propose that these errors can be explained by the misperception of the egocentric distance to the fixation point, or, in the absence of a fixation point, to a reference point that depends on the configuration of points. To test this idea we must consider how the pattern of disparities produced by the observer compares with the pattern of disparities corresponding to the physical configuration specified by the instructions. By pattern of disparities I mean the function that relates binocular disparity to direction. The left side of Figure 3 shows this function for physically fronto-parallel planes (PFPP) at different distances and the right side shows the same function for AFPP at different distances. If all the error in the AFPP settings is due to the misperception of the distance to the fixation point, then the function for an AFPP should be identical to the function for a PFPP, but generally this will be a PFPP at another distance. This is what the experiments show. For example, an AFPP at 1.2 m has less disparity than a PFPP at 1.2 m, but corresponds to the same disparity pattern as a PFPP at 1.45 m. Patterns of disparities obtained in the AEDC task also correspond closely with disparities produced physically by EDCs at other distances. Thus, the experimental settings can be accounted for by the hypothesis that the observer misperceives the egocentric distance to the configuration and produces the pattern of disparities appropriate to the misperceived distance.

This hypothesis has several important implications. First, the fact that the pattern of disparities changes with the distance to the fixed point implies that there is an egocentric distance signal related to the vergence of the eyes, and that this egocentric distance signal is not accurate. Second, effective binocular disparity equals binocular disparity. This is illustrated in Figure 1. In general, the distance to point r will be misperceived. But if r is misperceived, any other point i will also be misperceived, so that the difference between the effective binocular parallaxes equals the difference between the binocular parallaxes. I call this the effective disparity invariance principle.

The data from relative distance tasks may be used to infer the perceived distance to the fixation point or to the reference point. The simplest way to conceptualize this is to imagine a more complete set of functions on both sides of Figure 3. Then, for each pattern on the right, we find the matching pattern on the left. The distance on the right is the physical distance that corresponds to the perceived distance on the left. This perceived distance is a concave downward function of physical distance, as is shown by the solid line on the left side of Figure 4. When both physical distance and perceived distance are transformed to parallaxes, their relation becomes linear, as is shown by the solid line on the right side of this Figure. I call the curved function on the left the reference distance function and the linear function on the right the reference parallax function.

Egocentric distance tasks

Next consider a different class of tasks—egocentric distance tasks. An egocentric distance task is one in which an observer indicates the distance from herself or himself to visual targets (Foley, 1977, 1985). Several different indicators have been used, but I have relied on two, verbal reports of perceived distance and pointing with an unseen hand. In the pointing experiments a horizontal board just beneath the targets prevents the observer from seeing his or her hand or arm. I will describe two simple experiments.

In the first experiment the stimulus is a single light point in dark surroundings. It is straight ahead. Pointed distances and reported distances from such experiments are shown in Figure 4. The smooth curves shown have parameters that are close to the average values fitted to the data of five observers (Foley, 1977). On the left, indicated distance is plotted against physical distance, and on the right, the same values are plotted as binocular parallaxes. The functions on the left have the same form as the reference distance function; those on the right, the same form as the reference parallax functions.

But there is a complication: verbal and manual indicators do not agree, and neither, in general, agrees with the function inferred from the relative distance tasks, which tends to lie between the verbal and manual functions. Since the indicators do not agree, both cannot correspond to perceived distance. I have defined perceived distance as the distance inferred from the

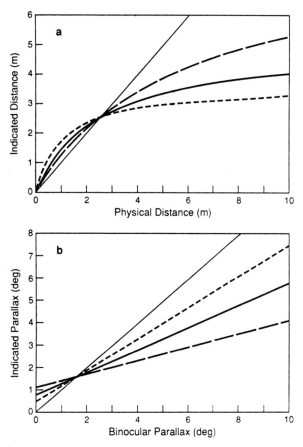

Figure 4. (a) Perceived (or indicated) distance as a function of target distance. Bold solid line, perceived distance inferred from relative distance tasks; long dashes, perceived distance indicated by manual pointing; short dashes, perceived distance indicated by verbal report. (b) The same three functions expressed as parallaxes.

relative distance tasks. When expressed as parallaxes, this value and the values indicated by pointing and verbal reports are all linearly related. This means that egocentric distance tasks can be used to test the implications of the theory. It is very important, however, to distinguish between perceived distance and indications of it. In Figure 4 only the solid lines derived from the relative distance tasks correspond to perceived distance and reference parallax; the other lines describe indicated distance and indicated parallax.

When the eyes move freely, there is one point the perceived distance of which is given by the reference distance function. I call this point the reference point. Perceived distances of all other points are determined by their disparities relative to this point. There are several ways to determine the reference point. The most obvious is to measure the effective parallax of each

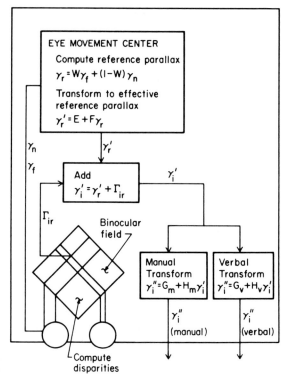

Figure 5. Diagram summarizing the formal operations of the model in a way that suggests underlying structures and processes (from Foley, 1985).

point in the configuration and then determine how these are related to the reference parallax function. This analysis has been carried out only for the case of two-point configurations (Foley, 1985). Here the parallax of the reference point is a weighted average of the parallaxes of the points, with the farther point tending to receive the greater weight. Thus the reference point need not correspond to any point of the configuration, although sometimes it may.

Discussion

Figure 5 is a schematic diagram illustrating the process of binocular distance perception. The visual system generates both binocular parallax and binocular disparity signals in response to the optic array. The binocular parallax signals determine a single reference point and its corresponding value of effective binocular parallax. Here this is shown as an outflow from an eye movement control center. For each point i, the disparity of i relative to the reference point is added to the effective reference parallax to give the effective parallax of the point. This value undergoes an indicator-specific linear transform to yield the indicated binocular parallax, which, in turn, determines the response.

When multiple cues are present, including perspective cues, distance perception is more accurate; however, the evidence indicates that there are systematic errors in distance perception under most cue conditions. There are several studies that have examined apparent distance bisection under such conditions. Although results have varied widely, no study has found consistently accurate bisection over a wide range of distances. The most common result is that the farther interval is set larger than the nearer one. There are also several studies that have obtained verbal reports of perceived distance under multiple cue conditions. The data are often fitted with a power function and the power is generally less than 1. An experiment limited to distances less than 70 cm yielded an accelerating verbal report function and a decelerating pointing response function (Foley, 1977). When the inverse output transforms derived from binocular experiments are applied to these data, both verbal and manual responses yield the same parallax function with a slope of about 0.8. The conclusion is that distance perception is generally inaccurate, even in the presence of multiple cues.

How can we perform accurately with respect to distance when distance perception is inaccurate? I can only answer this speculatively because the experiments needed to answer it scientifically have not been done. I hypothesize that we learn to behave accurately on the basis of feedback. This learning cannot be once and for all because the errors that it compensates for vary continuously with changing cue conditions. I hypothesize that the output transforms that I have proposed to explain open-loop performance are modified by feedback to compensate for perceptual errors.

What implications does this have for the design of visual displays? I would expect that most visual displays evoke erroneous distance percepts. I expect this because even a three-dimensional scene with multiple cues evokes erroneous percepts, and most displays both eliminate cues and introduce cue conflicts, both of which are associated with increasing errors. In principle, it might be possible to create a display that would evoke accurate percepts, at least in some limited domain, but I doubt the wisdom of attempting this. The perceptual-motor system is designed to make rapid compensation for certain forms of error, especially those that can be described by linear transforms of the reference parallax function. Displays that produce errors of this form should suffice to direct behavior. But every time a display is used to direct behavior in the real three-dimensional space, performance with feedback is necessary to calibrate the output transforms, just as performance with feedback is necessary when a three-dimensional scene directs behavior.

References

Foley, J.M. (1977). Effect of distance information and range on two indices of visually perceived distance. *Perception*, **6**, 449–460.

Foley, J.M. (1978). Primary distance perception. In Held, R., Leibowitz, H. and Teuber, L. (Eds) *Handbook of Sensory Physiology*, **8**, 181–213. Berlin: Springer-Verlag.
Foley, J.M. (1980). Binocular distance perception. *Psychological Review*, **87**, 411–434.
Foley, J.M. (1985). Binocular distance perception: Egocentric distance tasks. *Journal of Experimental Psychology: Human Perception and Performance*, **11**, 133–149.

38

Paradoxical monocular stereopsis and perspective vergence

J.T. Enright

*Scripps Institution of Oceanography,
La Jolla, California*

Summary

The question of how to convey depth most effectively in a picture is a multifaceted problem, both because of potential limitations of the chosen medium (stereopsis? image motion?), and because "effectiveness" can be defined in various ways. Practical applications usually focus on "information transfer," i.e., effective techniques for evoking recognition of implied depth relationships, but this issue depends on subjective judgments which are difficult to scale when stimuli are above threshold. Two new approaches to this question are proposed here which are based on alternative criteria for effectiveness.

Paradoxical monocular stereopsis is a remarkably compelling impression of depth which is evoked during one-eyed viewing of only certain illustrations; it can be unequivocally recognized because the feeling of depth collapses when one shifts to binocular viewing. An exploration of the stimulus properties which are effective for this phenomenon may contribute useful answers for the more general perceptual problem.

Perspective vergence is an eye-movement response associated with changes of fixation point within a picture which implies depth; it also arises only during monocular viewing. The response is directionally "appropriate" (i.e., apparently nearer objects evoke convergence, and vice versa), but the magnitude of the response can be altered consistently by making relatively minor changes in the illustration. The cross-subject agreement in changes of response magnitude would permit systematic exploration to determine which stimulus configurations are most effective in evoking perspective vergence, with quantitative answers based upon this involuntary reflex. It may well be that "most effective" pictures in this context will embody features which would increase "effectiveness" of pictures in a more general sense.

Introduction

One of the central issues involved in spatial display is the question, "What is the most effective way to convey three-dimensional depth in a pictorial representation?" This article deals only with a very restricted approach to that question, being confined to representations without stereopsis and without image motion; and so the problem addressed here should probably be rephrased, "What is the *third* most effective way of conveying depth in pictures?" Such rephrasing seems appropriate because there can be little doubt that the most effective representations of the third dimension are those which involve stereopsis; and that the second most effective way to convey a feeling for depth is through use of image motion: optical flow patterns, image shear, motion parallax and the like. When both stereopsis and image motion are excluded, one is dealing with no more than third best; and the rephrased question is in some ways like asking what is the best way to participate in a foot race, subject to the precondition that the runner's feet be tied together by his shoelaces.

Nevertheless, the question of how best to convey the third dimension in a static pictorial representation has been of central concern to artists for many hundreds of years; and the result of that interest is an organized body of technique, collectively known as perspective, to deal empirically with that problem. One might well ask, then, whether there is any hope for deriving new answers to this question—if thousands of artists, throughout their careers, have been experimenting for centuries with just this objective in mind. The honest reply is that this article has no new answers to offer, no new tricks to suggest. Instead, it focuses upon two interesting phenomena involving the perception of and response to depth in illustrations—phenomena which seem to me to have the potential of providing more quantitative answers to the question, "How can depth be more effectively represented?" These phenomena suggest research programs for the future, which would address this question within certain restricted contexts, and it is conceivable that the answers might be applicable to other, more general contexts as well. The hope is that such research might provide general, quantitative rules for optimizing the depth impression which is conveyed by the stimulus field in an illustration.

Paradoxical monocular stereopsis

The first of the phenomena of interest here is a remarkable and relatively little-known sort of depth perception which was described by the French visual scientist, Claparède, in a brief article published in 1904; he christened this visual experience "paradoxical monocular stereopsis". The essence of Claparède's message is that if certain pictures which illustrate a three-dimensional scene—drawings, paintings or photographs—are carefully examined *with one eye covered*, a truly compelling sense of depth can sometimes be obtained, an effect nearly as striking as looking into a stereoscope. Once this sort of perception has been achieved, it can be sustained while continuing to inspect the picture, and one might suspect that it results simply from thinking about and focusing attention on the illustrated subject matter. It is easy to demonstrate, however, that something unusual is involved, because the moment that the other eye is opened, to see the picture binocularly, the anomalous 3D effect vanishes; the picture flattens out just as suddenly and completely as when one *closes* one eye while looking into a stereoscope.

High-quality, well-printed color photographs of outdoor scenes, of the sort found in magazines like *National Geographic and Arizona Highways*, often provide good material for demonstrating this sort of depth perception, but one of the most interesting aspects of paradoxical monocular stereopsis is how difficult it is to predict whether a given illustration will be effective in evoking the response. The compelling impression of depth is not simply a response to monocular viewing of all illustrations which show a three-dimensional scene, but to certain configurations of stimuli. The question therefore arises, "What is the most effective way to evoke paradoxical monocular stereopsis with an illustration?" This is, of course, a much more limited question than asking what is the most effective way to convey depth in a picture, but it may be more tractable. One has available the clear-cut criterion, "Does the (supplementary) depth impression flatten out, when switching over to binocular viewing?" Furthermore, although the best stimuli for paradoxical monocular stereopsis may not turn out to be fully congruent with the stimuli which are optimal for conveying a three-dimensional impression during binocular viewing, preliminary evidence suggests that if a picture is effective in evoking paradoxical stereopsis, it will at least give a satisfying and convincing impression of depth during binocular viewing.

A search of the published literature indicates that there have apparently been no systematic investigations of which kinds of pictures best evoke paradoxical stereopsis; and in fact, I have encountered less than a dozen references, in the entire 80-year interval since Claparède's (1904) initial description of the phenomenon, in which this sort of depth perception is even mentioned (e.g., Pirenne, 1970; Schlosberg, 1941; Ames, 1925; Streiff, 1923; and the references cited there). Qualitative preliminary testing indicates that there is good agreement among subjects, in the sense that certain pictures seem to be very effective stimuli for everyone, so the project of exploring stimulus optimization should be relatively easy to carry through, with a relatively modest number of subjects. And if the illustrations which are to be used were to be carefully selected, it seems very likely that an organized body of rules will emerge which characterize the optimal stimuli.

Perspective vergence

In the brief article in which Claparède (1904) described this unusual sort of depth perception, he also proposed an interesting hypothesis about the mechanisms responsible. He speculated that during monocular inspection of a picture, the covered eye would be free to make vergence movements which might correspond to the relative distances implied by the illustration (converging, then, for apparently near objects and diverging for more remote ones), just as changes in vergence accompany binocular inspection of a real, three-dimensional scene. He pointed out that vergence changes of this sort could not take place during binocular viewing of a picture because of the demand for fusion; and he further proposed that this sort of postulated vergence movement might be responsible for the compelling sense of depth evoked during monocular viewing. Apparently there has been no test of Claparède's hypothesis, nor even any restatement of it, in the subsequent 80 years; a recently initiated research program, however, has provided compelling evidence that Claparède was essentially correct in his speculation about eye movements (Enright, 1987a, b). Vergence changes of the sort he postulated do, indeed, take place when inspecting a picture of a three-dimensional scene with one eye covered—though whether those eye movements are responsible for paradoxical stereopsis remains an open question, and one which will be much more difficult to investigate.

Methods

The experimental equipment which was used in this eye-movement research is extremely simple, both in principle and in practice (Figure 1). The subject sits with head held firmly in place by a bite board and headrest while two video cameras monitor eye position from somewhat below the line of sight. The output of the cameras is combined with an image splitter and recorded for subsequent analysis; the sum of the two distances between iris margins and the image-splitting line is an index for vergence state. The illustrations to be viewed are mounted at about 30 cm from the subject's eyes, and an obstruction is placed a few centimeters in front of the nondominant eye, at a level which hides the picture from that eye, but permits the camera to record eye position. While viewing the picture monocularly, the subject changes fixation at intervals of 2 to 3 sec, between points which are at different implied distances away. Single-measurement precision of the recording method is about 6 arcmin for each evaluation of eye position, and averaging results over repeated tests can further reduce the influence of random measurement error; but the between-trial variability within a given test session for a given subject and target is sufficiently large that a more precise monitoring technique could not appreciably improve the reliability of the estimates of average response; the variability in the eye movements from one

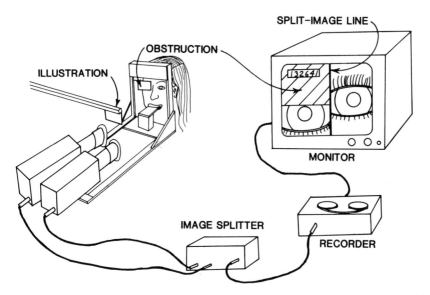

Figure 1. Diagram of the equipment and setup used for recording eye position while viewing illustations.

refixation to the next limits precision of the estimates, as reflected in the standard errors.

Results

An excerpt from a longer recording is shown in Figure 2, made while a subject changed fixation from the upper front corner to the upper back corner of the perspective drawing of a small box (target illustrated in Figure 3). Concurrent with the recording, a three-position switch, which was connected to two tone generators, was activated by the subject to indicate the fixation point; the timing of those signals is shown as open and solid bars in Figure 2. It is, then, quite clear that convergence occurred while fixating on the apparently nearer corner of the box, and divergence while fixating on the farther corner. A simple summary value for the typical vergence-change response can be obtained from such a recording, based on measuring one value of vergence state for each steady-state fixation, and then calculating differences between successive values; in this case, the average change in vergence, over 20 fixations, was 68 arcmin ±8 arcmin. In Figure 3, this summary value is shown for Subject 1, along with five other values for her, each with this same target, each recorded on a different day; and values of average vergence change are also shown there for another eight subjects with this target. Average vergence change, based on the method of calculation, could in principle also be negative (i.e., contrary to the perspective implication of the drawing); in fact, however, all 24 measured values are positive,

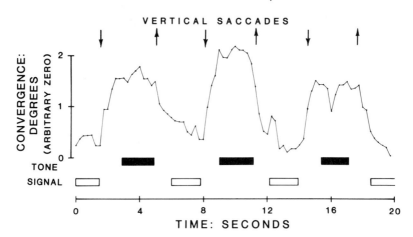

Figure 2. Excerpt from a recording made while Subject 1 alternated monocular fixation between apparently nearer and apparently farther topside corners in a line drawing of a small cubical box (picture shown in Figure 3 and as "Standard" in Figure 4). Bars beneath graph correspond to the timing of tone signals; solid bars represent fixation on "near" corner, open bars represent fixation on "far" corner. (Reprinted with permission from Vision Research **27**, J.T. Enright, "Perspective vergence: oculomotor response to line drawings", Copyright 1987, Pergamon Journals Ltd.)

and all except one of the results are statistically significant, most of them at the 0.01 level. In other words, the subjects all showed consistent vergence changes during changes in fixation point in this drawing; and those vergence changes corresponded in direction with the relative distances implied by the perspective of the drawing. For those who may be concerned about the reliability of this simple and unconventional method of recording eye movements, it is worth mentioning that the basic result of Figure 3 has now been replicated for other subjects in two other laboratories, each of them using a fundamentally different and more familiar measurement technique. I have proposed (Enright, 1987a) that these oculomotor responses to pictorial representations be called "perspective vergence".

Before considering additional details of the responses which have been measured for other kinds of illustrations, it seems worthwhile to try to place perspective-vergence responses into some sort of broader context. A phenomenon which is now called "proximal vergence" has long been known to visual physiologists, an eye-movement response which has been attributed to "knowledge of nearness" (Maddox, 1893). Although vergence responses to perspective representations have not been previously studied, it is probably appropriate to consider perspective vergence to be a subcategory of "proximal vergence" (Hokoda and Cuiffreda, 1983). It is important, however, to distinguish between these responses and another subcategory known as "voluntary vergence": some trained subjects can cross or uncross their eyes

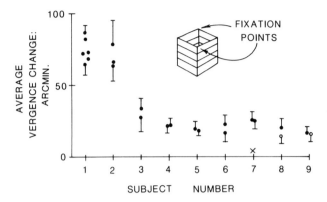

Figure 3. Summary of average vergence changes made by 9 subjects in conjunction with changes in fixation on the line drawing of a small cubical box; each point represents average value during a separate test session, with standard errors based on N of 10 (20 changes in fixation).

at will, even in total darkness. Many lines of evidence indicate, however, that the eye-movement responses to perspective illustrations are instead the result of an involuntary reflex. It is conceivable—even likely—that training or an "act of will" might enhance the responses, but fully naive, untrained subjects also show comparable behavior in their first test session—even subjects who are fully unaware that convergence is the appropriate response to objects which are nearby. They show this response even though they are uninformed about the purpose of the experiment, even though they have no visual feedback or other clues to tell them whether vergence has changed—much less whether the response was "as intended". Perspective vergence is an automatic response to components of the visual stimulus field—truly a reflex. Furthermore, at least certain components of the stimulus field which evoke this kind of response are apparently not a reflection of learning or prior experience, but instead represent built-in constraints on the visual system—although it seems likely that "learning" may also play a role—that prior visual experience with our three-dimensional world may build upon and supplement those components which are "hard-wired" into the system. Because of the reflex nature of the responses, an evaluation of illustrations, in terms of the magnitude of the vergence responses evoked, represents something far more substantial than can be achieved by asking for subjective opinions about picture quality.

An experimental program has been initiated, designed to determine what features of an illustration enhance or inhibit this oculomotor response. The results of Figure 4 summarize some of the kinds of data which have been obtained, with modest variations on the compositional theme of a single rectangular box. Despite the large inter-subject differences in response magnitude for a given picture, as shown in Figure 3, there are remarkably

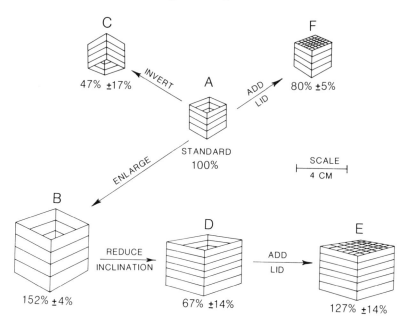

Figure 4. Cross-subject mean values and their standard errors for 100 times the ratio: "average vergence change for a given drawing", divided by the same-subject value of "average vergence change for 'standard' illustration". $N = 3$ for part B, $N = 9$ for all other parts.

consistent cross-subject *changes* in response magnitude for particular alterations in the picture; hence, the ratio of response for a given picture to the same subject's response for a standard, represents a reliable way of demonstrating the relative effectiveness of various representations in evoking perspective vergence. Doubling the size of the picture in all dimensions, for example, reliably led to an increase of about 50 per cent in response magnitude (Figure 4 *vs.* Figure 4B); inverting the picture led to a reduction in response (Figure 4A *vs.* Figure 4C), with 7 of 9 subjects showing smaller vergence changes. A reduction in the inclination of the box (with only minor other modifications in line spacing) led to a drastic reduction in response magnitude (Figure 4B *vs.* Figure 4D); for 8 of the 9 subjects, the response was even smaller than that to the "standard" picture, which shows a box half the size (Figure 4A). When a cross-hatched lid was superimposed upon a box which was in the relatively ineffective orientation, response magnitude increased for all 9 subjects (Figure 4D *vs.* Figure 4E), but when a similar lid was superimposed on a box with more effective orientation, it tended to reduce the response (Figure 4A *vs.* Figure 4F; 8 subjects out of 9). In all cases, there was remarkably good cross-subject agreement in the way in which a given change in the drawing affected magnitude of the response (details in Enright, 1987a).

One other closely related kind of target has been tested, which is not shown in this figure; three-dimensional cardboard models of the boxes shown in Figures 4A and 4D were constructed and photographed from 30 cm with illumination which produced a distribution of light and shadow, and prints of those photos, at appropriate scaling, were tested as targets. The rationale for this approach is that shading might enhance the resulting vergence changes. In these tests there was indeed a slight but significant increase in response for the box shown with suboptimal orientation (Figure 4D), but no significant change—in fact a slight decrease—for the more optimally oriented box (Figure 4A).

The vergence responses of this same group of 9 subjects have also been tested with a set of more complex pictorial representations: photographs which reproduce five classical paintings and an etching; and those experimental results have offered further hints about the kinds of stimuli which can be effective in evoking perspective vergence. By using a portrait by Rembrandt, for example, statistically significant vergence changes in the appropriate direction (nearly as large as those for the "small-box" drawing [Figure 3]), were evoked in all 9 subjects by a change in fixation from the nose to the ear of the portrayed philosopher and back again, although no suggestion of *linear* perspective was evident in the picture, and the implied difference in distance between the fixation points was quite small (ca. 10 cm, at a distance of 2 to 3 m from the viewer). One landscape scene evoked strong responses in every subject tested, and another outdoor scene, in which linear perspective was conspicuous, did not lead to statistically significant results for *any* of the subjects. Again, then, there was very good cross-subject agreement, in terms of which artworks were effective stimuli and which were not.

Discussion

The cross-subject consistency in terms of response magnitude demonstrates that in measuring perspective vergence we are dealing with relatively general characteristics of the oculomotor response system; but the experiments conducted so far do no more than define a few of the dimensions of the multidimensional coordinate system implied in the question, "What is the optimal stimulus for this response?" There seems to be clear non-additivity (a cross-hatched surface between fixation points enhances a response, or it does not, depending on context), which considerably complicates the exploration of these dimensions. Furthermore, it is by no means clear that the rules which might be derived from a line drawing of a cubical box can be generalized to other sorts of figures; nor do the available data define an optimum point in any stimulus dimension. Consider, for example, the conspicuous effect of tilt of the opening on responsiveness (Figure 4B *vs.* 4D): while it seems clear that a 22° tilt (4B) is much more effective than an 11° tilt (4D), there is presumably a continuous function relating responsiveness to inclination in the illustrated box, with a maximum someplace between 0° and 90°; and it may well be that 22° is far removed from that optimum tilt. The necessary experiments to explore this dimension should be enlightening—but the existence of nonlinearities cautions against overgeneralization.

The consistently positive responses to the Rembrandt portrait demonstrate that the dimensions which must be explored in any complete attempt to define optimal stimuli go far beyond the systems of lines and angles which constitute linear perspective. The opportunity to explore the question of stimulus optimization offers exciting promise for the future, but it is self-evident that the available data do not even adequately define the dimensions of the problem. Beyond the issue of stimulus optimization, the intriguing possibility exists that perspective vergence responses may provide an objective metric for evaluating the general effectiveness of an attempt to convey depth in a picture: that oculomotor responsiveness may prove to be well correlated with subjective perceptual responsiveness to pictorial implications of depth. Such a correlation would be a necessary—but not a sufficient—condition for establishing the validity of Claparède's most interesting speculation: that perhaps vergence movement itself contributes to the perception of paradoxical monocular stereopsis.

Acknowledgement

Research supported by Grant Number BNS 85-19616 from the National Science Foundation.

References

Ames, A., Jr. (1925). The illusion of depth from single pictures. *Journal of the Optical Society of America*, **10**, 137–148.
Claparède, E. (1904). Steréréoscopie monoculaire paradoxale. *Annales d' Oculistique*, **132**, 465–466.
Enright, J.T. (1987a). Perspective vergence: oculomotor responses to line drawings. *Vision Research*, **27**, 1513–1526.
Enright, J.T. (1987b). Art and the oculomotor system: perspective illustrations evoke vergence changes. *Perception*, **16**, 731–746.
Hodoka, S.C. and Cuiffreda, K.J. (1983). Theoretical and clinical importance of proximal vergence and accommodation. In Schor, C.M. and Cuiffreda, K.J. (Eds), *Vergence Eye Movements; Basic and Clinical Aspects*, 75–97. London: Butterworths.
Maddox, E.E. (1893). *The Clinical Use of Prisms; and the Decentering of Lenses*. Second Edition. Bristol: John Wright and Sons.
Pirenne, M.H. (1970). *Optics, Painting and Photography*, Cambridge, England: Cambridge University Press.
Schlosberg, H. (1941). Stereoscopic depth from single pictures. *American Journal of Psychology*, **54**, 601–605.
Streiff, J. (1923). Die binoculare Verflachung von Bildern, ein vielseitig bedeutsames Sehproblem. *Klinische Monatsblaetter fuer Augenheilkunde*, **70**, 1–17.

39

The eyes prefer real images

Stanley N. Roscoe

ILLIANA Aviation Sciences Limited,
Las Cruces, New Mexico

For better or worse, virtual imaging displays are with us in the form of narrow-angle combining-glass presentations, head-up displays (HUD), and head-mounted projections of wide-angle sensor-generated or computer-animated imagery (HMD). All of our military and civil aviation services and a large number of aerospace companies are involved in one way or another in a frantic competition to develop the best virtual imaging display system. The success or failure of major weapon systems hangs in the balance, and billions of dollars in potential business are at stake. Because of the degree to which our national defense is committed to the perfection of virtual imaging displays, a brief consideration of their status, an investigation and analysis of their problems, and a search for realistic alternatives are long overdue.

Current status

All of our currently operational tactical fighter aircraft are equipped with HUDs. Helicopters are navigated and controlled, and their weapons are delivered, with a variety of imaging displays including, in addition to HUDs, both panel-mounted and head-mounted image intensifiers and forward-looking infrared (FLIR) and low-light TV displays. Even some strategic aircraft and a few commercial airliners contain virtual imaging displays. A new generation of remotely piloted vehicles (RPV) is intended to be flown by reference to wide-angle but relatively low-resolution sensor imagery presented stereoscopically by head-mounted binocular displays. And Detroit is about to offer HUDs for cars.

The trouble with HUDs and HMDs

As for the operational problems, about 30 per cent of tactical pilots report that using a HUD tends to cause disorientation, especially when flying in and out of clouds (Barnette, 1976; Newman, 1980). Pilots frequently experience confusion in trying to maintain aircraft attitude by reference to the HUD's artificial horizon and "pitch-ladder" symbology, particularly at night and over water, and there are documented cases of airplanes becoming inverted without the pilots' awareness (Kehoe, 1985). Pilots have also reported a tendency to focus on the HUD combining glass instead of the outside real-world scene (Jarvi, 1981; Norton, 1981). The resulting myopia is a special case of the more general anomaly known as "instrument myopia" (Hennessy, 1975).

Misaccommodation of the eyes

Whatever the cause, it is a repeatedly observed experimental fact that our eyes do not automatically focus at optical infinity when viewing collimated virtual images, but lapse inward toward their dark focus, or resting accommodation distance, at about arm's length on average (Hull et al., 1982; Iavecchia et al., 1988; Norman and Ehrlich, 1986; Randle et al., 1980). The perceptual consequence of positive misaccommodation is that the whole visual scene shrinks in apparent angular size. This shrunken appearance causes distant objects to be judged farther away than they are, and anything below the line of sight, such as the surface of the terrain or an airport runway, appears higher than it really is relative to the horizon (Roscoe, 1984, 1985).

The effect of the HUD optics is illustrated in Figure 1. The experiment was conducted by Joyce and Helene Iavecchia at the Naval Air Development Center in Pennyslvania. A HUD was set up on one rooftop and a "scoreboard" assembly with selectively lighted numerals of various sizes was mounted on top of another building 182 m away and of about the same height. Observers were asked to read scoreboard numbers as they appeared and also numbers presented by the HUD on half the trials. Concurrently, the eye accommodation of the observers was measured with a polarized vernier optometer.

Figure 1 shows the average focal responses to the scoreboard numerals and the background terrain beyond the scoreboard, with the HUD turned off and with it turned on. In either case the observers' focal responses were highly dependent on their individual dark focus distances; in fact, knowing each individual's dark focus accounted for 88 per cent of the variance in focal responses under all conditions of the experiment. Excluding Observer 9, whose dark focus was almost three diopters (D) beyond infinity, the average for the remaining nine emmetropes was 1.06 D, or just short of 1 m.

But the striking result shown in Figure 1 is the fact that when the HUD was turned on, for all 10 observers, focus shifted inward from an average of 0.02 D, or 50 m, to an average of 0.20 D, or 5 m. Once again excluding Observer 9, the average inward shift was from 0.27 D, about 4 m, to 0.47 D,

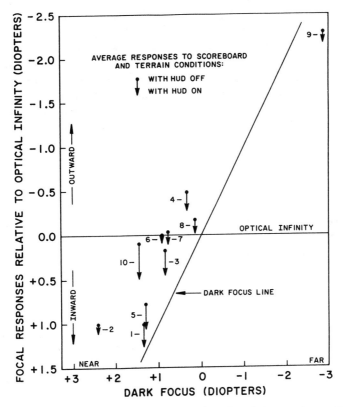

Figure 1. Average focal responses to the scoreboard and the terrain conditions with HUD on and off, plotted against each individual's dark focus.

about 2 m. Although such shifts have little effect on the apparent clarity of the visual scene, they have tremendous effects on the apparent size, distance, and angular direction of terrain features.

Accommodation and apparent size

Despite wide individual differences among observers, the average apparent size of objects is almost perfectly correlated ($r > 0.9$) with the distance at which the eyes are focused (Benel, 1979; Hull et al., 1982; Iavecchia et al., 1983; Roscoe et al., 1976; Simonelli, 1979). Thus, the positive misaccommodation induced by collimated HUD symbology can partially account for the fact that pilots flying airplanes or flight simulators by reference to virtual imaging systems make fast approaches, round out high, and land long and hard (Campbell et al., 1955; Palmer and Cronn, 1973).

Such biased judgments also partially account for the fact that helicopter pilots flying with imaging displays frequently collide with trees and other surface objects and the fact that the US Air Force between 1980 and 1985 lost 73 airplanes in clear weather because of pilot misorientation, resulting in controlled flight into the terrain (54), or disorientation resulting in loss of control (19) while flying by reference to collimated HUDs (Morphew, 1985). When flying by reference to panel-mounted or head-mounted imaging displays, helicopter pilots approach objects slowly and tentatively, and still they are frequently surprised when an apparently distant tree or rock suddenly fills the wide-angle sensor's entire field of view.

Fixed-wing airplane pilots flying with HUDs also judge a target to be farther away and the dive angle shallower than they are, resulting in almost-always-fatal "controlled-flight-into-the-terrain" accidents. In the US Air Force, such accidents have continued to occur at the rate of about one per month since HUDs came into general use at the beginning of this decade. Two months ago (June 1987) an F-16 left a smoking hole in the ground, and last month it was an F-111. The Navy's experience has been essentially the same.

Optical minification

Misorientation and disorientation with panel-mounted and some head-mounted imaging displays are exacerbated by the fact that limited display size and the need to display the widest practical outside visual angles typically result in drastic optical minification, which adds to the perceptual minification caused by the misaccommodation. If the display area were not so limited and could be varied to accommodate the wide individual differences in dark focus distances, images of the outside world could be magnified by appropriate amounts to neutralize each individual's perceptual bias. The average magnification required would be $\times 1.25$ (Roscoe, 1984; Roscoe et al., 1966), but this value would be correct for only a portion of the population, possibly requiring stricter pilot selection.

Image quality

Display minification and perceptual biases are two sources of error in human judgments of size, distance, and angular location, but there are other sources of error as well, namely, the variable errors associated with adverse ambient viewing conditions (atmospheric attenuation and reduced illumination), the limited resolution of cameras and display devices, and the further loss of resolution with image intensification. All of these factors serve to reduce contrast and detail, the principal components of image quality, and the accuracy with which people can extract positions, rates, and accelerations relative to outside objects in the visual environment.

Display alternatives

Because of the adverse effects of virtual images on eye accommodation, as well as the optical minification and poor image quality typically associated with sensor-generated displays, our judgments of spatial relations are simply not good enough to support complex flight missions as safely or effectively as we need. To date the advocates of virtual image displays have adamantly refused to acknowledge the implication of misaccommodation in the misorientation and disorientation of pilots flying with HUDs. Instead they have attributed the problems primarily to the limited fields of view afforded by the combining glasses used with current systems.

To address the limited-field-of-view problem, each of our military services, including the Marines, is spending millions of dollars a year—to say nothing of the IR&D funds invested by private companies—to develop wide-angle, head-mounted imaging displays, in many cases coupling camera line-of-sight to head or eye orientation. Still clinging to the assumption that the eyes will focus collimated images at optical infinity, the advocates of head-mounted displays and head-coupled sensors now promise that a pilot will be able to maintain geographic orientation and make veridical judgments of distances, rates of closure, and angular directions to visible navigation points and targets.

To dispel any doubt that such promises will come true, designers of some sensor and display systems are delivering imagery from two cameras independently to the two eyes to provide stereoscopic viewing (or even hyperstereo by exaggerating the interocular distance between the cameras). Many are convinced that stereo viewing will create an illusion of "remote presence" and thereby improve judgments of size, distance, and angular location sufficiently to make it unnecessary to provide automatic sensors of vehicle positions and rates for navigation and obstacle avoidance. Experience with head-mounted displays, whether binocular or biocular (both eyes receiving the same images), does not warrant these wishful thoughts.

Evidence from a variety of experimental and operational contexts indicates that binocular judgments of size and distance are not markedly better than monocular judgments, except at very short distances (as in threading a needle). In fact, Holway and Boring (1941) found monocular size judgments to be more nearly veridical than binocular judgments when good distance cues are present. In any case, the large bias errors in size, distance, and angular position judgments caused by misaccommodation to virtual images would more than cancel any minor benefits of disparate images to the two eyes.

In the absence of some striking breakthrough in human genetic engineering, the long-range prognosis for head-mounted displays is not good. Not only do our eyes refuse to behave as display designers would like to believe, but the illusion of vection induced by the "streaming" of objects near the periphery of wide-angle views often leads to motion sickness, particularly with head-coupled sensors and the consequent smearing of the images with head movements. Unfortunately our sole dependence on virtual imaging displays for tactical missions (HUDs now and HMDs in the future) has resulted in almost total suppression of research and development of more easily optimized direct-view displays of sufficient angular size to provide the needed fields of view with appropriate magnification.

What can be done

If we dismiss the genetic engineering approach, there are still several reasonable courses of action. In the short run, these include (1) trying to "fix" the HUD optics to compensate for the misaccommodation that leads to misorientation, and (2) modifying the ambiguous HUD symbology that leads to attitude reversals and subsequent disorientation. In the longer run, abandon the virtual image approach and concentrate on large, integrated forward-looking and downward-looking direct-view displays in which computer-animated flight attitude, guidance, and prediction symbology is superposed on sensor-generated real-world imagery.

Fixing the HUD

To induce pilots to focus at optical infinity when viewing virtual images, Norman and Ehrlich (1986) in Israel introduced a negative focal demand of -0.5 D with the desired result, although there were wide individual differences in responses as a function of individual dark-focus distances. Thus, the first experimental fix should be the addition of variable optical refraction to offset each individual pilot's inward focal lapse induced by the HUD's virtual images. Turning the HUD on would require a key coded to select the pilot's specific correction based on the dark focus. At this time, no one can be sure how successful this fix will be, but it must be tried.

Almost as important is the complete redesign of HUD symbology. Just how complicated and confusing it is can be appreciated from the estimate of an army instructor pilot that an average student helicopter pilot requires 200 hours of simulator and flight training to master the gaggle of symbols (personal communication). Furthermore, the attitude presentation in fixed-wing airplanes is conducive to horizon and pitch-ladder control reversals that result in disorientation and "graveyard spirals" at night and in marginal weather. At the very least, a frequency-separated predicted flightpath "airplane" symbol that banks and translates in immediate response and in the same direction as control inputs should replace the present velocity vector and acceleration symbology (Roscoe, 1980, Chapter 7; Roscoe and Jensen, 1981).

Presenting the big picture

If head-mounted, wide-angle imaging displays are ever to be safe and successful, the apparent minification of the outside world will have to be compensated for by individually selectable optical magnification, or the eyes will have to be induced to focus at or near optical infinity, as in the case of HUDs. Neither approach will be simple. Furthermore, the whole virtual image display concept depends on a gross reduction, rather than any increase, in the weight of any head-mounted device to be used in a high-G environment. All things considered, it is surely premature to give up on direct-view, panel-mounted displays.

Large, integrated, direct-view displays offer many advantages in terms of visual performance as well as ease of achievement and lower cost. Eyes focus real images more accurately than virtual images (Hull et al., 1982; Iavecchia et al., 1988; Randle et al., 1980). Although many with 20/20 vision cannot focus out to optical infinity, all emmetropes can focus at the distance of cockpit instrument panels. Thus, although magnification of sensor-generated or computer-animated images of the outside world will be required, as it is with direct-view projection periscopes (Roscoe, 1984; Roscoe et al., 1966), a single, fixed-magnification factor of about ×1.25 will suffice for most emmetropes.

To make room for large forward-looking and downward-looking (and possibly sideways-looking) displays, a lot of single-variable dedicated instruments and controls will have to be replaced by insets that appear selectively on the large displays as a function of the mission phase, aircraft configuration, mode of operation, weather and traffic, system malfunctions, and in the case of military aircraft, weapon selection. Furthermore, with the ever-increasing complexity of aircraft systems and military missions, many future airplanes—despite their high degrees of automation—will require at least two pilots with a redistribution of functions and available information.

In the military there will always be a heavy premium on being able to take advantage of whatever is visible to the naked eye. However, trying to combine synthetic imagery with contact visibility compromises both, and a strong case can be made for distributing operational functions and information sources between an "inside" pilot and an "outside" pilot. The inside pilot would normally do all the flying in instrument meteorological conditions (IMC) and most of the flying under visual meteorological conditions (VMC), using a direct-view, wide-angle projection periscope and the large, panel-mounted pictorial displays surrounding the pilot deep inside the airplane. The outside pilot would use his or her eyes to supplement the imaging sensors, do most of the communicating and procedural housekeeping, and fly any maneuver that requires direct contact visibility.

References

Barnette, J.F. (1976). Role of Head-up Display in Instrument Flight. IFC-LR-76-2, Randolph Air Force Base, TX: Instrument Flight Center.

Benel, R.A. (1979). "Visual accommodation, the Mandelbaum effect, and apparent size." BEL-79-1/ AFOSR-79-5, New Mexico State Univ., Behavioral Engineering Laboratory, Las Cruces. Dissertation Abst. Intern., 40 (10B), 1980, 5044; Univ. Microfilm No. 80-08974.

Campbell, C.J., McEachern, L.J. and Marg, E. (1955). Flight by Periscope. WADC-TR-55-142, Wright-Patterson Air Force Base, OH: Wright Air Development Center, Aero Medical Laboratory.

Hennessy, R.T. (1975). Instrument myopia. *Journal of the Optical Society of America*, **65** (10), 1114–1120.

Holway, A.H. and Boring, E.G. (1941). Determinants of apparent size with distance variant. *American Journal of Psychology*, **54** (1), 21–37.

Hull, J.C., Gill, R.T. and Roscoe, S.N. (1982). Locus of the stimulus to visual accommodation: where in the world, or where in the eye? *Human Factors*, **24** (3), 311–319.

Iavecchia, J.H., Iavecchia, H.P. and Roscoe, S.N. (1983). The Moon illusion revisited. *Aviation, Space and Environmental Medicine*, **54** (1), 39–46.

Iavecchia, J.H., Iavecchia, H.P. and Roscoe, S.N. (1988). Eye accommodation to head-up virtual images. *Human Factors*, **30**(6), 689-702.

Jarvi, D.W. (1981). Investigation of Spatial Disorientation of F-15 Eagle Pilots. ASD-TR-81-5016, Wright-Patterson Air Force Base, OH: Aeronautical Systems Division.

Kehoe, N.B. (1985). Colonel Kehoe's spatial disorientation (SDO) incident in a F-15 during VMC. In McNaughton, G.B. (Ed.), *Aircraft Attitude Awareness Workshop Proceedings*, Wright-Patterson Air Force Base, OH: Flight Dynamics Laboratory, 1-5-1 to 1-5-4.

Morphew, G.R. (1985). Transcript of open forum session. In McNaughton, G.B. (Ed.), *Aircraft Attitude Awareness Workshop Proceedings*, Wright-Patterson Air Force Base, OH: Flight Dynamics Laboratory, 3-8-1.

Newman, R.L. (1980). Operational Problems with Head-up Displays During Instrument Flight. AFAMRL-TR-80-116, Wright-Patterson Air Force Base, OH: USAF Aerospace Medical Research Laboratory.

Norman, J. and Ehrlich, S. (1986). Visual accommodation and virtual image displays; Target detection and recognition. *Human Factors*, **28** (1), 135–151.

Norton, P.S. (Moderator) and members of the SETP Cockpit Design Subcommittee (1981). *Proceedings from the Society of Experimental Test Pilots Aviation Safety Workshop*, New York: AIAA, 19–47.

Palmer, E. and Cronn, F.W. (1973). Touchdown performance with a computer graphics night visual attachment. *Proceedings AIAA Visual and Motion Simulation Conference* (AIAA paper 73–927), New York: AIAA.

Randle, R.J., Roscoe, S.N. and Petitt, J. (1980). *Effects of Magnification and Visual Accommodation on Aimpoint Estimation in Simulated Landings with Real and Virtual Image Displays.* NASA TP-1635.

Roscoe, S.N. (1980). *Aviation Psychology*. Ames: Iowa State University Press.

Roscoe, S.N. (1984). Judgments of size and distance with imaging displays. *Human Factors*, **26** (6), 617–629.

Roscoe, S.N. (1985). Bigness is in the eye of the beholder. *Human Factors*, **27** (6), 615–636.

Roscoe, S.N. and Jensen, R.S. (1981). Computer-animated predictive displays for microwave landing approaches. *IEEE Transactions on Systems Man, and Cybernetics*, **SMC-11** (11), 760–765.

Roscoe, S.N., Hasler, S.G. and Dougherty, D.J. (1966). Flight by periscope: Making takeoffs and landings; the influence of image magnification, practice, and various conditions of flight. *Human Factors*, **8** (1), 13–40.

Roscoe, S.N., Olzak, L.A. and Randle, R.J. (1976). Ground-referenced visual orientation with imaging displays: Monocular versus binocular accommodation and judgments of relative size. *Proceedings AGARD Conference Visual Presentation of Cockpit Information Including Special Devices for Particular Conditions of Flying*, Neuilly-sur-Seine, France: NATO, A5.1–A5.9.

Simonelli, N.M. (1979, 1980). 'The dark focus of accommodation: Its existence, its Measurement, its effects,' BEL-79-3/AFOSR-79-7, New Mexico State Univ., Behavioral Engineering Laboratory, Las Cruces. Dissertation Abstr. Intern., **41** (02B), 722; Univer. Microfilms No. 80-17984.

Index

Page numbers in bold refer to complete chapters. Page references in italics indicate illustrations.

accommodation 511, 513
  and 3D perception 453
  and apparent size 579–80
  in depth perception *527–33*, 536, 538, 541–2
  effect of virtual images 581
  and HUD displays 578, *579*
  in pictorial displays 473
  and stereoscopic displays 270
  in video displays 549–50
accommodative-convergence (AC/A) ratio 550
actors and acting 3–4, 6, 8, 10, **159–70**
acuity 508, 517–18
  grating 334
  letter 334
  Snellen 120
  stereoacuity 549, *550*
  Vernier 549, *550*
adaptation 232
  color 335
  motion perception 336
  motion-sickness 365, 372–3
  problem solving 164–5
  sensory rearrangement 366
  sensory-motor, time delay effects *240–4*
  in teleoperations 193
  and telepresence 238
  variable prismatic displacement 295–304
  visual-motor 165–8, 292
additive superposition (suppression) 384–5, *386–7*, 388
aerospace industry 103
afternystagmus, optinokinetic (OKAN) 359–60
agnosia, clinical 330–1
air-perspective contrast 527, *528–9, 533*, 536, 538

aircraft cockpit *30–2*
airsickness 362
algebra, visual metaphors 144, 151–3
alignment 44
  distortions *67–9*, 70–1, 73
ambient array 340–1
ambiguity, Necker-type 452
amblyopia (lazy eye) 334–5
Ames demonstrations 48–9, 51–3, 163
analog clock 26, *27*, 29
anamorphoser, slit-line 88
animation *see* computer animation
Annenberg/CPB Project 138
anthropomorphism 234–5, 249
antimotion-sickness drugs 365
artificial horizon 202
aspect blindness 330–1
attitude *see* orientation
attitude director indicator (ADI) *30*, 174–5, 178, *200*
Aubert phenomenon 379
Aubert-Fleischel effect 360
autokinesis 70, 360
automatization 305
autopilot 172–4
axial lighting *107*

Bernoulli effect 332
binocular conflict 520, 523
bioengineering 108
biofeedback 374
"blind sight" 306
blindness 329–31
blindness, color 334
blobs 67–8, 71, 337
brightness 333–5
  in depth perception 527, *528*, 530, 532, *533*, 536, 538–9

CADD workstation *104*
calculus 153, *154*, 155
camera
 fixed 182, 184, 248–9
 in graphic displays 280
 and perception 488
 remote video 266
 resolution 580
 steering-coupled 182, 184, 188, 191
 stereo 281
 in telemanipulator systems 251–3
 in teleoperations 182, 184, 188, *189*, 191
 video 570
 wide-angle lens 270
carsickness 362
Cartesian coordinates 268, 343
cartograms *89–92*
cartography 23, 44, **76–93**
CAT (computer assisted translation) scanner 35, 131, 135
center of projection (COP) *33*
central nervous system (CNS) 238–9, 363–4, 366, 373, 382, 388
cerebral ischemia, acceleration-induced 363
cerebral ischemia, motion-induced 362
cerebrospinal fluid (CSF) 364
Challenger space shuttle 15
chemoreceptive trigger zone (CTZ) 364
children and perception 332–3
chromaticity 46
cinema, efficacy of **486–92**
cinerama sickness 362–3, 371
circularvection 348, 350–1, *352–3*
clinical agnosia 330–1
cognitive science 99, 309
cognitive system 317–19
collision-avoidance systems 305, 312
color *45*
 anomaly 334
 blindness 334
 contrast 333–4, 336
 in depth perception 527, *528–9*, 538
 and graphic design 141–3
 harmony 99
 in medical art 98, 101
 in obstacle detection 188, *189*, 191
 perception 329, 333–4, 336–7
 space 46
 temperature 99
 vision 333–4
 and visual metaphor 146–9
Columbia space shuttle 13, 15, *16*

communication viii
 graphic 91
 health 97
 optimization 5
 pictorial 10, **22–38**
 satellite *20*, 21
computer animation **138–55**
 algebraic 144, 151–3
 calculus 153, *154*, 155
 character 144
 Disney 145–6
 molecular dynamics 146
 and perception 336
 physics 146–53, *154*, 155
computer command languages 5
computer graphics 83
 control of 256
 design, dynamic 138–9, 143–6
 design, static 139–43
 and graphical perception 111–12
 in medical art 97–8, 103
 in teleoperations 184
computer graphics system 88, **196–205**
computer graphics technology 43
computer science 99, 486
computer technology in medical art 103, 108–9
congruence 165, 178, 474
constancy
 position 505, *506*, 507, 510–11, 520
 shape 486, 504–5
 spatial layout 480, *481*, 484
 stereo-depth 551
 in virtual space of picture 473–4
control
 action 160–1
 automatic, in teleoperations 183
 Cartesian position 267
 dynamic complications 209
 end point 247–8, 251, 253–4, 262
 error 160–1, 163
 feedback 178–9
 fingertip 251
 geometric 204
 hand 208, 251, 253
 "inner loop" 172
 inside-out 183, *185*, 192
 isometric 248, 254
 joint by joint 247–8, 252
 kinetic complications 208–9
 manipulative 164–5
 manual 172, 174, 178, 183
 mode 249–50
 models 160–1

nonlinearities 207
on-off (bang-bang) 248
in orbital environment 210, 215
outside-in 183–4, *185*, 186, 192
over 164, 193–4, 248, 254
problems 9
resolution 208–9
resolved rate 248, 251–3, 263
  algorithm 256–7
reversal 186
robot *257*
and situation awareness 178–80
stations for teleoperations 184, 186, *187*
supervisory 240
translation-rotation 20–1
vehicular 162–4
viewpoint 197, *222*
of visualization by user 199, *200*, 201
visuomotor 167
control devices 136
  attitude instrument *30*, 174–5, 178, *200*
  autonomous and semiautonomous 247
  CAE 260, *261*, 262
  design 247
  displacement 260, 262
  exoskeletal 248
  and eye-hand coordination 165–8
  hand 164–5, 247–8, 258, *259*, 262
  isometric 260, *261*, 262
  master/slave 257, 263
  multi-axis 165, 247–63
  on-off (bang-bang) 255
  in orbital planning 218
  Rotational Hand Control (RHC) 165, 251, 253
  spatial 486
  standardization 165
  Translational Hand Control (THC) 165, 251, 253
control system theory 160–3
control systems
  aeronautical 160
  flight 247, 249, 251
  manual 209
  master/slave 248, 250, 253
  motor 307
  neurological 208
  in orbital planning 221
  "output feedback" optimal 367
  reaction 216
  robotic 249–51

SIMFAC *252*, 253, 262
for space station 196–7
visual-motor 159, 284, 290–1, 293
convergence
  and 3D perception 453–4
  in depth perception 527, *528–9*, 530, *533*, 536, 539, 541–2
  and direction error 507
  in distance perception 558
  in pictorial displays 473
  and slant 496–7, 499, 502
  and vergence 567, 571, *572*, 573
  in video displays 547, 549
Cooper-Harper ratings 258, 260, *261*
Coriolis force 21
corneal vascularization 120
Crewman Optical Alignment Sight (COAS) 213
crystalline lens 527–8
cues
  and 3D perception 450
  color 142, 191
  conflicts 565
  convergence 86
  depth 9, 142
    in 3D representation *107*
    binocular 266, 513
    cognitive 280
    dynamic 546
    and eye level judgment 392
    in medical art *109*
    monocular 280, 454, 457, 513
    perception 527, *528–9*, 530, 536, 538, *539*, 540–1, *542*, 543
    pictorial 51–2
    static 546
    in vection 350–1
    visual-enhancement 267, 273, 281
    see also accommodation; convergence; stereopsis
  in direction error 508, 516–17, 520–1, 523
  force 249
  gravity 169
  intensity 85
  motion 198, 451
  nonvisual 162
  perceptual 70
  perspective 49
  proprioceptive/kinesthetic 164
  sensory 363–4
  solid modeling 49
  stereo 32, 165
  surface segregation 48

symmetry 61
tactile 256
task-related 250
texture 451
vestibular 164
in video displays 546–7, 558, 565
visual 150, 161–2, 169
  and form perception 383, 386
  and orientation perception 378–9, 381
cyclofusion limits 554
cyclopean eye 345

DaVinci Project *104*, 108–9
depth 24, 488, 521
  and 3D 135, 267, 312, 452, 568–9
  apparent 453–4, 492
  cues *see* cues, depth
  gradient 452–3, 455
  illusion 86
  kinetic depth effect (KDE) 52
  measurable 453
  perception 185, 266, 270, 335, 449–50
    in pictures **527–44**
    in telemanipulator systems 249
    and three-dimensionality *454–8*
  probe, stereo 453
  sensitivity 527, 530–2, *533*, 534, *535*, 536, 538–41, 547
  stereoscopic 329, 335, 449
    in video displays 546, *547*, 549–50, *553*, 554–5
  and vection 350
  and viewpoint 463, *464*, *468*, *469*, *472–3*
depth-order ambiguity 48, *49–50*
derivative machine 154–6
design theory 97
"Desktop Metaphor" 8–9
diagnostic imaging 103–4, 108–9
Dietzel-Roelofs effect 396
difference of Gaussians (DOG) *548*, 549, 551, *552–3*
digital elevation data 87
Digital Elevation Model (DEM) 88
Digital Terrain Model (DTM) 88
diplopia 549, 551, *552–3*, 554
direction
  cardinal 71
  compass 79
  egocentric *35*, 391
  exocentric 32–3, 391, 505, *506*, 507, 518, 523

and eye level 393, 402
and frame of reference 338, 340, 342, 344
gaze 236, 341
gravity 338
headcentric 345
in induced visual motion 354–5, 360
judgment error **504–23**
meridional 341
motion 108, 284–6, 292–3, 546–7
in oculogyral illusion 346
perception 329
slant 452
tilt 452
in vection 348, 352, *353*
and visualization *91*, 199
director, ADI (attitude director indicator) *30*, 175, 178, *200*
director, flight 172–4
Discovery spacecraft 13
discrimination 112–14, 117–18, *119–21*, 122–3, *125*, 126
disorientation 36, 577–8, 580–2
disparity *see* stereopsis
displacement 48
  adaptation *242*, 243
  binocular optical 298
  control 253–4
  control devices 260, 262
  and eye level judgment 396, 402
  fixed prismatic 295–8, *299–303*
  gaze 345
  joysticks 267–8
  in motion displays *54*, 55, *56*
  in motion sickness 367–8
  prism 241, 344
  random dot patterns 335
  in saccadic suppression 316–19
  in static displays 53, *54*
  in telemanipulation 248
  variable prismatic **295–304**
  viewpoint 468, *469*–71, 474, 499
display(s)
  alphanumeric 199, *200*
  ambiguous 51
  animated 48, *49*, 50–1, 57
  attitude director indicator (ADI) *30*, 174–5, 178, *200*
  CCTV 253
  conventional 218
  cross-and-square 209
  CRT 30, 283, *285*, *287*, 500, 509
  CRT map 174, *175*

Index

2D 249, 265, 268
3D 223, 249, 312
design 47–8, 55, 93, 326
distance and vection 350–1, *352–3*
downward-looking 582–3
feedback 249, 283
force-torque 266
forward-looking 175, 218, 582–3
forward-looking infrared (FLIR) 577
frame of reference 180
full-field 352, *353*
geometry 9
graphic 83, 93
   memory distortions *62–73*
   and perception 112, 129, 450
   in situation awareness 181
   statistical *90*
   of supercomputer images 103
   and visual enhancement 266
head-mounted (HMD) *36*–7, 159, 163, **577–83**
   and eye level judgment 391
   and orientation 168–70
   in teleoperations 184, 192
   in telepresence 236
head-up (HUD) 103, 159, **577–82**
helmet-mounted 159, 312
high-definition 527, 542
high-resolution graphics 101
interactive, in medical art **97–109**
interactive systems 103, 232, 235–6, 239
isoluminant 334
liquid crystal 546
low-light television 577
luminance-coded *111–29*
map **76–93**, 178
memory distortions for **62–73**
monoscopic 265–70
motion 49–53, 57–8
multiple orthogonal 249
multiscreen, panoramic 184, *186*
numerical 112, 116, *119–29*
on-board 196–8
orbit position indicator (OPI) *200*
orbital maneuvering 159, 163–4, **207–31**
panel-mounted 36, 580
peripheral 351, *352*
perspective 33, *34*, 165
primary flight (PFD) 172, *173*, 175
pseudoperspective *167*
real-time dynamic 267
sensor-generated 581
side-looking (profile) 175–6, 178, 583
"smart" 266
for spacecraft in orbit 196–8
spatial 26, 30, 46, 180, 196
static 48, *49*, 51–3, *54*, 56, *57*, 58
stereo-camera 549
stereoscopic 249, 265–8, 527, 539
surface form 92
symbolic 181
teleoperation 163–4
traffic, for commercial aircraft *32*
vector 265, 267
vertical situation 175, *176*
video 391, **546–55**
virtual environment 362, 374
virtual image 36, 577, 579, 581–3
virtual reality (VR) 312
volumetric 137
wide-angle television 527, 542–3
distal visual stimulus *see* field of view (FOV)
distance 46, 142, 199, 472, 547
   absolute 451–4, 464, *529*
   in depth perception 540–1
   and direction error 507–8
   egocentric 449, 562, *563*, 564
   estimation 163–4
   and eye level judgment 392
   and frame of reference 340–1
   gradient 452
   interocular (IOD) 271
   judgment 513–14
   perception 87, 168, 331
      binocular 558, *559*
      egocentric 561
      stereoscopic **558–65**
      in teleoperations 185–6, 188–9, 192
      in vection 351
   relative 452, *529*, *560*, 561–2, 572
   rotational 292
   and slant 497, *498*, 499–500
   in vergence 572, 575
   viewing 450, 518
      in depth perception 527, 529–30, 532, *533–5*, *542*, 543
      and geometry of pictures *481*, 484
   viewpoint 202, *203*, 468
distortion 52, 159
   in computer animation 144
   conceptually-induced *70*, 71
   depth 456, 554
   and direction error 517
   displayed data 326
   geometric 36, 137

592 *Index*

and geometry of virtual space 487–8
irreversible 165
magnification 507
in maps 81
in memory **61–73**
non-invertible 237
oculocentric 344–5
optical 133
perception 303, 336
perceptually-induced 70, 71
perspective 336–7
in pictorial displays 474–6
and screen slant 491–2
size 336
spatial 32, 336–7
systematic 165, 167
and viewpoint 465–7, 469–70, 471, 472
distortion-free optical system 341
Donder's law 516
dynamics 55–6, 57, 58
 counter-intuitive 210
 of environment 3–4, 6, 8
 environmental properties 48
 inverse 207, 220
 laws of 48, 51, 55–7
 Newtonian 4
 orbital 204, 212, 214, 216, 226
 and orbital planning 208–9
 pictorial environments 9–10

eccentricity 311, 341, 354, 518, 549, 550, 555
ecoholonomic matches 311
Ecologically Insulated Event Input Operations (EIEIOs) 168, 309, 310, 311–13
electron microscope 101, 107
electronic textbook 101, 102
engineering 99, 108, 138, 197, 235
 control 160–2, 366–7
 software 139
 video 139
 and visual metaphor 148–9
enhancement
 combined geometric and symbolic 35
 computational 36
 control 209
 display 165
 dynamic 36–7, 38, 164, 216
 geometric 30, 32–3, 36, 46, 164, 216
 in pick-and-place tasks **265–81**
 symbolic 33–5, 81, 86, 164, 216

environment
 artificial 3, 10
 communications 5
 content (objects) 3–4, 6, 8–10
 definition 3–4
 earthbound 6
 exotic and alien viii
 extreme ix
 frictionless 8
 highly structured 310–11
 imaginary 233
 inertial 5–6
 inertially fixed 210
 low structured 310–11
 multi-spacecraft 221–3
 and optimization 3–8
 orbital 5–6, 8, 209–11, 212–15
 pictorial, for system interfaces 8–10
 properties **47–58**
 radioactive 266
 self in 9
 sensory 303
 space 169, 266
 space-station traffic 215
 spatial 10
 spatial maps 159
 synthetic 8, 207, 235
 underwater 266, 303
 unnatural 159
 virtual 159, 163, 165, 168–9, 233
 zero-g 263
error *see* distortion
estimators 161–3
Euler angles 15
Euler-Hill solution 229–30
Euler's equations 15
Exploratorium 329
Exploratory 329, 332
extravehicular activity (EVA) 204, 247
eye level
 gravitationally referenced (GREL) 390, 391–7
 head-referenced (HREL) 390, 391–2, 400–2, 403
 judgment determinants **390–403**
 and orientation 169–70
 in prismatic displacement 297
 surface-referenced (SREL) 390, 391–2, 396–7, 398–9, 401
eye movement
 and induced visual motion 360
 misregistration 354
 in oculogyral illusion 347
 optokinetic pursuit 351

perspective vergence 567, 570, *571*, 572–3
saccadic 306, 316–17, 319, 325
and video displays 547
eye-hand coordination 165–8, 240–1

Fechner's law 530
feedback
　and adaptation 241
　biofeedback 374
　control 178–9
　delays 239, 241, 243
　direct, in motion sickness 367
　direct force 249–50, 262–3
　direct view 248–9
　in direction error 519
　display 249, 283
　in distance perception 558, 565
　force-reflecting 253, 262–3
　kinesthetic 194, 248
　loop *239*, 247–8
　operator control 234, *239*
　operator control action 211, 216
　operator design action 216, 218
　positional 252
　steering-wheel torque 194
　tactile 253
　in teleoperations 182–4, *185*, *188*, *190*, 192–4, 232
　visual 249, 253, 262, 296, 573
　　error-corrective 297, 302
Fick coordinate system 342
field of view (FOV) 164, 319
　from within space suit 19
　and image displays 580–2
　in monoscopic displays 270
　in obstacle detection 188, 191, 193
　and perspective parameters 273, *276–7*, 279
　and visualization control 199–200
figure, detection 61, *70*, 73
figure, disparity 377
Filehne illusion 360
fluid dynamics, computational 131
forced-choice staircase procedure 126, *127*
form
　apparent 169
　idealized 108
　and orientation interaction **377–89**
　perception 159, 169, 377, 383–6, *387*, 388
　range and standard 108
　slant-tilt 452

in video displays 546
see also structure; three-dimensionality (3D)
frame of reference 279, 336
　3D object centered 308
　bodycentric (torsocentric) 338, *339*, 342–3
　conceptual 61
　in direction error 509
　Earth-fixed 200, 204
　egocentric 338, *339*, 340–3, 359–61
　exocentric 338, *339*, 343–4, *347–56*, 359–61
　for figures 61
　geodetic 198
　headcentric 338, *339*, 342, 344–5, 352, 354, *355*, 356
　inertial 200–2
　local vertical local horizontal (LVLH) 6, 198–202
　in memory distortions *70*, 71–3
　oculocentric 342, 344–5, 352, 354, *355*, 356, 359
　off-centre 325
　in orbital environment 210
　of orbiting spacecraft 45
　of picture 473
　proprioceptive 338, *339*, 347
　retinocentric 338, *339*, 340–2
　rotation 21, 165
　semi-exocentric *339*, 344, 348
　shifting 205
　and slanted image 488
　spacecraft "body" 200–1
　in spatial orientation 169
　station-point 338, *339*, 342
　in video displays 555
　for visual representations 318–19, *320–2*, 325

Gaussian distributions 111–17, *125–6*
Gaussian random numbers 123
Gaussian surfaces 148
gaze 236, 341, 345, 354, 360
genetic engineering 581–2
Geographical Information Systems (GIS) 88
geometry 9, 25
　and 3D perception 454
　in 3D representation *107*
　in direction error 507, 515, 523, *524*, 525
　of environment 3–4, 6
　extrinsic and intrinsic 343

fractal 85
and frame of reference 338, 340–1
map 76
and map perspective 83
in pictorial displays 474
of pictures **480–5**, *492*, 493
projection 76, *78*
reconstructive *487*
and shape constancy 504
stimulus 484, 508–9, 518
viewing 518
and viewpoint *463–71*
of virtual space 487–8
graphics 26
  computer 30, 83
  interactive 99–100
  orbital maneuvering display 211, 215
  orbital planning display 221, *222*
  in scale resolution 176, *177–9*
  software system 132
  symbols 76, *82*
  in visual enhancement 266
graphs
  analysis, visual perception in **111–129**
  and memory distortions 61, 65, 68, *70–3*
  parallel processing 121–3, 129
  plotting 140, 143
  preattentive processing 121–2, 129
  serial processing 122–3, 127, 129
graticule in map design 81, *82*
grating acuity 334
gravity 6
  cues 169
  direction 338
  and eye level judgments **390–403**
  and form perception 386
  law of 229
  and orientation 169–70, 381
  perception 333
  and spatial vision 347
  and vection *349*
guidance systems 367
gyroscope 332

hand-eye coordination 295–6, 301, 303, 317
head-mounted displays (HMD) *see* display(s)
head-up displays (HUD) 103, 159, **577–82**
Helmholtz coordinate system 342
Hermann grid 455, *456*
heterophoria 313

histology 101–3
hologram 131, 529
Hooke's law *145*
hyperacuity 547
Hypertext 93
hysteresis 250

illumination *392–5*, *398*, *400*, 543, 575
illusion
  body tilt 348, *349*, 350
  Café Wall 336
  Craik-O'Brien Cornsweet 555
  depth 86
  disturbance 344
  figures 453
  Filehne 360
  geometric 344
  induced-motion 318–19, 325–6, 344
  of inertially stable space 220–1
  Muller-Lyer 336
  oculogyral 169, 344, 346–7, 354
  and perception 62, 328, 331
  perspective distortion 336–7
  Poggendorf 336
  Roelofs effect 168, 316, 319, 321, *322–4*, 325–6
  self motion (vection) 169, 348, 354, 356
  self rotation (circularvection) 348, 350–1, *352–3*
  slant perception 496
  spatial 344
  visual direction shifts 345
  wallpaper 550
  see also vection
illustration, medical 44, 97, *98*, 100, 108
illustration, surgical 97, 99, 105, *109*
image bank 108
inference, unconscious 451
interactive digital video (IDV) interface 101–3, *105*
interactive "hands-on" science centers 329
isoluminance 334–7
isomorphism 234

Kalman estimator 161–2
Kepler-induced windage 19
Keplerian orbits 142
Kepler's laws 6, 142, 152, 230
kinematic laws 52–3
kinematics 208–9, 515–16
kinematics, inverse 262, 268
kinesthetic/tactual system 232

kinesthetics in telepresence 237
kinetic depth effect (KDE) 52
knowing 10, **43–6**

La Gournerie's paradox 486, *487*, 489
Legendre functions 15
Lennard-Jones atomic motion simulation 142
letter acuity 334
light source 99, *107*, 134, 143, 340–1
linear momentum conservation laws 55
Listing coordinate system 342
Listing's law 341, 516
luminance 46, 336–7, 455, *529*, 538, *552*

Mach bands 455–6
magnetic resonance imaging (MRI) 35, 103, 131
man-in-the-loop (teleoperator) systems 247
Man-Vehicle Systems Research Facility *31*
manipulator, real-time simulation 267–8
manipulator, simulated cylindrical 265, 267
Manned Maneuvering Unit (MMU) 3, 6, *17–8, 20–1*, 202
  simulator *19*, 255–6, *257*
map(s) ix, 26
  artificial 67–8
  automobile strip 89
  characteristics *76–83*, 92
  clarity 87
  coastline 81
  cognitive 168, 316, 325
  and computer graphics 44
  computer-supported 86
  3D 81, *83–8*, 505
  design 79, 81
  displays **76–93**, 178
  distortions 61, 89
  elevation *82*
  familiar 284, *285*, 287, 291–2
  land 81, *82*
  mapping problems 93
  and memory distortions 65, *66–73*
  memory for 61
  Mercator world 77
  in monoscopic displays 268, 270
  motor 316, 325
  non-transformed 283
  orientation 87, 505
  planar digitizing tablet 283–4, *285*, *287–8*, 291

planimetric *85*, 86
projections 76, *77–83*, 93
reading 92
real 68, *69*
road *90*
schematic 89, *90*
semantic similarities 46
space 46
and spatial instruments 29
and statistics 93
T-in-O 89
technology 81
in teleoperations 193–4
texture 140
topographic 32, 308, 317
transformed visual-motor 283–94
types and uses 93
using IDV interface 103
of visual space 316, 325–6
visuo-motor 165–8
world 77, 79
mathematics 144–5, 152
Mechanical Universe, The 44–5, **138–55**
media technology 97–8
medical art, interactive displays in **97–109**
medical diagnostics ix
medical education 108
medical illustration 44, 97, *98*, 100, 108
medical imaging 108, 131
medical sculpture 101
memory distortions **61–73**
misaccommodation 578, *579*, 580–2
misorientation 580–2
mobile servicing center (MSS) 260
model
  binocular distance perception *564*
  breadboard 253
  conceptual 135
  control 160–1
  Cutting's perceived orientation 480, 482–3
  3D 98, *100*, 101
  Digital Elevation (DEM) 88
  Digital Terrain (DTM) 88
  graphic 136
  induced visual motion 360
  internal CNS dynamic 367, *368–9*, 370, 373
  Kalman estimator 161–2
  Kalman filter 366–7
  McRuer's "crossover" 160
  man-machine 160–1

mathematical, of sensory conflict 366–70, 373–4
mental structural 133
Mittelstaedt's vector combination 401
motion sickness 362, 366–70, 374
optical 134
optimal control 366
parallel distributed processing (PDP) computational 309, 312
perception plus transformation 307–8, 310
position constancy 511
predictive 240, 250
Reason's neural mismatch 366–7, 369–70
of remote system dynamics 250
Shuttle Remote Manipulation System (SRMS) 262
slant perception 496–7
slant underestimation 497–502
solid 49, 51
"source attenuation" 131
space-station 224
spatial layout 308, 462
summation 457
for symptom dynamics 370, 371, 372–3
theoretical, of subjective vertical 169
tracking 161
transformed world 238, 239
view vector 516–17, 520
visual representation 308
visualizability of scientific 43
visualization 134
Monte Carlo simulations 123
morphometric analysis 108
motion
  absolute 54, 55, 56
  after effect 335–6, 345
  apparent 169, 335–6
  Aristotelean 332
  common 54, 55
  cue for navigation 198
  in depth perception 542–3
  direction in video displays 546–7
  display 49–53, 57–8
  double-framed 145
  endogenous 363
  equations of spacecraft 229–31
  exocentric 346
  exogenous 363–5, 368, 369–70, 374
  and frame of reference 338, 340
  and graphic design 142
  headcentric 346–7, 350–1

induced visual 168–9, 316, 352, 354, 355, 356, 359–61
information **47–58**
inverse 208–9
kinematic 6
long-range 336
opposite-direction 335
orbital 210–11, 212–13, 215, 229
overlapped 144
perceived 54, 325
perception 47, 136, 335–6
real 335–6
relative 54, 55, 227, 228, 229–31
and rigidity 486, 488–9
rotational 163
self 169, 348, 354, 356
short-range 336
single-framed 145
stroboscopic induced 316, 318–19
to communicate structure 134–5
viridical 325
and visual metaphor 146, 151
see also illusion, oculogyral; vection
motion parallax 9, 280, 473, 504, 546
  and 3D perception 452, 457
  in depth perception 527, 528–34, 536, 541, 568
  in direction error 513, 521
motion sickness 159, 161, 163, 169, 581–2
  sensory conflict in **362–75**
motion-detecting system, oculocentric 344
motion-induced cerebral ischemia 362
myopia, instrument 578

Nap of the Earth (NOE) 159, 258, 260
NASA Aviation Safety Reporting System 174
NASA TCV program 179
Navie orbital planning system 38
navigation viii
  cartograms 89–92
  inertial 161
  for manned space flight 196, 198
  and map projections 78, 79, 93
  maps as instruments 77
  orbital 30, 31
  skills training 305
  systems 367
  in teleoperations 193–4
  using IDV interface 103
Necker cube 168, 331
neurology 306

neurophysiology 308–9, 362, 388
Newton's Laws 8, 17, 152, 229
noise
  data 112
  internal 117, 126–8, 237
  intrinsic 113
  measurement 507
  in motor systems 325
  in multi-axis control devices 254
  quantization 116–17
  random 165
  sensor 366–7, *369*, 370
  system 159, 161–2
  visual 113, 122
numerical tables 111–12
nystagmus, optinokinetic (OKN) 347,
    354–5, 359

observer theory of motion sickness
    362–75
occlusion 9, 48, *49–50*, 51, 54, 280, 457
oculocentric motion-detecting system
    344
oculogyral illusion *see* illusion
oculomotor system 309, 318, 325,
    575–6
Operation OVERLORD 79
optic array 340
  and 3D perception 451
  distance perception 564
  lateral viewpoint 468
  perspective structure *462–7*, 472
  pictorial compensation 475
  picture plane 473–4
  shape constancy 505
  slanted image *489*
  spatial layout of environment 461
optic atrophy 120
optical minification 580–1, 583
optimization of environments 3–8
optokinetic afternystagmus (OKAN)
    359–60
optokinetic nystagmus (OKN) 347,
    354–5, 359
orbit position indicator (OPI) *200*
orbital maneuvering vehicle (OMV) *200*,
    202
orbital maneuvering visual display aid
    207–31
orbital planning system 37–8, 45, 207
orientation viii, 159, 168–70
  in direction error 517
  in displays 268, 581
  eye 340–2

and eye level 393, *395*, 396–7, *398–9*,
    400–2
and form interaction **377–89**
and frame of reference 338, *339*, 340
in motion sickness 362–7, 374
perceived 480–1, *482–3*, 484–5
perception 337
  effect of form on *377–85*
  in motion sickness 366–7, *368*, 370
research 105
retinocentric 341
and shape constancy 504
and slant 497
of spacecraft 199–200, 202
spatial 161–2
surface 449–50, 452, 454
tactile cues 256
in teleoperations 193
in vection *349*
and vergence 574–5
viewer 105
and viewpoint 465, *466–7*, 469–70,
    471
and virtual space of pictures 472

"Pandora's Box" (Gregory) 453
Panum's fusion 551, *553*
Panum's fusional area (PFA) 551, 554
parallax
  binocular 280, 527, *528–9*
    in depth perception *530–43*
    in distance perception *558–9*, 561–5
    in video displays 547
  zero 340
  see also motion parallax
perception 43
  cognitive processes 333
  distortions 64, 66, 68
  graphical **111–29**
  motor response mismatches 167–8
  organization 64–5, 68, *70–3*
  research 47
  serial analysis 122–3
  theories 61
  visual representations for **316–26**
perception-action discrepancies 305–7
perception-action skills 306
Perrone slant underestimation model
    497–502
perspective 9, 134, 467, 488
  aerial *107*
  angular *85*, 86
  3D form 51
  3D map 83, *84–5*, 86–8

3D representation 568
and direction of movement 108
linear *107*, 486
monoscopic displays 265–71, *272*, 273, *274–8*, 279–81
multiple 49
oblique 3D 81
parallel *85*, 86
parameters *272*, 273, *276–8*, 279–81
in pick-and-place performance 265
planimetric 88
projection 270, 280, 496
 and 3D perception 453–4
 in direction error 517, 523, *524*
 map displays 83, 87
 object-transformation method 268
 and synthetic universe 23, *24*, 33
 viewpoint-transformation method 268
stereoscopic displays 265–7, 270, *271*, 273, 275, *278*, 279–81
structure *462–7*, 469, 472, 475
transformation 488–9, 493, 509
vergence 567, 570, *571–5*, 576
pharmacology of motion sickness 362, 374
photography 99, 201, 486, *487*, 491–2
 black and white 98, 333
 spatial display 26
 and vergence 575
physics 56–7, 138, 145
 plasma 21
 visual metaphors 146–53, *154*, 155
physiology 328–9, 572
 of motion sickness 362, 364–5, 367, 373–4
Piaget's decentering task 505, *506*, 522–3
pick-and-place tasks, visual enhancements **265–81**
pictograph *24*, 381–2
pictorial communication 10, **22–38**
see also picture(s)
picture(s) ix, *22–6*
 definition 22–3
 depth perception in **527–44**
 dual nature of 473–4
 geometry of **480–5**, *492*, 493
 and judgment making 57
 paradoxical monocular stereopsis 568–9
 perception 52, 451, 484–5, 497, 517
 perspective vergence 567, *574*
 static 36
 stereoscopic 539, 543

surface segregation in 48
virtual space **460–76**
pitch 347, 507
 overestimation 514, *515*, 516–17, 523
 in vection 348, 350
pitchbox and eye level judgment *394–402*
pixel 10, 102–3, 116, 281
planar digitizing tablet 283–4, *285*, *287–8*, 291
plasma physics 21
plastic surgery 104
previsualization 98
primary flight display (PFD) 175
prism
 binocular 297–8
 displacement adaptation **295–304**
 downward-displacing 297, 299
 spectacles 240
 upward-displacing 297, 299, 301
 wedge 295, 297
projection
 geometric 76, *78*, *85*, 508
 map 76, *77–83*, 93
 orthographic 453–4
 perceived 480
 Plate carre *29*
 retinal 308
 surface 132, 488, *489*
 transparent 133
 volumetric 132–3
 see also perspective
proximity operations planning *209–23*
psychology
 applied 61
 and direction error 517–18
 experimental 160
 of motion sickness 367, 374
 and perception 52, 335, 450, 452, 486, 488
 perceptual 99
 symmetry experiments 71
psychophysics 92, 328
 and 3D perception 451
 and depth perception *531–44*
 and direction error 518
 in form perception 386
 and induced visual motion 354
 of KDE 52
 and shape constancy 504
 studies 306
 and visual function 317

quantization error 116–17
quantum mechanics 149

quantum theory 5
QUEST procedure 117

R-bar 211, *212*, 213, 226, *228*, 229
randomized-staircase forced-choice experiment 453
Rayleigh criterion 551
real-time behavior 326
Reason's neural mismatch model 366–7, 369–70
receiver-operating characteristics (ROC) curves 114
reflection, nonuniformities 167
reflection, specular 133–4
reflection in transformed mapping *283–94*
reflectorscope (vue d'optique) *541*, 543
relative distance tasks *560*, 561–2
remapping, visual-motor 284
remotely piloted vehicle (RPV) 577
reorientation 364
RESCALE 202, *204*
resolution
  camera 580
  control 208–9
  and 3D perception 458
  display 116
  image, in graphic displays 101–3
  and map projection 81
  scale 176, *177–9*
  in teleoperations 189, *190*, 191, 193
  and telepresence 236
  of vector-display module 267
  in vertical awareness 175
  in video displays 551, 554
  visual 120
rhumb lines *28–9*, *78*, 79
Ricco-Piper's law 538
rigid structures 52–3, *486–92*
rigidity and physical realizability 291–2
rigidity principle 52, 486
roam capability 203–5
robot, slave 232, *233*, 234, 236–7
robotic systems 233–4
robotics 248
Roelofs effect 168, 316, 319, 321, *322–4*, 325–6
root mean squared (RMS) error 287–8, *289–91*, 293
rotation 17, 55, 148, 167
  of attitude 19
  azimuth, of aircraft 34, *35*
  differential 481–3

in direction error 510, 513, 515–16, 518
and displacement 54–5, *56*, 254
in distance perception 558
drift 18
error in maps 67–8, *69*, 70
eye 341–2
and eye level 402
and frame of reference 201, 340, 343
in hand controller 165
mental 105, 488
in monoscopic displays 268–9
in motion displays 52–3
and motion perception 335–6
in multi-axis control devices 262
in oculogyral illusion 346
in orbital environment 6, 8
and orientation perception 378
and perspective parameters 279
positive 229
rate 21
satellite's center of 16
and shape constancy 504
and slant 489, 499–500, 502
and spatial orientation 169
and spatial vision 347–8
in SRMS system 251
of structures in diagnostic imaging 104
in telemanipulation 247
in telepresence 236–7
in transformed mapping *283–94*
in vection 348, 350–1
in video displays 546
viewpoint 198
in visualization 134

saccadic suppression 316–18
sampling efficiency 126–7, 134
scale 10, 163–5, 464
  construction 235
  for depth perception 527, 530
  distortion 79, 86
  map 76, 81, 86
  in perspective displays 270–1
  resolution 176, *177–9*
  of rigidity *490*
  in telemanipulator systems 253
  texture 472–3
  and vergence 575
scanner 99
  CAT (computer assisted translation) 35, 131, 135
scatterplots *111–29*

seasickness 362
seeing by exploring **328–37**
segregation, surface **48–51**, 56
seismological exploration 131
sensor displacement 312
sensor, infra-red 221
sensor, noisy 366–7, *369*, 370
sensorimotor interaction in telepresence 236–7
sensorimotor subsystems viii
sensorimotor system 160, 234
sensorimotor tuning ix
sensory conflict in motion sickness **362–75**
sensory conflict theory 362, 365–6
sensory miscalibration ix
sensory rearrangement 363–6, *368*, 370–1, 374
Shuttle Remote Manipulator System (SRMS) 247, 251, *252*, 253, 256, 262–3
Shuttle Remote Manipulator System (SRMS) model 262
signal-detection theory 111–12, 114, 318
signal-to-noise (S/N) ratio 135
SIMFAC control system *252*, 253, 262
simulation/simulator 98, 582
  anatomical 108
  electronic textbook 101, *102*
  of environment 233, 235
  flight 163, 374, 579
  ground-based 197
  interactive *100*, 101
  Lennard-Jones atomic motion 142
  Manned Maneuvering Unit (MMU) *19*, 255–6, *257*
  Monte Carlo 123
  on-board 197
  projection surface *489*
  sickness 362–3, 371, 374
  Space Station Simulator Project 196–7, 201
  symptom dynamics model *371*
  and telemanipulation 247
  teleoperations 265, 267
  viewpoint 202
  and visualization control *200*
situation awareness 6, **172–81**
size 472, 546
  in depth perception 527, *528–9*, 536, *538–40*
  in direction error 521
  estimation 189, 192

and perceived orientation *483*
perception 92
scaling 331, 464
skeletal motor system 325
slant 452, 536
  overestimation 499–500, *501*, 502, 514, *515*, 523
  perceived 493–5
  perception 496–500, 502
  of pictures 486, 488, *489*
  screen 488, *489–92*
  surface 455, 488, 496–7, 499
  underestimation **496–503**
Snellen acuity 120
solid rendering 136
space
  exocentric 348
  extraterrestrial viii
  inertial 170, 198, 343, 347–8
  inertially stable 220–1
  manual 283
  perception 529
  photo-colorimetric *45*
  virtual 9
    and direction error 517, 523
    geometry of 487–8
    of pictures 25, **460–76**, 480, 482–4
    and slant 489, 493–5
  visual 167, 283, 286
space shuttle *13–16*, 199, 202, *203–4*, 251
space sickness 262–3
space station *204*, 205
  dual-keel 223–4
  manned 196–7
  and orbital planning 207, 211, *212–14*, 215–16
  simulator 196–7, 201
  visualization system for 198, *200*
space station coordinates, relative motion in 227, *228*, 229–31
Space Station Freedom 260
Space Station Simulator Project 196–7, 201
spacecraft in orbit visualization **196–205**
spatial ambiguities 168
spatial "envelopes" 221
spatial frequency 329, 549
  in video displays 550–1, *552–3*, 554–5
spatial instruments 22, *26–30*, 46–9, 159
  skills training **305–13**
  and visuomotor modularity 305, 308, 312
  see also control devices

spatial layout 43–4
  and direction error 508, 521
  of environment 460
  model 308, 462
  perceived 480, *481*, 484–5
  perception 467, 472
  pictorial display 473
  and shape constancy 505
spatial organization *70*
spatial vision, egocentric/exocentric
    frames of reference **338–56**, 359–61
spectacle sickness 362–3
spectral analysis 329
spectral density 126
Sputnik satellite 12–13
standardization 105, 108–9, 250, 262–3
stereo pairs 132, 135
stereo viewing, static 135–6
stereo-sensitivity 549, 554
stereoacuity 549, *550*
stereogram 270, 275, *455*–7, 550, 554
stereology 108
stereopsis 9, 135–6, 280, 312, 473
  and 3D perception 451, 453–5
  depth perception and 266, 543
  and direction error 513, 518, 521
  paradoxical monocular 567–70, 576
  and shape constancy 504
  and slanted image 488
  and vection 350
  and vergence **567–76**
  in video displays **546–55**
stereoscope 275, *276*, 335, 453, 569
stroboscope 335–6
structure(s) 10, 56
  clearance between 215, 223
  detection 136
  dual-keel space-station 223–4
  from nonspatial domains 46
  perspective *462*–7, 472
  rigid 52–3, *486–92*
  spacecraft 199–200, 202, 204–5
  surface 133, 546
  in surgical illustration *109*
  tissue 137
  transformation 234
  volumetric 133, 136
supercomputers 99, 108, 131
suppression (additive superposition)
    384–5, *386*–7, 388
surface
  continuous, 3D perception across
    449–58

curvature 449–50, 452, 454, 458
orientation 449–50, 452, 454
perception 497
projection 132, 488, *489*
segregation **48–51**, 56
slant 452, 455, 488, 496–7, 499
texture 451, 454, 472–3, 488, 497
  in depth perception 527, 536, 541
  and direction error 514, 517, 521
  in video displays 546, 550, 554
tilt 452, 455
visualization 132
surgical illustration 97, 99, 105, *109*
symbolization 81, 83, 88
symmetry 44–5
  detection of mirror 112, 118
  and direction error 510–11, *512*,
    513–14, *515*, 520, *521–2*
  distortions 61, *65*, 66, 68, *70*, 71
"synthetic window" 202, *203*

telemanipulation 159, *166*–7, **247–63**,
    266, 280
telemanipulator systems 247, 249
Teleoperated Mobile Antiarmor Platform
    (TMAP) Project 188
teleoperated systems in space 164–5
teleoperation 159, 234, 240, 258, 265–7,
    281
  of land vehicles 162, **182–94**
teleoperation systems 183–4, 190, 194,
    232, *233*, 234–5, 247
telephoto lens 270, 488, 492
telepresence viii, 159, 232, 234–7, *238–9*,
    244, 546
telerobotics *166*, 208–9, 266
telescope 221
television 266, 281, 488
thermodynamics 148
three-dimensionality (3D) 133
  apparent 450–3
  cursors 136
  form 48, 265, 554
    and computer technology 25–6
    in medical art *106*–7
    in motion displays 51–3
    realism 23, *24*
    visualization of 35, 46, **131–7**
  graphical design 142–3
  graphics 132
  imaging, noninvasive 104
  lattice 134
  maps 83, *84–5*, 86–8, *89*

models 98, *100*, 101
perception across continuous surfaces **449–58**
space 9, 32, *33*, 46, 140
tilt 359, 452, 555
   and form perception 383–4, 386, 388
   head 377–8
   image *378–80*
   surface 452, 455
   in vection 348, *349*
time delay 164–5, 194, 232, 250
   effects on sensory-motor adaptation 240–4
   pure 165
   and telepresence 237, 239–41, *242*, 243, *244*
   in visual representations *320–5*
   and VR displays 312
tomography 103, 133
torque and telemanipulation *166–7*, 169, 194
tracking 165, 239–40
   error 172–4, 265–6, 280
transformation
   by projection geometry 76
   computed 198–9
   inverse kinematic 268
   linear 284
   mapping 83, 85–6, *89*, **283–94**
   mathematical *78*, 79
   nonlinear 116–17
   nonparallel 88, *89*
   paradigm 285–6
   spatial *90*, 293
   topological 89
   using primary displays 172
transparency 132–6, 250

UNIX operating system 5, 7, 8

V-bar 211–12, *213*, 224, 226, 228–9
Van der Waals' equation 148
vection 169, *348–53*, 581–2
   forward linear 354
   induced-motion 352, 359–61
vector-sum error correction 292, *293*
velocity
   absolute 205
   approach 211, 215, 221, *222*
   between spacecraft 215
   in computer animation 144–6
   in depth perception 532, 534, *535*, 538
   estimation 209

flow fields 340
   and frame of reference 340–1
   and HUDs 582
   of induced motion 354–5
   in motion sickness 367
   in oculogyral illusion 347
   orbital 224
   in orbital environment 8
   perception 329
   relative 218, *219*, 220–1
   and situational awareness 174
   of spacecraft 199
   in spacecraft equations 229–31
   and spatial vision 348
   and vection 350, 359–60
   vector, orbital 210–11, 216–17, *228*
   vector, relative *222*, 224, *228*
vergence 513
   in distance perception 559, 561
   perspective 567, 570, *571–5*, 576
   proximal 572
   and stereopsis **567–76**
   in video displays 547, 549–50
   voluntary 572
Vernier acuity 549, *550*
vestibular disease 363
vestibular ocular reflex (VOR) 170, 346–7
vestibular system 163, 346–8, 364
vestibulo-ocular pathways 374
viewpoint 83, 105, *107*, 250, 488
   cinematic 489
   control *222*
   cross-sectional 104
   in direction error 509, 522
   effects on picture virtual space **460–76**
   fixed 249
   geometrically correct 460
   ideal 249
   in monoscopic displays 268–70
   oblique 104
   orbital maneuvering 164
   and position constancy 505
   and screen slant 490–1, 499
   of *self* in environment 3–4
   in visualization system 197–9, *200–4*, 205
   in volumetric visualization 133–4
"virtual cockpit" 205
virtual reality (VR) systems 305
virtual space *see* space, virtual
Vision and Ophthalmology, Association for Research in 113

Index

visual acuity *see* acuity
visual egocentre 345
visual imagery 43
visual metaphor *see* computer animation
visual ocular reflex 159
visual representation
  ecological theory 308–9, *310*, 311
  for perception **316–26**
  training and spatial displays 311–13
  for visually guided behavior **316–26**
visual sensitivity 5
visual system 159, 309, 311, 329
visual underestimation *92*
visual variables *91*, 92
visual-motor system 159, 284, 290–1, 293
visual-vestibular coordination 36
visualization *84*, 103, 132–3
  biomedical 44
  of 3D data 35, 46, **131–7**
  in DaVinci Project 108–9
  direction *91*, 199
  graphic 30, 45
  interactive 103, 131–2, 136
  land surfaces *82*
  map 79
  motion-based 136
  orbital maneuvering 211, 215
  orbital planning 218, 221, 223, 226
  orthopedic reconstructions 104

scientific 43
spatial 43–4
spatial relationships 101
structures and processes 97
system 133–4, 164
  for spacecraft in orbit **196–205**
  uses *201–5*
tools 99
transformed maps 83, *89*
of viewpoints 104
volumetric **131–7**
visually guided behavior 168, 285, 306, 309
  Roelofs effect 316, 319, 321, *322–4*, 325–6
  under transformed mapping **283–94**
  visual representations for **316–26**
visuomotor modularity **305–13**
Volume Rendering Technique 132

way-point
  editing 218, *219*, 220–1
  in orbital planning 211, *216–21*, 223, 225
Weber fraction 548

X-ray images 132

Zipf's Law 5, 7